Warming Mountains

Sachchidanand Tripathi
Rahul Bhadouria • Satish Chandra Garkoti
Editors

Warming Mountains

Implications for Livelihood and Sustainability

Springer

Editors
Sachchidanand Tripathi
Deen Dayal Upadhyaya College
University of Delhi
New Delhi, India

Rahul Bhadouria
Delhi College of Arts and Commerce
University of Delhi
New Delhi, India

Satish Chandra Garkoti
School of Environmental Sciences
Jawaharlal Nehru University
New Delhi, India

ISBN 978-3-031-62196-3 ISBN 978-3-031-62197-0 (eBook)
https://doi.org/10.1007/978-3-031-62197-0

This Springer imprint is published by the registered company Springer Nature Switzerland AG
The registered company address is: Gewerbestrasse 11, 6330 Cham, Switzerland

Contents

About the Editors

Sachchidanand Tripathi, Ph.D., is an Associate Professor in the Department of Botany, Deen Dayal Upadhyaya College, University of Delhi, Delhi, India. He obtained his doctoral degree from the Department of Botany, Banaras Hindu University, India. The areas of his interest are plant ecology, soil ecology of various ecosystems, ecophysiology, and urban ecology. He has published more than 60 publications (including research papers, books and book chapters, conference proceedings) with journals and publishers of international repute. He is serving as an editorial board member of several reputed international journals.

Rahul Bhadouria, Ph.D., is an Assistant Professor in the Department of Environmental Studies, Delhi College of Arts and Commerce, University of Delhi, New Delhi, India. He obtained his doctoral degree from the Department of Botany, Banaras Hindu University, Varanasi, India. He has published more than 25 papers, 27 book chapters, and 13 edited books in internationally reputed journals/publishers. His current research areas are management of soil C dynamics to mitigate climate change, a perspective on tree seedling survival and growth attributes in tropical dry forests under the realm of climate change, plant community assembly, functional diversity and soil attributes along the forest-savanna-grassland continuum in India, recovery of degraded mountains in central Himalayas, and urban ecology.

Satish Chandra Garkoti, Ph.D., is Professor of Environmental Sciences at the School of Environmental Sciences, Jawaharlal Nehru University, India. He is a Himalayan ecologist specializing in Forest Ecology, emphasizing forest productivity and nutrient cycling, tree ecophysiology, phenology, biodiversity characterization, plant invasion, and climate change adaptation and mitigation. He is among the pioneers contributing to the structure and functioning of high-elevation forest ecosystems leading to tree lines in the Himalayas. Plant diversity characterization in tropical forests of North Eastern India and estimation of ecosystem services in Himalayan Forest ecosystems are other areas where he has made substantial contributions.

Part I
Mountains: Ecosystem Functioning and Vulnerability

Warming Mountains: Challenges and Opportunities

Sachchidanand Tripathi, Rahul Bhadouria, and Satish Chandra Garkoti

1 Introduction

Mountain ecosystems are fragile (Schickhoff, 2011; Grêt-Regamey et al., 2012; Luo et al., 2023).

The adverse effects of climate change on these ecosystems are manifesting through extreme weather events and natural calamities (Heyojoo et al., 2017; Bolan et al., 2023). Consequently, it is imperative to comprehend the science and mechanisms underlying how climate change impacts mountain ecosystems. A thorough understanding of the factors driving changes in mountain environments and affecting its sustainability is vital for effective management and policy-making (Beniston, 2016). The retreat of mountain cryospheres due to global warming is threatening the physical, biological, and human systems in both the mountains and surrounding lowlands (Beniston et al., 2018; Rasul & Molden, 2019; Schickhoff et al., 2022). Therefore, a scientific understanding of this process in the current context is essential. Rising temperatures are also affecting biodiversity and ecosystem services in the mountains, posing significant challenges for climate adaptation (Pereira et al., 2022; Kattel, 2022). Additionally, increasing temperatures are also hampering crop productivity and hindering livestock rearing, directly affecting the livelihoods of the

S. Tripathi
Deen Dayal Upadhyaya College, University of Delhi, New Delhi, India
e-mail: drsachchidanandtripathi@ddu.du.ac.in

R. Bhadouria (✉)
Delhi College of Arts and Commerce, University of Delhi, New Delhi, India
e-mail: rahul.bhadouria@dcac.du.ac.in

S. C. Garkoti
School of Environmental Sciences, Jawaharlal Nehru University, New Delhi, India
e-mail: scgarkoti@mail.jnu.ac.in

S. Tripathi et al. (eds.), *Warming Mountains*,
https://doi.org/10.1007/978-3-031-62197-0_1

mountain communities (Praveen & Sharma, 2019; Bhadra et al., 2021). It is, therefore, imperative to focus on alternative livelihood opportunities.

In addition to changing climate, some prominent challenges to livelihood in mountainous landscapes are poor infrastructure, limited access to water and energy, poor market linkages, and limited know-how on the development of marketable products and post-harvest management, which have threatened the sustainability of mountain agriculture and traditional rural livelihoods (Shahzad et al., 2019; Dame et al., 2019; Dhakal & Kattel, 2019). Apart from handling the limitations mentioned above, the livelihood of local people may be improved by assistive and adaptive livestock rearing, improved processing and management of forest products, skill-based opportunities, and other developmental activities. Sustainable and adaptive agricultural practices have been adopted by farmers in mountainous regions across the globe. It is also equally significant to explore a green economy to strengthen the socioeconomic status of local inhabitants of the mountain (Unay-Gailhard & Bojnec, 2019; Ali et al., 2021). Further, there is an urgency to ensure food and nutrition security for mountain people in the backdrop of depleting natural resources, altering crop patterns (both agricultural and horticultural), and depleting sources to rear livestock under changing climate (Kumar et al., 2015; Romeo et al., 2021).

Increased intensity of disasters in mountain zones suggests employing state-of-the-art remote sensing techniques, more accurate modeling, and the latest techniques to avert or manage disasters in the near future effectively (Khan et al., 2020; Chen et al., 2023). The application of remote sensing in forecasting disasters may help to prevent the loss of human life and livestock (Oštir et al., 2003; Kirschbaum et al., 2019; Sharma et al., 2019; Gupta et al., 2022).

Mountains are also areas of conflict between economic growth and environmental protection. In the modern world, mountains have drawn huge economic investments for tourism, electricity, and communication lines, negatively harming natural ecosystems (Perlik, 2019; Zeng et al., 2022). Evidence suggests that the changing climate is causing considerable perturbations in mountain landscapes, which will undoubtedly impact the economics, livelihood, and sustainability of mountainous regions. Global warming will likely worsen the current conflict between commercial and environmental concerns. This warrants effective policy framing and implementation both at national and international levels. If adequately implemented, national policy frameworks, international agreements, and networks may be effective in achieving sustainable development goals (SDGs) in mountains (Leal Filho et al., 2019; Trane et al., 2023). There is an urgent need to achieve various SDGs identified for mountains, such as poverty eradication, ensuring food security and nutrition, sustainable agriculture, water security, energy security, forests management, disaster risk reduction, biodiversity conservation, and social and gender equality. This can only be achieved through targeted policies for mountain regions involving local communities/people who are immediate stakeholders. The chapter provides an overview of the content included in the book. It focuses on the emerging challenges in the sustainability of mountainous landscapes and associated livelihood opportunities involving possible adaptive and mitigatory measures.

2 Warming Mountains: A Global Overview

Approximately 10% of the world's population lives in the mountain regions, which cover about 25% of the earth's total land area (Kohler et al., 2010; Lu, 2021; Pepin et al., 2022). Between 1990 and 2100, global temperatures are expected to rise by an average of 1.1–6.4 °C, depending on the levels of greenhouse gas emissions worldwide (Kohler & Maselli, 2009). Since the shifting ecosystems in the mountains may provide a first indication of what could occur in lowland areas, many scientists believe that mountains function as early warning systems. Mountains are located in diverse parts of the world and vary in shape, length, altitude, plant cover, and temperature regime. As a result, climate change will affect them differently. The temperature in the mountains varies with elevation. For different altitudes, a warmer climate has varied effects. These mountains are vulnerable to climatic fluctuation and change (Shrestha & Aryal, 2011). The rise in anthropogenic greenhouse gas emissions has been attributed to the majority of the warming seen in the last few decades of the twentieth century (You et al., 2020).

Determining the observed variability and long-term changes in climate throughout the mountains is difficult due to a lack of consistent and adequate studies (Ren & Shrestha, 2017). Furthermore, accurate climatic forecasts over the mountains are essential for determining the effects of climate change. The need to understand why the consequences of global warming and climate change are often exacerbated in high mountains is becoming increasingly urgent.

Mountains are considered the world's water reservoirs and harbor enormous biological diversity at the genetic, species, and ecosystem levels, offering crucial resources to a significant portion of the global human population. Ecosystem services provided by the mountains constitute most of the livelihood opportunities in these areas, in addition to supporting people and organizations in lowland areas. Despite all the services these ecosystems render to humanity, they are the least well-documented ecosystems in this aspect.

3 Mountain Ecosystem Services Under Changing Climate

People who live in lowland and mountainous areas depend on the wide range of ecosystem services that mountain ecosystems offer for their well-being (Mina et al., 2017). The advantages provided by mountain ecosystems support around half of the world's population (Mina et al., 2017). The various ecosystem services of mountain ecosystems are given in Fig. 1. Climate change has developed as a global problem since it endangers ecosystems, biodiversity, and societal well-being in the mountains (Schirpke et al., 2017; Palomo, 2017). The geography of the mountains, the environmental gradient, and other elements together impact extreme and unexpected climatic events that are likely to exacerbate hazards like avalanches and landslides (Beniston, 2016; Jakob, 2022). The availability of food and water, the risk of

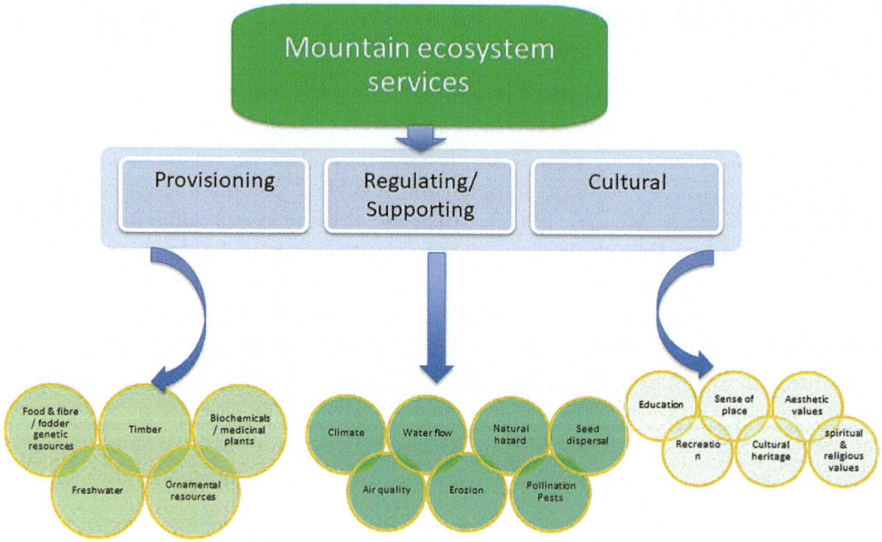

Fig. 1 Mountain ecosystem services

natural disasters, tourism, aesthetics, recreation, spirituality, and cultural distinctiveness are all negatively impacted by climate change, which also compromises the well-being of mountain inhabitants (Debarbieux et al., 2014; Aayog, 2018). Thus, it is crucial to concentrate on sustainable resource management and subsequently formulate programs/policies to protect them from the detrimental impacts of climate change, given the grave threat that climate change poses to the mountain ecosystem and the services that it provides.

4 Mountain Biodiversity and Climate Change

Mountain ecosystems are diversified, with one of the most diverse flora and fauna, and they provide a variety of ecosystem services to one-fourth of the population. In the current scenario, mountain ecosystems are confronting increased global and regional causes of biodiversity and ecosystem degradation. A high level of endemism is a characteristic of mountains. Major centers of plant biodiversity include the eastern Himalaya-Yunnan region, the Atlantic forests of Brazil, the tropical and subtropical highlands of Costa Rica and Panama, the tropical and subtropical eastern Andes, the northern Borneo, New Guinea, and East Africa. With their diverse livestock and crops suited to their particular regions, mountains are also important centers of agro-biodiversity. These resources, including their genetic diversity, are crucial for guaranteeing food security to the world's growing population. The lives of those in mountains and lowlands are also supported by mountain biodiversity, which provides vital environmental services like fresh water, timber, medicinal plants, and recreational possibilities for an increasingly urbanized world.

Mountain biodiversity is at risk due to warming linked to changing climate. Rampant urbanization and landuse change in mountainous landscapes have been frequently manifested in biodiversity loss (Jiménez-Olivencia et al., 2021). According to forecasts released by IPCC, the habitats for species acclimated to the cold will be reduced as temperatures rise. Because they have nowhere else to go, species that live in high-altitude zones will be in jeopardy. Rare species or those acclimated to the cold may be outcompeted in the long run. Additionally, mountain ecosystems are anticipated to experience more extreme occurrences, including intense rainstorms, severe pest and disease outbreaks, extended fire seasons, and more serious fires (Walsh et al., 2020).

Recently, it has been more widely acknowledged that managing mountain biodiversity is a worldwide obligation (Villeneuve et al., 2002; Şekercioğlu et al., 2011). Many mountains are inhabited by locals. Climate change may increase the urge for more conservation and intensive resource usage in the mountains. Thus, novel ideas and strategies are needed to balance development with biodiversity conservation. Conservation landscapes—which combine broad yet diversified small-scale farming with high human density—are becoming increasingly recognized for their ability to conserve high levels of biodiversity in regions where the creation or expansion of protected areas is hindered.

5 Ecosystem Functioning in Mountains: Challenges for Indigenous Communities Adaptation to Climate Change

Sustainable development in mountainous areas of the world requires effective resource management while also considering more general concerns like environmental degradation and climate change (Ramakrishnan, 2003; Sgroi, 2020). Indigenous peoples have long practiced and managed a variety of farming techniques that depend on forests, which has helped them build a strong knowledge base (Dawson et al., 2021; Sharma et al., 2020). However, the effects of climate change combined with human activity in the mountains are seriously endangering both human livelihoods and the functioning of ecosystems.

6 Shrinking Cryosphere: Impact of Warming in the Mountains

As a primary source of freshwater, snow cover is a crucial component of the cryosphere and one of the most significant types of land cover (Yongjian et al., 2020; You et al., 2020). For the ecosystems and people of the high-altitude mountains to replenish streams and groundwater, snowmelt is necessary. Crop production, tourism, and other industries may be impacted by water regime variations brought about by snowmelt fluctuations (Bormann et al., 2018; You et al., 2020; Notarnicola, 2020). The impact of many weather systems causes the snow cover peak to change throughout a range of periods (You et al., 2020). Studies conducted recently revealed that the global average temperature increases by twice as much in mountains, increasing with elevation (Notarnicola, 2020). In recent decades, numerous mountain locations have substantially used remotely sensed data, model simulations, and surface station observations to better understand these changes (Mote et al., 2018; Bormann et al., 2018).

7 Sustainability and Diversification in Livelihood Opportunities in the Mountains

7.1 Plant Phenological Shifts

Mountainous areas, home to several hotspots of biodiversity, are especially heavily impacted by the effects of climate change (Bellard et al., 2014; Trew & Maclean, 2021). In the global preview, mountainous regions have been particularly affected by global warming and its aftereffects, leading to phenological changes, early snow and ice melting, quicker deglacialization, and exceptional climatic occurrences (Tiwari et al., 2014; Schickhoff et al., 2022). In light of continuous environmental changes worldwide, investigating the relationships between plant phenology and other ecological aspects may provide important insights for adaptive management strategies. More study is needed to create effective conservation methods, protect biodiversity hotspots, and minimize the effects of climate change on these delicate mountain ecosystems.

8 Warming Mountains and Crop Productivity: Implications for Livelihood and Food Security

Climate change has a significant impact on people's lives and livelihoods, as well as on overall socio-economic growth. As per IPCC, 2013, over the period 1901–2012, the climate warmed by 0.89 °C, primarily due to anthropogenic activities. The time and amount of precipitation may be affected by rising temperatures, resulting in changes in water availability. Temperature increases at a faster pace at greater altitudes (Ghalambor et al., 2006; Smith et al., 2009). Crop productivity in the mountains is hampered by changing temperatures and irregular rainfall patterns.

The main determinants of agricultural productivity and production are climatic, and it is generally recognized that ideal crop yield depends on favorable climatic conditions. The effects of climatic fluctuation on the environment have been well-documented. These effects include a rise in temperature, precipitation, and the frequency and severity of extreme weather events. According to FAO research from 2003, up to 245 million, or one-third, of the 720 million people who live in rural mountain areas globally are either in danger of or already suffering from starvation. According to FAO estimates, 78% of the world's mountainous landscape is either unsuitable for agricultural cultivation or only moderately acceptable (Huddleston et al., 2003). Numerous mountain communities, many of whom depend on a single cash crop for their livelihoods, have recently changed their farming practices to adapt to the demands of a more developed market economy (Gentle & Maraseni, 2012; Gentle et al., 2018). The variety of agriculture has decreased due to the decline of indigenous knowledge about local foods and traditional farming methods. Due to this, food instability and malnutrition have been made worse in mountainous places. The effects of climatic fluctuation on the environment have been well-documented. These effects include a rise in temperature, precipitation, and the frequency and severity of extreme weather events. Global warming is the primary cause of temperature increases, which lower agricultural yields and promote the spread of pests (Nelson et al., 2010).

Additionally, the changes in precipitation patterns increase the likelihood of both short-term crop failure and long-term output reduction. There is an imminent risk of increasing crop failure and animal loss for rural mountain populations who rely on agriculture in a fragile environment. Research from Cameroon's Kilum-Ijim Mountain Range to assess rural livelihood options for climate change adaptation suggested that stakeholder synergies, nature-based solutions, and ecosystem-based adaptation can achieve sustainability. Further, the Afromontane landscape of Cameroon is becoming less sustainable due to rising temperatures. It suggests best practices like agroforestry, afforestation, reforestation, nature-based remedies, and ecosystem-based adaptation to mitigate the negative effects of rising temperatures on the Afromontane landscape. This book discussed the effects of climate change on the growth, development, and transmission of several vector-borne diseases, water-borne diseases, and foodborne diseases, along with effects on non-communicable diseases, injuries, and mental health in the Hindu Kush Himalaya. Further, the effect of climate change on the gender roles and livelihood of the people with mitigative measures was also highlighted.

9 Relatedness of Ecosystem Services and Livelihood in Mountains

Along with producing agricultural products like food and fiber, mountain agroecosystems also contribute to the preservation of biodiversity, water, carbon storage, and scenic amenities (Romeo et al., 2021). In combination with other stresses, climate change compromises the mountain agroecosystem's ability to supply the

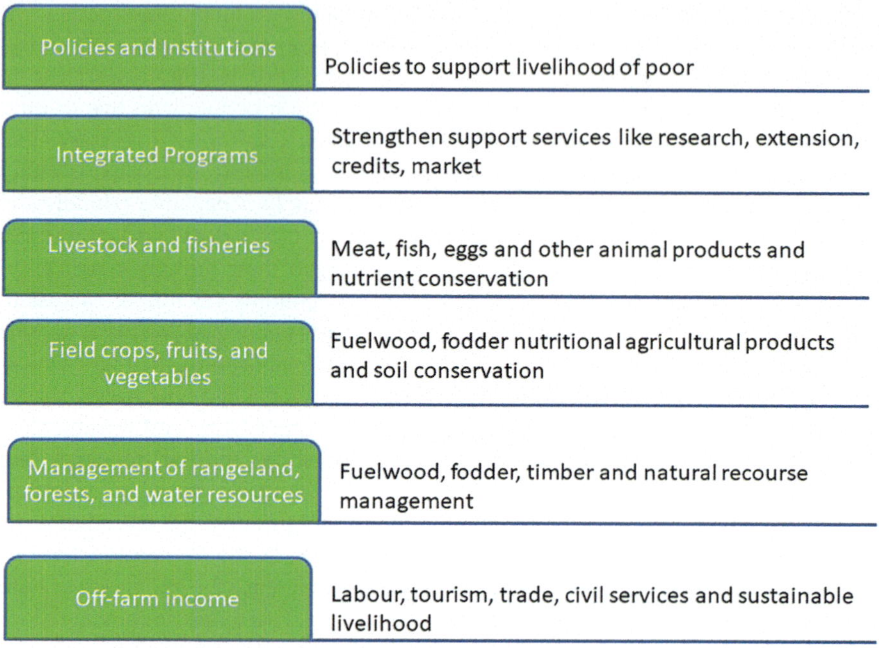

Fig. 2 Food security and alternative livelihood opportunities in mountains

ecosystem services essential for rural livelihood sustainability. It is difficult to accurately characterize poverty and vulnerability in the mountains due to the vast gaps in data that are unique to most mountain nations (Thakur et al., 2021). Approaches and programs to alleviate poverty developed at the national level are prone to overlook important subnational and local expressions of poverty (Thakur et al., 2021). In most cases, national measures have little influence on poverty reduction in the mountains, and they might even make mountain livelihoods more susceptible (Liu et al., 2018; Jodha, 2000). Without additional mountain-specific poverty-reduction programs, closing the economic gap between the lowlands and mountain areas will be impossible (Romeo et al., 2021). There is a need for diversification in livelihood opportunities for the people residing in the mountains in the backdrop of a changing climate. Food security and alternative livelihood opportunities in the mountains are given in Fig. 2.

Global environmental change is increasingly impacting mountain ecosystems, posing a threat to the people's pervasive agroecosystem-based livelihoods. In addition to producing agricultural products like food and fiber, mountain agroecosystems support the conservation of biodiversity, water resources, carbon sequestration, and aesthetic value of the landscape. However, mountain agroecosystems' ability to continue to supply the quality and quantity of ecosystem services necessary for sustained rural livelihoods is hampered by recent environmental change combined with additional stresses (Shrestha et al., 2012; Baral, 2013). According to global climate change predictions, ecosystems and their ecosystem services will be significantly impacted, substantially affecting communities' ability to support themselves, especially in developing nations (ICIMOD, 2010; van de Sand, 2012). Because of their isolation and lack of access to transportation and communication, mountain communities are frequently marginalized and more susceptible to environmental effects. Most poor people rely on ecosystem services for their survival, and they often lack the capacity to adapt to change, making them more vulnerable to climate change and other types of changes (ICIMOD, 2010). Because of the susceptibility of mountain ecosystems to warming, studies predicted that climate change will have a significant impact on them (IPCC, 2007). Mountain communities have a relatively poor ability to adjust to these changes since they have restricted access to other resources. These factors make the consequences of climate change on the supply of ecosystem services a more crucial field of research.

10 Managing Disasters in the Mountains: Challenges in the Era of Global Warming

Earthquakes, landslides, cloud bursts, and water-related calamities are hazardous in mountainous areas. Due to climate change and land degradation, mountains are experiencing more frequent flash floods, river-line floods, erosion, and wet mass movement during the monsoon season and drought during the non-monsoon season (Carrivick & Tweed, 2016).

The mountainous regions of the Andes, the European Alps, and the Himalayas are among those most vulnerable to various climate-related calamities (Kohler et al., 2014; Bhadouria et al., 2023). Snowfall and other natural disasters reduce the availability of water, wild food, fuel, wood, and biodiversity, impacting both human society and wildlife (Tse-Ring et al., 2010). Additionally, the habitat for both humans and animals is affected by changes in land use patterns brought on by climate change (Khanduri et al., 2018). The intensity and reach of socioecological consequences, such as deterioration of ecosystem services, scarcity of fuel wood, and population migration, can be increased by the rapid loss of forests due to fires (Maass et al., 2005; Roy et al., 2022). By altering their way of life, agricultural practices, and cultural traditions, people in the mountains have created coping mechanisms to adapt to the unpredictable nature of climate change, unpredictability, and extremes; they have even experienced major losses (Son et al., 2019; Sun et al., 2022). Building the adaptive capacities for shared water management, fair distribution of irrigation resources, livestock management, communal grazing, and infrastructure for flood control and water security requires bolstering social networks and local institutional support through sound planning and governance (Alfthan et al., 2009).

The analysis of community responses to disaster risks and disaster events in upstream and downstream locations in mountainous landscapes and an understanding of emerging risks as a result of climatic and land-use changes, as well as a wide range of other drivers contributing to increased disaster risk in the mountains, are required. Foreseeing how a natural calamity brought on by climate change may affect the mountains is difficult in the long run. Therefore, extensive research on snow, glaciers, landslides, earthquakes, floods, snow avalanches, and drought has to be done in these fragile ecosystems. Additionally, all-mountain locations share the intricacy of their topography, which results in the precipitation and temperature varying over even short distances, making projection studies challenging.

The components of risk, climate exposure, and susceptibility are included in risk information, which is used to analyze the risks associated with the several hazards in the mountain region.

Effective adaptation entails both adaptation (changing one's way of life, practices, and behavior to fit the new circumstances) and the development of a capacity for adaptation (knowledge, awareness, and governance) (Mirza & Burton, 2007).

Road building increases the risk of landslides and other disasters in mountainous areas. Numerous air pollutants are released by alternative technologies, such as diesel span that runs on fossil fuels. For mountain rural communities, alternative modes of transportation like the gravity goods ropeway provide a sustainable transportation option that also increases local economic activity. This book recommends solar power in small tar span technology for transportation to reduce carbon emissions and disasters in mountain valleys. The book also evaluated the effects of climate change and the challenges that they pose for the sustainable development of the populations living in the disaster-prone mountains of Uttarakhand's Almora district. The problems-facing mountainous areas of the world were also highlighted, along with potential natural remedies and adaptations from both developed and emerging nations.

11 Modelling the Future Impacts of Climate Change on Mountains

Future climate change projections indicate that natural disasters will become more frequent and severe, impacting our civilization and the ecosystem. Internationally, it is acknowledged that one of the most important steps in creating resilient societies is to be ready for and adapt to a variety of climatic disasters. Assessing the dynamics of multi-risk processes and climate change requires looking at the extent and frequency of different hazardous events over time and space, as well as how these extremes interact with dynamic social and economic fabrics and processes. Mountain regions are very sensitive to climatic change, resulting in changes in the water cycle (rainfall variability, ice melting, surface runoff), biodiversity loss and ecosystem services, damage to local economies, and human safety, according to future scenarios. They have been exposed to the effects of the present global warming, such as temperature regime shifts and permafrost thawing, because of their high degree of specialization in both natural and human systems (such as mountain species, valley population density, and tourism-based economy). The effects of climate change on the supply of ecosystem services are becoming a more pertinent area of study. The book discusses employing a fixed box model to forecast the ground-level concentration of Black Carbon at the city level. The Positive Matrix Factorization model was utilized to allocate sources of pollution and gain insight into the concentration, behavior, and seasonal patterns of aerosols and Black Carbon. This information can be utilized to ascertain the warming trend in Kullu, Himachal Pradesh, India.

12 Spatio-Temporal Ecological Vulnerability

The effects of climate change put greater pressure on ecological systems, including geographic ranges, evolutionary processes, and the phenology of plant and animal species, as well as societal systems like food production, water security, health, well-being, and settlements (IPCC, 2022). Furthermore, the stressors brought on by climate change have varying effects on different levels of biological structure. To provide a framework for implementing an ecosystem-based strategy for climate change mitigation and adaptation, the vulnerability of ecological systems must be assessed. Assessments of ecological vulnerability consider the interconnections between the interconnected human and ecological systems, as well as the early identification of potentially harmful changes in the ecosystem that protect its sustainability. Therefore, ecological vulnerability assessments are an essential instrument that may direct efficient policymaking and help decision-makers comprehend the various effects of factual and natural variables on the ecosystem. This book assessed ecological vulnerability at a spatiotemporal scale in the North-Eastern Himalayan region, which includes nine Indian states, and proposed that, in addition to climate change, anthropogenic challenges, such as soil erosion, forest degradation, and agriculture, were primarily the reason for the changes.

13 Livelihood Opportunities in the Mountains

Accurately characterizing poverty and vulnerability in the mountains is challenging since most mountain nations lack substantial data gaps pertaining to their mountains. Approaches and programs to alleviate poverty developed at the national level are prone to overlook important subnational and local expressions of poverty. In most cases, national measures have little influence on poverty reduction in the mountains and might even make mountain livelihoods more susceptible. Closing the economic gap between the lowlands and mountain areas will be impossible without additional mountain-specific poverty-reduction programs. There is a need for diversification in livelihood opportunities for the people residing in the mountains in the backdrop of a changing climate. A few alternative livelihood options that have been explicitly investigated or put into practice in Chyulu Hills, Kenya, are covered in the book, along with their actual effects on the environment, biodiversity, and community livelihoods.

14 Food and Nutrition Security in the Mountains

In the mountains, food insecurity and malnutrition are still major problems. In distant mountain areas, insecurity is even worse. The difficulties to food security in the plains are not the same as those in the mountains due to factors such as harsh biophysical conditions, restricted market access, limited economic possibilities, seasonality, inaccessibility, fragility, and rising temperatures (Jodha, 1996). Food and nutrition insecurity in the mountains is caused by several factors, including extreme poverty, the depletion of natural resources, climate change, limited market development, uncertain food aid, and a lack of institutional and governmental support (Jodha, 1996; Tiwari et al., 2014). Rapid socioeconomic and environmental changes, such as changing consumer tastes, a move toward monocropping and commodity crops, depletion of water resources, deterioration of the soil, and a decline in market value, are increasingly endangering traditional mountain food systems. Lower food yield and a detrimental effect on regional food systems result from mountain agriculture's declining competitiveness and the increased number of young people leaving the field (Devaux et al., 2020; Gliessman, 2021). These factors will ultimately affect the livelihood of the people residing in the mountainous landscapes. The book presents the depressing and intricate socio-economic effects of the vulnerability based on actual and empirical data. These observations were made from the Kosi basin in India, where farmers frequently experience land loss and soil infertility due to flooding, migrate in large numbers to distant locations in order to cope with the hardships, and grow commercial crops like maize in order to sell their produce to local buyers while securing food security with the help of subsidized public support. In the mountains, food insecurity and malnutrition are still major problems. In distant mountain areas, insecurity is even worse. The difficulties to food security in the plains are not the same as those in the mountains due to factors such as harsh biophysical conditions, restricted market access, limited economic possibilities, seasonality, inaccessibility, fragility, and rising temperatures. High poverty, natural resource degradation, climatic change, poor market development, unpredictable food assistance, and inadequate policy and institutional support are all issues that contribute to food and nutrition insecurity in the mountains. Traditional mountain food systems are now threatened by fast socioeconomic and environmental changes, such as shifting eating preferences, shifting to monocropping and commodity crops, loss of water supplies, soil degradation, and a drop in market value. Mountain agriculture is becoming less competitive, and young people are quitting agricultural careers in greater numbers, resulting in lower food output and a negative impact on local food systems. These factors will ultimately affect the livelihood of the people residing in the mountainous landscapes. In recent years, the notion of a green economy has been a focal point of policy debates (Clapp & Dauvergne, 2011). A green economy boosts economic growth, helps to reduce and adapt to climate change, provides jobs, and helps to eradicate poverty (Beniston, 2003; World Health Organization, 2012). Reduced carbon emissions and pollution, improved energy and resource efficiency, and preserving ecosystem services like biodiversity are characteristics of economic development in a green economy (Kohler et al., 2015; Zhang et al., 2022). Apart from providing sustainable livelihood opportunities and thereby strengthening the socio-economic status of locals, a green economy may also augment the prospect of mountain sustainability.

15 Soil Carbon Stocks, Ecological Security, and Livelihood

Since soil emits greenhouse gases and absorbs carbon, it is an essential climate system component. Large quantities of C are stored in soil (Ellert et al., 2007; Scharlemann et al., 2014). Among the many strategies, increasing soil carbon stocks in forest and agricultural soils has come to be seen as an important and effective one. To mitigate climate change and support ecological security, this technique entails increasing the quantity of carbon stored in the soil. One realistic way to reduce greenhouse gas emissions and decrease the negative effects of climate change is to increase the carbon content of soils. This will remove carbon dioxide from the atmosphere and store it in a stable form. Carbon-rich soils improve the overall health of plants, increase soil fertility, and sustain varied and resilient ecosystems (Song et al., 2023). These soils are more resilient to severe weather, preserve water quality, and offer vital ecosystem services. The book discussed the impact of climate change on Soil C stock in agricultural and forest soils for ecological security.

16 Gendered Implications for Livelihood

Climate change is accelerating the melting of glaciers, which is affecting mountain people's lives severely through river flooding, land erosion, droughts, and sedimentation. These situations have a wide range of consequences for people's lives and livelihoods. People who live in mountains and mountains themselves are often more susceptible to these occurrences and changes. These incidents have consequences that are not gender-neutral. Their preexisting distinctions based on caste, class, ethnicity, (dis)ability, and geography have diverse effects on women, men, and various groups of society. When preexisting roles, duties, circumstances, and places within the family, community, and wider society intersect with gender, people become either privileged or vulnerable during and after catastrophes and disasters. A chapter in the book examined the disparate effects of warming mountains on various societal groups, including those with varying gender identities, disabilities, and sexual orientations. The authors have included a few examples from the most vulnerable regions of South Asia and the Hindu Kush Himalayas to help readers think through concerns and suggest solutions.

17 Remote Sensing in Food Security and Livelihood

The incorporation of remote sensing technology into ecosystem services concepts and practices has the potential to provide practical advantages for biodiversity conservation and the development of sustainable use of Earth's natural resources. The tremendous growth of study on the issue of ecosystem services in the mountains during the last two decades has resulted in a plethora of scholarly papers. In a chapter, an analysis of Landsat images from 1991 to 2021 showed how the vegetative covering of Bangladesh's Chittagong Hill Tracts varied. Utilizing the NDVI model, land cover maps were produced with six classes: water, built-up areas, bare land, grasslands and agriculture, sparse vegetation, and thick vegetation. One remote sensing method for evaluating the density and health of vegetation is the Normalized Difference Vegetation Index (NDVI).

18 Communal Arrangements for Soil Quality Management

In mountainous regions, sustainable soil quality management and adaptation to climate change are critical challenges requiring coordinated communal efforts. Regular monitoring and evaluation are integral to the framework, enabling communities to track soil health, assess the effectiveness of adaptation measures, and adjust plans as needed. This participatory approach ensures that local knowledge and practices are incorporated, fostering resilience and sustainability. The book provides a case study in the Peruvian Central Andes for communal arrangements for sustainably managing soil quality and adapting to climate change.

19 Medicinal and Aromatic Plants

Through their positive effects on health and the economy, medicinal and aromatic plants enhance people's well-being. Over 25 percent of recently approved medications are made from natural sources. For most people in Asia, MAPs are their main source of medical supplies. The consequences of climate change on Medicinal and Aromatic Plants in Indian Himalayan Regions are described in the book.

20 International Agreements and Networks to Achieve SDGs in Mountains

Managing natural resources is crucial to the future of sustainable development of mountainous areas. Forests, water, biodiversity, and agriculture are among the region's numerous environmental resources. The management of these resources necessitates a complex network of policies, institutions, decision-making processes, and execution methods. Governments and environmental organizations must act immediately to enhance the link between knowledge, policy, and practice. Transboundary collaboration is essential for strengthening mountain environmental governance. Governments must build legal frameworks and local institutional structures to facilitate the development of ambitious programs to empower community action and encourage community-government collaborations.

Forests are essential to the lives of the mountain people. A specific goal to "ensure the conservation of mountain ecosystems, including their biodiversity, in order to enhance their capacity to provide benefits that are essential for sustainable development" is found in the UN's Agenda 2030 category Life on Land (SDG 15), which recognizes the importance of mountain ecosystems. Mountains must also be conserved in order to "guarantee the protection, restoration, and sustainable use of terrestrial and inland freshwater ecosystems and their services, in particular forests, wetlands, mountains, and drylands, in accordance with international conventions." Landslides in mountainous regions seriously threaten ecosystems, human settlements, and sustainable development. This study examines strategies for the recovery and restoration of landslide-affected areas in the mountains, emphasizing their contribution to Sustainable Development Goal (SDG) 15: Life on Land. By focusing on ecological recovery and resilience, the research demonstrates how effective landslide management supports the conservation of terrestrial ecosystems and promotes sustainable use of mountain landscapes. The proposed framework for landslide recovery includes immediate response actions such as debris removal and stabilization of slopes and infrastructure. Long-term restoration efforts are crucial and involve reforestation, soil stabilization, and the reintroduction of native plant species to restore biodiversity and ecosystem functions. Integrating community involvement and participatory planning is essential to ensure that local knowledge and needs are addressed in the restoration projects. This book highlights the importance of ongoing monitoring and adaptive management to respond to environmental changes and enhance the resilience of restored areas. By aligning landslide recovery efforts with SDG 15, this chapter advocates for a comprehensive approach that mitigates the impacts of landslides and promotes sustainable land use practices. This approach contributes to mountain ecosystems' overall health and stability, ensuring their long-term sustainability and resilience. The key elements of Sustainable Development Goals (SDGs) in mountains are given in Fig. 3. A chapter of this book focuses on developing various indicators for selected SDGs specific to the Western Himalayan region for the overall well-being of the dependent community, such as

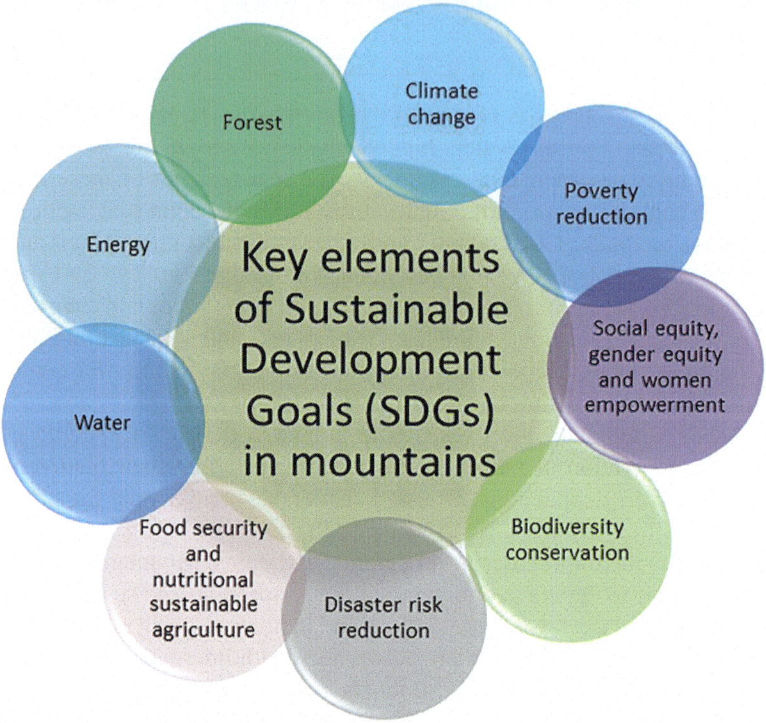

Fig. 3 Key elements of Sustainable Development Goals (SDGs) in mountains

social, economic, health, and environmental well-being based on focused group discussion and questionnaire-based survey.

21 Conclusion

In summary, the warming of mountain regions presents a complex array of challenges, opportunities, and necessary responses. The melting of glaciers, loss of biodiversity, heightened landslide risks, and shifts in water availability underscore the urgent need for adaptive and mitigative strategies. Addressing these challenges requires robust conservation efforts, comprehensive disaster management planning, sustainable tourism practices, and continuous research and monitoring to navigate these changing environments. Despite these difficulties, significant opportunities also arise. The shift towards renewable energy sources, the adoption of innovative agricultural practices, the diversification of tourism, and the deepening of scientific understanding offer pathways to mitigate the adverse effects of warming mountains and unlock new benefits. However, the challenges extend beyond immediate responses. Future research must focus on understanding the long-term impacts of climate change on mountain ecosystems, developing more accurate predictive models, and exploring the socio-economic implications of communities dependent on these regions. By embracing these opportunities and addressing the challenges through targeted research and informed strategies, we can foster resilient mountain communities and ecosystems capable of thriving amidst climate change. A comprehensive approach is crucial for ensuring mountain areas remain sustainable and that their valuable natural resources are preserved for future generations.

References

Aayog, N. I. T. I. (2018). *Sustainable Tourism in the Indian Himalayan Region*. Report of Working Group II, NITI Aayog, Government of India.

Alfthan, B., Klatze, F., Murray, A. B., Maharjan, D. R., & Thaku, A. K. (2009). *Local responses to too much and too little water in the Greater Himalayan Region.*

Ali, E. B., Anufriev, V. P., & Amfo, B. (2021). Green economy implementation in Ghana as a road map for a sustainable development drive: A review. *Scientific African, 12*, e00756.

Baral, H. (2013). *Ecosystem goods and services in production landscapes in south-eastern Australia* (Doctoral dissertation, University of Melbourne, Department of Forest and Ecosystem Science).

Bellard, C., Leclerc, C., Leroy, B., Bakkenes, M., Veloz, S., Thuiller, W., & Courchamp, F. (2014). Vulnerability of biodiversity hotspots to global change. *Global Ecology and Biogeography, 23*(12), 1376–1386.

Beniston, M. (2003). Climatic change in mountain regions: A review of possible impacts. *Climatic Change, 59*(1), 5–31.

Beniston, M. (2016). *Environmental change in mountains and uplands*. Routledge.

Beniston, M., Farinotti, D., Stoffel, M., Andreassen, L. M., Coppola, E., Eckert, N., Fantini, A., Giacona, F., Hauck, C., Huss, M., & Huwald, H. (2018). The European mountain cryosphere: A review of its current state, trends, and future challenges. *The Cryosphere, 12*(2), 759–794.

Bhadouria, R., Singh, S., Tripathi, S., & Singh, P. (Eds.). (2023). *Understanding soils of mountainous landscapes: Sustainable use of soil ecosystem services and management*. Elsevier.

Bhadra, R., Neupane, B., & Khadka, U. R. (2021). Induced impacts of climate change on livelihood and migration in the upper Himalayas: A case of Mustang, Nepal. In *Handbook of climate change management: Research, leadership, transformation* (pp. 2229–2269). Springer International Publishing.

Bolan, S., Padhye, L. P., Jasemizad, T., Govarthanan, M., Karmegam, N., Wijesekara, H., Amarasiri, D., Hou, D., Zhou, P., Biswal, B. K., & Balasubramanian, R. (2023). Impacts of climate change on the fate of contaminants through extreme weather events. *Science of the Total Environment, 909*, 168388.

Bormann, K. J., Brown, R. D., Derksen, C., & Painter, T. H. (2018). Estimating snow-cover trends from space. *Nature Climate Change, 8*(11), 924–928.

Carrivick, J. L., & Tweed, F. S. (2016). A global assessment of the societal impacts of glacier outburst floods. *Global and Planetary Change, 144*, 1–16.

Chen, Q., Li, X., Zhang, Z., Zhou, C., Guo, Z., Liu, Z., & Zhang, H. (2023). Remote sensing of photovoltaic scenarios: Techniques, applications and future directions. *Applied Energy, 333*, 120579.

Clapp, J., & Dauvergne, P. (2011). *Paths to a green world: The political economy of the global environment*. MIT press.

Dame, J., Schmidt, S., Müller, J., & Nüsser, M. (2019). Urbanisation and socio-ecological challenges in high mountain towns: Insights from Leh (Ladakh), India. *Landscape and Urban Planning, 189*, 189–199.

Dawson, N.M., Coolsaet, B., Sterling, E.J., Loveridge, R., Gross-Camp, N.D., Wongbusarakum, S., Sangha, K.K., Scherl, L.M., Phuong Phan, H., Zafra-Calvo, N. and Lavey, W.G., 2021. The role of indigenous peoples and local communities in effective and equitable conservation.

Debarbieux, B., Oiry Varacca, M., Rudaz, G., Maselli, D., Kohler, T., & Jurek, M. (2014). *Tourism in mountain regions: Hopes, fears and realities*. Department of Geography and Environment, University of Geneva; Centre for Development and Environment, University of Bern.

Devaux, A., Goffart, J. P., Petsakos, A., Kromann, P., Gatto, M., Okello, J., Suarez, V., & Hareau, G. (2020). Global food security, contributions from sustainable potato agri-food systems. In *The potato crop: Its agricultural, nutritional and social contribution to humankind* (pp. 3–35).

Dhakal, B., & Kattel, R. R. (2019). Effects of global changes on ecosystems services of multiple natural resources in mountain agricultural landscapes. *Science of the Total Environment, 676*, 665–682.

Ellert, B. H., Janzen, H. H., VandenBygaart, A. J., & Bremer, E. (2007). Measuring change in soil organic carbon storage. In *Soil sampling and methods of analysis* (pp. 25–38).

Gentle, P., & Maraseni, T. N. (2012). Climate change, poverty and livelihoods: Adaptation practices by rural mountain communities in Nepal. *Environmental Science & Policy, 21*, 24–34.

Gentle, P., Thwaites, R., Race, D., Alexander, K., & Maraseni, T. (2018). Household and community responses to impacts of climate change in the rural hills of Nepal. *Climatic Change, 147*, 267–282.

Ghalambor, C. K., Huey, R. B., Martin, P. R., Tewksbury, J. J., & Wang, G. (2006). Are mountain passes higher in the tropics? Janzen's hypothesis revisited. *Integrative and Comparative Biology, 46*(1), 5–17.

Gliessman, S. R. (2021). *Package price agroecology: The ecology of sustainable food systems*. CRC press.

Grêt-Regamey, A., Brunner, S. H., & Kienast, F. (2012). Mountain ecosystem services: Who cares? *Mountain Research and Development, 32*(S1), S23.

Gupta, M., Almomani, O., Khasawneh, A. M., & Darabkh, K. A. (2022). Smart remote sensing network for early warning of disaster risks. In *Nanotechnology-based smart remote sensing networks for disaster prevention* (pp. 303–324). Elsevier.

Heyojoo, B. P., Yadav, N. K., & Subedi, R. (2017). Assessments of climate change indicators, climate-induced disasters, and community adaptation strategies: A case from high mountain of Nepal. In *Land cover change and its eco-environmental responses in Nepal* (pp. 203–221).

Huddleston, B., Ataman, E., & d'Ostiani, L. F. (2003). *Towards a GIS based analysis of mountain environment and population* (Vol. 1, No. 10, pp. 2–4). FAO.

ICIMOD. (2010). In K. Tse-Ring, E. Sharma, N. Chettri, & A. B. Shrestha (Eds.), *Climate change vulnerability of mountain ecosystems in the Eastern Himalayas*. International Centre for Integrated Mountain Development.

IPCC. (2007). Climate change 2007: The physical science basis. In *Contribution of Working Group I to the fourth assessment report of the intergovernmental panel on climate change*. Cambridge University Press.

IPCC. (2013). *Climate change 2013: The physical science basis* (p. 1535). Cambridge University Press. https://doi.org/10.1017/CBO9781107415324

IPCC. (2022). *Climate change 2022: Mitigation of climate change. Working Group III contribution to the sixth assessment report of the intergovernmental panel on climate change (AR6). Accepted version subject to final edits*. Available online at: https://report.ipcc.ch/ar6wg3/pdf/IPCC_AR6_WGIII_FinalDraft_FullReport.pdf

Jakob, M. (2022). Landslides in a changing climate. In *Landslide hazards, risks, and disasters* (pp. 505–579). Elsevier.

Jiménez-Olivencia, Y., Ibáñez-Jiménez, Á., Porcel-Rodríguez, L., & Zimmerer, K. (2021). Land use change dynamics in Euro-mediterranean mountain regions: Driving forces and consequences for the landscape. *Land Use Policy, 109*, 105721.

Jodha, N. S. (1996). Enhancing food security in a warmer and more crowded world: Factors and processes in fragile zones. In *Climate change and world food security* (pp. 381–419). Springer Berlin Heidelberg.

Jodha, N. S. (2000). Poverty alleviation and sustainable development in mountain areas: Role of highland–lowland links in the context of rapid globalisation. In *Growth, poverty alleviation, and sustainable resource management in the mountain areas of South Asia* (pp. 541–570).

Kattel, G. R. (2022). Climate warming in the Himalayas threatens biodiversity, ecosystem functioning and ecosystem services in the 21st century: Is there a better solution? *Biodiversity and Conservation, 31*(8–9), 2017–2044.

Khan, A., Gupta, S., & Gupta, S. K. (2020). Multi-hazard disaster studies: Monitoring, detection, recovery, and management, based on emerging technologies and optimal techniques. *International Journal of Disaster Risk Reduction, 47*, 101642.

Khanduri, S., Sajwan, K. S., & Rawat, A. (2018). Disastrous events on Kelash-Mansarowar Route, Dharchula Tehsil in Pithoragarh District, Uttarakhand in India. *Journal of Earth Science & Climatic Change, 9*(4), 1–4.

Kirschbaum, D., Watson, C. S., Rounce, D. R., Shugar, D. H., Kargel, J. S., Haritashya, U. K., Amatya, P., Shean, D., Anderson, E. R., & Jo, M. (2019). The state of remote sensing capabilities of cascading hazards over High Mountain Asia. *Frontiers in Earth Science, 7*, 197.

Kohler, T., & Maselli, D. (2009). *Mountains and climate change. From understanding to action.* Geographica Bernensia.

Kohler, T., Giger, M., Hurni, H., Ott, C., Wiesmann, U., von Dach, S. W., & Maselli, D. (2010). Mountains and climate change: A global concern. *Mountain Research and Development, 30*(1), 53–55.

Kohler, T., Wehrli, A., & Jurek, M. (2014). *Mountains and climate chance. A global concern.* Centre for Development and Environment (CDE), Swiss Agency for Development and Cooperation (SDC) and Geographica Bernensia.

Kohler, T., Balsiger, J., Rudaz, G., Debarbieux, B., Pratt, D. J., & Maselli, D. (2015). *Green economy and institutions for sustainable mountain development: From Rio 1992 to Rio 2012 and beyond.* Centre for Development and Environment (CDE); Swiss Agency for Development and Cooperation (SDC); University of Geneva; Geographica Bernensia.

Kumar, A. M. I. T., Sharma, N. I. R. M. A. L., Ahmad, M., & Siddiqui, M. W. (2015). Climate change, food security, and livelihood opportunities in mountain agriculture. In *Climate dynamics horticultural science* (Vol. 28, pp. 349–360).

Leal Filho, W., Tripathi, S. K., Andrade Guerra, J. B. S. O. D., Giné-Garriga, R., Orlovic Lovren, V., & Willats, J. (2019). Using the sustainable development goals towards a better understanding of sustainability challenges. *International Journal of Sustainable Development & World Ecology, 26*(2), 179–190.

Liu, W., Xu, J., & Li, J. (2018). The influence of poverty alleviation resettlement on rural household livelihood vulnerability in the western mountainous areas, China. *Sustainability, 10*(8), 2793.

Lu, X. (2021). Mountain surface processes and regulation. *Scientific Reports, 11*(1), 5296.

Luo, Q., Bao, Y., Wang, Z., Chen, X., Wei, W., & Fang, Z. (2023). Vulnerability assessment of urban remnant mountain ecosystems based on ecological sensitivity and ecosystem services. *Ecological Indicators, 151*, 110314.

Maass, J. M., Balvanera, P., Castillo, A., Daily, G. C., Mooney, H. A., Ehrlich, P., Quesada, M., Miranda, A., Jaramillo, V. J., García-Oliva, F., & Martínez-Yrizar, A. (2005). Ecosystem services of tropical dry forests: Insights from long-term ecological and social research on the Pacific Coast of Mexico. *Ecology and Society, 10*(1), 23.

Mina, M., Bugmann, H., Cordonnier, T., Irauschek, F., Klopcic, M., Pardos, M., & Cailleret, M. (2017). Future ecosystem services from European mountain forests under climate change. *Journal of Applied Ecology, 54*(2), 389–401.

Mirza, M., & Burton, I. (2007). Climate change, adaptation and adaptive governance in water sector in South Asia. *Phys Sci Basis, 2007*, 1–19.

Mote, P. W., Li, S., Lettenmaier, D. P., Xiao, M., & Engel, R. (2018). Dramatic declines in snowpack in the western US. *npj Climate and Atmospheric Science, 1*(1), 2.

Nelson, G. C., Rosegrant, M. W., Palazzo, A., Gray, I., Ingersoll, C., Robertson, R., Tokgoz, S., Zhu, T., Sulser, T. B., Ringler, C., & Msangi, S. (2010). *Food security, farming, and climate change to 2050: Scenarios, results, policy options* (Vol. 172). Intl Food Policy Res Inst.
Notarnicola, C. (2020). Hotspots of snow cover changes in global mountain regions over 2000–2018. *Remote Sensing of Environment, 243*, 111781.
Oštir, K., Veljanovski, T., Podobnikar, T., & Stančič, Z. (2003). Application of satellite remote sensing in natural hazard management: The Mount Mangart landslide case study. *International Journal of Remote Sensing, 24*(20), 3983–4002.
Palomo, I. (2017). Climate change impacts on ecosystem services in high mountain areas: A literature review. *Mountain Research and Development, 37*(2), 179–187.
Pepin, N., Adler, C., Kotlarksi, S., & Palazzi, E. (2022). Mountains undergo enhanced impacts of climate change. *Eos Earth & Space Science News*, 103.
Pereira, P., Ignacio, M., Bogunivic, I., Francos, M., Barceló, D., & Zhao, W. (2022). Ecosystem services in mountain environments: Benefits and threats. *Pirineos, 177*, e068.
Perlik, M. (2019). *The spatial and economic transformation of mountain regions: Landscapes as commodities*. Routledge.
Praveen, B., & Sharma, P. (2019). A review of literature on climate change and its impacts on agriculture productivity. *Journal of Public Affairs, 19*(4), e1960.
Ramakrishnan, P. S. (2003). Linking natural resource management with sustainable development of traditional mountain societies. *Tropical Ecology, 44*(1), 43–54.
Rasul, G., & Molden, D. (2019). The global social and economic consequences of mountain cryospheric change. *Frontiers in Environmental Science, 7*, 91.
Ren, G., & Shrestha, A. B. (2017). Climate change in the Hindu Kush Himalaya. *Advance in Climate Change Research, 8*(3), 130–140.
Romeo, R., Manuelli, S., Geringer, M., & Barchiesi, V. (2021). *Mountain farming systems–seeds for the future: Sustainable agricultural practices for resilient mountain livelihoods*. Food & Agriculture Org.
Roy, P. S., Ramachandran, R. M., Paul, O., Thakur, P. K., Ravan, S., Behera, M. D., Sarangi, C., & Kanawade, V. P. (2022). Anthropogenic land use and land cover changes—A review on its environmental consequences and climate change. *Journal of the Indian Society of Remote Sensing, 50*(8), 1615–1640.
Scharlemann, J. P., Tanner, E. V., Hiederer, R., & Kapos, V. (2014). Global soil carbon: Understanding and managing the largest terrestrial carbon pool. *Carbon Management, 5*(1), 81–91.
Schickhoff, U. (2011). Dynamics of mountain ecosystems. In A. Millington, M. Blumler, & U. Schickhoff (Eds.), *Handbook of biogeography* (pp. 313–337).
Schickhoff, U., Bobrowski, M., Mal, S., Schwab, N., & Singh, R. B. (2022). The world's mountains in the Anthropocene. In *Mountain landscapes in transition: Effects of land use and climate change* (pp. 1–144).
Schirpke, U., Kohler, M., Leitinger, G., Fontana, V., Tasser, E., & Tappeiner, U. (2017). Future impacts of changing land-use and climate on ecosystem services of mountain grassland and their resilience. *Ecosystem Services, 26*, 79–94.
Şekercioğlu, Ç. H., Anderson, S., Akçay, E., Bilgin, R., Can, Ö. E., Semiz, G., Tavşanoğlu, Ç., Yokeş, M. B., Soyumert, A., Ipekdal, K., & Sağlam, İ. K. (2011). Turkey's globally important biodiversity in crisis. *Biological Conservation, 144*(12), 2752–2769.
Sgroi, F. (2020). Forest resources and sustainable tourism, a combination for the resilience of the landscape and development of mountain areas. *Science of the Total Environment, 736*, 139539.
Shahzad, L., Tahir, A., Sharif, F., Khan, W. U. D., Farooq, M. A., Abbas, A., & Saqib, Z. A. (2019). Vulnerability, well-being, and livelihood adaptation under changing environmental conditions: A case from mountainous region of Pakistan. *Environmental Science and Pollution Research, 26*, 26748–26764.
Sharma, T. P. P., Zhang, J., Koju, U. A., Zhang, S., Bai, Y., & Suwal, M. K. (2019). Review of flood disaster studies in Nepal: A remote sensing perspective. *International Journal of Disaster Risk Reduction, 34*, 18–27.

Sharma, I. P., Kanta, C., Dwivedi, T., & Rani, R. (2020). Indigenous agricultural practices: A supreme key to maintaining biodiversity. In *Microbiological advancements for higher altitude agro-ecosystems & sustainability* (pp. 91–112).

Shrestha, A. B., & Aryal, R. (2011). Climate change in Nepal and its impact on Himalayan glaciers. *Regional Environmental Change, 11*, 65–77.

Shrestha, U. B., Gautam, S., & Bawa, K. S. (2012). Widespread climate change in the Himalayas and associated changes in local ecosystems. *PLoS One, 7*(5), e36741.

Smith, W. K., Germino, M. J., Johnson, D. M., & Reinhardt, K. (2009). The altitude of alpine treeline: A bellwether of climate change effects. *The Botanical Review, 75*, 163–190.

Son, H. N., Chi, D. T. L., & Kingsbury, A. (2019). Indigenous knowledge and climate change adaptation of ethnic minorities in the mountainous regions of Vietnam: A case study of the Yao people in Bac Kan Province. *Agricultural Systems, 176*, 102683.

Song, M., Li, J., Gao, L., & Tian, Y. (2023). Comprehensive evaluation of effects of various carbon-rich amendments on overall soil quality and crop productivity in degraded soils. *Geoderma, 436*, 116529.

Sun, Y., Zhai, B., Saierjiang, H., & Chang, H. (2022). Disaster adaptation evolution and resilience mechanisms of traditional rural settlement landscape in Xinjiang, China. *International Journal of Disaster Risk Reduction, 73*, 102869.

Thakur, S., Negi, V. S., Dhyani, R., Satish, K. V., & Bhatt, I. D. (2021). Vulnerability assessments of mountain forest ecosystems: A global synthesis. *Trees, Forests and People, 6*, 100156.

Tiwari, P. C., Joshi, B., Grover, V. I., Borsdorf, A., Breuste, J., & Frangetto, F. W. (2014). Global change and mountains: Consequences, responses and opportunities. In *Impact of global changes on mountains: Responses and adaptation* (pp. 79–136).

Trane, M., Marelli, L., Siragusa, A., Pollo, R., & Lombardi, P. (2023). Progress by research to achieve the sustainable development goals in the EU: A systematic literature review. *Sustainability, 15*(9), 7055.

Trew, B. T., & Maclean, I. M. (2021). Vulnerability of global biodiversity hotspots to climate change. *Global Ecology and Biogeography, 30*(4), 768–783.

Tse-ring, K., Sharma, E., Chettri, N., Shrestha, A. (eds). (2010). *Climate change vulnerability of mountain ecosystems in the Eastern Himalayas; Climate change impact an vulnerability in the Eastern Himalayas – Synthesis report*. Kathmandu: ICIMOD.

Unay-Gailhard, İ., & Bojnec, Š. (2019). The impact of green economy measures on rural employment: Green jobs in farms. *Journal of Cleaner Production, 208*, 541–551.

van de Sand, I. (2012). Payments for ecosystem services in the context of adaptation to climate change. *Ecology and Society, 17*(1), 14.

Villeneuve, A., Talla, P., & Mekouar, M. A. (2002). *The legal framework for sustainable mountain management: An overview of mountain-specific instruments* (pp. 56–65). UNASYLVA-FAO.

Walsh, J. E., Ballinger, T. J., Euskirchen, E. S., Hanna, E., Mård, J., Overland, J. E., Tangen, H., & Vihma, T. (2020). Extreme weather and climate events in northern areas: A review. *Earth-Science Reviews, 209*, 103324.

World Health Organization. (2012). *Health in the green economy: Health co-benefits of climate change mitigation-transport sector*. World Health Organization.

Yongjian, D. I. N. G., Shiqiang, Z. H. A. N. G., & Rensheng, C. H. E. N. (2020). Cryospheric hydrology: Decode the largest freshwater reservoir on earth. *Bulletin of Chinese Academy of Sciences (Chinese Version), 35*(4), 414–424.

You, Q., Wu, T., Shen, L., Pepin, N., Zhang, L., Jiang, Z., Wu, Z., Kang, S., & AghaKouchak, A. (2020). Review of snow cover variation over the Tibetan Plateau and its influence on the broad climate system. *Earth-Science Reviews, 201*, 103043.

Zeng, L., Li, R. Y. M., Nuttapong, J., Sun, J., & Mao, Y. (2022). Economic development and mountain tourism research from 2010 to 2020: Bibliometric analysis and science mapping approach. *Sustainability, 14*(1), 562.

Zhang, L., Xu, M., Chen, H., Li, Y., & Chen, S. (2022). Globalization, green economy and environmental challenges: State of the art review for practical implications. *Frontiers in Environmental Science, 10*, 870271.

Emerging Threats to Mountain Ecosystem Services Under Changing Climate

Reema Mishra, Ridhi Khurana, Aparajita Mohanty, and Preeti Agarwal

1 Introduction

Climate change has emerged as an issue of global concern as it poses a threat to ecosystems, biodiversity and consequently the well-being of society. It refers to long-term variation in the temperature and weather forms that describes the regional, local and global climates of the earth. Climate change is primarily a result of emissions of greenhouse gases derived mainly from the burning of fossil fuels. Besides this, various natural processes such as cyclical ocean patterns (e.g. El Niño, La Niña and Pacific Decadal Oscillation), volcanic eruptions, variations in the sun's energy output, alterations in earth's orbit and earth's crust are other factors responsible for climate change (Climate Change, United Nations; https://www.un.org/en/climatechange/science/causes-effects-climate-change).

Ecosystems provide numerous economic and ecosystem services (ES). Economic services include food, fresh water, medicine, fuel wood and raw material for industries. It offers various ecological services including regulatory services such as climate regulation, water cycle, control of disease and pollinators for plants; cultural services such as spiritual value, recreational and aesthetic appeal; and supporting services such as soil formation, nutrient cycling and flood control. They also affect the well-being in terms of security, livelihood, health and social security. Thus,

R. Mishra · A. Mohanty · P. Agarwal (✉)
Department of Botany, Gargi College, University of Delhi, New Delhi, India
e-mail: preeti.agrawal@gargi.du.ac.in

R. Khurana
Department of Botany, Hansraj College, University of Delhi, New Delhi, India

S. Tripathi et al. (eds.), *Warming Mountains*,
https://doi.org/10.1007/978-3-031-62197-0_2

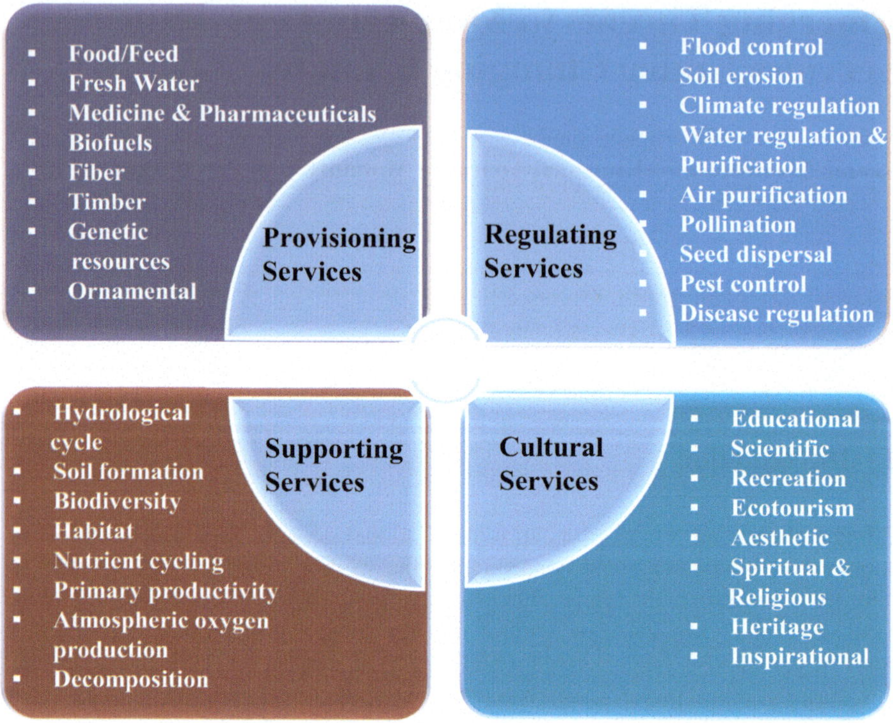

Fig. 1 Schematic representation of different types of mountain ecosystem services

diversity in communities and the functioning of ecosystems plays an essential role in ES sustenance and well-being of society (Palomo, 2017). Climate change disturbs the structure and function of ecosystems, and studies suggest that there exists a connecting link between climate change and ecosystem functioning, ES, biodiversity, social well-being and other drivers of global changes such as population growth and consumption; and changes in land use, energy use and pollution (Parmesan & Yohe, 2003; IPCC, 2007a, b). A schematic representation of types of ES has been depicted in Fig. 1.

Mountains are the most delicate, severely affected ecosystem and are primary indicators of climate change (Singh et al., 2010). Mountains cover 22% of the earth's surface and serve as home to 13% of the world's population (Sharma et al., 2019). They are broadly distributed covering all continents, ranging from arctic to tropical latitudes and maintain a large divergent biota, human population, livelihoods and ES. Further, mountains are rich repositories of biodiversity, genetic resources, water, food, medicinal plants, fibre, timber and non-timber products. They are home to many endemic and threatened species (Tse-Ring et al., 2010; Fort, 2015).

The climate change affects mountain ecosystem (ME) mostly via. effect on food and water availability, natural hazards control, culture, spirituality, aesthetics and recreation (Palomo, 2017). Mountains display significant environmental variations at both regional and local levels. It is expected that there will be a rise in both mean and extreme temperature; an increase in the frequency of natural catastrophes; regional variation in precipitation; dry areas will become drier; wet areas will turn wetter; and growing temperature will decrease precipitation in the form of snow. Mountains on high and medium latitudes will experience more changes (Egan & Price, 2017). The natural hazards are expected to increase in terms of both frequency and magnitude and will have a serious impact on both infrastructure and the livelihood of people in the mountains (Oppenheimer et al., 2015; Egan & Price, 2017). The impact is widespread; however, its response differs w.r.t. variation in exposure, susceptibility and ability to adapt (Kovach et al., 2019). A tabular representation of various conditions arising as a consequence of climate change and its impact on ME/MES and people surviving on it is provided in Table 1.

With the receding glaciers, there is a shift in snowlines in the upward direction and a change in the river flow (Preston et al., 2016; Palomo, 2017). The change in water flow will adversely affect hydropower generation, biodiversity, forestry and agriculture (Fort, 2015). Mountains are further exposed to other hazards as well and change in climate will intensify its vulnerability by increasing the probability of landslides, storms and avalanches in both frequency and intensity (Kohler et al., 2010). The distinctive environment that symbolize mountains such as remoteness, instability, heterogeneity and marginality may get affected in response to this increasing climate unpredictability. Thus, rapid changes in these fragile ecosystems as an outcome of climate change will impose unprecedented threats to wildlife, livelihoods, socio-economic condition and culture of local people and also to billions of those living downstream (Fort, 2015; Egan & Price, 2017).

In addition to climate change, there are other interacting drivers of change that is globally influencing the mountain system such as increasing population pressure, modification in land-use pattern, globalization and economic policy. These changes together signify that the impact of climate change on mountains is a matter of grave concern that needs to be globally addressed as it has far-reaching consequences. It further highlights the need to comprehend the multidimensional drivers of climate change, their response and feedback mechanisms (Kohler et al., 2010; Egan & Price, 2017).

In view of the importance of ME and its susceptibility to the climate change, this work will provide an insight into ME and its services and the impact of climate change on ME, its services and people residing in mountains. Thus, it will highlight the requisites to keep check on MES and develop adaptation method needed for locals and tourists. This study will also aid in spreading awareness among students and scholars, and thus will encourage research and innovation to develop and improve methods that will help to assess the impact of climate change. This would further help in developing strategy for mitigating the impact of climate change on mountains.

Table 1 Climate change induced parameters and their impact on the environment

S. no.	Parameters impacted by climate change	Impact on ME/MES	References
1	Extreme hot and dry summers	The amount of water recharging the aquifers decreases and thus reduces water availability	Vanham et al. (2009)
2	Reduced stream flow	Decreased accessibility of water leads to water shortage to meet the demand of water of tourists in Khumbu, Nepal	McDowell et al. (2013)
3	Reduced snow cover and varying patterns in precipitation	In high-elevation villages such as India and Nepal, it affects the subsistence agriculture	Ogra and Badola (2015) and Smadja et al. (2015)
4	Increased temperature	With rise in temperature, there is possibility of cultivation of new crops that could not be grown before in high mountain areas, which could lead to diversification in local diets	Chaudhary et al. (2011) and Konchar et al. (2015)
5	Permafrost melting; glacial lake outburst floods (GLOFs)	Permafrost melting results in landslides and rock fall; glacial lake outburst floods and extreme weather leads to virulent storms	Huggel et al. (2012) and Ritter et al. (2012)
6	Withdrawal of glaciers	It decreases the stability of land surface, which in turn affects the frequency of natural hazards	Keiler et al. (2010) and Nepal (2011)
7	Natural hazards such as landslides, rockfall, GLOFs and stream flooding	This leads to destruction of infrastructure and transportation. Destruction of roads restricts the connectivity among regions and also affects access to markets	Dame and Nüsser (2011) and Fort (2015)
8	GLOFs	It can affect the hydropower dams and thus pose threat to distribution of water and energy regulation. Many dams in Bhutan have been identified to be at risk	Hoy et al. (2016)
9	Droughts and other severe events	The yield and quality of crop decreases. The decrease in production can also affect the price of the crops	Zhao et al. (2017) and Gowda et al. (2018)
10	Rise in ocean temperature	Decreases the level of oxygen and may cause decline in average body size of fish by 14–24% by 2050	Cheung et al. (2013)
11	Rise in stream temperatures	In freshwater systems the increased temperature will adversely impact some of the harvested species	Isaak et al. (2012) and Crozier et al. (2019)
12	Increased temperature	Enhances the decomposition of soil organic matter, will cause increase in soil carbon losses and will also modify the balance of C:N	Davidson and Janssens (2006) and Crowther et al (2011)
13.	High loading of nutrient and increase in temperature	May increase frequency of harmful algal blooms that has a negative impact on human and animal health. It is projected that by the end of this century, conditions may favour blooms of *Alexandrium catanella*, a toxic alga that will further affect the aquaculture, recreation, and so on	Sandifer and Sutton-Grier (2014) and Hilborn and Beasley (2015)

2 Mountain Ecosystem (ME)

Mountains exhibit unique diversity and complexity. Despite sharing common characteristics of steep slopes and high relative relief, mountains display extreme diversity (Messerli & Ives, 1997). The physical and chemical characteristics of the soil generally differ across and within the mountain ranges, and furthermore, there is a lack of uniformity in the pattern of topography among the mountain regions (Rahbek et al., 2019a). The characteristics of soil along the mountain slopes are determined by local features such as their geographical location, topography and vegetation. Despite the significant variation in the topography of mountains from one region to another, the fall in mean air temperature is steady as the elevation rises. The rate of fall in temperature also decreases with the elevation that is further determined by the location and slope of the mountain. Other abiotic factors such as relative humidity, precipitation and solar radiation also differ in mountains and do not show unidirectional trends with elevation (Körner, 2007; Tito et al., 2020). Thus, mountains are characterized by their steep topography, and biological and climatic gradient with marked seasonal contrasts, which further supports the extreme geomorphic and climatic events that may consequently affect the ecological as well as human environment (Huggel et al., 2010; Körner, 2013; Tito et al., 2020).

Owing to the high degree of differences in topography and climate, mountains in general display a high degree of endemism and diversity at all levels (genetic, species and ecosystem) (Mittermeier et al., 2011). The structure, composition and richness of both plant and animal communities show variation with elevation. Though they cover 12% of the land surface, they serve as hosts for greater than 85% of the world's amphibian, mammal and bird species and for about one-third of terrestrial plant species (Körner, 2004, 2007; Körner et al., 2011; Rahbek et al., 2019b). They are home for certain world's endemic and threatened species and, interestingly, one half of the worldwide biodiversity hot spots are in the mountain regions. Although ME are essential components of the earth, they are also the most fragile and are under immense threat imposed by climate change and additional drivers associated with global climate change (Tse-Ring et al., 2010). Thus, mountains are considered as centers of biological diversity and are prioritized areas for conservation programmes (Tito et al., 2020). Further, in view of the significance of ME, the Convention on Biological Diversity initiated a Programme of Work on Mountain Biodiversity in 2004 that intended to minimize the decline of mountain diversity at regional, global and national levels by 2010. Mountains hold international importance as world's most important major rivers and several smaller rivers originate from mountains (Mountain Agenda, 1998). They are referred as 'water towers' as they stock and supply fresh water to the downstream areas for domestic use, irrigation, production of food, hydropower energy, etc. (de Jong et al., 2009; Viviroli et al., 2011). Agenda 21—Chapter 13, UNCED, Rio de Janeiro—1992 (UNEP, 1992), clearly identifies mountains and uplands as a main part of the worldwide environment (Barnett et al., 2005; Beniston, 2005; Tse-Ring et al., 2010).

The complex topography of mountains makes it more challenging to understand the climatic conditions in mountains. Climate and ecosystems are closely associated with each other, the climate being an integral component of ME. Climate influences the environment and its characteristics in mountains and likewise mountains (via. altitude, latitude, topography) also influence the climatic variables such as relative humidity, wind patterns, solar radiation and cloud conditions. Thus, ME displays tremendous variation in patterns of local climatic conditions that can be observed in the form of types of vegetation and cover and hydrological and geomorphic landscapes (Tse-Ring et al., 2010; Palomo, 2017).

In view of the significance of ME and the growing threats imposed by ongoing climate change in the mountains, The International Centre for Integrated Mountain Development (ICIMOD) conducted research in collaboration with the Eastern Himalayas from 2007 to 2008. They analysed the influence of climate change on Eastern Himalayas and observed its susceptibility to climate change because of its ecological instability and economic marginality (Tse-Ring et al., 2010). A national adaptation programme (UN Framework Convention on Climate Change) aims to build a methodology (harmonized policy) to combat the ecological and socio-economic impacts of climate change (Egan & Price, 2017). The National Mission for Sustaining the Himalayan Ecosystem (NMSHE) is one of the missions of the Indian government (under the NAPCC—National Action Plan on Climate Change) administered by the Ministry of Science and Technology. It strives to address certain significant issues associated with the Himalayan glaciers; conservation and protection of biodiversity and wildlife; livelihoods; societies for traditional knowledge; and planning for sustainable of Himalayan ecosystem (https://dst.gov.in/sites/default/files/NMSHE_Mission_document.pdf). The National Mission on Himalayan Studies (NMHS), a scheme launched by MoEF&CC (Ministry of Environment, Forest & Climate Change), supports the attempt of NMSHE. This is dedicated to provide holistic understanding of component of system and address major issues associated to conservation and sustainable management of natural resources in IHR—Indian Himalayan Region (http://nmhs.org.in/pdf/publication/Mission_Documents/MIssion_Document.pdf). The Mountain Partnership is a United Nations (UN) association of partnership that targets to improve the lives of people in mountains and safeguard the mountain environments globally (https://www.fao.org/mountain-partnership/about/en/). Another is the IPCC—The Intergovernmental Panel on Climate Change (IPCC)—founded by the WMO (World Meteorological Organization) and the United Nations Environment Programme to estimate relevant data at technical, scientific and socio-economic to gain an insight into the influence of climate change and to find out possibilities for mitigation and adaptation. The WG II report approves that mountains among others and people depending on them are most susceptible to the impacts of climate change. The WMO (World Meteorological Organization), in partnership at global level would provide support in capacity building, education, training and awareness at various levels to address problems associated with the climate change. It further provides an active assistance to the accomplishment of the UN Millennium Development Goals (MDGs) in eliminating hunger and poverty and in guaranteeing environmental sustainability. Adaptation at Altitude programme funded by the Swiss Agency of Development of Cooperation aims to increase the accessibility and use of mountain observation data and data on climate change-associated impacts in mountains; incorporate mountain climate change adaptation into planning and policy procedures; support and apply information on climate change adaptation solutions in mountains; confirm that key global policy procedures take account of climate change adaptation in mountains. The first phase of programme will start from 2020 to 2023 (http://www.sd-caucasus.com/en/projects/project_item/208). Thus, the inherent environmental instability of the ME and the vulnerability of socio-economic conditions of people residing on the mountains have pushed the issues associated with ME to the top agenda of international sustainability.

3 Mountain Ecosystem Services (MES)

ME offer a wide range of ES to both their residents as well as tourists. In Millennium Ecosystem Assessment report (MEA, 2003), MES have been categorized into provisioning, regulating, supporting and cultural services. Important provisioning services consist of food, medicines, fibre, ornamental plants, wood, fodder, habitat and genetic resources. Regulating and supporting services have been further subdivided into physical (climate regulation, erosion, natural threats, quality of air and water flow of water) and biological factors. Supporting services include control of pathogen and diseases, pollination, purification of water and dispersal of seeds. Cultural services consist of education, beautiful landscapes, religious and spiritual values, leisure and tourism, sacredness and cultural heritage (Egan & Price, 2017). Some of the important MES are discussed in the following sections.

3.1 Biodiversity

Mountains have long been known to be key regional and worldwide hubs of biodiversity. They are recognized as ‘hot spots of biological diversity’. Due to extreme climatic conditions and topography, mountain regions frequently show the presence of many endemic species and tremendous biological variability at genetic, ecosystem and species levels. The abundant biodiversity offers food, medicines, timber and fibres to the mountain people. Many essential crops such as barley, maize and potatoes and domestic animals such as sheep, yaks and goats also support the livelihood of mountain people. Mountains also form a significant gene reservoir for numerous medicinal plants, wild relatives of crop plants and ornamental plants (Egan & Price, 2017).

3.2 Climate Regulation

Physical MES includes climate regulation, quality of air, soil erosion, movement of water and natural calamities (MEA, 2003; Egan & Price, 2017). By influencing atmospheric circulation, large mountain ranges may have an even greater impact on the spread of total ozone in the atmosphere (Han et al., 1998). The hydro-biosphere of the mountain is made up of the number of intricately interconnected systems that eventually influence the availability and quantity of water. Healthy functioning ME also provide protection against natural hazards and the effects of extreme events, mainly floods and droughts. The above-mentioned MES are critical, especially to downstream areas, where the influence of such events are often more severe, and sometimes extends to several hundreds of kilometres away (Egan & Price, 2017).

3.3 Cultural Services

Cultural ES are primarily non-materialistic in nature and include aesthetics, natural landscapes, spiritual, religious, educational values, ecotourism and recreation (Wester et al., 2019; Li & Cai, 2022). ME offers cultural values to both their inhabitants and tourists. Cultural ES are very important for supporting the psychological facets of human well-being. The remote and inaccessible nature of mountains has enabled the preservation of rituals, distinctive indigenous culture, religious beliefs and related traditional knowledge and influenced the biophysical environment of ME. Sacred groves, forests, lakes and holy mountain peaks are important landscapes, which have significant tourism and biodiversity values (Anthwal et al., 2010; Egan & Price, 2017; Wester et al., 2019).

3.4 Recreation and Tourism

In many mountainous regions around the globe, tourism and leisure are the main drivers of local economies and frequently make sizable assistance to national economies. A growing industry for creating jobs and revenue is tourism. For thousands of years, pilgrimage, a form of tourism, has been centred around mountains (Pohner et al., 2009). The majority of ME offer a wide range of leisure opportunities, particularly for winter sports, trekking and climbing (Debarbieux et al., 2014). While several lowland people still associate mountains with 'purity' and a close connection between inhabitants and the environment, today's mountain regions actually draw the majority of tourists, frequently from very different cultural backgrounds (Egan & Price, 2017).

4 MES Impacted by Climate Change

The Himalayan mountain ranges, their ecosystems and services that have borne the impact of climate change have been specifically discussed.

4.1 Habitat

Ecosystems in the Himalayan ranges are exposed to a multitude of human activities influencing and altering the environment at various levels. The vulnerability of local ecosystems to climate change increases as a result of the disruption of habitat for both flora and fauna from land-use and land-cover changes. Changing weather conditions over the recent decades have enormously impacted frozen water sources such as snow and glaciers, reduced precipitation and subsequent spring water depletion. The green cover has been impacted as well with shifts in vegetation patterns in terms of geographical locations and an overall reduction of the biological diversity in the region.

Studies on the western Himalayan mountain ranges have shown an increase in temperature from 1 to 2.8 °C, an overall decrease in precipitation, reduced cloud cover and a reduced number of snowfall days during peak snow season from January to March (Shekhar et al., 2010). The change in temperature and precipitation in turn impacts the phenological aspects of the ecosystem. Studies show advancement as well as delays in the start of the growing season (Shrestha et al., 2012). A higher proportion is indicative of advancements of the growing season, as much as a month earlier is reported for rhododendrons (Xu et al., 2009). Selective studies have been conducted for various mountain ranges, but an overall comparison of changing trends still needs to be consolidated with studies approaching a wider geographical domain along with spreading awareness on the matter at hand (Shekhar et al., 2010; Dimri & Dash, 2012).

4.2 Biodiversity and Their Distribution

The Himalayas along with the Andes and the Alps comprise the major mountain biodiversity hotspots of the world. Of these, the Himalayas are the largest mountain ecosystem harbouring one of the richest reserves of mountain flora along with supporting the indigenous human population (Xu et al., 2009). More than 10,000 vascular plant species alone inhabit the Himalayan ranges, of these approximately 31.6% are endemic to the region in addition to other plant species (Mittermeier et al., 2004). Studies indicate high vulnerability for the states of Jammu and Kashmir and Himachal Pradesh occupying a significant proportion of the biodiversity hotspots in the IHR owing to the rising temperature conditions (Chaturvedi et al.,

2011). The rich diversity of flora and fauna in the IHR provides the livelihood for the rural sector from both crop and livestock farming (Gairola et al., 2009). The receding glaciers are impacting freshwater ecosystems severely, along with risking the survival of freshwater species such as molluscs, fishes and dragonflies (IUCN, 2022; https://www.iucnredlist.org/). The Indian agrarian economy is dependent on these livestock, crop and horticultural farming activities. Long-term as well as short-term strategies need to be imposed for maintaining the resilience of the IHR ecosystems.

Harvey et al. (2023) have extensively reviewed the impact of anthropogenic stress coupled with a changing climate on insect biodiversity globally in alpine, temperate and tropical ecosystems. Insects form an integral part of both overground and underground ecosystems where they contribute to various ES such as pollination, controlling pests and regulating nutrient uptake and availability (Schowalter et al., 2018). Climatic fluctuations along with the ability to track climatic shifts are essential for determining species' distributions (Hewitt, 2000). Biodiversity hotspots, both terrestrial and aquatic, have been shown to bear negative impact from the changing climate and most severely impacted are endemic species across various geographical areas (Chapin & Díaz, 2020; Manes et al., 2021).

The changing climate poses threats to both the montane flora and fauna. These need to be implemented by prioritizing the structural organization and understanding the various networks underlying these rich and diverse ecosystems in a climate-smart manner.

4.3 Water Resources

Frozen glaciers across the globe are crucial for sustenance of the world. Climate change influences generation of hydroelectric power either directly or indirectly: directly, through changing precipitation patterns and river inflows and indirectly, through the ever-increasing demand for power (Maran et al., 2014). Thus, the climate change-induced hastened melting of the glaciers poses a major strain to hydroelectric power generation along with catastrophic floods in surrounding regions (Laghari, 2013). The Himalayas in particular are drained by ten major river systems. Of these, the three major river systems important for India include the Indus, the Ganges and the Brahmaputra. The hydrologic cycle of these is immensely impacted by the changing climate conditions. The most widely reported manifestation of the same is the rapid recession of glaciers with its implications for future downstream water supplies (Solomon, Intergovernmental Panel on Climate Change and Intergovernmental Panel on Climate Change, 2007). Precipitation, temperature, snowfall and area covered by glaciers influence the river flow and the freshwater springs. Efficient water utilization and conservation strategies must be implemented by the people residing in the IHR (Bhutiyani et al., 2008; Bhambri et al., 2011). Prominent and frequent glacier recession has added risks to human life in the vicinity of these glaciers, particularly to those located downstream of potentially dangerous glacial lakes (PDGLs).

Studies indicate towards a non-uniform change in the Hindu-Kush Himalaya cryosphere owing to changing climatic conditions across different river basins (Miller et al., 2012). Over and above the dumping of waste into water bodies further impacts the water quality and aquatic biodiversity of the same (Bouzas-Monroy et al., 2022; Wilkinson et al., 2022). Although we have highlighted the drain, dependence and possible impact of climate change on the water resources in the IHR, similar studies have been reported on and from other mountain systems of the world (Maran et al., 2014; Berga, 2016; Mishra et al., 2020; Wei et al., 2020; Hunt et al., 2022).

4.4 Human Health and Livelihood

Human health and livelihood in the IHR are at risk from the detrimental impact of changing weather on vegetation and wildlife in the region. The changing climate has vastly increased the frequency of landslides in the IHR (Chalise, 2001). Additionally, studies have also reported the degradation of agricultural land resulting from heightened human activities in the Himalayan region (Mishra et al., 2022). Soil health has declined owing to soil erosion from increased human activity, increase incidence of landslides and loss of green cover coupled with a lack of management and sustainable development regimes.

Moreover, the native populations inhabiting the IHR have a largely untapped traditional knowledge system (TKS) for health care, farming and the use and management of natural resources. The shifting climate patterns are leading to changes in socio-economic and altering domestic gender-biased roles in the IHR (Bhadwal et al., 2019). Such vast reservoir of traditional knowledge passed over from generations suffers today due to the loss of not only biological but linguistic and cultural and diversity as well. Additionally, youth migration out of the IHR in search of better, stable and secure opportunities also poses threats to the TKS in the region.

Tourism is a significant source of livelihood in mountainous regions across the globe. The primary tourist attractions here are largely climate sensitive—the natural biodiversity, landscapes and adventure sports. The changing climate has forced locals to migrate towards alternate livelihoods and also move to other locations impacting not only themselves but the economy at large at the level of employers as well as workers and labourers. Changing climate will eventually bring about changes in tourism patterns in geographical locations globally (Steiger et al., 2022). Understanding and improving the impact of climate change on the IHR requires the development of ecological calendars or plans in sync with the TKS in place in the local communities. IHR boasts of rich biodiversity naturally resilient to crop diseases. Baranaja, a mixed cropping strategy involving 12 crops (Baranaja), is a good example of such diversity (Ghosh & Dhyani, 2004). The inherent ability of these crops to tolerate environmental risks and natural hazards imparts food and nutritional security to generations of farmers in the region.

Crop yields have seen drastic reductions in recent years as a result of climatic changes. The impact on agricultural practices in the hilly regions arises from the reduced irrigation due to lack of water, altered rain patterns and conditions of drought leading to failure in crop germination and fruiting, as well as invasion of weeds (Kc et al., 2022). Climate change also leads to a shift in the incidence, migration and viability of crop pests with temperature being a dominant factor. The faster growth and development of temperate zone insects are promoted by high summer temperatures, adding to the number of pest generations per year (Bale et al., 2002).

5 Future Directions

Climate change is impacting the entire world, but more so in countries such as India which are largely dependent on their agricultural resources. The ecosystem dynamics of the IHR pose challenges and opportunities alike. Studying, analysing and utilizing adaptive traits in traditional varieties of crops and livestock breeds could provide solutions in executing ecosystem resilience measures. Promotion of in situ conservation of the various indigenous genetic resources and expanding environmental impact awareness, as well as imparting climate-sensitive education are the need of the hour.

It therefore becomes essential to integrate research with outreach programmes to generate a two-way communication channel between the locals and concerned authorities and institutions for making a climate-smart future. Global, national and regional climate actions that foresight climate resilience per se is crucial for securing the future of the ecosystems harboured in the IHR.

6 Conclusions

Climate change is a growing and widespread threat to the ecosystem, biodiversity and ES. Mountain ecosystem, a global heritage, is under major threat imposed by climate change, anthropogenic activities and other drivers of global change, consequently exerting a catastrophic effect on infrastructure, livelihoods, well-being and life of people as well. It is difficult to assess the impacts of environmental changes owing to the diversity and intricacy of mountains, and the prediction of future climate trend depends on the huge network of meteorological stations and the modelling outputs derived from satellite data. To address and combat issues associated with climate change impacts, it requires research in several areas.

- In view of the significance of mountains and the growing threat of climate change in mountains, there is a need to develop and implement mitigation and adaptive methods more effectively.
- Effective ecosystem management involving proactive and flexible methods can help to promote adaptation to climate change. It must focus on climate-associated risks as well as socio-economic advancement. This ecosystem-based adaptation procedure can provide a better way to safeguard and restore the mountain ecosystem by improving its resilience.
- This method demands sustainable and judicious use of biodiversity and other ES in order to strengthen the adaptive capability of the socio-ecological structures of the mountain under the expected impact of climate change.
- There is a lack of consistent and long-term data on climate and resources of the mountains that necessitates the development of monitoring methods that would provide exact predictions of climate change and impact assessment, as well as documentation is needed for prevailing adaptation approaches in and beyond the mountain areas.
- Further, there is the requirement for conservation policy at both regional and national levels as well as integrated research to combat climate issues and turn challenges into opportunities by improving and diversifying the livelihood and income of people on mountains. One such example is Adaptation at Altitude, a collective agenda that strive to escalate the adaptive capacity as well as resilience and of communities residing on mountains and ecosystems in response to climate change. Another is the 'UN Decade on Ecosystem Restoration' (launched by UN Environment Programme and allies) (2021), which is dedicated to restore ecosystems across the world. The Mountain Partnership is a UN partnership and international association dedicated for mountain and improving the lives of peoples on mountains across the world. Another is the IPCC (founded by the World Meteorological Organization and the United Nations Environment Programme), which evaluates relevant information at the technical, scientific and socio-economic levels to gain insight into climate change, its impact and the possibilities and opportunities for mitigation and adaptation in response of climate change.
- Education sector can play a significant role in providing an opportunity for successful mitigation and adaptation to climate change through information, knowledge and skill development and enhancement. Moreover, education can help in building and improving the capacity of individuals and thereby communities to make cognizant decisions and take suitable actions for developing climate resilient and sustainable world.

References

Anthwal, A., Gupta, N., Sharma, A., Anthwal, S., & Kim, K. H. (2010). Conserving biodiversity through traditional beliefs in sacred groves in Uttarakhand Himalaya, India. *Resources, Conservation and Recycling, 54*(11), 962–971. https://doi.org/10.1016/j.resconrec.2010.02.003

Bale, J. S., Masters, G. J., Hodkinson, I. D., Awmack, C., Bezemer, T. M., Brown, V. K., Butterfield, J., Buse, A., Coulson, J. C., Farrar, J., Good, J. E. G., Harrington, R., Hartley, S., Hefin Jones, T., Lindroth, R. L., Press, M. C., Symrnioudis, I., Watt, A. D., & Whittaker, J. B. (2002). Herbivory in global climate change research: Direct effects of rising temperature on insect herbivores. *Global Change Biology, 8*(1), 1–16. https://doi.org/10.1046/j.1365-2486.2002.00451.x

Barnett, T. P., Adam, J. C., & Lettenmaier, D. P. (2005). Potential impacts of a warming climate on water availability in snow-dominated regions. *Nature, 438*(7066), 303–309. https://doi.org/10.1038/nature04141

Beniston, M. (2005). The risks associated with climatic change in mountain regions. In U. M. Huber, H. K. M. Bugmann, & M. A. Reasoner (Eds.), *Global change in mountain regions. An overview of current knowledge* (pp. 511–519). Springer. https://doi.org/10.1007/1-4020-3508-X_51

Berga, L. (2016). The role of hydropower in climate change mitigation and adaptation: A review. *Engineering, 2*(3), 313–318. https://doi.org/10.1016/J.ENG.2016.03.004

Bhadwal, S., Sharma, G., Gorti, G., & Sen, S. M. (2019). Livelihoods, gender and climate change in the Eastern Himalayas. *Environmental Development, 31*, 68–77. https://doi.org/10.1016/j.envdev.2019.04.008

Bhambri, R., Bolch, T., Chaujar, R. K., & Kulshreshtha, S. C. (2011). Glacier changes in the Garhwal Himalaya, India, from 1968 to 2006 based on remote sensing. *Journal of Glaciology, 57*(203), 543–556. https://doi.org/10.3189/002214311796905604

Bhutiyani, M. R., Kale, V. S., & Pawar, N. J. (2008). Changing streamflow patterns in the rivers of northwestern Himalaya: Implications of global warming in the 20th century. *Current Science, 95*(5), 618–626.

Bouzas-Monroy, A., Wilkinson, J. L., Melling, M., & Boxall, A. B. (2022). Assessment of the potential ecotoxicological effects of pharmaceuticals in the World's rivers. *Environmental Toxicology and Chemistry, 41*(8), 2008–2020. https://doi.org/10.1002/etc.5355

Chalise, S. R. (2001). An introduction to climate, hydrology, and landslide hazards in the Hindu Kush Himalayan region. In *Landslide hazard mitigation in the Hindu Kush Himalayas*. ICIMOD.

Chapin, F. S., III, & Díaz, S. (2020). Interactions between changing climate and biodiversity: Shaping humanity's future. *Proceedings of the National Academy of Sciences, 117*(12), 6295–6296. https://doi.org/10.1073/pnas.2001686117

Chaturvedi, R. K., Gopalakrishnan, R., Jayaraman, M., Bala, G., Joshi, N. V., Sukumar, R., & Ravindranath, N. H. (2011). Impact of climate change on Indian forests: A dynamic vegetation modeling approach. *Mitigation and Adaptation Strategies for Global Change, 16*, 119–142. https://doi.org/10.1007/s11027-010-9257-7

Chaudhary, P., Rai, S., Wangdi, S., Mao, A., Rehman, N., Chettri, S., & Bawa, K. S. (2011). Consistency of local perceptions of climate change in the Kangchenjunga Himalaya landscape. *Current Science, 101*(4), 504–513. http://www.jstor.org/stable/24078981

Cheung, W. W., Sarmiento, J. L., Dunne, J., Frölicher, T. L., Lam, V. W., Deng Palomares, M. L., Watson, R., & Pauly, D. (2013). Shrinking of fishes exacerbates impacts of global ocean changes on marine ecosystems. *Nature Climate Change, 3*(3), 254–258. https://doi.org/10.1038/nclimate1691

Crowther, T. W., Boddy, L., & Jones, T. H. (2011). Outcomes of fungal interactions are determined by soil invertebrate grazers. *Ecology Letters, 14*(11), 1134–1142. https://doi.org/10.1111/j.1461-0248.2011.01682.x

Crozier, L. G., McClure, M. M., Beechie, T., Bograd, S. J., Boughton, D. A., Carr, M., Cooney, T. D., Dunham, J. B., Greene, C. M., Haltuch, M. A., & Hazen, E. L. (2019). Climate vulnerability assessment for Pacific salmon and steelhead in the California Current Large Marine Ecosystem. *PLoS One, 14*(7), e0217711. https://doi.org/10.1371/journal.pone.0217711

Dame, J., & Nüsser, M. (2011). Food security in high mountain regions: Agricultural production and the impact of food subsidies in Ladakh, Northern India. *Food Security, 3*, 179–194. https://doi.org/10.1007/s12571-011-0127-2

Davidson, E. A., & Janssens, I. A. (2006). Temperature sensitivity of soil carbon decomposition and feedbacks to climate change. *Nature, 440*(7081), 165–173. https://doi.org/10.1038/nature04514

de Jong, C., Lawler, D., & Essery, R. (2009). Mountain hydroclimatology and snow seasonality-perspectives on climate impacts, snow seasonality and hydrological change in mountain environments. *Hydrological Processes, 23*(7), 955. https://doi.org/10.1002/hyp.7193

Debarbieux, B., Oiry Varacca, M., Rudaz, G., Maselli, D., Kohler, T., & Jurek, M. (2014). *Tourism in mountain regions: Hopes, fears and realities*. Department of Geography and Environment, University of Geneva; Centre for Development and Environment, University of Bern.

Dimri, A. P., & Dash, S. K. (2012). Wintertime climatic trends in the western Himalayas. *Climatic Change, 111*(3–4), 775–800. https://doi.org/10.1007/s10584-011-0201-y

Egan, P. A., & Price, M. F. (2017). *Mountain ecosystem services and climate change: A global overview of potential threats and strategies for adaptation.* http://unesdoc.unesco.org/images/0024/002487/248768e.pdf

Fort, M. (2015). Impact of climate change on mountain environment dynamics. An introduction. *Journal of Alpine Research, 103*(2), 1–7. https://doi.org/10.4000/rga.2877

Gairola, S., Sharma, C. M., Ghildiyal, S. K., Suyal, S., Rana, C. S., & Butola, D. S. (2009). Biodiversity conservation and sustainable rural development in the Garhwal Himalaya. *Report and Opinion, 1*(4), 6–12.

Ghosh, P., & Dhyani, P. P. (2004). Baranaaja: The traditional mixed cropping system of the central Himalaya. *Outlook on Agriculture, 33*(4), 261–266. https://doi.org/10.5367/0000000042664756

Gowda, P., Steiner, J. L., Olson, C., Boggess, M., Farrigan, T., & Grusak, M. A. (2018). Agriculture and rural communities. In *Impacts, risks, and adaptation in the United States: Fourth national climate assessment* (Vol. 2, pp. 391–437). U.S. Global Change Research Program.

Han, Z., Yongqi, G., & Libo, Z. H. O. U. (1998). Ozone low and surface heating over large scale topography. *Climate and Environmental Research, 3*, 18–26.

Harvey, J. A., Tougeron, K., Gols, R., Heinen, R., Abarca, M., Abram, P. K., Basset, Y., Berg, M., Boggs, C., Brodeur, J., & Cardoso, P. (2023). Scientists' warning on climate change and insects. *Ecological Monographs, 93*(1), e1553. https://doi.org/10.1002/ecm.1553

Hewitt, G. (2000). The genetic legacy of the Quaternary ice ages. *Nature, 405*(6789), 907–913. https://doi.org/10.1038/35016000

Hilborn, E. D., & Beasley, V. R. (2015). One health and cyanobacteria in freshwater systems: Animal illnesses and deaths are sentinel events for human health risks. *Toxins, 7*(4), 1374–1395. https://doi.org/10.3390/toxins7041374

Hoy, A., Katel, O., Thapa, P., Dendup, N., & Matschullat, J. (2016). Climatic changes and their impact on socio-economic sectors in the Bhutan Himalayas: An implementation strategy. *Regional Environmental Change, 16*, 1401–1415. https://doi.org/10.1007/s10113-015-0868-0

Huggel, C., Salzmann, N., Allen, S., Caplan-Auerbach, J., Fischer, L., Haeberli, W., Larsen, C., Schneider, D., & Wessels, R. (2010). Recent and future warm extreme events and high-mountain slope stability. *Philosophical Transactions of the Royal Society A: Mathematical, Physical and Engineering Sciences, 368*(1919), 2435–2459. https://doi.org/10.1098/rsta.2010.0078

Huggel, C., Clague, J. J., & Korup, O. (2012). Is climate change responsible for changing landslide activity in high mountains? *Earth Surface Processes and Landforms, 37*(1), 77–91. https://doi.org/10.1002/esp.2223

Hunt, J. D., Jurasz, J., Zakeri, B., Nascimento, A., Cross, S., ten Caten, C. S., de Jesus Pacheco, D. A., Pongpairoj, P., Leal Filho, W., Tomé, F. M. C., & Senne, R. (2022). Electric Truck Hydropower, a flexible solution to hydropower in mountainous regions. *Energy, 248*, 123495. https://doi.org/10.1016/j.energy.2022.123495

IPCC. (2007a). *Fourth assessment report*. Intergovernmental Panel on Climate Change Secretariat.

IPCC. (2007b). *Climate change 2007: Climate change impacts, adaptation and vulnerability*. In *Contribution of working group II contribution to the Fourth Assessment Report of the Intergovernmental Panel on Climate Change*. IPCC. www.ipcc.ch/SPM13apr07.pdf. Accessed June 2007.

Isaak, D. J., Wollrab, S., Horan, D., & Chandler, G. (2012). Climate change effects on stream and river temperatures across the northwest US from 1980–2009 and implications for salmonid fishes. *Climatic Change, 113*, 499–524. https://doi.org/10.1007/s10584-011-0326-z

IUCN. (2022). *The IUCN red list of threatened species* (Version 2022-2). https://www.iucnredlist.org

Kc, K. B., Tzadok, E., & Pant, L. (2022). Himalayan ecosystem services and climate change driven agricultural frontiers: A scoping review. *Discover Sustainability, 3*(1), 35. https://doi.org/10.1007/s43621-022-00103-9

Keiler, M., Knight, J., & Harrison, S. (2010). Climate change and geomorphological hazards in the eastern European Alps. *Philosophical Transactions of the Royal Society A: Mathematical, Physical and Engineering Sciences, 368*(1919), 2461–2479. https://doi.org/10.1098/rsta.2010.0047

Kohler, T., Giger, M., Hurni, H., Ott, C., Wiesmann, U., von Dach, S. W., & Maselli, D. (2010). Mountains and climate change: A global concern. *Mountain Research and Development, 30*(1), 53–55. https://doi.org/10.1659/MRD-JOURNAL-D-09-00086.1

Konchar, K. M., Staver, B., Salick, J., Chapagain, A., Joshi, L., Karki, S., Lo, S., Paudel, A., Subedi, P., & Ghimire, S. K. (2015). Adapting in the shadow of Annapurna: A climate tipping point. *Journal of Ethnobiology, 35*(3), 449–471. https://doi.org/10.2993/0278-0771-35.3.449

Körner, C. (2004). Mountain biodiversity, its causes and function. *AMBIO: A Journal of the Human Environment, 33*(sp13), 11–17. https://doi.org/10.1007/0044-7447-33.sp13.11

Körner, C. (2007). The use of 'altitude' in ecological research. *Trends in Ecology & Evolution, 22*(11), 569–574. https://doi.org/10.1016/j.tree.2007.09.006

Körner, C. (2013). Alpine ecosystems. In S. A. Levin (Ed.), *Encyclopedia of biodiversity* (Vol. 1, 2nd ed., pp. 148–157). Academic Press.

Körner, C., Paulsen, J., & Spehn, E. M. (2011). A definition of mountains and their bioclimatic belts for global comparisons of biodiversity data. *Alpine Botany, 121*, 73–78. https://doi.org/10.1007/s00035-011-0094-4

Kovach, R. P., Dunham, J. B., Al-Chokhachy, R., Snyder, C. D., Letcher, B. H., Young, J. A., Beever, E. A., Pederson, G. T., Lynch, A. J., Hitt, N. P., Konrad, C. P., Jaeger, K. L., Rea, A. H., Sepulveda, A. J., Lambert, P. M., Stoker, J., Giersch, J. J., & Muhlfeld, C. C. (2019). An integrated framework for ecological drought across riverscapes of North America. *Bioscience, 69*(6), 418–431. https://doi.org/10.1093/biosci/biz040

Laghari, J. (2013). Climate change: Melting glaciers bring energy uncertainty. *Nature, 502*(7473), 617–618. https://doi.org/10.1038/502617a

Li, G., & Cai, J. (2022). Spatial and temporal differentiation of mountain ecosystem service trade-offs and synergies: A case study of Jieshi Mountain, China. *Sustainability, 14*(8), 4652. https://doi.org/10.3390/su14084652

Manes, S., Costello, M. J., Beckett, H., Debnath, A., Devenish-Nelson, E., Grey, K. A., Jenkins, R., Khan, T. M., Kiessling, W., Krause, C., & Maharaj, S. S. (2021). Endemism increases species' climate change risk in areas of global biodiversity importance. *Biological Conservation, 257*, 109070. https://doi.org/10.1016/j.biocon.2021.109070

Maran, S., Volonterio, M., & Gaudard, L. (2014). Climate change impacts on hydropower in an alpine catchment. *Environmental Science & Policy, 43*, 15–25. https://doi.org/10.1016/j.envsci.2013.12.001

McDowell, G., Ford, J. D., Lehner, B., Berrang-Ford, L., & Sherpa, A. (2013). Climate-related hydrological change and human vulnerability in remote mountain regions: A case study from Khumbu, Nepal. *Regional Environmental Change, 13*, 299–310. https://doi.org/10.1007/s10113-012-0333-2

Messerli, B., & Ives, J. (Eds.). (1997). *Mountains of the word. A global priority* (p. 496). Parthenon Publishing.

Millennium Ecosystem Assessment–MEA (2003) Ecosystems and Human Well-Being: A Framework forAssessment. Island Press, Washington DC.

Miller, J. D., Immerzeel, W. W., & Rees, G. (2012). Climate change impacts on glacier hydrology and river discharge in the Hindu Kush–Himalayas. *Mountain Research and Development, 32*(4), 461–467. https://doi.org/10.1659/MRD-JOURNAL-D-12-00027.1

Mishra, S. K., Veselka, T. D., Prusevich, A. A., Grogan, D. S., Lammers, R. B., Rounce, D. R., Ali, S. H., & Christian, M. H. (2020). Differential impact of climate change on the hydropower economics of two river basins in high mountain Asia. *Frontiers in Environmental Science, 8*, 26. https://doi.org/10.3389/fenvs.2020.00026

Mishra, P. K., Rai, A., Abdelrahman, K., Rai, S. C., & Tiwari, A. (2022). Land degradation, overland flow, soil erosion, and nutrient loss in the Eastern Himalayas, India. *Land, 11*(2), 179. https://doi.org/10.3390/land11020179

Mittermeier, R. A., Mittermeier, R. A., & Cemex, S.A. de C. V (Eds.). (2004). *Hotspots revisited* (First English ed.). Cemex (Cemex books on nature).

Mittermeier, R. A., Turner, W. L., Larsen, F. W., Brooks, T. M., & Gascon, C. (2011). Global biodiversity conservation: The critical role of hotspots. In F. E. Zachos & J. C. Habel (Eds.), *Biodiversity hotspots: Distribution and protection of conservation priority areas* (pp. 3–22). Springer.

Mountain Agenda. (1998). *Mountains of the world: Water towers for the 21st Century*. Mountain Agenda group, Centre for Development and Environment (CDE), University of Berne.

Nepal, S. K. (2011). Mountain tourism and climate change: Implications for the Nepal Himalaya. *Nepal Tourism and Development Review, 1*(1), 1–14. https://doi.org/10.3126/ntdr.v1i1.7367

Ogra, M. V., & Badola, R. (2015). Gender and climate change in the Indian Himalayas: Global threats, local vulnerabilities, and livelihood diversification at the Nanda Devi Biosphere Reserve. *Earth System Dynamics, 6*(2), 505–523. https://doi.org/10.5194/esd-6-505-2015

Oppenheimer, M., Campos, M., Warren, R., Birkmann, J., Luber, G., O'Neill, B., Takahashi, K., Brklacich, M., Semenov, S., Licker, R., & Hsiang, S. (2015). Emergent risks and key vulnerabilities. In *Climate change 2014 impacts, adaptation and vulnerability: Part A: Global and sectoral aspects* (pp. 1039–1100). Cambridge University Press. https://doi.org/10.1017/CBO9781107415379.024

Palomo, I. (2017). Climate change impacts on ecosystem services in high mountain areas: A literature review. *Mountain Research and Development, 37*(2), 179–187. https://doi.org/10.1659/MRD-JOURNAL-D-16-00110.1

Parmesan, C., & Yohe, G. (2003). A globally coherent fingerprint of climate change impacts across natural systems. *Nature, 421*(6918), 37–42. https://doi.org/10.1038/nature01286

Pohner, T., Berki, T., & Rátz, T. (2009). Religious and pilgrimage tourism as a special segment of mountain tourism. *Journal of Tourism Challenges and Trends, 2*(1), 27–42.

Preston, D. L., Caine, N., McKnight, D. M., Williams, M. W., Hell, K., Miller, M. P., Hart, S. J., & Johnson, P. T. J. (2016). Climate regulates alpine lake ice cover phenology and aquatic ecosystem structure. *Geophysical Research Letters, 43*(10), 5353–5360. https://doi.org/10.1002/2016GL069036

Rahbek, C., Borregaard, M. K., Antonelli, A., Colwell, R. K., Holt, B. G., Nogues-Bravo, D., Rasmussen, C. M., Richardson, K., Rosing, M. T., Whittaker, R. J., & Fjeldså, J. (2019a). Building mountain biodiversity: Geological and evolutionary processes. *Science, 365*(6458), 1114–1119. https://doi.org/10.1126/science.aax0151

Rahbek, C., Borregaard, M. K., Colwell, R. K., Dalsgaard, B. O., Holt, B. G., Morueta-Holme, N., Nogues-Bravo, D., Whittaker, R. J., & Fjeldså, J. (2019b). Humboldt's enigma: What causes global patterns of mountain biodiversity? *Science, 365*(6458), 1108–1113. https://doi.org/10.1126/science.aax0149

Ritter, F., Fiebig, M., & Muhar, A. (2012). Impacts of global warming on mountaineering: A classification of phenomena affecting the alpine trail network. *Mountain Research and Development, 32*(1), 4–15. https://doi.org/10.1659/MRD-JOURNAL-D-11-00036.1

Sandifer, P. A., & Sutton-Grier, A. E. (2014). Connecting stressors, ocean ecosystem services, and human health. *Natural Resources Forum, 38*(3), 157–167. https://doi.org/10.1111/1477-8947.12047

Schowalter, T. D., Noriega, J. A., & Tscharntke, T. (2018). Insect effects on ecosystem services—Introduction. *Basic and Applied Ecology, 26*, 1–7. https://doi.org/10.1016/j.baae.2017.09.011

Sharma, E., Molden, D., Rahman, A., Khatiwada, Y. R., Zhang, L., Singh, S. P., Yao, T., & Wester, P. (2019). Introduction to the Hindu Kush Himalaya assessment. In P. Wester, A. Mishra, A. Mukherji, & A. Shrestha (Eds.), *The Hindu Kush Himalaya Assessment*. Springer. https://doi.org/10.1007/978-3-319-92288-1_1

Shekhar, M. S., Chand, H., Kumar, S., Srinivasan, K., & Ganju, A. (2010). Climate-change studies in the western Himalaya. *Annals of Glaciology, 51*(54), 105–112. https://doi.org/10.3189/172756410791386508

Shrestha, U. B., Gautam, S., & Bawa, K. S. (2012). Widespread climate change in the Himalayas and associated changes in local ecosystems. *PLoS One, 7*(5), e36741. https://doi.org/10.1371/journal.pone.0036741

Singh, S. P., Singh, V., & Skutsch, M. (2010). Rapid warming in the Himalayas: Ecosystem responses and development options. *Climate and Development, 2*(3), 221–232. https://doi.org/10.3763/cdev.2010.0048

Smadja, J., Aubriot, O., Puschiasis, O., Duplan, T., Grimaldi, J., Hugonnet, M., & Buchheit, P. (2015). Climate change and water resources in the Himalayas. Field study in four geographic units of the Koshi basin, Nepal. *Journal of Alpine Research|Revue de géographie alpine, 103*(2), 1–23. https://doi.org/10.4000/rga.2910

Solomon, S., & Intergovernmental Panel on Climate Change and Intergovernmental Panel on Climate Change (Eds.). (2007). *Climate change 2007: The physical science basis: Contribution of Working Group I to the Fourth Assessment Report of the Intergovernmental Panel on Climate Change. Cambridge*. Cambridge University Press.

Steiger, R., Knowles, N., Pöll, K., & Rutty, M. (2022). Impacts of climate change on mountain tourism: A review. *Journal of Sustainable Tourism*, 1–34. https://doi.org/10.1080/09669582.2022.2112204

Tito, R., Vasconcelos, H. L., & Feeley, K. J. (2020). Mountain ecosystems as natural laboratories for climate change experiments. *Frontiers in Forests and Global Change, 3*, 38. https://doi.org/10.3389/ffgc.2020.00038

Tse-Ring, K., Sharma, E., Chettri, N., & Shrestha, A. B. (Eds.). (2010). *Climate change vulnerability of mountain ecosystems in the Eastern Himalayas*. International centre for integrated mountain development (ICIMOD).

UNEP. (1992). Earth Summit: Agenda 21 (Chapter 13: Managing Fragile Ecosystems: Sustainable Mountain Development). United Nations Conference on Environment and Development (UNCED), 3–14 June 1992, Rio deJaneiro, Brazil.

Vanham, D., Fleischhacker, E., & Rauch, W. (2009). Impact of an extreme dry and hot summer on water supply security in an alpine region. *Water Science and Technology, 59*(3), 469–477. https://doi.org/10.2166/wst.2009.887

Viviroli, D., Archer, D. R., Buytaert, W., Fowler, H. J., Greenwood, G. B., Hamlet, A. F., Huang, Y., Koboltschnig, G., Litaor, M. I., López-Moreno, J. I., & Lorentz, S. (2011). Climate change and mountain water resources: Overview and recommendations for research, management and policy. *Hydrology and Earth System Sciences, 15*(2), 471–504. https://doi.org/10.5194/hess-15-471-2011

Wei, L., Jiheng, L., Junhong, G., Zhe, B., Lingbo, F., & Baodeng, H. (2020). The effect of precipitation on hydropower generation capacity: A perspective of climate change. *Frontiers in Earth Science, 8*, 268. https://doi.org/10.3389/feart.2020.00268

Wester, P., Mishra, A., Mukherji, A., & Shrestha, A. B. (2019). *The Hindu Kush Himalaya assessment: Mountains, climate change, sustainability and people* (p. 627). Springer Nature.

Wilkinson, J. L., Boxall, A. B., Kolpin, D. W., Leung, K. M., Lai, R. W., Galbán-Malagón, C., Adell, A. D., Mondon, J., Metian, M., Marchant, R. A., & Bouzas-Monroy, A. (2022). Pharmaceutical pollution of the world's rivers. *Proceedings of the National Academy of Sciences, 119*(8), e2113947119. https://doi.org/10.1073/pnas.2113947119

Xu, J., Grumbine, R. E., Shrestha, A., Eriksson, M., Yang, X., Wang, Y. U. N., & Wilkes, A. (2009). The melting Himalayas: Cascading effects of climate change on water, biodiversity, and livelihoods. *Conservation Biology, 23*(3), 520–530. https://doi.org/10.1111/j.1523-1739.2009.01237.x

Zhao, C., Liu, B., Piao, S., Wang, X., Lobell, D. B., Huang, Y., Huang, M., Yao, Y., Bassu, S., Ciais, P., Durand, J. L., Elliott, J., Ewert, F., Janssens, I. A., Li, T., Lin, E., Liu, Q., Martre, P., Müller, C., Peng, S., Peñuelas, J., Ruane, A. C., Wallach, D., Wang, T., Wu, D., Liu, Z., Zhu, Y., Zhu, Z., & Asseng, S. (2017). Temperature increase reduces global yields of major crops in four independent estimates. *Proceedings of the National Academy of Sciences United States of America, 114*, 9326–9331. https://doi.org/10.1073/pnas.1701762114

Ecosystem Functioning in the Chattogram Hill Tracts, Bangladesh: Scopes and Challenges for Indigenous Communities Adaptation to Climate Change

Tasneem Siddiqui, Ekhtekharul Islam, and Tamim Billah

Abbreviations

BINA	Bangladesh Institute of Nuclear Agriculture
CBD	Convention on Biological Diversity
CHT	Chattogram Hill Tracts
HKH	Hindu Kush Himalayan
ICIMOD	International Centre for Integrated Mountain Development
IPCC	Intergovernmental Panel on Climate Change
REDD+	Reducing Emissions from Deforestation and Forest Degradation and Enhancing Forest Carbon Stocks
SDG	Sustainable Development Goals
SDI	Shannon's Diversity Index
SEI	Shannon Evenness Index
USF	Unclassed State Forest
VCF	Village Common Forest

Thematic Area: Biodiversity and Ecosystem services in mountains: challenges towards adapting to climate change.

T. Siddiqui (✉)
Department of Political Science, University of Dhaka, and Founding Chair, Refugee and Migratory Movements Research Unit (RMMRU), Dhaka, Bangladesh

E. Islam
Department of Environmental Science and Management, School of Environment and Life Sciences, Independent University, Dhaka, Bangladesh
e-mail: ekhtekhar@iub.edu.bd

T. Billah
Refugee and Migratory Movements Research Unit (RMMRU), Dhaka, Bangladesh

S. Tripathi et al. (eds.), *Warming Mountains*,
https://doi.org/10.1007/978-3-031-62197-0_3

1 Introduction

Resource management in mountainous regions worldwide is vital for sustainable development while acknowledging other overarching issues such as climate change and environmental degradation. The unique landscapes of mountains cover one-fifth of the world's land surface that host exclusive biological and cultural diversities. Roughly half of the world's biodiversity hotspots and 30% of the key biodiversity areas are found in the mountainous region (Payne et al., 2020). Mountain ecosystems are habitats for approximately 87% of amphibians, birds, and mammals found in the world (Rahbek et al., 2019). Globally, an estimated number of 650 million people live in mountainous rural or peri-urban regions (Dach & Peyré, 2021). Mountains have substantial importance for mountainous communities as well as for people residing on the lowland. The mountain ecosystem and its functions are indispensable as they provide livelihood necessities (Bär et al., 2017). The tropical mountainous regions are extraordinarily different and have the highest species richness due to unique climatic conditions. The mountains of the Hindu Kush Himalayan (HKH) region have the largest ice cover right after what is there in the two poles and are home to ten river basins (GRID Arendal, 2018). It stretches over an area of 3500 km^2 across eight countries including Bangladesh. These river basins support 1.3 billion people living in mountainous regions and downstream by providing support such as drinking water, irrigation, energy production, industry, tourism, and sanitation (ICIMOD, 2021). In developing countries such as Bangladesh, natural resources in the mountains represent the wealth of the poor, providing crucial livelihood support. Chattogram (formerly known as Chittagong) Hill Tracts (CHT) region, a part of the HKH, is the habitat for 12 indigenous communities and contains nearly half of the total forested lands in Bangladesh (Haque et al., 2021).

Mountain ecosystems are undeniably underprotected despite their significance for human wellbeing and it is reported that the global community is not on track to meet the sustainable development goals (SDG) of "*ensuring the conservation of mountain ecosystems, including their biodiversity, to enhance their capacity to provide benefits which are essential for sustainable development*" (Silveira et al., 2019). Unplanned anthropogenic activities in the mountainous regions worldwide are jeopardizing the ecosystem and human resilience while climate change effects are adding more into the problem. In relation to changing climatic conditions, CHT is reported to be the least-climate resilient place in Bangladesh (Bandara et al., 2021). Due to their intricate relationship with nature, CHT communities are at the forefront of the crises associated with climate change and perilous anthropogenic interventions. In addition, scholarly communities often fall into silos and do not take the risks of reporting the exploitation and systematic marginalization of indigenous communities (Newell et al., 2021). At this outset, this chapter attempts to review the anthropogenic and environmental drivers threatening mountain ecosystem functioning. Discussion on adaptation pathways, strategies, and challenges for mountainous communities in the Chattogram Hill Tracts (CHT) is also included. How existing institutions and policies incorporate ecosystem-based adaptation ideas explicit to CHT indigenous communities is also examined in detail. Before writing the book chapter, we made a visit to remote CHT areas including Thanchi of Bandarban district, Matiranga of Khagrachari district, and Borkol of Rangamati district. During the visit, we informally discussed many issues with indigenous communities and key informants about livelihood and climate change.

2 CHT Region in Bangladesh: Historical Routes to Environmental Degradation

Dr. Emil Riebeck (1885) while travelling in the Chattogram Hill Tracts (CHT) region in 1882 expressed how he observed the CHT region and we quote "*nevertheless such is the intricate character of the land, consisting of nearly parallel mountain ranges running close together, mainly north and south, often densely wooded, and inhabited by a multiplicity of semi-independent hill tribes in almost every stage of social culture, that the broad physical features both of the country and its inhabitants had hitherto been but imperfectly understood, while few of the details had been fully worked out*". CHT region located in the southeast of Bangladesh administratively comprised of three districts namely Bandarban, Rangamati, and Khagrachari, covering a total area of 12,395 km^2 (Gain, 2013). One-third of Bangladesh's forests (roughly 40%) belong to this region and are part of the Hindu Kush Himalayan (HKH) region, which is also manifested by mountains and river basins (Hussain et al., 2019). HKH region is alternatively and popularly known as the "Water Tower of Asia," "the Third Pole," and "the Planet's Pulse" because of its dynamic features (The Third Pole, 2023). Thirty-six percent of the CHT forests are reserved forests with moderate to dense tree cover and completely restricted, whereas 64% are unclassed state forests with low or no tree cover. The tropical to subtropical climate in the CHT is characterized by a mean monthly temperature between 25 and 35 °C and the average annual rainfall record is between 2032 and 3910 mm (Biswas et al., 2012). The CHT is home to most of the ethnic population of Bangladesh who primarily rely on forest resources for their livelihoods (Rasul, 2007). Barring Bengali settlers, 11 different indigenous communities of Sino-Tibetan descent, known as Chakma, Marma, Tripura, Tanchangya, Bawm, Panghua, Lusai, Khumi, Achik, Chak, and Khyang (Zaman, 1982), also reside there. Chakma forms the largest community, followed by Marma and Tripura as second and largest, respectively. The other communities are small compared to these three communities (Panday & Jamil, 2009). These people have a wide array of cultural, social, economic, and environmental diversities. During the 1947 partition, they became part of Pakistan against their will, based on the "two-nation" theory (Siddiqui & Chakma, 2016).

Before the colonial rule, the indigenous communities were fully political and cultural nations, living their life with indigenous norms while determining their self-political destiny (Levene, 2010). However, a greater level of autonomy also existed during the colonial rule since the CHT region was declared an excluded area and several stringent policies were adopted to restrict others to settle in this region (Datta, 2019). The Chittagong Hill Tracts Manual 1900 by the British rulers introduced the region's autonomous administrative processes while making the indigenous communities' representatives responsible for resource management. In the postcolonial period, rapid political-administrative changes were adopted without giving attention to indigenous rights and local ecology (Chakma, 2010). Such changes are marked by the government-supported development projects, government-initiated internal migration and internal colonialism, and the pervading control of the security forces, all of which culminated in the severe curtailment of indigenous rights (Ahsan & Chakma, 1989; Chakma, 2010; D'Costa & Ahmed, 2023). The construction of the Kaptai Dam in the 1960s is a textbook example of how government intervention can promote "internal colonialism" and "internal environmental imperialism" in the name of development and threaten indigenous lives. Construction of the dam has permanently inundated 40% of the Rangamati district's land and displaced more than 1,00,000 indigenous people (Parveen & Faisal, 2002). Following Bangladesh's independence in 1971, the colossal security efforts taken by the military are jeopardizing the unique biodiversity through dangerous and extractive processes of exploiting vast natural resources, initiating deforestation, and depleting indigenous sociocultural practices in the name of economic development (D'Costa & Ahmed, 2023).

Over the last few decades, environmental degradation in CHT has become more striking due to swidden farming, timber extraction, illegal wildlife trading, deforestation, commercial plantation, land grabbing, and chemical-based agricultural practices (Ahammad et al., 2021). A modeling study on landscape diversity and probability distribution indicates a decrease in Shannon's diversity index (SHDI) and Shannon evenness index (SHEI) from 1988 to 2018, which characterizes that the ecosystem heterogeneity and evenness have decreased (Mamnun & Hossen, 2021). A plethora of studies also suggest that there has been an exponential rise in forest fragmentation in the CHT since 1990 (International Centre for Integrated Mountain Development, 2013). Such analysis has also pointed out that forest fire incidences have also increased and happened in areas where deforestation is unprecedented. In a nutshell, many of the historical reforms and unsustainable changes in the CHT region have contributed to the gradual increase in human and ecosystem vulnerability and exaggerated the crisis in a changing climatic condition (Islam et al., 2020). The above-mentioned case of historical exploitation and experimentation is illustrated (see Fig. 1) as a multistage environmental and psychosocial process that happens in four stages: threats, distortion, rigidification, and collusion (Sensarma & Okada, 2006).

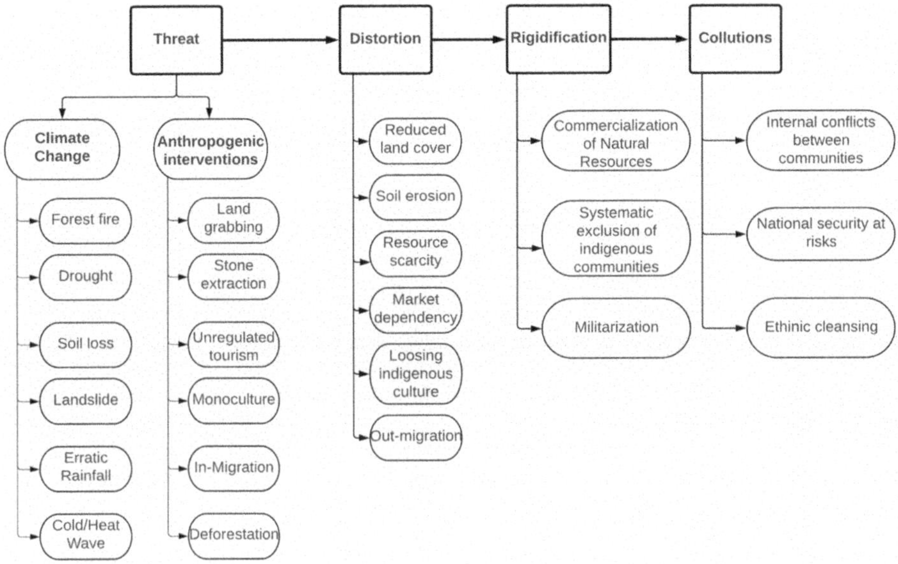

Fig. 1 Drivers and the underlying reasons behind the conflict in the indigenous CHT region

3 Value of Mountain Ecosystem: Focusing on CHT Region

The United Nations in Agenda 21 emphasized on essential aspects of the mountain ecosystem for biodiversity conservation and human sustainability (Dong et al., 2022). The value of a mountainous ecosystem is not only represented by ecological factors but largely by sociocultural dimensions. Therefore, a strategic plan that included 20 targets (Aichi Targets) under the Convention on Biological Diversity was formulated by global communities in 2010. It was done to "*take effective and urgent action to halt the loss of biodiversity in order to ensure that by 2020 ecosystems are resilient and continue to provide essential services, […] securing the planet's variety of life and contributing to human well-being, and poverty eradication*" (Roche & Campagne, 2017). Over the last four decades, a conceptual understanding of ecosystem services has gained momentum. The concept refers to the direct and indirect benefits that humans gain from ecosystem functioning. A good number of studies revealed that the delivery and performance of ecosystem services strongly correlated with the conditions of the ecosystem (Grizzetti et al., 2019). The condition of the mountain ecosystem is characterized based on certain factors. In mountain ecosystems, the recently developed ecological factors to accurately define the ecosystem include the altitude and slope of mountains and the topographic roughness (Foggin, 2016). CHT, due to its ecological and cultural richness, is called a hub for biocultural diversity and is part of the Indo-Burma biodiversity hotspot. Half the

Table 1 Ecosystem services from CHT region

CHT mountain ecosystem resources	Ecosystem services	Source
Land	Crop production: 27% of the CHT land are used for horticulture, 16% for plough land, and 15.5% for Jum cultivation	Rasul (2009) and Hossain et al. (2020)
	Crop variety transformation	Rasul and Thapa (2003)
	Recreational hiking	Siraz et al. (2023)
Forest	Firewood and biomass fuel supply	Miah et al. (2012)
	49 types of wild edible fruits, edible fungus, vegetables including bamboos, sun grass, and rattan come from forests that provide nutritional support to indigenous communities	Paul et al. (2020) and Siddiqi (1998)
	Carbon sequestration: 0.9 ton of carbon are absorbed by per hectare forest per year in CHT	Bangladesh Forest Department (2020)
	Air quality regulation: 46.51 kg of fine dust absorbed by per hectare forest per year	
	Forest provides spiritual benefits to indigenous communities	Ahammad et al. (2022)
River basin	Energy production: More than 40,000 MW of electricity is produced in the river basin areas of HKH, only 230 MW is produced in CHT river basin area	GRIDA Arendal (2018)
	Sangu and Matamuhuri river basins are main sources of freshwater flow to southeastern part of Bangladesh	Rudra and Alam (2022)

world's population depends on three basic resources from mountainous ecosystems: land, forest, and river basin (Price, 2015). From this perspective, illustrating the ecosystem services of the CHT mountainous region (Table 1) can be a source of inspiration to realize the value of the ecosystem by understanding ecosystem integrity and ecosystem functioning.

Land use by indigenous communities utilize shifting cultivation in CHT as customary rights for subsistence farming. Shifting cultivation is a form of agricultural practice or a cultivation system in which an area of ground is cleared of vegetation and cultivated for a few years and then abandoned for a new area until its fertility has been naturally restored (Zonayet & Kumar, 2020). Such cultivation practices provide food for subsistence and a little amount of earnings. The land preparation starts in April when communities utilize the benefits of precipitation and the harvesting takes place in December (Gafur et al., 2013). Though there is limited plain land for cultivation, the Bengali settlers introduced plough cultivation in CHT during the 1870s (GoB and FAO, 2013). Since then, the extent of shifting cultivation has gradually decreased and the communities have started delving into food cosmopolitanism (Jhala, 2022).

In CHT, forests are classified into four categories: unclassed state forests (USF) managed by the Forest Department, unclassed state forests (USF) managed by the district administration, reserved forests managed by the Forest Department, and village common forests utilized and managed by the indigenous communities. Reserve forests are restricted, and no one is allowed to enter these areas. A large portion of the CHT forests is declared reserve forest, restricting indigenous communities' access to these resources and triggering higher competition for resources outside the reserve forests (Chowdhury et al., 2018). The management of village common forests (VCFs) has been going on for centuries following certain indigenous practices. Such management has ensured the sustainability of VCF forest resources, whereas the other forest resources managed under state rules and regulations are severely degraded (Nath et al., 2016). This validates traditional indigenous knowledge's argument to be considered an ecosystem service since such knowledge contributes to greater ecosystem integrity and resilience. Numerous wild edible fruit plants and medicinal plants grown in CHT forests provide nutritional support for poor indigenous communities (Paul et al., 2020). Common forest-based plants include bamboo shoots, Turkey berries, Boken bones tree, Indian trumpet flower, Tree of Damocles, mushroom, golden flowered zinger, and Indian pennywort. The CHT region consists of six main river valleys, namely Bakkhali, Chengi, Kassalong, Maini, Matamuhuri, and Sangu (Mahmud et al., 2020). Other than these rivers, springs and wells also provide fresh water to indigenous communities and beyond (Zahan, 2022; Adnan, 2022). Upland communities in CHT fully rely on water for drinking and cooking from the springs.

4 Climate Change and Looming Threats for Mountainous Communities in CHT

Some of the world's poorest and most disadvantaged communities reside in the mountainous regions for whom climate change is an existential threat. In this section, we elucidate the changing climatic conditions in the mountainous region of CHT and how those eventually affect human lives.

4.1 Effects of Climate Change

Mountain biocultural diversity has been transformed by a wide variety of anthropogenic and environmental factors. Rapid changes in the climatic conditions coupled with exploitative actions by humans have put the ecosystem under serious threat (Payne et al., 2020; Silveira et al., 2019). Unusual and rapid changes in our ecosystem have already occurred due to anthropogenic climate change, and the number of weather climate extreme events have increased manifold (IPCC, 2023). It is already well-established from scientific works that the climate in the HKH region has changed dramatically and is anticipated to continue to change in the coming decades

Table 2 Effects of climate change in HHK and CHT

Effects of climate change	Significance for Hindu Kush Himalayan region (Krishn et al., 2019)	Significance for Chattogram Hill Tract
El-Nino Enso influence on weather and climate	Well-established	Well-established (Lalmalsawma et al., 2023; Kamruzzaman et al., 2022)
Increase in extreme warm days and nights	Well-established	Well-established (Baul et al., 2022; Lalmalsawma et al., 2023)
Changes in precipitation trend	Established but inconclusive	Established (Lalmalsawma et al., 2023; Islam et al., 2021a, b)
Future precipitation extremes	Inconclusive	Established (Imran et al., 2023)
Warming of 1.7–2.4 °C between 2036 and 2065	Well-established	Data not available

(Krishn et al., 2019). A 1.5 °C rise in temperature will accelerate the risks of species extinction, increase the frequency and magnitude of extreme events, changing rainfall patterns, and will enhance ice melting. According to scientific studies that have utilized climate data dated from 1900, downscaling it to local perspectives and warming in mountainous regions are faster compared to lowlands (Pepin et al., 2022; Huss & Hock, 2018). Observation based on data from the past five to six decades shows rising incidences of warm extreme events such as heat/cold waves in the HHK, including the CHT region. Table 2 shows observed climate change effects in HHK and CHT regions with their level of significance based on a detailed review of the literature.

Analysis of longitudinal data concluded that there has been a dramatic increase in the number of warm days and nights in the CHT region, whereas, in reality, the number of cold days and nights decreased (Abdullah et al., 2020). In terms of precipitation, CHT has been experiencing an increasing trend. In addition, the frequency of heavy rain days, very heavy rain days, and extreme have rain days have slightly increased in the southeastern hilly Chattogram region of Bangladesh (Imran et al., 2023). Over the last 40 years, consecutive warm or dry days have increased in the Chattogram region (Abdullah et al., 2020). In addition to changes in the climatic parameters, extreme weather events have also become more frequent in CHT and other parts of Bangladesh.

The loose silt clay soil of CHT is naturally susceptible to landslides (Islam et al. 2021a, b). There has been a sharp escalation of landslide incidences in CHT due to a combination of anthropogenic and environmental factors (Ahmed, 2015). Unplanned urbanizations dried up springs, reduced forest cover, torrential rainfall, and increasing temperature are the factors behind landslide incidents. Since indigenous communities consider hills part of their cultural identity, they make fewer disturbances and are well-versed with the survival process in mountainous regions in the context of landslide hazards (Ahmed, 2015).

4.2 Impacts on the Ecosystem and Communities

In CHT region, not only are the communities facing historical marginalization and exploitation, but the land is also marginalized with natural resources receiving little attention from the decision-makers (Foggin, 2016). It is evident that the intrinsic value of the CHT ecosystem is severely undermined in national policy architecture and hence the readiness to withstand climate change effects is poor. The three major threats for the CHT ecosystem are the loss or degradation of biodiversity and ecosystem services, accelerating risks of natural hazards induced by climate change, and the loss of cultural and traditional knowledge. Impacts of climate change on the CHT ecosystem and communities spread across all segments of society as well as nature.

4.2.1 Reduced Plants Availability

Several decades before, there was richness and abundance of plants collected from forests by indigenous people for daily consumption. During the last 10–15 years, the forest-based consumption behavior among indigenous people has decreased due to plant unavailability (Morgubatul et al., 2018). According to the participants in the interviews, the hardly found plants from forests include, but not limited to, *Bambusa* spp., *Musa* spp., *Colocasia* spp., *Zingiber chrysanthum*, *Centella asiatica*, and *Ampelopteris prolifera*. Therefore, the communities get dependent on volatile markets for food supply and due to unprecedented poverty, they often cut down their daily food intake. The CHT population suffers most since they have the highest rate of malnutrition, food insecurity, and poverty in Bangladesh (Adnan, 2022).

4.2.2 Public Health Impacts

Climate change has dramatically altered the ecological landscape of the planet while putting billions of people at risk of infectious and noncommunicable diseases (Baker et al., 2021). There is very little evidence of how climate change effects create a toll on the health of indigenous communities in CHT. Malaria is a very common disease and endemic in 13 out of 64 districts in Bangladesh. (Haque et al., 2014). Despite a steep decrease of 93% in malaria incidence, 13 districts including three CHT districts contribute to nearly 90% of the current malaria cases (Clancy, 2022). Such high endemicity of the disease in the CHT is attributed to the changing climate parameters—temperature, humidity, and precipitation (Lalmalsawma et al., 2023). The minimum temperature in the CHT has a consistently increasing trend over the last four decades, creating a favorable condition for malaria transmission not just in Bangladesh but also across the Indian border. Studies indicated that indigenous people over 50 and children below 15 are more at risk of malaria in relation to climate change (Emeto et al., 2020). Some literature also argue that climate change effects have profound impacts on the mental well-being of indigenous communities, but the overall scenario is not comprehensively documented (Raza & Khan, 2022). On the supply side, healthcare services are often unavailable, inaccessible, and expensive for indigenous communities, making them more vulnerable to diseases associated with climatic conditions (Uddin et al., 2013).

4.2.3 Impacts on River Basins and Streams

It is reported in myriad studies that natural springs in the CHT get dried during summer and communities face severe water crises. Changing in the rainfall pattern, prolonged summers and frequent heat waves are the few major causes behind water inadequacy and drying up of water in the streams (WaterAid., 2013). The WaterAid assessment of the spring revealed that 11 out of 18 streams in CHT got reduced or severely reduced and three streams died predominantly due to reduced rainfall, stone extraction, monoculture, and deforestation. Human-caused environmental degradation compounds the climate change impact, making indigenous life more challenging. The soil erosion process gets faster when deforested areas get exposed to erratic and torrential rainfall events (Chisty, 2014). Higher sediment load in the water eventually destroys the small springs and creeks in the CHT. Since the fallow period of Jum land has substantially decreased in recent decades, frequent Jum cultivation causes six times higher the regular soil loss in the watersheds in CHT (Gafur et al., 2003).

4.2.4 Food Security Issues

Food security is primarily a serious problem in developing countries and is gaining attention in global political discussions. Food insecurity may arise directly from climate change effects or indirectly through its compounding effects. Shifting cultivation is the primary source of food for most of the indigenous people in CHT. One of the key threats for shifting cultivation in CHT is the increasing weed, pest, wild boar, rat, and disease attacks. Since 2010, a number of climate change-induced disasters including flash floods and landslides have aggravated the situation. The other threats include landslides, flash floods, and erratic rainfall. All these contribute to episodic incidents of food shortages in the CHT region (IPC, 2016). A plethora of studies correlates the outbreak of rodents and other animal attacks on croplands with increasing temperature (Santos, 2018; Xu et al., 2015). The local term for the explosion of the rat population is called *indur bonna* and it happens during the flowering season of certain species of bamboo in CHT. The introduction of commercial cash crops for greater economic gains is another potential threat to food security in the CHT. For instance, tobacco cultivation is promoted by external agencies in the CHT. Though it offers higher earnings, it has severe health and ecological consequences (Bala & Hossain, 2013). A commercial crop cultivation expansion in CHT will make indigenous communities more dependent on products from markets that

Changes in Agricultural Land Use		Major Driving Factors	
Shifting cultivation util 1970s	**Shifting cultivation after 1970**	**Anthropogenic factors**	**Climate change factors**
Predomnant agricultural practice	Fallow period: 2-3 years	Deforestation	Changes in rainfall pattern
Fallow period: >10 years	Fewer crops produced	Excludability of forest resources	Loss of soil nutrients
Diverse crop production	Use of chemical fertilizers	Population growth and inward migration	Changing seasonal patterns

Fig. 2 Major driving factors behind agricultural changes in CHT

are usually rife with volatile pricing mechanisms (Adnan, 2022). The risks of starvation, malnutrition, and other diseases will be on the rise. Figure 2 combines anthropogenic and climate change factors behind declining crop production by indigenous communities.

4.2.5 Impacts on Soil Quality

Increased soil erosion and runoff due to changing climatic conditions lead to decreased availability of soil phosphorus, potassium, and organic matter content in the CHT forest areas (Hossain et al., 2014). Declining shifting cultivation has forced indigenous communities to switch to traditional agricultural practices to grow commercially profitable crops. Such cultivation techniques heavily rely on the usage of chemical fertilizers, pesticides, and insecticides, which eventually create a nutrient imbalance in the soil. Excessive rainfall has increased the likelihood and occurrences of landslides in CHT due to loosely packed soils induced from land cover changes (Sarker & Rashid, 2013).

4.2.6 Displacement and Migration

Mountainous communities worldwide are at the forefront of the most profound impacts of climate change since they rely heavily and inextricably on natural resources for their livelihood. Therefore, they are at high risk of displacement. There is consensus among learned communities that displacement occurs due to the various vulnerabilities created by climate change (Jayawardhan, 2017). A handful number of studies through analyzing decadal data on migration and displacement revealed that there is a positive correlation between climatic factors, such as rainfall, and temperature with domestic and international migration (Chatterjee, 2023). However, migrating to a new area is a very difficult choice to make for indigenous communities due their greater level of attachment to the area (Ford et al., 2020). CHT is one the most vulnerable region where indigenous communities face recurrent displacement due to climate change impacts (Siddiqui, 2023). Displacement induced by food shortages is common in CHT, and indigenous people move from one part to another or move deep into the forest or sometimes cross the border illegally (Mahmud, 2021). Figure 3 shows the recurrent displacement process in the remote CHT areas partly influenced by climate change effects.

It is indicated that internal migration from CHT to other parts of the country is more common, whereas international migration from the plain land of Chattogram is too frequent. Migration decision is based on a combination of social, environmental, economic, and political factors, not just climate change effects (Siddiqui et al., 2014). In addition, migration from other climate change-affected areas of Bangladesh (e.g., low-lying coastal areas) to CHT has also occurred, realizing the growing economic potential in the region. This has led to escalated and unhealthy competition for resources, leading to more conflicts and, subsequently, forced displacement of indigenous communities (Rafi & Chowdhury, 2001).

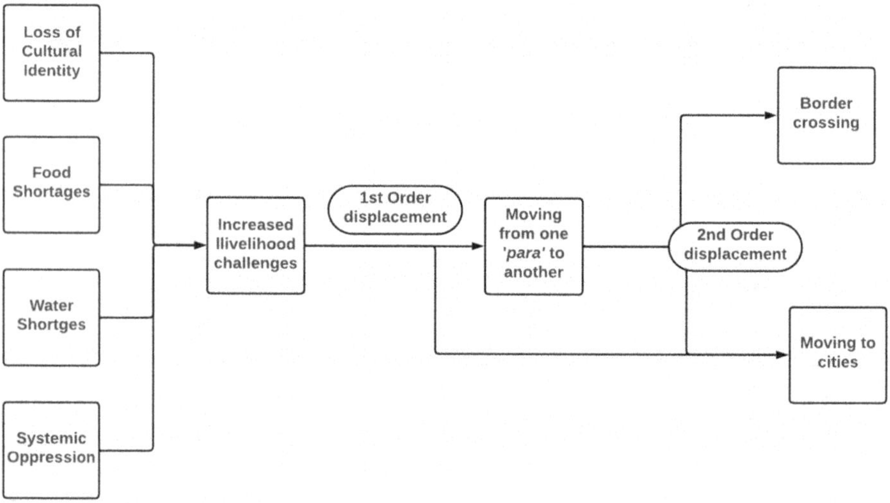

Fig. 3 Displacement in CHT induced by climatic and non-climatic factors

5 Climate Change Adaptation for Indigenous Communities in CHT

A deeper understanding of climate change adaptation concepts has been developed in the post-Paris climate policy architecture. Adaptation focuses on a three-tier process: enhancing adaptive capacity, strengthening resilience, and reducing the factors of vulnerability to climate change (Singh et al., 2022). Indigenous knowledge and techniques in response to climate variabilities all over the world have prepared mountainous communities to thrive in challenging environments. In this section, we identify and compile the current adaptation practices by the indigenous communities. Our analysis further extended to why negative coping mechanisms are undertaken.

5.1 Indigenous Adaptation Practices

Since shifting cultivation is becoming increasingly more challenging, many indigenous communities have adopted plain land cultivation methods and homestead gardening for nutritional crops (Nath & Inoue, 2009). Planting crops on the contours of the steep slopes in the mountain region ensures moisture retention, maintains soil nutrients, and prevents soil erosion. We discussed in the previous sections that warmer days and prolonged dry seasons have significantly increased in the CHT region. Drought resistance crop variety is an effective adaptation strategy, and it is imperative for supporting indigenous livelihood. At this outset, the Bangladesh Institute of Nuclear Agriculture (BINA) has developed a locally appropriate rice variety from the African NERICA rice variety for the CHT region and is preparing for commercialization. The variety is drought tolerant, have uniform flowering time, early maturity, and higher yield (Islam, et al., 2011). Orchards and agroforestry have emerged as an alternative income source for many indigenous communities in CHT considering the greater economic gains. In a participatory study, indigenous communities have expressed their satisfaction as higher income gains and diverse food production were ensured through fruit cultivation (Kibria et al., 2015).

Indigenous knowledge of the mountainous communities that have developed through centuries of practices is also important for climate change adaptation as the novel methodologies from indigenous practices can advance the understanding of climate adaptation (UNESCO, 2020). The collaborative management of forests by the indigenous communities has a lot to offer for forest resource management in other parts of the country. A study found that 42 and 18 diverse plant species were planted in the orchard and agroforest projects of CHT, and many of which were planted by the indigenous communities without any institutional support (Kibria et al., 2013). Consequently, the faunal diversity in these project areas have overwhelmingly increased (Dewi et al., 2013). Such positive outcomes have tremendous potential to reduce the dependency on shifting cultivation. In the CHT context, indigenous communities have developed a deep understanding of the medical properties of many plants and have built up medical practices based on those plants from remote forests. A total of 204 plant species were reported to be used by the CHT indigenous communities for the treatment of nearly 200 diseases and ailments (Rudra et al., 2021). Some plants such as bamboo are taken by the indigenous communities as regular food but they also offer a host of other health benefits (Misbahuzzaman & Smith-Hall, 2015). Such benefits from plants harvested from village common forests in CHT indicate sustainable management of common pool resources, which can have the potential to be replicated elsewhere for marginalized communities (Table 3).

Table 3 Most common indigenous adaptation practices in CHT

Sector	Indigenous adaptation practice	Type	Reference
Agriculture and Food Security	Sustainable management of common pool resources as a novel methodology	Institutional	Misbahuzzaman and Smith-Hall (2015) and Baul et al. (2022)
	Locally appropriate NERICA rice variety	Institutional	Islam et al. (2021b)
	Leaves are burnt under trees to create ash to prevent pest attacks on fruits	Autonomous	Bala and Hossain (2013)
	Adult members of the families moved to deep forest for wild plants	Autonomous	Misbahuzzaman and Smith-Hall (2015) and Roy (2000)
	Paddy production on fringe lands	Autonomous	
	Homestead vegetable gardening	Autonomous	Jhala (2022)
	Changing planting and harvesting time and cultivating short rotational crops	Institutional and autonomous	Bandara et al. (2021) and Misbahuzzaman and Smith-Hall (2015)
	Cultivating fruits and vegetables in interrow spaces	Institutional and autonomous	Kibria et al. (2015)
	Fruits cultivation through agroforestry project on degraded lad	Institutional	Kibria et al. (2015)
	Using organic fertilizer and compost	Autonomous	Paul et al. (2020)
	Crop diversification and intercropping	Institutional and autonomous	Ahammad et al. (2021) and Paul et al. (2020)
Health	Leaves, root, and stems of 204 plant species used for health purposes. Some widely used plants are *Bambusa bambos*, *Bambusa*, *Begonia roxburghii*, *Cheilocostus speciosus*, *Colocasia esculenta*, *Boehmeria glomerulifera*, *Hoya parasitica*, *Baccaurea ramiflora*, *Bischofia javanica*, *Lagerstroemia speciosa*, and *Brownlowia elata*	Autonomous	Rudra et al. (2021), (Ahammad et al. (2019) and Chowdhury et al. (2014a, b)

(continued)

Table 3 (continued)

Sector	Indigenous adaptation practice	Type	Reference
Sociocultural	Indigenous men weave baskets, mats, and other indigenous utensils from bamboo and rattans	Autonomous	Rahman and Jannatul (2012)
	Female members of indigenous communities weave their clothes by themselves from *Indigofera* sp. plant	Autonomous	Dewan (2002)
Forest Management	Indigenous practices and knowledge as a source of information for sustainable natural resources management	Autonomous	Mukul et al. (2002)
	Regenerative forest practices through VCF	Institutional	
	VCF communities value conservation of native species	Autonomous	Baten et al. (2010) and Saha (2010)
	Engaging in furniture business by collecting woods from the forest	Autonomous	
Water	Rainwater harvesting by excavating ponds and reservoirs, making cross dam in the springs.	Autonomous	Chakma et al. (2020) and Hossen et al. (2016)
	Bamboo pitch to collect water from rocky hill slopes	Autonomous	Hossen et al. (2016)
Others	Fuel from cow dung and agricultural residues	Autonomous	Bhuiyan et al. (2021)
	Handicraft business	Autonomous	Bunten (2010) and Anik and Khan (2012)
	Social media-based small enterprises to sell fruits and vegetables grown in hilly areas	Institutional	From interviews
	Planting deep-routed trees to protect houses from wind speed	Autonomous	

5.2 *Triple Dividend of Adaptation in CHT*

Development practitioners worldwide use the concept of triple dividend to highlight and appraise the benefits of climate change adaptation. The triple dividend of adaptation is a tactic that considers three types of benefits: losses in the communities avoided, economic benefits of adaptation, and socio-environmental benefits of adaptation (Heubaum et al., 2022). Figure 4 shows the triple dividends from climate change adaptation in the context of CHT region.

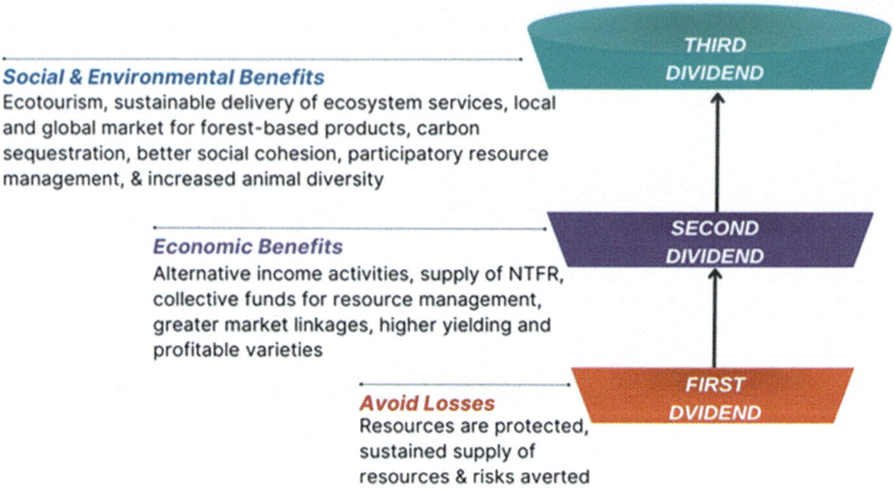

Fig. 4 Triple benefits of adaptation initiatives in the context of CHT

5.3 Negative Coping Mechanism

Decreasing fertility and productivity of land is forcing indigenous communities to adopt (Abdullah et al., 2020) certain negative coping mechanisms. Such coping mechanism includes the indiscriminate use of chemical fertilizers on land to increase productivity from Jum cultivation, reducing the land fallow period to 2–3 years further aggravating the soil quality and enhancing the deforestation process. There was a belief among the indigenous farmers that the loss from rat and pig attacks could be recovered by increasing yield through increased usage of chemical fertilizers (Miah et al., 2011). Consequently, soil quality degrades due to the presence of high organic matter and faster nutrient depletion (Kibria et al., 2013). The diet pattern of indigenous communities has changed, reflecting the decrease of plant availability. Indigenous communities hunt wild animal species to fulfill their food demands. Episodic food shortages due to climatic and non-climatic factors were explained in the preceding section. To adapt with such a situation, indigenous people often reduce the number of meals per day, increase their consumption of wild animals, and start consuming wild forest food (Misbahuzzaman & Smith-Hall, 2015). Tobacco cultivation has been expanding rapidly in recent years since it offers better financial incentives for indigenous communities. A modeling study found that at the current rate of conversion, all the crop lands in CHT will convert to tobacco land in 110 years (Bala & Hossain, 2013). Another recent study based on satellite image analysis concluded that there has been a staggering 95% increase of tobacco lands in CHT from 1989 to 2012. Such transition of crop lands will cause severe environmental degradation on top of human health hazards. According to a newspaper report, 134,010 mounds of fuelwood is burned each year in more than 20,000 tobacco kilns in the CHT region (Hussain, 2018). According to the communities, there has been a significant decline in faunal diversity due to environmental degradation in recent years. Turner et al. (2010) described such ecological encroachment as "*the natural systems upon which people depend, already under direct assault from climate change, are further threatened by how we respond to climate change*" and that "*human adaptation to climate change will have far-reaching effects, ricocheting between different biomes and affecting biodiversity through globally connected economies and markets ... biodiversity is likely to be undermined by a broad and complex array of human responses to climate change*".

Regarding agricultural management, in order to prevent attacks of wild animals to agricultural lands, they also kill animals. The most common wild animals they hunt include *Viverra zibetha*, *Canis aureus*, *Macaca mulatta*, *Elephas maximus*, *Ursus thibetanus*, *Callosciurus pygerythrus*, *Suncus murinus*, and *Herpestes auropunctatus* (Uddin, 2020). Such behavior is a direct threat for the ecological balance of the CHT mountainous region and beyond. According to a study, 8 out of 45 wild animal species are at risk of disappearance due to wildlife consumption by Mro communities (Chowdhury et al., 2014a, b). The same study also highlighted that in the name of making crafts to ensure earning, some indigenous communities kill wild animals and use their body parts to make necklaces, bracelets, and anklets. The topographic and cultural features of CHT offer ample opportunities for tourism. In recent years, the domestic tourism section has expanded at an unprecedented rate. Despite security threats, people from all over Bangladesh visit the natural beauty of CHT, which offers alternative income sources for indigenous communities. Indigenous youths work in the tourism sector as local guides and the traditional paras offer accommodation for tourists. However, the recent influx of tourists in the CHT has seriously jeopardized the local ecosystem, biodiversity, and culture (Ahamed et al., 2022). A recent study documented that built-up area to support tourism has increased mostly in an unplanned manner by 139% in Bandarban, 91% in Rangamati, and 73% in Khagrachari from 2010 to 2020 (Anjum et al., 2021). Due to over-tourism, the use of nontraditional items, including plastics, has increased significantly. A recent estimate shows that the river basins of Chattogram and Chattogram Hill Tracts region drain out 1087.18 billion microbeads to the Bay of Bengal every month (Chowdhury et al., 2021).

5.4 *Barriers and Scopes for Climate Change Adaptation in CHT*

5.4.1 Barriers to Climate Change Adaptation

After reviewing a plethora of scholarly works, we attempt to identify the barriers to climate change adaptation in the CHT region. This section, therefore, highlights some specific and significant barriers to adaptation for indigenous communities in CHT. IPCC defines adaptation barriers as "*factors that make it harder to plan and implement adaptation actions or that restrict options*". The barriers of adaptation are not static and they change over time. To better comprehend, we tend to adopt the classification of barriers by Ekstrom and Moser (Ekstrom & Moser, 2014) for this book chapter. According to them, there are four types of barriers to adaptation: institutional, behavioral, financial, and political. Institutional barriers stem from governance structure and frameworks. In the postliberation period, the governments allowed migration and resettlement of Bengali families to hilly areas and allowed to grab lands and natural resources, all leading to severe environmental degradation (Dhali, 2008). Ecosystem integrity is thus enormously affected in the CHT region. Environmental degradation amplifies the vulnerability of indigenous communities to climate change and limits the scope for adaptation (Webb et al., 2017). In many cases, elite groups and industrialists receive patronization from state agencies to encroach on reserve forest lands for commercial and industrial plantations (Dhali, 2017). Stone extraction from streams in CHT is another example of state-supported environmental degradation. Though stone extraction is illegal, local road construction works rely on supplies of stones from the streams in CHT (Chakma, 2022b). It is widely reported in the national newspapers that the elite groups who are regulating the stone extraction in CHT are from the ruling political party and hence the law-and-order agencies always remain silent and do not take drastic actions. Behavioral barriers to adaptation refer to the thought process of individuals and communities that influence the responses to climate change stimuli (Jones & Boyd, 2011). It is reported that trust toward the government inhibits stronger adaptive behavior in marginalized communities (Valkengoed & Steg, 2019). Historical records of exploitation, rigidification, experimentation, and marginalization by state agencies have created a severe lack of trust among the indigenous communities of CHT toward the government. Ineffective government control on resources coupled with a lack of trust has created a tragedy of the common's situation in the CHT region, which has reduced the adaptive capacity of the indigenous communities (Rasul, 2007). A wide range of anthropogenic activities has substantially altered the ecosystem functioning of the CHT region, which readily affects indigenous life. Therefore, communities focus more on human causes behind the degradation and destruction and are not able to visualize the climate change effects. Results from several studies revealed that perceived susceptibility and severity have influenced adaptation behavior (Semenza et al., 2011). This lack of understanding allows state and non-state agencies to exploit natural resources in the name of climate change adaptation. For instance, NGOs development projects in CHT largely ignored the local ecology and promoted monoculture of commercial plantations, which jeopardized the link between nature and indigenous people (Gain, 2000). Moreover, in the financial support for indigenous communities, incompatible microcredit policies were introduced in the CHT. The difficult repayment process designed by NGOs is extremely difficult for indigenous communities to maintain (Chakma, 2022a). Large-scale alienation of hilly lands through a complex political process can also be considered as a barrier to nature-based and locally led adaptation in CHT. Land grabbing under state support occurs in three distinct ways: conversion of private lands to state-owned lands, coercive occupation of hilly lands by Bengali business elites and commercial groups, and privatization of lands (Adnan & Dastidar, 2011).

Such changes in land ownership limit the adaptation opportunities of indigenous communities through "enclosing the commons" since much of their livelihoods are dependent on forest lands. Myriad studies argue that context-sensitive adaptation responses are required for community resilience (Anik & Khan, 2012).

5.4.2 Window of Adaptation Opportunities

Adaptation to climate change in the CHT is an overwhelmingly complex process and is well-connected with the development process of the region. Indigenous communities of CHT region are the most vulnerable population and yet their rights and privileges are not well-addressed in the national climate policies of Bangladesh (The Daily Star, 2021). There is an urgent need to thoroughly investigate and document the existing adaptation practices among the indigenous communities in CHT. Scaling up of those practices with adequate institutional and financial support and aligning those to the development process will empower indigenous communities and beyond. As we discussed earlier, the overwhelming reliance on natural resources and lack of resource ownership place the indigenous communities at the forefront of the climate crisis. Despite these facts, they have created examples of far better sustainable management practices of forests that have the tremendous potential to replicate elsewhere. Shrinking natural resources have dramatically reduced the income opportunities for indigenous communities in CHT. Therefore, sustainable tourism in CHT can potentially offer a window of alternative livelihood opportunities if managed under a proper policy framework and institutional arrangements. A thorough assessment of ethnic tourism, indigenous tourism, and heritage tourism in CHT for cultural exchange and learning can be a potential adaptation strategy. Such methods have proven to be effective to alleviate the poverty of the indigenous population (Bunten, 2010). It is mentioned earlier that forest resources in the CHT play a pivotal role in the livelihood and cultural activities of the indigenous communities. CHT has a vast opportunity to include nature-based solutions as adaptation strategies (Hossain & Saika, 2020). For instance, the topographical features of CHT such as mountains, waterfalls, forests, and cultural features can be part of sustainable, participatory, and inclusive tourism and can offer livelihood opportunities to local people. Nature-based solutions are a set of collective approaches to deal with climate change mitigation and adaptation. More than 100 nations have incorporated nature-based solutions into their climate adaptation and mitigation policies and actions in the post-Paris era. Nature-based solutions offer substantially higher adaptation benefits in response to the impacts of climate change (Mercer, 2022). Reducing emissions from deforestation and forest degradation and enhancing forest carbon stocks (REDD+) have been proven to be effective methods for climate mitigation (Streck, 2020). However, it is increasingly realized that such initiatives in the context of CHT have a lot of adaptation benefits where indigenous communities improve their livelihoods through sustainable practice as well as through participatory governance for forest management (Miah et al., 2014). However, a large part of the forest area is already restricted and more forests, if brought under REDD+, will further deteriorate the land scarcity crisis. It is speculated that a silent revolution of fruit cultivation is happening in the CHT region. It has been gradually becoming a hub for exotic fruit production (Zaman & Marma, 2022). Forecasting on the future development of the fruits production sector in CHT can offer huge adaptation opportunities for the communities. In addition, there is significant potential for economic development in the areas of high-value agricultural products, livestock, fisheries, agro-based microenterprise, and forestry (Rasul et al., 2016). Though labor migration as a livelihood strategy is very common from Chattogram, such a strategy is almost nonexistent among indigenous communities in hilly areas (Siddiqui et al., 2019). Internal and international labor migration can thus be an important adaptation strategy for households from the CHT region (Siddiqui et al., 2014).

6 Conclusion and Recommendations

The vulnerability of Bangladesh to climate change is well recognized in the literature. This chapter has discussed the impacts of climate change on indigenous communities and mountainous ecosystem in the CHT region. It has complied with the existing adaptation strategies undertaken by the communities and associated challenges. The discussion further extended to the potential areas of adaptation to climate change in the context of CHT to achieve sustainable development goals. Scientific studies have established the argument that warmer days will increase, summer will be prolonged, and rainfall patterns will change in the CHT region. Such changes will dramatically alter the livelihoods of hilly communities in terms of food, shelter, security, nutrition, and employment. Strengthening institutional capacities to incorporate indigenous knowledge, rights, and privileges are key to remove the physical, cultural, and economic exclusion pathways. At the outset, this book chapter makes several recommendations as follows:

- There is persistence threat of losing control on VCF resources by the indigenous communities since there is no formal land tenure for them. Hence, there is a need to establish a sustainable mechanism for resources ownership among indigenous communities to harness the benefits of their practices at a greater scale. It is found in many studies that community-managed forests in CHT are more effective than government-managed forests. Hence, more forests can be shifted to community ownership for greater benefits. However, the indigenous practice of exotic species consumption might put this initiative at risk, which can be minimized through better biodiversity conservation initiatives while including indigenous people.
- Micro-enterprises are good examples of ensuring alternative livelihoods for indigenous communities. Creating a financial architecture with support from both state and non-state actors will open avenues for growth and prosperity among indigenous communities. It is argued that crops and fruits from the CHT

region have higher nutritional values. Such aspects can be capitalized to create a sustainable supply and market chain for high-value crops from CHT.

- There is an urgent need to create a "knowledge bank" to document the indigenous knowledge and practices of all ethnic groups residing all over Bangladesh. Further research needs to be undertaken to scale up these practices for greater adaptation benefits.
- The current infrastructural development in the CHT heavily concentrated on gaining economic benefits from tourism and establishing more security controls. If such developments are aligned with indigenous needs and priorities, this can offer ample economic, social, and cultural benefits for indigenous communities as well as for the whole nation.
- This can be said undoubtedly that CHT has tremendous potential for tourism if managed by adopting sustainable tourism practices. Relevant state agencies can assess the tourism potential and devise a sustainable tourism plan for the CHT by engaging indigenous communities. In addition, foreign tourism opportunities can also be pursued in near future.
- A national repository of medicinal plants and a seed bank can be established to preserve the plant-based indigenous practices.
- It is mentioned by indigenous people in many participatory studies that the repayment mechanism of NGO run microcredit is brutal since income scopes are extremely limited. "One-size-fits-all" development approach by non-state actors needs to be revisited for the CHT region and context-sensitive methods need to be undertaken.
- Choice of internal and international livelihood migration can be facilitated as one of the strategies of in-situ adaptation for some indegenious households who have blick employment opportunities in the hills and would like to send one or more members to urban or international locations of short-term contract migration, but are unable to do so due to different socio-political and economic barriers.
- The magnitude and frequency of landslides are likely to increase in the near future due to a combination of climatic and anthropogenic factors. Hence, an improved, localized, and impact-based early warning mechanism needs to be developed for CHT. A combination of various methods such as acoustic emission slope alarms, moisture sensors, application of inclinometer and extensometers, GPS, and microelectromechanical systems are proven to be effective for landslide early warning.

References

Abdullah, A., Bhuian, M., Kiselev, G., Dewan, A., Hassan, Q., & Rafiuddin, M. (2020). Extreme temperature and rainfall events in Bangladesh: A comparison between coastal and inland areas. *International Journal of Climatology, 42*(6), 3253–3273.

Adnan, S. (2022). Livelihoods and food security of the indigenous peoples of the Chittagong Hill Tracts of Bangladesh: Factors of change and future prospects. In H. Khondker, O. Muurlink,

& A. Ali (Eds.), *The emergence of Bangladesh – Interdisciplinary perspectives* (pp. 229–250). Palgrave Macmillan.

Adnan, S., & Dastidar, R. (2011). *Alienation of the lands of indige-nous peoples in the Chittagong Hill Tracts of Bangladesh*. Chittagong Hill Tracts Commission and International Work Group for Indigenous Affairs.

Ahamed, F., Islam, M., & Nasrullah, A. (2022). *Rethinking development in South Asia: Issues, perspectives and practices*. Cambridge Scholars.

Ahammad, R., Stacey, N., & Sunderland, T. (2019). Use and perceived importance of forest ecosystem services in rural livelihoods of Chittagong Hill Tracts, Bangladesh. *Ecosystem Services, 35*, 87–98.

Ahammad, R., Stacey, N., & Sunderland, T. (2021). Assessing land use changes and livelihood outcomes of rural people in the Chittagong Hill Tracts region, Bangladesh. *Land Degradation and Development, 32*(13), 3626–3638.

Ahammad, R., Stacey, N., Sunderland, T., & Sangha, K. (2022). Land use preference for ecosystem services and well-being in Chittagong Hill Tracts of Bangladesh. *Forests, 13*(12), 2086.

Ahmed, B. (2015). Landslide susceptibility modelling applying user-defined weighting and data-driven statistical techniques in Cox's Bazar Municipality, Bangladesh. *Natural Hazards, 79*, 1707–1737.

Ahsan, S.-A., & Chakma, B. (1989). Problems of national integration in Bangladesh: The Chittagong Hill Tracts. *Asian Survey, 29*(10), 959–970.

Anik, S., & Khan, M. (2012). Climate change adaptation through local knowledge in the north eastern region of Bangladesh. *Mitigation and Adaptation Strategies for Global Change, 17*, 879–896.

Anjum, F., Zilany, H., Rakin, S., Hoque, M., Ania, A.-N., Zaman, M., et al. (2021). Land use/land cover change analysis due to tourism in the Chittagong Hill Tracts of Bangladesh. In *Advances and trends in artificial intelligence. From theory to practice: 34th International Conference on Industrial, Engineering and Other Applications of Applied Intelligent Systems, IEA/AIE 2021* (pp. 173–184). Springer International Publishing.

Baker, R., Mahmud, A., Miller, I., Rajeev, M., Rasambainarivo, F., Rice, B., et al. (2021). Infectious disease in an era of global change. *Nature Reviews Microbiology, 20*, 193–205.

Bala, B., & Hossain, M. (2013). Modeling of ecological footprint and climate change impacts on food security of the Hill Tracts of Chittagong in Bangladesh. *Environmental Modeling & Assessment, 18*, 39–55.

Bandara, T., Marambe, B., Pushpakumara, G., Silva, P., Punyawardena, R., Premalal, S., et al. (2021). Climate resilience of farming systems in steep mountain terrain of selected regions in South Asia. *APN Science Bulletin, 11*(1), 89–111.

Bangladesh Forest Department. (2020). *Ecosystem info*. Retrieved from Bangladesh Forest Information System: http://bfis.bforest.gov.bd/esvd/public/view-user-data/ESV%2029504320

Bär, R., Heinimann, A., & Ehrensperger, A. (2017). Assessing the potential supply of biomass cooking fuels in Kilimanjaro region using land use units and spatial Bayesian networks. *Energy for Sustainable Development, 40*, 112–125.

Baten, M., Khan, N., Ahammad, R., & Misbahuzzaman, K. (2010). *Village common forests in Chittagong Hill Tracts, Bangladesh: Balance between conservation and exploitation*. Unnayan Anneshan.

Baul, T., Peuly, T., Nandi, R., Kar, S., & Karmakar, S. (2022). Role of homestead forests in adaptation to climate change: A study on households' perceptions and relevant factors in Bandarban Hill District, Bangladesh. *Environmental Management, 69*, 906–918.

Bhuiyan, M., Mamur, H., & Begum, J. (2021). A brief review on renewable and sustainable energy resources in Bangladesh. *Cleaner Engineering and Technology, 4*, 100208.

Biswas, S., Swanson, M., & Vacik, H. (2012). Natural resources depletion in hill areas of Bangladesh: A review. *Journal of Mountain Science, 9*, 147–156.

Bunten, A. (2010). More like ourselves: Indigenous capitalism through tourism. *American Indian Quarterly, 34*(3), 285–311.

Chakma, B. (2010). The post-colonial state and minorities: Ethnocide in the Chittagong Hill Tracts, Bangladesh. *Commonwealth & Comparative Politics, 48*(3), 281–300.

Chakma, B. (2022a). Everyday politics of Dadan contracts in the Chittagong Hill Tracts, Bangladesh. *Development and Change, 54*(1), 143–167.

Chakma, S. (2022b, August 4). *Development projects are illegally removing stones from streams in Bandarban, Bangladesh.* Earth Journalism Network.

Chakma, U., Hossain, A., Islam, K., Hasnat, G., & Kabir, M. (2020). Water crisis and adaptation strategies by tribal community: A case study in Baghaichari Upazila of Rangamati District in Bangladesh. *International Journal of Disaster Risk Management, 2*, 37–46.

Chatterjee, S. (2023). Global Climate Change and Environmental Refugees. In P. Singh, B. Ao, & A. Yadav (Eds.), *Global climate change and environmental refugees: Nature, framework and legality* (pp. 185–198). Springer.

Chisty, K. (2014). Landslide in Chittagong city: A perspective on hill cutting. *Journal of Bangladesh Institute of Planners, 2075*, 9363.

Chowdhury, M., Halim, A., Muhamed, N., Koike, M., & Biswas, S. (2014a). Indigenous knowledge in natural resource management by the Hill People: A case of the Mro Tribe in Bangladesh. *Forests, Trees and Livelihoods, 19*(2), 129–151.

Chowdhury, M., Izumiyamaa, S., Nazia, N., Muhammed, N., & Koike, M. (2014b). Dietetic use of wild animals and traditional cultural beliefs in the Mro community of Bangladesh: An insight into biodiversity conservation. *Biodiversity, 15*(1), 23–38.

Chowdhury, M., Zahra, F.-T., Rahman, M., & Islam, K. (2018). Village common forest management in Komolchori, Chittagong Hill Tracts, Bangladesh: An example of community based natural resources management. *Small Scale Forestry, 17*, 535–553.

Chowdhury, G., Koldewey, H., Duncan, E., Napper, I., Niloy, M., Nelms, S., et al. (2021). Plastic pollution in aquatic systems in Bangladesh: A review of current knowledge. *Science of The Total Environment, 761*, 143285.

Clancy, E. (2022). *Is Bangladesh entering the era of malaria elimination?* Speaking of Medicine and Health.

D'Costa, B., & Ahmed, H. (2023). Under the state's securitized gaze – The indigenous peoples of the Chittagong Hill Tracts. In I. S. Parvez & M. Rahman (Eds.), *The politics of terrorism and counterterrorism in Bangladesh* (pp. 128–147). Routledge.

Dach, S., & Peyré, F. (2021). *Vibrant mountain communities regional development in mountains: Realizing potentials, tackling disparities*. Bern Open Publishing (BOP).

Datta, R. (2019). Implementation of Indigenous environmental heritage rights: An experience with Laitu Khyeng Indigenous community, Chittagong Hill Tracts, Bangladesh. *AlterNative: An International Journal of Indigenous Peoples, 15*(4), 309–320.

Dewan, A. (2002). *Woven textiles as art: An examination of the revival of weaving in the Chittagong Hill tracts*. Concordia University.

Dewi, S., Noordwijk, M., Ekadinata, A., & Pfund, J.-L. (2013). Protected areas within multifunctional landscapes: Squeezing out intermediate land use intensities in the tropics? *Land Use Policy, 30*, 38–56.

Dhali, H. (2008). Deforestation and its impacts on indigenous women: A case from the Chittagong Hill Tracts in Bangladesh. *Gender, Technology and Development, 12*(2), 229–246.

Dhali, H. (2017). Deforestation and its impacts on indigenous women: A case from the Chittagong Hill Tracts in Bangladesh. *Gender, Technology and Development, 12*, 229–246.

Dong, J., Jiang, H., Gu, T., Liu, Y., & Peng, J. (2022). Sustainable landscape pattern: A landscape approach to serving spatial planning. *Landscape Ecology, 37*(1), 31–42.

Ekstrom, J., & Moser, S. (2014). Identifying and overcoming barriers in urban climate adaptation: Case study findings from the San Francisco Bay Area, California, USA. *Urban Climate, 9*, 54–74.

Emeto, T., Adegboye, O., Rumi, R., Khan, M.-U., Adegboye, M., Khan, W., et al. (2020). Disparities in risks of malaria associated with climatic variability among women, children and elderly in the Chittagong Hill Tracts of Bangladesh. *Environmental Research and Public Health, 17*(24), 9469.

Foggin, J. (2016). Conservation issues: Mountain ecosystems. In *Reference module in earth systems and environmental sciences*. Elsevier.

Ford, J., King, N., Galappaththi, E., Pearce, T., & Harper, S. (2020). The resilience of indigenous peoples to environmental change. *One Earth, 2*(6), 532–543.

Gafur, A., Jensen, J., Borggaard, O., & Petersen, L. (2003). Runoff and losses of soil and nutrients from small watersheds under shifting cultivation (Jhum) in the Chittagong Hill Tracts of Bangladesh. *Journal of Hydrology, 274*(1–4), 30–46.

Gafur, A., Borggaard, O., Jensen, J., & Petersen, L. (2013). Changes in soil nutrient content under shifting cultivation in the Chittagong Hill Tracts of Bangladesh. *Geografisk Tidsskrift-Danish Journal of Geography, 100*(1), 37–46.

Gain, P. (2000). *The Chittagong Hill Tracts: Life and nature at risk*. Society for Environment and Human Development.

Gain, P. (2013). *The Chittagong Hill Tracts: Man-nature Nexus Torn*. Society for Environment and Human Development.

GoB and FAO. (2013). *Support to preparation of an integrated project for environment friendly agriculture in the Chittagong Hill Tract*. GoB and FAO.

GRID Arendal. (2018). *The Hindu Kush Himalayan region*. Retrieved from GRID Arendal: https://www.grida.no/resources/12806

GRIDA Arendal. (2018). *Hydropower in the Hindu Kush Himalaya region*. Retrieved from GRIDA Arendal: https://www.grida.no/resources/12790

Grizzetti, B., Liquete, C., Vigiak, O., Zulian, G., Bouraoui, F., Roo, A., & Cardoso, A. (2019). Relationship between ecological condition and ecosystem services in European rivers, lakes and coastal waters. *Science of the Total Environment, 671*, 452–465.

Haque, D., Overgaard, H., Clements, A., Norris, D., Islam, N., Karim, J., et al. (2014). Malaria burden and control in Bangladesh and prospects for elimination: An epidemiological and economic assessment. *The Lancet Global Health, 2*(2), e98–e105.

Haque, M. N., Baroi, A., Gomes, J., Toppo, A., Dasand, R., & Hossain, M. (2021). Food security In Chittagong Hill Tracts (Cht), Bangladesh and way forward for achieving through sustainable agricultural practices. *International Journal of Information Research and Review, 8*(9), 7330–7340.

Heubaum, H., Brandon, C., Tanner, T., Surminski, S., & Roezer, V. (2022). *The triple dividend of building climate resilience: Taking stock, moving forward*. World Resources Institute.

Hossain, M., & Saika, U. (2020). An exploration into sustainable hill-tourism potential of the Bandarban Region of Bangladesh. *Tourism Today, 19*, 88–103.

Hossain, I., Osman, K., Kashem, A., & Sarker, A. (2014). Correlations of available phosphorus and potassium with pH and organic matter content in the different forested soils of Chittagong Hill Tracts, Bangladesh. *International Journal of Forest, Soil and Erosion (IJFSE), 4*(1), 7–10.

Hossain, M., Riyadh, Z., Ferdousi, J., Rahman, M., & Saha, S. (2020). Crop agriculture of Chittagong Hill Tracts: Reviewing its management, performance, vulnerability and development model. *International Journal of Agriculture and Environmental Research, 6*(5), 707–727.

Hossen, M., Muhibbullah, M., Hasnat, M., & Rahman, M. (2016). Water crisis, adaptation practices and management strategies in Chittagong Hill Tracts (CHT) area: A case study on Naniarchar Upazila. *Social Change, 6*(1–2), 69–90.

Huss, M., & Hock, R. (2018). Global-scale hydrological response to future glacier mass loss. *Nature Climate Change, 8*, 135–140.

Hussain, A. (2018, September 30). 134,010 maunds of fuelwood burned in CHT tobacco kilns, yearly. *The Dhaka Tribune*.

Hussain, A., Sarangi, G., Pandit, A., Ishaq, S., Mamnun, N., Ahmad, B., & Jamil, M. (2019). Hydropower development in the Hindu Kush Himalayan region: Issues, policies and opportunities. *Renewable and Sustainable Energy Reviews, 107*, 446–461.

ICIMOD. (2021, April 22). *Major river basins in the Hindu Kush Himalaya (HKH) region*. Retrieved from Regional Database System: https://rds.icimod.org/home/datadetail?metadataid=2732#:~:text=The%20dataset%20provides%20first%20level,the%20Yellow%20River%20(Huanghe).

Imran, H., Kala, J., Uddin, S., Islam, A. K., & Acharya, N. (2023). Spatiotemporal analysis of temperature and precipitation extremes over Bangladesh using a novel gridded observational dataset. *Weather and Climate Extremes, 39*, 100544.

International Centre for Integrated Mountain Development. (2013). *Chittagong Hill Tracts ecological forecasting*. NASA DEVELOP National Program.

IPC. (2016). *Chronic food security situation in 10 districts of Bangladesh—December 2015–2018/20*. Ministry of Food, Food Planning and Monitoring Unit.

IPCC. (2023). *Synthesis report of the IPCC sixth assessment report (AR6)- Summary for the policymakers*. IPCC.

Islam, R., Schech, S., & Saikia, U. (2020). Climate change events in the Bengali migration to the Chittagong Hill Tracts (CHT) in Bangladesh. *Climate and Development, 13*(5), 375–385.

Islam, M. A., Islam, M. S., & Jeet, A. A. (2021a). A geotechnical investigation of 2017 Chattogram landslides. *Geosciences, 11*(8), 337.

Islam, M., Begum, S., & Gupta, R. (2021b). High yielding NERICA mutant rice for upland areas and hope for Bangladeshi farmers. In S. Sivasankar, N. Ellis, L. Jankuloski, & I. Ingelbrecht (Eds.), *Mutation breeding, genetic diversity and crop adaptation to climate change* (pp. 53–64). CABI.

Jayawardhan, S. (2017). Vulnerability and climate change induced human displacement. *Consilience, 17*, 103–142.

Jhala, A. (2022). Farming and eating in an indigenous Asian Borderland: Histories of botany, agriculture, and food in the Chittagong Hill Tracts, East Bengal. *Global Food History, 8*(1), 34–55.

Jones, L., & Boyd, E. (2011). Exploring social barriers to adaptation: Insights from Western Nepal. *Global Environmental Change, 21*(4), 1262–1274.

Kamruzzaman, M., Almazroui, M., Salam, M. A., Mondol, M. A., Rahman, M. M., Deb, L., et al. (2022). Spatiotemporal drought analysis in Bangladesh using the standardized precipitation index (SPI) and standardized precipitation evapotranspiration index (SPEI). *Scientific Reports, 12*, 20694.

Kibria, M., Ahammad, M., & Osman, K. (2013). Soil fertility status of some mango orchards in Chittagong Hill Tracts. *Chittagong University Journal of Biological Science, 6*(1–2), 187–197.

Kibria, A., Inoue, M., & Nath, T. (2015). Analysing the land uses of forest-dwelling indigenous people in the Chittagong Hill Tracts, Bangladesh. *Agroforestry Systems, 89*(4), 663–676.

Krishnan, R., Shrestha, A. B., Ren, G., Rajbhandari, R., Saeed, S., Sanjay, J., et al. (2019). Unravelling climate change in the Hindu Kush Himalaya: Rapid warming in the mountains and increasing extremes. In P. Wester, A. Mukherji, A. Mishra, & A. Shrestha (Eds.), *The Hindu Kush Himalaya assessment: Mountains, climate change, sustainability and people* (pp. 57–97). Springer.

Lalmalsawma, P., Balasubramani, K., James, M., Pautu, L., Prasad, K., Sarma, D., & Nina, P. (2023). Malaria hotspots and climate change trends in the hyper-endemic malaria settings of Mizoram along the India–Bangladesh borders. *Scientific Reports, 13*, 4538.

Levene, M. (2010). The Chittagong Hill Tracts: A case study in the political economy of 'creeping' genocide. *Third World Quarterly, 20*(2), 339–369.

Mahmud, F. (2021, January 7). *The silent climate migration on Bangladesh's unforgiving high altitudes*. TRT World.

Mahmud, M., Amin, S., Rashed, M., & Mahmud, R. (2020). Water, sanitation and hygiene practices among ethnic communities in Chittagong Hill Tracts, Bangladesh. *International Journal of Current Research, 12*, 13269–13275.

Mamnun, M., & Hossen, S. (2021). Spatial alteration of fragmented landscape in evergreen and semievergreen rainforest: A case study in Chittagong Hill Tracts, Bangladesh. *Asian Journal of Forestry, 5*(1), 17–27.

Mercer, L. (2022, November 15). *What are nature-based solutions to climate change?* Retrieved from London School of Economics: https://www.lse.ac.uk/granthaminstitute/explainers/what-are-nature-based-solutions-to-climate-change/#:~:text=Nature%2Dbased%20solutions%20include%3A%20avoiding,singular%20species%3B%20improving%20management%20practices

Miah, M., Chakma, S., Koike, M., & Muhammed, N. (2011). Contribution of forests to the livelihood of the Chakma Communiy in the Chittagong HIll Tracts of Bangladesh. *Journal for Science, 17*(6), 449–457.

Miah, M., Chakma, S., Koike, M., & Muhammed, N. (2012). Contribution of forests to the livelihood of the Chakma community in the Chittagong Hill Tracts of Bangladesh. *Journal of Forest Research, 17*(6), 449–457.

Miah, M., Akther, S., Shin, M., & Koike, M. (2014). Scaling up REDD+ strategies in Bangladesh: A forest dependence study in the Chittagong Hill Tracts. *Forest Science and Technology, 10*(3), 148–156.

Misbahuzzaman, K., & Smith-Hall, C. (2015). Role of forest income in rural household livelihoods: The case of village common forest communities in the Chittagong Hill Tracts, Bangladesh. *Small-Scale Forestry, 14*, 315–330.

Morgubatul, J., Hossain, M., Uddin, M., HossainIn, M., & Kamruzzaman, M. (2018). People's dependency on forest resources and contributions of forests to the livelihoods: A case study in Chittagong Hill Tracts (CHT) of Bangladesh. *International Journal of Sustainable Development & World Ecology, 25*(6), 554–561.

Mukul, S., Rashid, A., & Uddin, M. (2002). Indigenous knowledge for appropriate farming systems in Chittagong Hill Tracts. *Grassroots Voice*, 91–95.

Nath, T., & Inoue, M. (2009). Forest-based settlement project and its impacts on community livelihood in the Chittagong Hill Tracts, Bangladesh. *The International Forestry Review, 11*(3), 394–407.

Nath, T., Jashimuddin, M., & Inoue, M. (2016). *Community-Based Forest Management (CBFM) in Bangladesh*. Springer.

Newell, P., Srivastava, S., Naess, L., Contreras, G., & Price, R. (2021). Toward transformative climate justice: An emerging research agenda. *WIREs Climate Change, 12*(6), e733.

Panday, P., & Jamil, I. (2009). Conflict in the Chittagong Hill Tracts of Bangladesh: An unimplemented accord and continued violence. *Asian Survey, 49*(6), 1052–1070.

Parveen, S., & Faisal, I. (2002). People versus power: The geopolitics of Kaptai Dam in Bangladesh. *International Journal of Water Resources Development, 18*(1), 197–208.

Paul, A., Alam, M., & Alam, A. (2020). Assessment of wild edible fruits consumed through the tribal people of Chittagong Hill Tracts (CHTs), Bangladesh. *Indian Journal of Traditional Knowledge, 19*(3), 598–603.

Payne, D., Spehn, E., Prescott, G., Geschke, J., Snethlage, M., & Fischer, M. (2020). Mountain biodiversity is central to sustainable development in mountains and beyond. *One Earth, 3*(5), 530–533.

Pepin, N., Arnone, E., Gobiet, A., Haslinger, K., Kotlarski, S., Notarnicola, C., et al. (2022). Climate changes and their elevational patterns in the mountains of the world. *Reviews of Geophysics, 60*(1), e2020RG000730.

Price, M. (2015). *Mountains: A very short introduction*. Oxford University Press.

Rafi, M., & Chowdhury, A. (2001). *Counting the hills assessing development in Chittagong*. University Press Limited.

Rahbek, C., Borregaard, M. K., Colwell, R. K., Dalsgaard, B., Holt, B. G., Morueta-Holme, N., et al. (2019). Humboldt's enigma: What causes global patterns of mountain biodiversity? *Science, 365*(6458), 1108–1113.

Rahman, M., & Jannatul, F. (2012). Indigenous plant utilization and farming system of Garo tribe in North-East Bangladesh: A means of sustainable biodiversity conservation. *Journal of Forest and Environmental Science, 28*(2), 84–96.

Rasul, G. (2007). Political ecology of the degradation of forest commons in the Chittagong Hill Tracts of Bangladesh. *Environmental Conservation, 34*(2), 153–163.

Rasul, G. (2009). Ecosystem services and agricultural land-use practices: A case study of the Chittagong Hill Tracts of Bangladesh. *Sustainability: Science, Practice, and Policy, 5*(2), 15–27.

Rasul, G., & Thapa, G. (2003). Shifting cultivation in the mountains of South and Southeast Asia: Regional patterns and factors influencing the change. *Land Degradation and Development, 14*(5), 495–508.

Rasul, G., Bikram, N., & Tripura, K. (2016). *Achieving the sustainable development goals in Chittagong Hill Tracts: Challenges and opportunities*. International Centre for Integrated Mountain Development.

Raza, W., & Khan, A. (2022). *Water and health: Impact of climate change in Bangladesh*. World Bank.

Riebeck, E. (1885). The Chittagong Hill tribes. *Nature, 32*, 169–170.

Roche, P., & Campagne, C. (2017). From ecosystem integrity to ecosystem condition: A continuity of concepts supporting different aspects of ecosystem sustainability. *Current Opinion in Environmental Sustainability, 29*, 63–68.

Roy, R. (2000). *Land rights of the indigenous peoples of the Chittagong Hill Tracts, Bangladesh*. IWGIA.

Rudra, A., & Alam, A. (2022). Streamflow characteristics of Sangu-Matamuhuri watershed in the southeastern part of Bangladesh. *Heliyon, 9*(3), e14559.

Rudra, S., Islam, K., Rahman, M., & Uddin, S. (2021). Medicinal plant diversity and their therapeutic uses in selected village common forests in Chittagong Hill Tracts, Bangladesh. *Journal of Herbs, Spices & Medicinal Plants, 27*, 83–107.

Saha, P. (2010). Parbattya Chattagramer Mouza Ban: PrachinPraggyar Arek Rup. In P. Gain (Ed.), *11th issue, an occasional SEHD magazine (Bangla)*. Society for Envi-ronment and Human Development (SEHD).

Santos, G. (2018). Characteristics of rodent outbreaks in the Low San Francisco Sergipano (Sergipe, Brazil) and influence of anomalies on sea surface temperature on temperatures in this region. *International Journal of Design & Nature and Ecodynamics, 13*(2), 156–165.

Sarker, A., & Rashid, A. (2013). Landslide and flashflood in Bangladesh. In R. Shaw, F. Mallick, & A. Islam (Eds.), *Disaster risk reduction approaches in Bangladesh* (pp. 165–189). Springer.

Semenza, J., Ploubidis, G., & George, L. (2011). Climate change and climate variability: Personal motivation for adaptation and mitigation. *Environmental Health, 10*, 1–12.

Sensarma, S., & Okada, N. (2006). *Multi stage conflict resolution process in the case of community disaster risk management: A retrospective analysis*. IEEE Xplore.

Siddiqi, N. (1998). Ethnobotany of non timber forest products of Chittagong Hill Tracts. *Applied Ethnobontany*, 52–55.

Siddiqui, T. (2023). Climate change and displacement: Locating the most vulnerable groups. In H. Khondker, O. Muurlink, & A. Ali (Eds.), *The emergence of Bangladesh – Interdisciplinary perspectives* (pp. 259–272). Palgrave Macmillan.

Siddiqui, M., & Chakma, A. (2016). State-building, identity crisis and ethnic conflict: The case of Chittagong Hill Tracts (CHT) of Bangladesh. *Journal of Sociology, 8*(2), 59–73.

Siddiqui, T., Faisal, M., Masud, M., & Billah, T. (2014). *The role of labour migration in climate change adaptation in Bangladesh with a special focus on the Chittagong Hill Tracts region*. RMMRU.

Siddiqui, T., Bhagat, R., Banerjee, S., Liu, C., Sijapati, B., Memon, R., et al. (2019). Migration in the Hindu Kush Himalaya: Drivers, consequences, and governance. In P. Wester, A. Mishra, A. Mukherji, & A. Shrestha (Eds.), *The Hindu Kush Himalaya assessment: Mountains, climate change, sustainability and people* (pp. 517–544). Springer.

Silveira, F., Barbosa, M., Beiroz, W., Callisto, M., Macedo, D., Morellato, L., et al. (2019). Tropical mountains as natural laboratories to study global changes: A long-term ecological research project in a megadiverse biodiversity hotspot. *Perspectives in Plant Ecology, Evolution and Systematics, 38*, 64–73.

Siraz, M. M., Roy, D., Dewan, M. J., Alam, M. S., AM, J., Rashid, M. B., ... & Yeasmin, S. (2023). Vertical distributions of radionuclides along the tourist-attractive Marayon Tong Hill in the Bandarban district of Bangladesh. Environmental monitoring and assessment, 195(3), 382.

Singh, C., Iyera, S., New, M., Few, R., Kuchimanchid, B., Segnon, A., & Morchaing, D. (2022). Interrogating 'effectiveness' in climate change adaptation: 11 guiding principles for adaptation research and practice. *Climate and Development, 14*(7), 650–664.

Streck, C. (2020). REDD+ and leakage: Debunking myths and promoting integrated solutions. *Climate Policy, 21*(6), 843–852.

The Daily Star. (2021, September 22). Inclusion of indigenous peoples in sustainable development and realisation of rights. *The Daily Star*.

The Third Pole. (2023, February 7). *Opinion: The hidden waters of the Himalayas are key to mitigating disasters*. Retrieved from The Third Pole: https://www.thethirdpole.net/en/climate/opinion-the-hidden-waters-of-the-himalayas-are-key-to-mitigating-disasters/

Turner, W., Bradley, B., Estes, L., Hole, D., Oppenheimer, M., & Wilcove, D. (2010). Climate change: Helping nature survive the human response. *Conservation Letters, 3*(5), 304–312.

Uddin, A. (2020). Man, medicine and foods in Chittagong Hill Tracts, Bangladesh. In K. Mukhopahyay (Ed.), *Food and power: Expressions of food-politics in South Asia* (pp. 109–126). Sage.

Uddin, J., Hossin, M. Z., Mahbub, F., & Hossain, M. Z. (2013). Healthcare-seeking behavior among the Chakma ethnic group in Bangladesh: Can accessibility and cultural beliefs shape healthcare utilization? *Community Health Equity Research & Policy, 33*(4), 375–389.

UNESCO. (2020). *Mobilizing indigenous and local knowledge solutions: Addressing climate impacts and vulnerabilities, a perspective from the Caribbean region*. UNESCO.

Valkengoed, A., & Steg, L. (2019). Meta-analyses of factors motivating climate change adaptation behaviour. *Nature Climate Change, 9*, 158–163.

WaterAid. (2013). *Assessing the natural springs of Chittagong Hill Tracts (CHT)*. WaterAid.

Webb, N., Marshall, N., Stringer, L., Reed, M., Chappell, A., & Herrick, J. (2017). Land degradation and climate change: Building climate resilience in agriculture. *Entries in Ecology and the Environment, 15*(8), 450–459.

Xu, L., Schmid, B., Liu, J., Si, X., Stenseth, N., & Zhang, Z. (2015). The trophic responses of two different rodent–vector–plague systems to climate change. *Biological Sciences, 282*(1800), 20141846.

Zahan, N. (2022). Water scarcity and sufferings of the tribal people: A study on Bandarban hilly area. *ABC Research Alert, 10*(1), 31–38.

Zaman, M. (1982). Crisis in Chittagong Hill Tracts: Ethnicity and integration. *Economic and Political Weekly, 17*(3), 75–80.

Zaman, M., & Marma, M. (2022, December 15). High-value exotic fruits make a splash. *The Daily Star*.

Zonayet, M., & Kumar, P. (2020). Effect of fertilizer on the growth and yield of jhum crop under the slush-mulch condition for sustainable crop production in the Chittagong Hill Tracts (CHTs) of Bangladesh. *International Journal on Bioresource and Stress Management, 11*(5), 456–464.

Long-Term Monitoring of Snow Cover Extent and Land Surface Temperature to Understand the Warming of Mountains in a Semi-urban Location of Northwestern Himalayas

Dharam Chand, Renu Lata, and Isha Thakur

1 Introduction

A common feature of high-altitude regions is snow cover, particularly in the Himalayan mountain ranges, which are the least accessible to people due to their rough terrain. Snow cover is an essential part of the cryosphere and is one of the most important types of land cover as its major source of freshwater (Kulkarni, 2007). The Western Himalayas is widely known for having a greater amount of snow cover due to its higher average elevation and the influence of winter westerly winds (Bookhagen & Burbank, 2010). Due to the influence of several weather systems, the snow cover peak fluctuates in different time periods (Kripalani et al., 2003). The socioeconomic progress of the states in northern India and its society is directly influenced by the snow cover of the Himalayas (Kulkarni et al., 2007). Snowmelt dominates the hydrological regime of the northern rivers, and it also regulates the thermal regime of the land (Koster, 1991). The term "land surface temperature" (LST) refers to the radiative temperature of the land surface as measured by geostationary satellites, utilizing brightness temperature at the top of the atmosphere form infrared spectral channels. LST becomes one of the most effective instruments for observing the temporal variation in earth surface heat. It is now a significant environmental problem (Sahoo, 2013). It is related to how surface energy is physically processed and how different types of land use and land cover affect it (LULC). In order to identify urban heat islands and ecological comfort zones, many times, land surface temperature (LST) derived from multiple remotely sensed data (Weng, 2009; Fu & Weng, 2015; Hao et al., 2016; Tran et al., 2017). A large homogeneous land surface or even a relatively small heterogeneous urban region can

D. Chand · R. Lata (✉) · I. Thakur
G.B. Pant National Institute of Himalayan Environment, Himachal Regional Centre, Mohal, Kullu, Himachal Pradesh, India

S. Tripathi et al. (eds.), *Warming Mountains*,
https://doi.org/10.1007/978-3-031-62197-0_4

experience significant LST changes (Chen et al., 2006; Guha et al., 2019; Hou et al., 2010). Every LST-related work uses normalized difference vegetation index (NDVI) since it is a key component of LST derivation methods (Smith & Choudhury, 1990; Hope & McDowell, 1992; Julien et al., 2006; Yuan et al., 2017; Mondal et al., 2011). Since NDVI is directly utilized to calculate land surface emissivity, it plays a key role in estimating LST (Carlson & Ripley, 1997; Sobrino et al., 2004). The ideal threshold limits in various physical environments are also used to determine the LULC classifications (Chen et al., 2006). Since it measures vegetation, NDVI is heavily influenced by seasonal changes (Guha & Govil, 2021). As a result, LST is influenced by the changing of the seasons. Therefore, a crucial step in LST mapping and monitoring, particularly in an urban context, is the seasonal evaluation of LST and NDVI. The link between LST and NDVI is interesting (Gutman & Ignatov, 1998; Goward et al., 2002; Voogt & Oke, 2003; Weng et al., 2004). The present study deals with the snow cover extent. Climate change has had a significant impact on the Indian Himalayan glaciers in recent decades (Kothawale et al., 2010). Changes in LST patterns and snow cover extent can help identify regions that are warming or changing their precipitation patterns by pointing to changes in regional climate patterns. Monitoring snow cover is essential to conserving water resources, especially in areas where snowmelt is a significant source of freshwater. Monitoring of LST and snow cover helps us better understand the dynamics of ecosystems. They aid in identifying regions that may experience ecological disruptions as a result of variations in temperature and snowmelt patterns. Assessing the effects on vegetation, wildlife habitats, and biodiversity is made easier by monitoring these parameters. The present study will help researchers working in the Indian Himalayan region to understand the snow melting and land surface temperature of the region.

2 Study Area

Himachal Pradesh is located in northern India. It is located in the Western Himalayas between latitudes 30°22′N and 33°12′N and longitudes 75°47′E and 79°04′E. It covers an area of 55,673 km^2. It is located in the mountainous Himalaya region, which is abundant in natural resources. Himachal's drainage system is made up of both rivers and glaciers. The perennial rivers are fed by snowmelt and rainfall. The climate in Himachal Pradesh varies greatly because of the extreme elevation variation. Himachal is regarded as the nation's fruit basket due to the abundance of orchards. It has a diverse range of forests. The elevation ranges from 465 m to more than 7000 m above sea level. The state of Himachal Pradesh has 12 districts, which are further subdivided into three divisions: Shimla, Kangra, Mandi. Agriculture generates 9.4% of the state's GDP and is the primary source of income and employment in Himachal. Approximately 90% of the population in Himachal is directly dependent on agriculture, which employs 62% of the state's total workers. The region begins with the Shivalik mountain range. One of the largest Indian states experiencing rapid urbanization is Himachal Pradesh (Fig. 1).

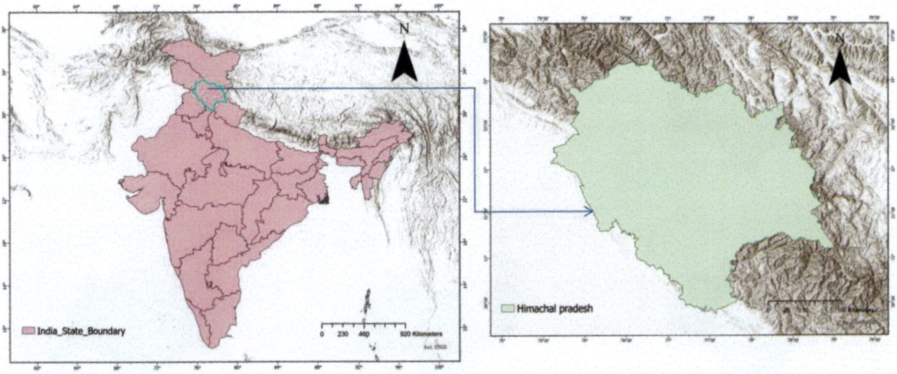

Fig. 1 Location of the study area

3 Methodology

3.1 Data Collection

In the present study, the snow cover maps were created or analyzed on an annual basis in October from 2011 to 2021, which represents the comparison of snow cover. Data preparation and analysis were performed using the Google Earth Engine, Arc GIS Pro, processing software. The Snow Cover Daily Global 500 m (MOD10A1.006) was downloaded from Google datasets using Google Earth Engine. The MODIS's Snow Cover daily Global 500 m (MOD10A1.006) product is used for snow cover, not for snow volume. MODIS's product (MODIS/061/MOD11A1) was used to calculate the land surface temperature of the study area. For calculating the LST in °C, the formula used was as follows:

$$C = K - 273.15$$

After the correction of image, the resultant LST map was exported into chart form.

4 Results

4.1 Snow Cover Extent

Only October month data for snow cover is used because October gives us information about the snow accumulation. The present analysis has shown that an average area percentage of 11.14, 14.51, 11.78, 12.22, 18.15, 6.41, 9.62, 29.36, 18.83, 5.82, and 11.21 observed as snow covered during the period 2011, 2012, 2013, 2014, 2015, 2016, 2017, 2018, 2019, 2020, 2021, respectively (Fig. 2). According to our

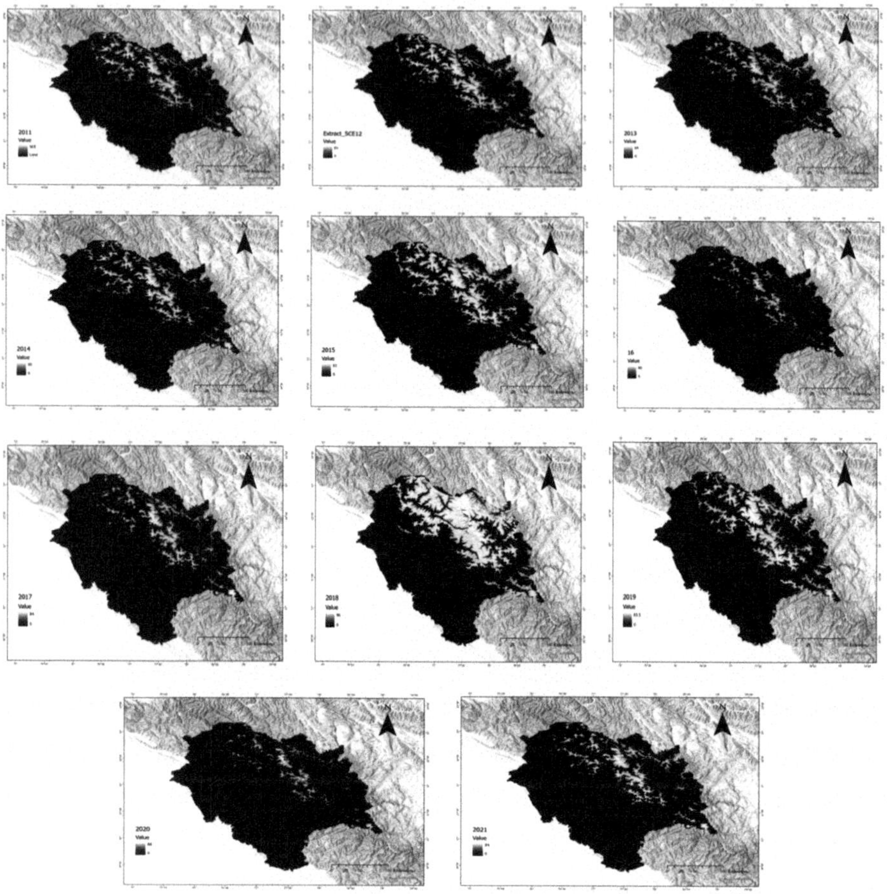

Fig. 2 Year-wise changes in snow cover extent of Himachal Pradesh

findings, the year 2018 had the highest snow cover of 29.36% and the year 2020 shows the least amount of snow cover in Himachal Pradesh with 5.82% coverage. Throughout the study period (2011–2021), the overall snow cover extent in Himachal Pradesh is 13.55%. The overall positive trend in snow cover extent has been observed during the period 2016–2018 (Fig. 3). According to the results of the current study, the year 2017–2018 shows a significant increase in snow cover extent from 9.62% to 29.36%. However, a sudden decrease was observed in the year 2016, where the extent of snow cover directly decreased from 18.15% to 6.41%, representing an 11.71% decrease in the extent of snow cover.

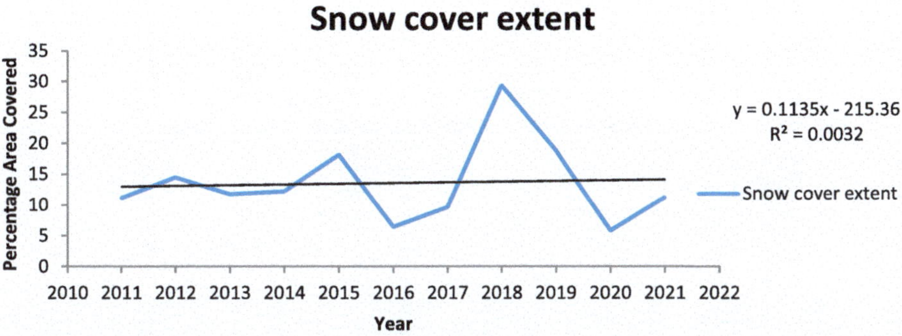

Fig. 3 Year-wise snow cover extent of Himachal Pradesh

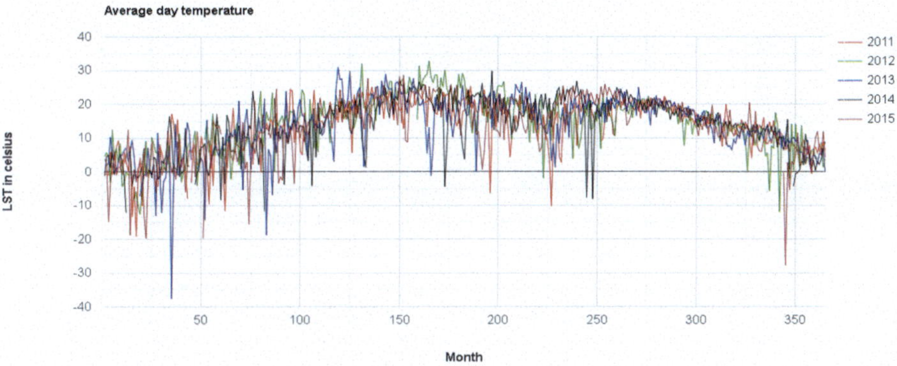

Fig. 4 Annual average land surface temperature from 2011 to 2015

4.2 *Land Surface Temperature*

The land surface temperature (LST) was analyzed annually on a daily basis from 2011 to 2021. From the results of the analysis and processing of data, the LST of the study area ranging from 12.14 to 15.05 in 2011. During the summer, the average daytime LST is higher than the winter. The average day land surface temperature was 13.26 °C, whereas in 2012, 2013, 2014, 2015, 2016, 2017, 2018, 2019, 2020, and 2021, it is 13.92 °C, 13.43 °C, 13.25 °C, 12.91 °C, 15.05 °C, 14.33 °C, 13.81 °C, 12.14 °C, 13.58 °C, and 13.77 °C, respectively (Figs. 4 and 5). As per the results, the average annual land surface temperature ranges from 12.14 °C to 15.05 °C.

4.3 *Relation Between Snow Cover Extent and Land Surface Temperature*

A scatter diagram of LST and snow cover extent (SCE) was created to further analyze the relationship between the LST and SCE. The result of linear regression analysis shows the significance between the two variables. The correlation

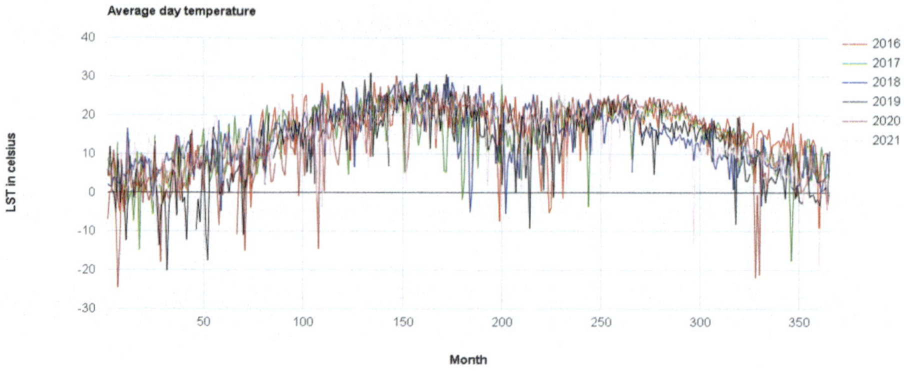

Fig. 5 Annual average land surface temperature from 2016 to 2021

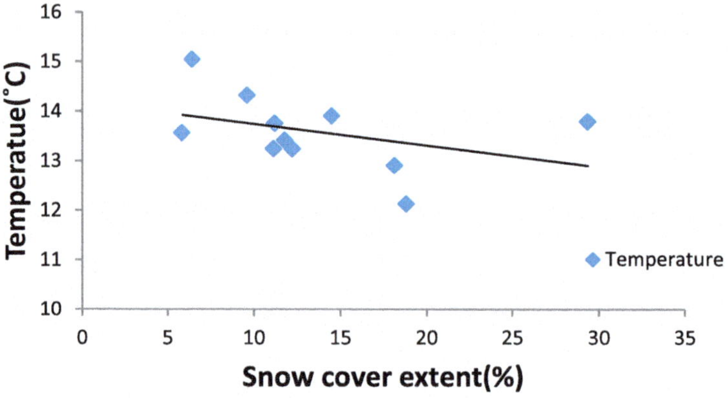

Fig. 6 Snow cover extent and its relation to the land surface temperature

coefficient between annual land surface temperature and snow cover extent was −0.382 (Fig. 6). This shows the negative correlation between the two variables. A negative correlation between LST and SCE shows us that as the value of LST increases, the snow cover extent decreases.

5 Discussion

The pattern of land cover in mountains is significantly influenced by snow cover. The loss of mass from glaciers is primarily caused by warming, which is primarily caused by the CO_2 concentration in the atmosphere (Burkhat & et al., 2017). Climate change has a significant impact on how quickly snowpacks melt. Snowmelt and stream runoff patterns may be impacted by an increase in atmospheric temperature

(Kulkarni et al., 2002). Due to its connection to the terrestrial climate system and the fact that it is a hydrological variable that is extremely sensitive to climate changes, snow cover has a significant impact over the earth's surface (Frei & Robinson, 1999). Temperature and water vapor gradients within the snowpack, as well as between the snow cover, the atmosphere, and the underlying soil, cause snow metamorphosis and melting. Accounting for energy fluxes to the snow pack allows one to estimate the temperature profile of the pack as well as account for snow mass loss due to sublimation and melting. The primary operational application of remote sensing of snow properties has been to map the area of snow cover. Anthropogenic activities such as increasing the number of artificial materials on the earth's surface and using more energy, both of which have high heat capacities and conductivities, are responsible for the increase in land surface and atmospheric temperatures. These activities are also linked to a decline in vegetation and water surfaces, two key components that lower surface temperatures through evapotranspiration (Kaiser, 2014). The temperature felt at the surface–atmosphere interface when long-wave radiation and turbulent heat fluxes interact is known as LST (Myneni et al., 2011). Land surface temperature is a mixture of vegetation and bare soil temperature. Vegetation and soil temperature both respond to a change in incoming solar radiation due to the cloud cover and variance in the illumination. Land surface temperature is an important variable in defining the climatological and hydrological processes that operate in a glaciated environment (Rawat et al., 2021). The Himalayan glaciers are being impacted significantly by climate change. To understand the warming of mountains, it is essential to understand the snow cover extent and its relation with the land surface temperature. For the monitoring of snow cover in Himachal Pradesh, maps for snow cover extent were constructed by using MODIS's data. In Himachal Pradesh, the minimum snow cover was found in the year 2020 and 2016. Average snow cover extent in Himachal Pradesh is 13.55% from 2011 to 2021 in October. As per the results, a general increase in snow cover extent from 2017 to 2018 from 9% snow cover to 29% snow cover extent was observed. Snow cover extent was recorded as maximum in the year 2018. The variation in land surface temperature throughout the study period can be seen in the annual average land surface temperature, which ranges from 12.14 °C to 15.05 °C. Land surface temperature has a significant impact on ecosystems, plant growth, and animal behavior. A sudden decline was seen in 2016 when the amount of snow cover decreased from 18.15% to 6.4%. This might be brought on by factors such as decreased snowfall; earlier snowmelt brought on by higher temperatures. Our findings indicate that the year 2016 had the highest land surface temperature (15.05 °C), which could be the most likely cause of this drop in snow cover. The mean land surface temperature from 2011 to 2021 was observed as 13.59 °C and the snow cover extent and the annual land surface temperature had a −0.382 correlation coefficient. According to this negative correlation coefficient, greater amounts of snow cover are linked to cooler land surface temperatures, while lesser amounts of snow cover are linked to warmer land surface temperatures.

6 Conclusion

The current study provides useful information on the snow cover extent in relation to LST. The assessment of SCE and LST over 11 years between 2011 and 2021 in Himachal Pradesh was carried out. In Himachal Pradesh, the percentage of snow cover extent ranges from 5.82% to 29.63%. From 2011 to 2021, the average snow cover extent is 13.55%. Seven years out of 11 had snow cover extents that were less than the average snow cover extent of 13.58%. Based on our study, the average Land surface temperature during our study period is 13.59 °C. The highest LST values were seen in the year 2020 (5.82 °C) followed by 2016 (5.82 °C), while snow cover extent was lowest for both the years, respectively. Since there is a negative correlation between LST and SCE, which demonstrated that as the value of land surface temperature increases, the extent of snow cover decreases. A rise in LST could speed up the snowmelt this would be a serious threat to people who live in mountains. Therefore, we should plant more trees and install green roofs and vertical gardens to increase the amount of vegetation cover. Cooler land surface temperatures are caused by vegetation because it provides shade, absorbs solar radiation, and releases moisture through evapotranspiration. Continual geospatial monitoring of snow cover patterns will aid in identifying key regions that are particularly vulnerable to climate change. The current study will assist researchers working in the Indian Himalayan region to comprehend snow melting and the region's land surface temperature.

References

Bookhagen, B., & Burbank, D. W. (2010). Toward a complete Himalayan hydrological budget: Spatiotemporal distribution of snowmelt and rainfall and their impact on river discharge. *Journal of Geophysical Research, 115*, F03019. https://doi.org/10.1029/2009jf001426

Burkhart, P. A., Alley, R. B., Thompson, L. G., Balog, J. D., Baldauf, P. E., & Baker, G. S. (2017). Savor the Cryosphere. GSA Today, 4–11. https://doi.org/10.1130/gsatg293a.1

Carlson, T. N., & Ripley, D. A. (1997). On the relation between NDVI, fractional vegetation cover, and leaf area index. *Remote Sensing of Environment, 62*(3), 241–252. https://doi.org/10.1016/S0034-4257(97)00104-1

Chen, X. L., Zhao, H. M., Li, P. X., & Yin, Z. Y. (2006). Remote sensing image-based analysis of the relationship between urban heat Island and land use/cover changes. *Remote Sensing of Environment, 104*(2), 133–146. https://doi.org/10.1016/j.rse.2005.11.016

Frei, A., & Robinson, D. A. (1999). Northern Hemisphere snow extent: Regional variability 1972 to 1994. *International Journal of Climatology, 19*, 1535–1560.

Fu, P., & Weng, Q. (2015). Temporal dynamics of land surface temperature from Landsat TIR time series images. *IEEE Geoscience and Remote Sensing Letters, 12*, 1–5.

Goward, S. N., Xue, Y. K., & Czajkowski, K. P. (2002). Evaluating land surface moisture conditions from the remotely sensed temperature/vegetation index measurements. *Remote Sensing of Environment, 79*(2–3), 225–242. https://doi.org/10.1016/S0034-4257(01)00275-9

Guha, S., & Govil, H. (2021). An assessment on the relationship between land surface temperature and normalized difference vegetation index. *Environment, Development and Sustainability, 23*(2), 1944–1963. https://doi.org/10.1007/s10668-020-00657-6

Guha, S., Govil, H., & Diwan, P. (2019). Analytical study of seasonal variability in land surface temperature with normalized difference vegetation index, normalized difference water index, normalized difference built-up index, and normalized multiband drought index. *Journal of Applied Remote Sensing, 13*(2), 024518. https://doi.org/10.1117/1.JRS.13.024518

Gutman, G., & Ignatov, A. (1998). The derivation of the green vegetation fraction from NOAA/AVHRR data for use in numerical weather prediction models. *International Journal of Remote Sensing, 19*(8), 1533–1543. https://doi.org/10.1080/014311698215333

Hao, X., Li, W., & Deng, H. (2016). The oasis effect and summer temperature rise in arid regions-case study in Tarim Basin. *Scientific Reports, 6*(1), 35418. https://doi.org/10.1038/srep35418

Hope, A. S., & McDowell, T. P. (1992). The relationship between surface temperature and a spectral vegetation index of a tall grass prairie: Effects of burning and other landscape controls. *International Journal of Remote Sensing, 13*(15), 2849–2863. https://doi.org/10.1080/01431169208904086

Hou, G. L., Zhang, H. Y., Wang, Y. Q., Qiao, Z. H., & Zhang, Z. X. (2010). Retrieval and spatial distribution of land surface temperature in the middle part of Jilin province based on MODIS data. *Scientia Geographica Sinica, 30*, 421.

Julien, Y., Sobrino, J. A., & Verhoef, W. (2006). Changes in land surface temperatures and NDVI values over Europe between 1982 and 1999. *Remote Sensing of Environment, 103*(1), 43–55. https://doi.org/10.1016/j.rse.2006.03.011

Kaiser, M. F. (2014). Impact of the human activities on the local climate and environment of the suez City in Egypt. *International Journal of Geosciences, 5*(7), 700–709). Scientific Research Publishing, Inc. https://doi.org/10.4236/ijg.2014.57063

Koster. (1991). *Assessment of climate change impact in high-latitudinal regions* (Vol. 103, pp. 3–13). E.N. Terra.

Kothawale, D. R., Munot, A. A., & Krishna, K. K. (2010). Surface air temperature variability over India during 1901–2007, and its association with ENSO. *Climate Research, 42*(2), 89–104.

Kripalani, R. H., Kulkarni, A., & Sabade, S. S. (2003). Indian monsoon variability in a global warming scenario. *Natural Hazards, 29*, 189–206. https://doi.org/10.1023/A:1023695326825

Kulkarni, A. V. (2007). Effect of global warming on Himalayan cryosphere. *Jalvigyan Sameeksha, 22*, 93–108.

Kulkarni, A. V., Mathur, P., Rathore, B. P., Alex, S., Thakur, N., & Kumar, M. (2002). Effect of global warming on snow ablation pattern in the Himalayas. *Current Science, 83*(2), 120–123.

Kulkarni, A. V., Bahuguna, I. M., Rathore, B. P., Singh, S. K., Randhawa, S. S., Sood, R. K., & Dhar, S. (2007). Glacial retreat in Himalayas using Indian remote sensing satellite data. *Current Science, 92*, 69–74.

Mondal, A., Guha, S., Mishra, P. K., & Kundu, S. (2011). Land use/land cover changes in Hugli Estuary using Fuzzy C-Mean algorithm. *International Journal of Geomatics and Geosciences, 2*(2), 613–626.

Myneni, C. J., Chapman, L., Thornes, J. E., & Baker, C. (2011). Remote sensing land surface temperature for meteorology and climatology: A review. *Meteorological Applications, 18*(3), 296–306.

Rawat, M., Karwariya, S., Raushan, R., Kanga, S., Taloor, A. K., & Thapliyal, A. (2021). Snow cover and land surface temperature assessment of Mana basin Uttarakhand India using MODIS satellite data. In *Water, cryosphere, and climate change in the Himalayas* (pp. 159–174). Springer International Publishing. https://doi.org/10.1007/978-3-030-67932-3_10

Sahoo, S. (2013). Monitoring urban land use land cover changeably multi-temporal remote sensing information in Howrah City, India. *International Research Journal of Earth Sciences, 1*(5), 1–6. Retrieved from http://www.isca.in. Retrieved from http://www.isca.m

Smith, R. C. G., & Choudhury, B. J. (1990). On the correlation of indices of vegetation and surface temperature over south-eastern Australia. *International Journal of Remote Sensing, 11*(11), 2113–2120. https://doi.org/10.1080/01431169008955164

Sobrino, J. A., Jimenez-Munoz, J. C., & Paolini, L. (2004). Land surface temperature retrieval from Landsat TM5. *Remote Sensing of the Environment, 9*, 434–440. https://doi.org/10.1016/j.rse.2004.02.003

Tran, D. X., Pla, F., Latorre-Carmona, P., Myint, S. W., Caetano, M., & Kieu, H. V. (2017). Characterizing the relationship between land use land cover change and land surface temperature. *ISPRS Journal of Photogrammetry and Remote Sensing, 124*, 119–132. https://doi.org/10.1016/j.isprsjprs.2017.01.001

Voogt, J. A., & Oke, T. R. (2003). Thermal remote sensing of urban climates. *Remote Sensing of Environment, 86*(3), 370–384. https://doi.org/10.1016/S0034-4257(03)00079-8

Weng, Q. (2009). Thermal infrared remote sensing for urban climate and environmental studies: Methods, applications, and trends. *ISPRS Journal of Photogrammetry and Remote Sensing, 64*(4), 335–344. https://doi.org/10.1016/j.isprsjprs.2009.03.007

Weng, Q. H., Lu, D. S., & Schubring, J. (2004). Estimation of land surface temperature-vegetation abundance relationship for urban heat Island studies. *Remote Sensing of Environment, 89*(4), 467–483. https://doi.org/10.1016/j.rse.2003.11.005

Yuan, X. L., Wang, W., Cui, J., Meng, F., Kurban, A., & De Maeyer, P. (2017). Vegetation changes and land surface feedbacks drive shifts in local temperatures over Central Asia. *Scientific Reports, 7*(1), 3287. https://doi.org/10.1038/s41598-017-03432-2

Part II
Mountain's Sustainability: Perspectives for Biodiversity and Livelihood

Plant Phenological Shifts in the Indian Himalayan Region

Aditi Mishra, Harshit Pant Jugran, K. Chandra Sekar, and Gautam Talukdar

Abbreviations

EOS	End of the Season
HKH	Hindu Kush Himalayas
IBP	International Biological Program
IHR	Indian Himalayan Region
IUCN CEM	International Union for Conservation of Nature and Natural Resources Commission on Ecosystem Management
LGS	Length of the Growing Season
NPN	National Phenology Network
SOS	Start of the Season

A. Mishra (✉)
Centre for Biodiversity Conservation and Management, Govind Ballabh Pant National Institute of Himalayan Environment, Almora, Uttarakhand, India

H. P. Jugran
Centre for Socio-Economic Development, Govind Ballabh Pant National Institute of Himalayan Environments, Almora, Uttarakhand, India
e-mail: harshit.pantj@gbpihed.nic.in

K. C. Sekar
G.B. Pant National Institute of Himalayan Environment, Garhwal Regional Centre, Srinagar, Uttarakhand, India
e-mail: kcsekar@gbpihed.nic.in

G. Talukdar
Protected Area Network, Wildlife Management and Conservation Education, Wildlife Institute of India, Dehradun, Uttarakhand, India
e-mail: gautam@wii.gov.in

S. Tripathi et al. (eds.), *Warming Mountains*,
https://doi.org/10.1007/978-3-031-62197-0_5

1 Introduction

The mountains cover 24% of the land surface on Earth (IUCN CEM) and are a source of vital goods and services to the residing communities. They provide habitat to 12% of the world's population (GTOS, 2008). The livelihoods of nearly 40% of people at the global level are influenced by the mountains (Singh et al., 2010). Numerous scientific studies provide substantial evidence, indicating that mountainous regions are encountering some of the most elevated rates of climate warming (Pepin et al., 2015; Steinbauer et al., 2018).

The Indian Himalayan Region (IHR) lies between 21° 57″–37° 50″N and 72° 40″–97° 25″E, covering an area of 5.3 lakh km^2 (Mehta et al., 2023). The foothills of the forests of the Indian Himalayan Region are dominated by tropical dry deciduous forests, whereas the alpine meadows are found above the timber line. The Indian Himalayan Region boasts an abundance of high-value biodiversity elements (Mehta et al., 2021) and is one of the 34 global biodiversity hotspots (Sri & Rao, 2023). These areas are known for their unique flora and fauna and for showing high endemism (Sharma et al., 2010). These high-value biodiversity elements in the Indian Himalayan Region include several medicinal and aromatic plants, as well as woody tree species that hold economic importance, among others. The ongoing climate change has put these high-value biodiversity elements in the Indian Himalayan Region under constant threat. The Himalayan region is not only ecologically significant but also shows great deal of cultural, economic, and social significance.

Climate change exerts a profound impact on the physiology and seasonal behaviors of plants, leading to significant alterations in their life processes and seasonal activities (Vitasse et al., 2021). Phenology is a field of study concerned with the timing and recurring patterns of biological events or "phenophases," such as the initial flowering of plants, the emergence of buds on trees, and the migration of birds to different areas (Chmielewski & Roetzer, 2001; Fitter & Fitter, 2002; Walther et al., 2012). These events are often recorded as Julian dates within the calendar year (Cook et al., 2008). The International Biological Program (IBP) defined phenology as "the study of the timing of recurrent biological events, the causes of their timing with regard to biotic and abiotic forces, and the interrelation among phases of the same or different species" (Richardson et al., 2013). For centuries, agricultural practices have relied on phenological observations and calendars, as evidenced by their utilization in ancient Chinese and Roman civilizations (Hopp, 1974). In the past 15 years, there has been a growing recognition of the importance of phenology in the global scientific community. The first flowering dates have often been used to examine the changes in flowering phenology (Fitter & Fitter, 2002; Bradley et al., 1999). Phenology serves as a valuable tool for tracking the biological responses to climate change (Beaubien & Johnson, 1994). Phenology exhibits considerable variability (Schmidt & Lotan, 1980) and exhibits heightened sensitivity to prolonged climate shifts (Sparks & Carey, 1995). Since the 1970s, the onset of spring has advanced by 2.3–5.2 days per decade (IPCC AR-4, 2007), and Rosenzweig et al. (2008) described phenology as one of the simplest methods of monitoring alterations in species ecology as a response to climate change. In order to anticipate the future course of global change, it is essential to comprehend the climate-induced changes in plant phenology and their influence on climate change feedback (Ma & Zhou, 2012). The plants shift their biological events in order to adapt to the changes in the environmental cues (Jerome et al., 2021).

Phenological shifts can vary across species and geographical regions, but the consistent trends of earlier spring onset, delayed autumn events, and extended growing seasons are evident. This may be the result of anthropogenic global changes. According to Cleland et al. (2007), increased CO_2 levels accelerated flowering of forbs but had a delaying effect on grasses. The delay in flowering can result in a shortened flowering period, which may lead to reduced seed and fruit production (Rafferty et al., 2016). The composition of species at the community level is determined by the individual responses of each species to temperature. The decline in the native species is linked to weak responses to temperature in flowering and vice versa. Modifications in the biophysical attributes of the land surface also impact evapotranspiration and phenology (IPCC, 2018). The shifts in phenological phases can impact terrestrial ecosystems, agriculture, and economies (Ma & Zhou, 2012). Most plants have a clear definite phenology, which is linked to the phenology of other organisms (Pau et al., 2011). Most research on the early arrival of spring due to increasing temperatures has been conducted in North America and Europe, where distinct cold and warm seasons exist and vegetation is not limited by moisture; however, studies on phenology in the arid and semiarid areas where moisture turns out to be the limiting resource are scarce (Crimmins et al., 2013). Phenological study is essential for comprehending the ecological interactions in both natural and modified ecosystems, which are integral to human society's dependence on the same (Denny et al., 2014). Phenology can serve as a useful tool to indicate changing patterns in climate and can therefore serve as a marker of climate change as noted by Parmesan and Yohe (2003), Root et al. (2003), and Willis et al. (2017). Moreover, it plays a significant role in the assessment of different aspects of climatic change patterns on the studied tree species. It is very crucial to study the phenology of the Himalayan vegetation in order to come up with a conservation strategy. The purpose of this study is to provide an overview of the changes in plant phenology across the Indian Himalayan Region (IHR) and to assess the extent of this shift. In this chapter, we will delve into the fascinating world of plant phenology and explore the factors that drive shifts in plant phenological events. In addition to examining the drivers of shifts in plant phenology, the chapter presents case studies from the Indian Himalayan Region, showcasing observed changes in the phenological events of various plant species in response to climate change.

2 Drivers of Shifts in Plant Phenology

Plant phenological changes are influenced by a multitude of factors. These factors can include fluctuations in air temperature, rainfall and precipitation, snowmelt timing, and photoperiod, amongst others. In order to gain a comprehensive understanding of global drivers and patterns in phenology, a further investigation into species-specific mechanisms of phenological change is essential (Gallinat et al., 2021). Climate-induced shifts in phenology have extensive impacts on ecological interactions and the abundance of species (Song et al., 2021). The shifts in plant phenology occur as a consequence of their adaptive responses to the changing environmental conditions. Furthermore, an illustrative visual representation of the interplay among these factors is shown in Fig. 1 to enhance understanding of this intricate process.

2.1 *Temperature*

Temperature stands as the primary abiotic factor accountable for alterations in plant phenology. Warming induces advancement in phenological events (Menzel et al., 2006; Wolkovich et al., 2012). The timing of flowering and how plants respond to various thermal inductions varies between different species and it reflects how the

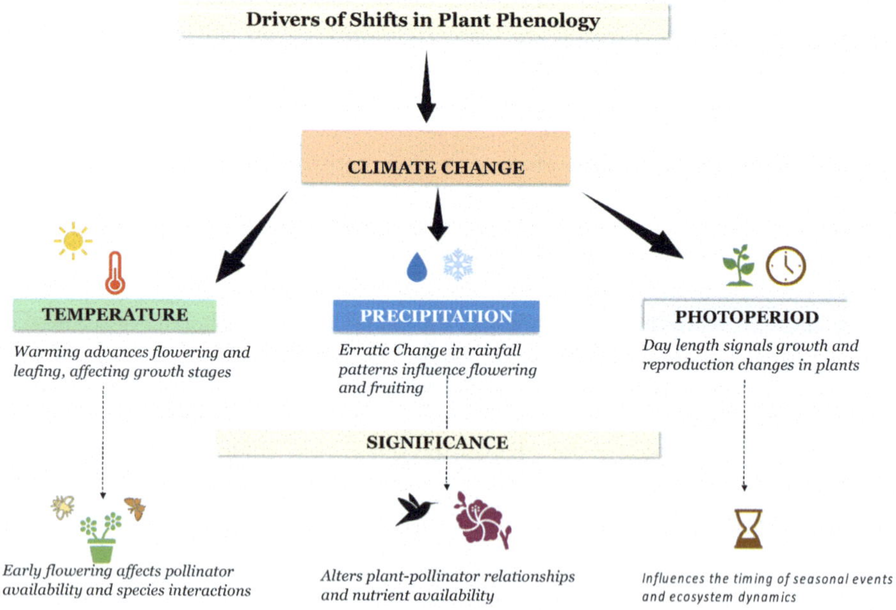

Fig. 1 Illustration to depict the drivers of plant phenology and their significance

plants adapt to their native environment (Capovilla et al., 2014). The timing of flowering in plants is strongly influenced by temperature, with the effect of prolonged cold temperatures, known as vernalization, playing a crucial role as well as the thermal incline or decline in the atmosphere. Elevated temperatures can accelerate plant growth across different developmental stages, causing them to shorten, as reported in studies by Bernier et al. (1981), Halevy (1985), and Kinet et al. (1985). In areas with milder winters and lower latitudes, temperature plays a significant role in regulating autumnal phenology (Estiarte & Penuela, 2015). Doi and Takahashi (2008) have proposed that leaves in the upper crown that are exposed to more sunlight tend to senesce more slowly due to higher temperatures, while leaves at higher latitudes tend to senesce faster due to colder temperatures. Even slight alterations in air temperature, caused by climate change, can have significant impacts on both resource allocation patterns and biotic factors (Stenstrom & Jonsdbttir, 2006). Plant species composition can be changed when the mean daily air temperature rises by 3.5 °C, as shown in the long-term ecological research in the Alaskan Arctic region (Chapin et al., 1995). Climate warming can cause temperature stress and reductions in soil moisture, which can lead to reduced nutrient availability due to increased evapotranspiration (Rustad et al., 2001). During the past three to five decades, there has been a notable shift in the onset of spring phenology in the middle and high latitudes of the Northern Hemisphere. On average, there has been a shift toward an earlier onset of spring by approximately 2.5 days per decade, primarily attributed to changes in temperature (Kariyeva et al., 2012). Studies have indicated that plants that undergo early spring flowering are more susceptible to the impacts of warming compared to those that flower later in the year (Garcia-Mozo et al., 2010). A study carried out in Spain by Garcia-Mozo et al. (2010) have reported that an increase in temperature is the primary factor causing an advance in foliation, flowering, and fruiting, as well as a delay in leaf fall dates. Increased temperatures in spring lead to an advancement in spring phenology, whereas increasing temperatures in winter delays the requirement of chilling temperatures and hence delays the start of spring (Yu et al., 2010). Temperature in itself is not the sole driver responsible for the shifts in phenology; however, photoperiod plays a vital role as a critical cue, triggering changes in growth and reproduction across diverse species (Ettinger et al., 2021). Shen et al. (2022) in their study have reported that the temperature responses of phenophases are influenced by soil water availability, with the start and end of the season showing greater temperature sensitivity under wetter conditions.

Since 1986, a notable warming trend has been observed in India, with average mean, maximum, and minimum temperatures showing a recorded increase of 0.15 °C, 0.15 °C, and 0.13 °C per decade, respectively (Sanjay et al., 2020). The temperature irregularities also have a substantial impact on regulating the phenology of plants. Sabin et al. (2020) in their study indicated that by the end of the twenty-first century, the Hindu Kush Himalayan (HKH) region is expected to experience a temperature increase ranging from 2.6 °C to 4.6 °C. This significant warming trend is projected to contribute to a further decline in snowfall and the melting of glaciers in the region. This, in turn, will result in significant hydrological and agricultural consequences for the aforementioned area.

2.2 Rainfall

Unlike temperature, the impacts of rainfall and precipitation on phenology are less well understood (Siegmund et al., 2016). Based on recent climate projections, it is expected that high-altitude and high-latitude regions will encounter a rise in extreme rainfall and precipitation events if temperatures continue to increase (IPCC, 2018). Understanding the effects of these rainfall and precipitation patterns is crucial as it can indirectly influence the phenology of many species by reducing temperatures, while in some cases, it can directly influence the phenology of various species (Dai et al., 2012). Herbaceous species were seen to be more affected by the changes in rainfall and precipitation patterns in comparison to the tree species (Garcia-Mozo et al., 2010). As a result of milder winters, the advancement of leaf unfolding may be slowed down if the chilling requirements for breaking dormancy are not met, potentially resulting from milder winter seasons in the future (Meier et al., 2021).

The experiments conducted by Olivares et al. (2009) on *Avena barbata* link showed that although the amount and distribution of subsequent rainfall and precipitation affect the dry matter production as well as the number of seeds, the vegetative and reproductive development is facilitated by the late distribution of rain, which neither leads to a change in the generative capacity nor seed viability. According to Chambers and Keatley (2010), summer phenological events showed significant correlations with rainfall, with minimal or no lag period. Increased incidences of rainfall and precipitation in December and, to a certain extent, during summer led to a postponement of phenological events. However, the timing of autumn or winter phenology was found to be less impacted by rainfall. According to a report by Collins et al. (2013), there would be changes in the spatial distribution of rainfall, some regions might experience an increase in rainfall, whereas others might have a decrease but the higher latitude and higher altitudes along with the tropical regions are likely to experience greater amount of rainfall and precipitation. North Central and Western Europe regions for instance are expected to have increased winter rainfall (Scaife et al., 2012). The alterations in these rainfall and precipitation patterns may cause changes to the timing of various phenological stages and are in turn going to drastically impact the plant–pollinator interactions (Lawson & Rands, 2019). Unprecedented events of rain disrupt the pollen transfer and hence leads to a hindrance in the reproductive efforts of flowering plants (Lawson & Rands, 2019).

2.3 Snowmelt

Winter snow is undergoing the most rapid changes in the context of ongoing climate change (IPCC, 2007). The initiation and duration of the feasible growing period in high-altitude ecosystems are determined by the timing, extent, and duration of the snow cover, hence it is very crucial (Inouye & Wielgolaski, 2003). Future projections indicate that winter incidences of rainfall and precipitation is expected to increase, primarily in the form of rain rather than snow (IPCC, 2007). As a consequence, thinner and less insulating snow cover will be formed, characterized by delayed onset and earlier melting (Wipf et al., 2009), leading to significant changes in the snow cover formation, duration, and depth. This will result in colder winters in the mountains, but also an extended and warmer growing season. The melting of snow, in conjunction with temperature and photoperiod, plays a fundamental role in shaping plant phenology in Arctic and Alpine regions (Hülber et al., 2010; Keller & Körner, 2003; Wipf et al., 2009). Research has documented a robust correlation between the timing of snowmelt and the initiation of the growing season (e.g., Bjorkman et al., 2015; Ernakovich et al., 2014). The structure of communities in the Alpine and Arctic ecosystems is highly dependent on the spatial patterns of snow distribution (Odland & Munkejord, 2008). Therefore, the characteristics of snow, including its depth, melt timing, and duration, significantly influence the dynamics of these ecosystems. The composition of vegetation and the growth of plants are immediately affected by the changes in the snowmelt timings (Wipf et al., 2009). One can expect that the phenology of plants in the Arctic will advance because of the phenomenon of earlier snowmelt and warmer temperatures (Bjorkman et al., 2015), whereas the phenology in the Alpine regions is expected to be moderate because of the limiting effect of photoperiod (Ernakovich et al., 2014). The individual species vary in their responses to snowmelt timing (Winkler et al., 2018). Due to climate warming, earlier snowmelt occurs and that ultimately leads to warmer winters and hence also an earlier onset of spring, thus advancing the spring phenology. According to Dunne et al. (2003), the timing of flowering in early-flowering species exhibits a stronger correlation with snowmelt compared to the timing of flowering in later-flowering species. According to a study by Inouye et al. (2002), early snowmelt led to a decline in the number of inflorescences in *Delphinium barbeyi* Huth and, consequently, there was a reduction in the total duration of the flowering period within the studied plot. Any change in the timing of snowmelt can have a severe impact on the surrounding air temperatures, therefore it drastically impacts spring phenology (Vitasse et al., 2016). The continued climate change can produce further severe impacts (Ettinger et al., 2022) (Table 1).

Table 1 Checklist of plant phenology: drivers of shift in plant phenology and observed phenological shifts

Category	Elements	Exemplar papers reporting the elements
Spring phenology	Advanced	Advanced flowering by 11.8 days/°C. "a significant advancement in the flowering phenology of an early spring flowering plant *Senecio vernalis* Waldst. & Kit. over four decades in this Himalayan region." (Hassan et al., 2021)
	Delayed	"The flowering period delayed from 2 to 10 days. It is judged that the flowering date is highly related to the temperature" (Kim et al., 2012)
Autumn phenology	Advanced	Advanced leaf senescence in *Quercus lamellose* Sm. by 1 month (Chand et al., 2017)
	Delayed	"Leaf Colouration Date was found to have been delayed by 4.56 days" (Dai et al., 2012)

3 Impact of Climate Change on Plant Phenological Events

The rapid and readily observable response of plant phenology to global climate change makes it a crucial indicator, with significant implications for the flora and fauna reliant on periodic availability of plant resources (Corlett & Lafrankie Jr, 1998). In recent decades, climate change has had an impact on plant phenology by advancing the first flowering dates as winter and spring temperatures rise, in reaction to the changing climate (Cayan et al., 2001; Defila & Clot, 2001). Human activities are a prominent driver of climate change, playing a substantial role in influencing changes in the timing of biological events and disturbances to biodiversity (Ranjitkar et al., 2013). At higher latitudes, warming has altered the phenology and caused plant-pollinator and pest mismatch in the region (IPCC, 2022). Flowering occurs earlier than usual, implying an earlier pollen season, and the timing of spring and summer phenology is influenced by climate and local weather. The duration of the growth period, which is determined by the timing of both the beginning of growth and the onset of senescence, plays a vital role in influencing the yearly carbon assimilation in terrestrial ecosystems (Du et al., 2019; Li et al., 2019). Wolkovich et al. (2012) found that in response to rising annual temperatures, woody species in temperate regions across the globe are adjusting the timing of their leaf out, advancing it earlier by an average of 4.6 days/°C. Cherry blossom (*Prunus jamasakura* Makino) flowering dates have been recorded for centuries in Japan, for example, and this has revealed that due to the impacts of warming and urbanization, the flowering of Cherry trees has advanced than at any time in the previous 712 centuries (Wenden et al., 2016). The IPCC's fourth assessment report presented clear compelling evidences, indicating that human activities have caused significant alterations in the Earth's climate (IPCC, 2007). Climate change is the primary factor responsible for alterations in temperature and precipitation patterns, leading to changes in phenology, CO_2 exchange, and annual carbon uptake across various ecosystems (Richardson et al., 2013).

The IPCC Special 1.5° report had indicated the influence of climate change on the timing of plant biological events. The following is quoted from the report:

- "In recent decades, the Northern Hemisphere ecosystems have experienced an advance in spring phenology by approximately 2.8 days per decade, which is attributed to the impact of climate change" (Settele et al., 2014).
- "A projected 1.5°C increase in global temperature, as opposed to 2°C, would result in a 10-day advancement of temperate forest phenology" (Roberts et al., 2015).
- "Mediterranean species are experiencing shifts in phenology and range contraction due to decreasing precipitation and increasing temperatures" (Settele et al., 2014).
- "Anthropogenic climate change strongly impacts plant life cycles and species interactions."

- "Climate change causes biome transformations, species range losses, increased extinction risks, changes in phenology, and an increase in extreme weather events. All of these changes can damage ecosystem services and functions."

3.1 Spring Phenology

Spring phenology pertains to the timing of biological events, such as the emergence of the first leaves on trees or the emergence of early flowering plants like crocuses, that mark the beginning of spring. Springtime phenology interests the researchers the most because of the prominent phenological changes such as bud bursts or the emergence of new foliage. The blooming of Cherry blossoms in Japan indicates that spring is on its way. Spring phenology records the initiation of photosynthetic processes as the season progresses (Wang et al., 2015). Temperature plays a dominant role in shaping the spring phenology of temperate and boreal vegetation in the Northern Hemisphere (Zhang et al., 2004). Changes such as plants blooming and buds shooting earlier, migrant birds arriving earlier, butterflies appearing earlier, amphibians exhibiting earlier choruses and spawning, and earlier breeding or first singing of birds are commonly observed due to shifts in the onset of spring (Walther et al., 2012). Key metrics of vegetation phenology such as the start of the growing season (SOS) and end of the growing season (EOS) are highly vulnerable to climate changes (Fu et al., 2015). In recent decades, global temperature increases have led to an earlier initiation of spring plant growth (Allstadt et al., 2015). The dominant factor controlling spring phenology appears to be increasing temperature in the spring (Flynn & Wolkovich, 2018). Winter chilling (Chuine et al., 2000; Ghelardini et al., 2010) and photoperiod (Caffarra et al., 2011) are additional factors that can contribute to breaking bud dormancy and inducing bud bursts in plants. The initial appearance of spring foliage is a crucial indicator of climate variability, as it demonstrates the most pronounced response to temperature changes, according to research conducted by Cayan et al. (2001) and Chuine and Beaubien (2001). In recent decades, studies have reported that winter and spring have experienced greater warming than any other season (Schwartz et al., 2006). Climate variability can have a strong impact on the timing of the first leaf and first flower phenophases, which occur following a period of winter dormancy (Cayan et al., 2001; Schwartz et al., 2006). Climate change has wide-ranging consequences that are most pronounced in the spring, and there are typically more observations available for plant phenophases during this season compared to others. Spring phenology also affects the length of the growing season (LGS) and holds significant importance in determining autumn phenology (Fu et al., 2018).

3.2 Autumn Phenology

The onset of foliar senescence marks the autumn phenology. Advances in spring phenology have been well-marked all over the world but the shifts in autumn phenology have shown more mixed results. The autumn events, such as leaf senescence, fruit ripening, bird and insect migration, as well as the initiation of hibernation and diapause, hold significant importance. Deciduous trees undergo visible changes during the transition from the active to dormant stages, as the leaves undergo a color transformation from green to yellow or red, indicating the occurrence of leaf senescence. The timing of foliage coloration also impacts the ecotourism industry in the United States, which generates billions of dollars in revenue (Rustad et al., 2012). One way to quantify the phenology of leaf yellowing is to observe and note the date of the initiation of senescence. The leaf senescence is not just a matter of 1 day but it is an entire process. Leaf abscission can serve as an indicator to evaluate potential delays in the timing of leaf senescence, given that it is the subsequent step following the senescence process (Thakur et al., 2016). Autumn phenophases affect tree development and growth potential for the next year autumn events are quite important ecologically and evolutionarily. Autumn phenology acts as indicators for animal behavior, including bird migration (Ellwood et al., 2015). Despite its importance in the environment, autumn phenology remains poorly understood (Gallinat et al., 2015). Not much attention has been paid to autumn phenology and the reason for this can be its very narrow reach, limited only to deciduous plants.

Monitoring of autumn phenology in evergreen plants is a very difficult task and requires extreme precision and skill. The shift from the active phase to the dormant stage in deciduous trees is signaled by leaf senescence (Estiarte & Penuela, 2015). Studies have shown that there has been a delay in leaf senescence by a rate of 1–2.5 days per decade (Menzel & Fabian, 1999; Matsumoto et al., 2003; Gordo & Sanz, 2010). The major driver responsible for the delayed autumn phenology is climate warming (Menzel et al., 2001; Chen et al., 2005; Rutishauser et al., 2008). Menzel et al. (2006) in their study indicated that the impact of a 0.1 °C temperature increase on the delay in autumn phenology was much weaker when compared to the advancement in spring phenology. Studies have indicated that the timing of spring can affect autumn phenology (Fu et al., 2014). According to Walther et al. (2012), the change in leaf color in autumn has been delayed by 0.3–1.6 days per decade in Europe due to climate warming. The timing of leaf emergence and senescence in temperate deciduous forests controls the ecosystem properties such as nutrient cycling and productivity (Fridley, 2012).

3.3 Case Studies of Shifting Plant Phenology in the Indian Himalayan Region

Gaira et al. (2014) had investigated the changes in the phenology of *Rhododendron arboreum* Sm. (Image 1) in the Central Himalayan region at an altitude of 1800–2100 m above sea level (a.s.l.). *Rhododendron* is an important tree species in the Indian Himalayan region whose flowering is considered as an indicator of spring. They utilized real-time field observation data (2009–2011) along with past herbarium records (1893–2003) to study the phenology of *Rhododendron arboreum* Sm. The results clearly showed peak flowering from early February to mid-March. The study observed an increase in both seasonal and annual mean maximum temperatures. According to the generalized additive model (GAM), as a consequence of the rising annual mean maximum temperature, a forecast of early flowering by 88–97 days has been made over the past century.

Quercus is another important broad-leaved species from the Himalayan region with several economic importance. A study conducted by Chand et al. (2017) in Phulchowki Hill, Lalitpur, Nepal found that the phenology of three species, namely *Quercus semecarpifolia* Sm., *Quercus lamellose* Sm., and *Quercus glauca* Thunb., had shifted as a result of climate change. The study recorded phenological observations from 2015 to 2017. There was an advancement in spring phenology for *Quercus lamellosa* Sm. and *Quercus glauca* Thunb., whereas a delay in the case of *Quercus semecarpifolia* Sm. However, there was an advancement in leaf senescence in *Quercus lamellosa* Sm. and *Quercus glauca* and a delay in autumn phenology for *Quercus semecarpifolia* Sm. These shifts in phenology have been attributed to erratic rainfall patterns.

Image 1 *Rhododendron arboreum* in full bloom at the Great Himalayan National Park

Negi et al. (2021) conducted a study on seven different plant species, namely *Shorea robusta* Roth., *Pinus roxburghii* Sarg., *Myrica esculenta* Buch.-Ham. ex D. Don, *Quercus leucotrichophora* A. Camus, *Rhododendron arboreum* Sm., *Quercus floribunda* Lindl. ex A. Camus, and *Machilus duthiei* King ex Hook.f., which are distributed across an altitude range of 300–2200 m a.s.l. The study analyzed the phenological records for two distinct time periods, 1985–1987 and 2014–2016, and found that bud break and leafing exhibited an advancement by 0.20 days per year. This shift has been linked to the increase in atmospheric temperature (0.038 °C per year) in the study area over the years.

Mohandas et al. (2015) investigated the change in flowering phenology for three alpine ginger *Roscoea* species of the Central Himalayas from 1913 to 2011 using herbarium records. Due to an upward trend in minimum and spring temperatures, the study found that the flowering time of *Roscoea* species has been either advanced or delayed. These findings indicate that the phenology of these plants has been influenced by climate warming.

Nautiyal et al. (2020) conducted a study aimed to examine the influence of climate change on the phenology of Apples (*Malus pumila* Borkh) in the lower hills of Uttarakhand Himalayas. The advancement of flowering due to rising temperatures, as mentioned earlier, has also led to a potential vulnerability of apple flowers to sustain frost damage during spring frost occurrences, as observed by Blanke and Kunz (2011). The study revealed that the flowering and bearing pattern of apple trees in the lower hills of Uttarakhand Himalayas has been affected by the changing climate, specifically the temperature rise and rainfall and precipitation decline. These trees experienced notable effects on both fruit yield and quality as a result of these changes.

A phenological study was conducted by Masoodi et al. (2018) on *Abies pindrow* Royle in the Uttarakhand Himalayas 2200–3300 m above mean sea level (m.s.l.) for the years 2012 and 2013. By comparing phenological data from 2 years, it was observed that the leaf emerged earlier than usual, the cone maturity period was extended, and seed dispersal took place earlier than expected in the year 2013. These alterations have been ascribed to the alterations in temperature and precipitation that occurred along the altitudinal gradient.

4 Existing Research Gaps and Way Forward

Climate change has the potential to induce alterations in the timing of flowering, leading to exposure to new biotic and abiotic environments, which can harm fruit and seed production due to reduced synchronization with pollinators and adverse conditions (Rafferty et al., 2016). Enhancement in the length of the growing season (LGS) exerts an influence on the productivity and nutrient cycling dynamics within forest ecosystems (Miki & Doi, 2016). Climate change has a direct impact on local livelihoods through its influence on crop phenology, the emergence of new pests, and the infestation of new weeds in agricultural practices (Chaudhary et al., 2011).

A study conducted by Pant (2017) on the perspective of the local communities on climate change revealed the concern of the forest-dwelling communities in the Uttarakhand Himalayas. According to this study, the dry spell of 2015 has delayed the ripening of economically important species, for example, *Myrica esculenta* Buch.-Ham. ex D. Don, a species that makes a significant contribution to the local economy and *Berberis asiatica* Roxb. ex DC., an important medicinal plant in this area. A delayed ripening followed by a shortened growing season impacted the marketing and hence the livelihood of the locals.

The IHR, a biodiversity hotspot in India that provides critical ecosystem services to the region (Bulbul et al., 2022), is highly fragile and susceptible to the impacts of climate change. Despite the ecological significance of plant phenology in the IHR, there is insufficient comprehensive and dependable information available on this subject. This data deficiency in IHR is ultimately a hindrance to the conservation efforts in the region. Gaining a deeper comprehension of phenology requires the integration of multiple disciplines, including ecology, evolutionary biology, climate science, and remote sensing, in a comprehensive manner. The consolidation of these disciplines should be facilitated by the utilization of long-term monitoring data collected across various spatial scales.

The involvement of the citizen science approach can serve as an important data collection method. The National Phenology Network (NPN) in the United States has a citizen science network called Nature's Notebook, which has more than 15,000 naturalists involved. It requires no prior experience in the area. Similarly, Canada has Plant Watch and the United Kingdom has Nature's Calendar as its citizen science database for phenological recordings. Season Watch is a citizen science program run by the National Centre for Biological Sciences (NCBS), which is the biological arm of the Tata Institute of Fundamental Research (TIFR) in India. Steps should be taken to broaden the horizon of these databases to the Himalayan regions, which, despite being severely impacted by climate change, are data deficient (IPCC, 2007). Future work should also focus on the assessment of the exigencies hampering the phenological functions along with the involvement of AI techniques for the measurement and monitoring of major phenological events. It is also important to study the community perception of phenological events in the Himalayan region along with its impact on the forest-based ecosystem services and their resilience towards it so that the policy-based suggestions can be provided for the enhancement of these services along with conservation and sustenance of forest resource-based livelihoods for the long run in the ongoing race of climate change.

Acknowledgments I would like to express my sincere appreciation to the Directors of the Wildlife Institute of India, the Director of GBPNIHE (G.B. Pant National Institute of Himalayan Environment), Kosi-Katarmal, Almora, and Forest Research Institute, Dehradun, for their generous support and collaboration throughout my research. Their guidance, encouragement, and access to resources have been instrumental in the successful completion of this study. I am grateful for the opportunity to be associated with these esteemed institutions and for their contributions to my academic career.

References

Allstadt, Andrew, J., Vavrus, S. J., Heglund, P. J., Pidgeon, A. M., Thogmartin, W. E., & Radeloff, V. C. (2015). Spring plant phenology and false springs in the conterminous US during the 21st century. *Environmental Research Letters, 10*, 104008. https://doi.org/10.1088/1748-9326

Beaubien, E. G., & Johnson, D. L. (1994). Flowering plant phenology and weather in Alberta, Canada. *International Journal of Biometeorology, 38*, 23–27.

Bernier, G., Kinet, J. M., & Sachs, R. M. (1981). *The physiology of flowering* (Vol. 1, p. 231). CRC Press.

Bjorkman, A. D., Elmendorf, S. C., Beamish, A. L., Vellend, M., & Henry, G. H. R. (2015). Contrasting effects of warming and increased snowfall on Arctic tundra plant phenology over the past two decades. *Global Change Biology, 21*, 4651–4661.

Blanke, M. M., & Kunz, A. (2011). Effects of climate change on pome fruit phenology and precipitation. *Acta Horticulturae, 922*, 381–386.

Bradley, N. L., Leopold, A. C., Ross, J., & Huffaker, W. (1999). Phenological changes reflect climate change in Wisconsin. *Proceedings of the National Academy of Sciences of the United States of America, 96*, 9701–9704.

Bulbul, M., Ankit, Y., Chirakkal, A., Anoop, A., & Mishra, P. K. (2022). Understanding the present and past climate-human-vegetation dynamics in the Indian Himalaya: A comprehensive review. In *Advances in remote sensing technology and the three poles* (pp. 247–256). John Wiley & Sons Ltd.

Caffarra, A., Donnelly, A., Chuine, I., Jones, M., & B. (2011). Modelling the timing of *Betula pubescens* bud-burst. I. Temperature and photoperiod: A conceptual model. *Climate Research, 46*, 147–157.

Capovilla, G., Schmid, M., & Posé, D. (2014). Control of flowering by ambient temperature. *Journal of Experimental Botany, 66*(1), 59–69.

Cayan, D. R., Kammerdiener, A. S., Dettinger, M. D., Caprio, J. M., & Peterson, D. H. (2001). Changes in the onset of spring in the western United States. *Bulletin of the American Meteorological Society, 82*(3), 399–416.

Chambers, L. E., & Keatley, M. R. (2010). Phenology and climate—early Australian botanical records. *Australian Journal of Botany, 58*, 473–484.

Chand, D. B., Poudyal, K., & Jha, P. K. (2017). Shifts in leaf phenology of three Himalayan oak species: Role of wood water properties. *Ecoprint: An International Journal of Ecology, 24*, 29–36.

Chapin, F. S., Shaver, G. R., Giblin, A. E., Nadelhoffer, K. J., & Laundre, J. A. (1995). Responses of arctic tundra to experimental and observed changes in climate. *Ecology, 76*, 694–711.

Chaudhary, P., Rai, S., Wangdi, S., Mao, A., Rehman, N., Chettri, S., & Bawa, K. S. (2011). Consistency of local perceptions of climate change in the Kangchenjunga Himalaya landscape. *Current Science, 101*, 504–513.

Chen, X., Hu, B., & Yu, R. (2005). Spatial and temporal variation of phenological growing season and climate change impacts in temperate eastern China. *Global Change Biology, 11*, 1118–1130.

Chmielewski, F. M., & Roetzer, T. (2001). Responses of tree phenology to climate change across Europe. *Agricultural and Forest Meteorology, 108*, 101–112.

Chuine, I., & Beaubien, E. (2001). Phenology is a major determinant of tree species range. *Ecology Letters, 4*, 500–510.

Chuine, I., Cambon, G., & Comtois, P. (2000). Scaling phenology from the local to the regional level: Advances from species-specific phenological models. *Global Change Biology, 6*, 943–952.

Cleland, E. E., Chuine, I., Menzel, A., Mooney, H. A., & Schwartz, M. D. (2007). Shifting plant phenology in response to global change. *Trends in Ecology & Evolution, 22*(7), 357–365.

Cook, B. I., Cook, E. R., Huth, P. C., Thompson, J. E., Forster, A., & Smiley, D. (2008). A cross-taxa phenological dataset from Mohonk Lake, NY and its relationship to climate. *International Journal of Climatology, 28*, 1369–1383.

Collins, M., Knutti, R., Arblaster, J., Dufresne, J. L., Fichefet, T., Friedlingstein, P., Gao, X., Gutowski, W. J., Johns, T., Krinner, G., Shongwe, M. (2013). Long-term climate change: projections, commitments and irreversibility.

Corlett, R. T., & Lafrankie, J. V., Jr. (1998). Potential impacts of climate change on tropical Asian forests through an influence on phenology. *Climatic Change, 39*(2–3), 439–453.

Crimmins, T. M., Crimmins, M. A., & Bertelsen, C. D. (2013). Spring and summer patterns in flowering onset, duration, and constancy across a water-limited gradient. *American Journal of Botany, 100*(6), 1137–1147.

Dai, J., Wang, H., & Ge, Q. (2012). Multiple phenological responses to climate change among 42 plant species in Xi'an, China. *International Journal of Biometeorology, 57*, 749–758.

Defila, C., & Clot, B. (2001). Phytophenological trends in Switzerland. *International Journal of Biometeorology, 45*, 203–207.

Denny, E. G., Gerst, K. L., Miller-Rushing, A. J., Tierney, G. L., Crimmins, T. M., Enquist, C. A., Guertin, P., Rosemartin, A. H., Schwartz, M. D., Thomas, K. A., & Weltzin, J. F. (2014). Standardized phenology monitoring methods to track plant and animal activity for science and resource management applications. *International Journal of Biometeorology, 58*(4), 591–601.

Doi, H., & Takahashi, M. (2008). Latitudinal patterns in the phenological responses of leaf colouring and leaf fall to climate change in Japan. *Global Ecology and Biogeography, 17*, 556–561.

Dunne, J. A., Harte, J., & Taylor, K. J. (2003). Subalpine meadow flowering phenology responses to climate change: Integrating experimental and gradient methods. *Ecological Monographs, 73*, 69–86.

Du, Q., Liu, H., Li, Y., Xu, L., & Diloksumpun, S. (2019). The effect of phenology on the carbon exchange process in grassland and maize cropland ecosystems across a semiarid area of China. *Science of the Total Environment, 695*, 133868.

Ellwood, E. R., Gallinat, A., Primack, R. B., & Lloyd-Evans, T. L. (2015). Autumn migration of North American landbirds. In *Phenological synchrony and bird migration: Changing climate and seasonal resources in North America* (pp. 193–205). CWC Press.

Ernakovich, J. G., Hopping, A., Berdanier, A. B., Simpson, R. T., Kachergis, E. J., Steltzer, H., & Wallenstein, M. D. (2014). Predicted responses of arctic and alpine ecosystems to altered seasonality under climate change. *Global Change Biology, 20*, 3256–3269.

Estiarte, M., & Penuela, J. S. (2015). Alteration of the phenology of leaf senescence and fall in winter deciduous species by climate change: Effects on nutrient proficiency. *Global Change Biology, 21*, 1005–1017.

Ettinger, A. K., Buonaiuto, D. M., Chamberlain, C. J., Morales-Castilla, I., & Wolkovich, E. M. (2021). Spatial and temporal shifts in photoperiod with climate change. *New Phytologist, 230*(2), 462–474.

Ettinger, A. K., Chamberlain, C. J., & Wolkovich, E. M. (2022). The increasing relevance of phenology to conservation. *Nature Climate Change, 12*(4), 305–307.

Fitter, A. H., & Fitter, R. S. R. (2002). Rapid changes in flowering time in British plants. *Science, 296*, 1689–1691.

Flynn, D. F. B., & Wolkovich, E. M. (2018). Temperature and photoperiod drive spring phenology across all species in a temperate forest community. *The New Phytologist, 219*, 1353–1362.

Fridley, J. D. (2012). Extended leaf phenology and the autumn niche in deciduous forest invasions. *Nature, 485*(7398), 359–362.

Fu, Y. S. H., Campioli, M., Vitasse, Y., Boeck, H. J. D., Berge, J. V., Elgawad, H. A., Asard, H., Piao, S., Deckmyn, G., & Janssens, I. A. (2014). Variation in leaf flushing date influences autumnal senescence and next year's flushing date in two temperate tree species. *Proceedings of the National Academy of Sciences, 111*(20), 7355–7360.

Fu, Y. H., Zhao, H., Piao, S., Peaucelle, M., Peng, S., Zhou, G., Ciais, P., Huang, M., Menzel, A., & Peñuelas, J. (2015). Declining global warming effects on the phenology of spring leaf unfolding. *Nature, 526*(7571), 104–107.

Fu, Y., He, H. S., Zhao, J., Larsen, D. R., Zhang, H., Sunde, M. G., & Duan, S. (2018). Climate and spring phenology effects on autumn phenology in the Greater Khingan Mountains, Northeastern China. *Remote Sensing, 10*(3), 449.

Gaira, K. S., Rawal, R. S., Rawat, B., & Bhatt, I. D. (2014). Impact of climate change on the flowering of *Rhododendron arboreum* in central Himalaya, India. *Current Science, 106*, 1735–1738.

Gallinat, A. S., Primack, R. B., & Wagner, D. L. (2015). Autumn, the neglected season in climate change research. *Trends in Ecology & Evolution, 30*, 169–176.

Gallinat, A. S., Ellwood, E. R., Heberling, J. M., Miller-Rushing, A. J., Pearse, W. D., & Primack, R. B. (2021). Macrophenology: Insights into the broad-scale patterns, drivers, and consequences of phenology. *American Journal of Botany, 108*(11), 2112–2126.

Garcia-Mozo, H., Mestre, A., & Galán, C. (2010). Phenological trends in southern Spain: A response to climate change. *Agricultural and Forest Meteorology, 150*, 575–580.

Ghelardini, L., Santini, A., Samuelsson, S. B., Myking, T., & Falusi, M. (2010). Bud dormancy release in elm (Ulmus spp.) clones—a case study of photoperiod and temperature responses. *Tree Physiology, 30*(2), 264–274.

Gordo, O., & Sanz, J. J. (2010). Impact of climate change on plant phenology in Mediterranean ecosystems. *Global Change Biology, 16*, 1082–1106.

GTOS. (2008). *Terrestrial observation of our planet: Biennial report 50*. GTOS Secretariat, FAO.

Halevy, H. (1985). *CRC handbook of flowering*. CRC Press.

Hassan, T., Hamid, M., Wani, S. A., Malik, A. H., Waza, S. A., & Khuroo, A. A. (2021). Substantial shifts in flowering phenology of *Sternbergia vernalis* in the Himalaya: Supplementing decadal field records with historical and experimental evidences. *Science of the Total Environment, 795*, 148811.

Hopp, R. J. (1974). Plant phenology observation networks. In *Phenology and seasonality modeling* (pp. 25–43). Springer-Verlag.

Hülber, K., Winkler, M., & Grabherr, G. (2010). Intraseasonal climate and habitat-specific variability controls the flowering phenology of high alpine plant species. *Functional Ecology, 24*, 245–252.

Inouye, D. W., & Wielgolaski, F. E. (2003). High altitude climates. In M. D. Schwartz (Ed.), *Phenology: An integrative environmental science* (pp. 195–214). Kluwer.

Inouye, D. W., Morales, M. A., & Dodge, G. J. (2002). Variation in timing and abundance of flowering by *Delphinium barbeyi* Huth (Ranunculaceae): The roles of snowpack, frost, and La Nin ˜a, in the context of climate change. *Oecologia, 130*, 543–550.

IPCC. (2007). *Climate change 2007: Synthesis report*. Contribution of Working Groups I, II and III to the fourth assessment report of the Intergovernmental Panel on Climate Change (Core Writing Team, R. K. Pachauri, & A. Reisinger, Eds.; 104 pp). IPCC.

IPCC. (2018). *Global warming of 1.5°C*. An IPCC Special Report on the impacts of global warming of 1.5°C above pre-industrial levels and related global greenhouse gas emission pathways, in the context of strengthening the global response to the threat of climate change, sustainable development, and efforts to eradicate poverty (V. Masson-Delmotte, P. Zhai, H. O. Pörtner, D. Roberts, J. Skea, P. R. Shukla, A. Pirani, W. Moufouma-Okia, C. Péan, R. Pidcock, S. Connors, J. B. R. Matthews, Y. Chen, X. Zhou, M. I. Gomis, E. Lonnoy, T. Maycock, M. Tignor, & T. Waterfield, Eds.). IPCC.

IPCC. (2022). Impacts, adaptation and vulnerability. Contribution of Working Group II to the sixth assessment report of the Intergovernmental Panel on Climate Change (H.-O. Pörtner, D. C. Roberts, M. Tignor, E. S. Poloczanska, K. Mintenbeck, A. Alegría, M. Craig, S. Löschke, V. Möller, A. Okem, B. Rama, Eds.; pp. 3–33). Cambridge University Press. https://doi.org/10.1017/9781009325844.001

Jerome, D. K., Petry, W. K., Mooney, K. A., & Iler, A. M. (2021). Snow melt timing acts independently and in conjunction with temperature accumulation to drive subalpine plant phenology. *Global Change Biology, 27*(20), 5054–5069.

Kariyeva, J., van Leeuwen, W. J., & Woodhouse, C. A. (2012). Impacts of climate gradients on the vegetation phenology of major land use types in Central Asia (1981–2008). *Frontiers of Earth Science, 6*(10), 1007.

Keller, F., & Körner, C. (2003). The role of photoperiodism in alpine plant development. *Arctic, Antarctic, and Alpine Research, 35*, 361–368.

Kim, B. D., Yu, S. T., tak Shin, H., hoon Yi, M., won Yoon, J., Song Kim, G., & won Sung, J. (2012). A study on the plants for phenology of the Mt. Palgongsan Provincial Park. *Journal of Korean Nature, 5*(4), 287–292.

Kinet, J. M., Bernier, G., & Sachs, R. M. (1985). *The physiology of flowering* (Vol. III). CRC Press.

Lawson, D. A., & Rands, S. A. (2019). The effects of rainfall on plant–pollinator interactions. *Arthropod-Plant Interactions, 13*, 561–569.

Li, B., Huang, F., Qin, L., Qi, H., & Sun, N. (2019). Spatiotemporal variations of carbon use efficiency in natural terrestrial ecosystems and the relationship with climatic factors in the Songnen Plain, China. *Remote Sensing, 11*(21), 2513.

Ma, T., & Zhou, C. (2012). Climate-associated changes in spring plant phenology in China. *International Journal of Biometeorology, 56*, 269–275.

Masoodi, H. U., Rana, D., Thapliyal, M., & Singh, V. R. R. (2018). A study on phenological traits of Abies pindrow (Royle) Spach. in the different sites of Garhwal Himalayas, India. *Journal of Applied and Natural Science, 10*(1), 178–184.

Matsumoto, K., Ohta, T., Irasawa, M., & Nakamura, T. (2003). Climate change and extension of the Ginkgo biloba L. growing season in Japan. *Global Change Biology, 9*, 1634–1642.

Mehta, P., Bisht, K., & Sekar, K. C. (2021). Diversity of threatened medicinal plants of Indian Himalayan Region. *Plant Biosystems-An International Journal Dealing with all Aspects of Plant Biology, 155*(6), 1121–1132.

Mehta, P., Bisht, K., Sekar, K. C., & Tewari, A. (2023). Mapping biodiversity conservation priorities for threatened plants of Indian Himalayan Region. *Biodiversity and Conservation, 32*, 1–37.

Meier, M., Vitasse, Y., Bugmann, H., & Bigler, C. (2021). Phenological shifts induced by climate change amplify drought for broad-leaved trees at low elevations in Switzerland. *Agricultural and Forest Meteorology, 307*, 108485.

Menzel, A., & Fabian, P. (1999). Growing season extend in Europe. *Nature, 397*(6721), 659.

Menzel, A., Estrella, N., & Fabian, P. (2001). Spatial and temporal variability of the phenological seasons in Germany from 1951 to 1996. *Global Change Biology, 7*, 657–666.

Menzel, A., Sparks, T. H., Estrella, N., Koch, E., Aasa, A., Ahas, R., Alm-Kübler, K., Bissolli, P., Braslavská, O., Briede, A., Chmielewski, F. M., Crepinsek, Z., Curnel, Y., Dahl, Å., Defila, C., Donnelly, A., Filella, Y., Jatczak, K., MÅge, F., Mestre, A., Nordli, Ø., Peñuelas, J., Pirinen, P., Remišová, V., Scheifinger, H., Striz, M., Susnik, A., Van Vliet, A. J. H., Wielgolaski, F.-E., Zach, S., & Zust, A. (2006). European phenological response to climate change matches the warming pattern. *Global Change Biology, 12*, 1969–1976.

Miki, T., & Doi, H. (2016). Leaf phenological shifts and plant–microbe–soil interactions can determine forest productivity and nutrient cycling under climate change in an ecosystem model. *Ecological Research, 31*, 263–274.

Mohandass, D., Zhao, J. L., Xia, Y. M., Campbell, M. J., & Li, Q. J. (2015). Increasing temperature causes flowering onset time changes of alpine ginger Roscoea in the Central Himalayas. *Journal of Asia-Pacific Biodiversity, 8*(3), 191–198.

Nautiyal, P., Bhaskar, R., Papnai, G., Joshi, N., & Supyal, V. (2020). Impact of climate change on apple phenology and adaptability of Anna variety (low chilling cultivar) in lower hills of Uttarakhand. *International Journal of Current Microbiology and Applied Sciences, 9*(9), 453–460.

Negi, G. C. S., Singh, P., & Singh, S. P. (2021). Atmospheric warming-associated phenological earliness does not increase the length of growing season in Himalayan trees. *Forest Science, 67*(6), 694–700.

Odland, A., & Munkejord, H. K. (2008). Plants as indicators of snow layer duration in southern Norwegian mountains. *Ecological Indicators, 8*(1), 57–68.

Olivares, A., Johnston, M., & Calderón, C. (2009). Effect of rainfall regimes on seed production and quality of *Avenabarbata. Cienciaeinvestigaciónagraria, 36*(1), 68–76.

Pant, H. (2017). Case study of Uttarakhand in perspective of extreme climatic events: Fire, ecosystem and livelihoods. In *Climate change research at universities: Addressing the mitigation and adaptation challenges* (pp. 519–531). Springer.

Parmesan, C., & Yohe, G. (2003). A globally coherent fingerprint of climate change impacts across natural systems. *Nature, 421*, 37–42.

Pau, S., Wolkovich, E. M., Cook, B. I., Davies, T. J., Kraft, N. J., Bolmgren, K., Betancourt, J. L., & Cleland, E. E. (2011). Predicting phenology by integrating ecology, evolution and climate science. *Global Change Biology, 17*(12), 3633–3643.

Pepin, N., Bradley, R. S., Diaz, H. F., Baraer, M., Caceres, E. B., Forsythe, N., Fowler, H., Greenwood, G., Hashmi, M. Z., Liu, X. D., Miller, J. R., Ning, L., Ohmura, A., Palazzi, E., Rangwala, I., Schöner, W., Severskiy, I., Shahagedanova, M., Wang, M. B., Williamson, S. N., & Yang, D. Q. (2015). Elevation-dependent warming in mountain regions of the world. *Nature Climate Change, 5*, 424–430.

Rafferty, N. E., Bertelsen, C. D., & Bronstein, J. L. (2016). Later flowering is associated with a compressed flowering season and reduced reproductive output in an early season floral resource. *Oikos, 125*(6), 821–828.

Ranjitkar, S., Luedeling, E., Shrestha, K. K., Guan, K., & Xu, J. (2013). Flowering phenology of tree rhododendron along an elevation gradient in two sites in the Eastern Himalayas. *International Journal of Biometeorology, 57*, 225–240.

Richardson, A. D., Keenan, T. F., Migliavacca, M., Ryu, Y., Sonnentag, O., & Toomey, M. (2013). Climate change, phenology, and phenological control of vegetation feedbacks to the climate system. *Agricultural and Forest Meteorology, 169*, 156–173.

Roberts, A. M., Tansey, C., Smithers, R. J., & Phillimore, A. B. (2015). Predicting a change in the order of spring phenology in temperate forests. *Global Change Biology, 21*(7), 2603–2611.

Root, T. L., Price, J. T., Hall, K. R., Schneider, S. H., Rosenzweig, C., & Pounds, J. A. (2003). Fingerprints of global warming on wild animals and plants. *Nature, 421*, 57–60.

Rosenzweig, C., Karoly, D., Vicarelli, M., Neofotis, P., Wu, Q. G., Casassa, G., Menzel, A., Root, T. L., Estrella, N., Seguin, B., Tryjanowski, P., Liu, C. Z., Rawlins, S., & Imeson, A. (2008). Attributing physical and biological impacts to anthropogenic climate change. *Nature, 453*(7193), 353–357.

Rustad, L. E., Campbell, J. L., Marion, G. M., Norby, R. J., Mitchell, M. J., Hartley, A. E., Cornelissen, J. H. C., & Gurevitch, J. (2001). A meta-analysis of the response of soil respiration, net nitrogen mineralization, and aboveground plant growth to experimental ecosystem warming. *Oecologia, 126*, 543–562.

Rustad, L., Campbell, J., Dukes, J. S., Huntington, T., Lambert, K. F., Mohan, J., & Rodenhouse, N. (2012). *Changing Climate, Changing Forests: The impacts of climate change on forests of the northeastern United States and eastern Canada*. Gen. Tech. Rep. NRS-99 (p. 49). U.S. Department of Agriculture, Forest Service, Northern Research Station.

Rutishauser, T., Luterbacher, J., Defila, C., Frank, D., & Wanner, H. (2008). Swiss spring plant phenology 2007: Extremes, a multi-century perspective and changes in temperature sensitivity. *Geophysical Research Letters, 35*, 1–5.

Sabin, T. P., Krishnan, R., Vellore, R., Priya, P., Borgaonkar, H. P., Singh, B. B., & Sagar, A. (2020). Climate change over the Himalayas. In *Assessment of climate change over the Indian region: A report of the Ministry of Earth Sciences (MoES), Government of India* (pp. 207–222). Springer Nature.

Sanjay, J., Revadekar, J. V., Ramarao, M. V. S., Borgaonkar, H., Sengupta, S., Kothawale, D. R., et al. (2020). Temperature changes in India. In *Assessment of climate change over the Indian region: A report of the Ministry of Earth Sciences (MoES), Government of India* (pp. 21–45). Springer Open.

Scaife, A. A., Spangehl, T., Fereday, D. R., Cubasch, U., Langematz, U., Akiyoshi, H., Bekki, S., Braesicke, P., Butchart, N., Chipperfield, M., Gettelman, A., Hardiman, S., Michou, M., Rozano, E., & Shepherd, T. (2012). Climate change projections and stratosphere-troposphere interaction. *Climate Dynamics, 38*, 1–9.

Schmidt, W. C., & Lotan, J. E. (1980). *Phenology of common forest flora of the Northern Rockies – 1928 to 1937* (USDA Forest Service Intermountain Forest and Range Experiment Station Research Paper INT-259). Ogden, UT.

Schwartz, M., Ahas, R., & Aasa, A. (2006). Onset of spring starting earlier across the Northern Hemisphere. *Global Change Biology, 12*, 343–351.

Settele, J. C. B., Barros, V. R., Dokken, D. J., Mach, K. J., Mastrandrea, M. D., Bilir, T. E., Chatterjee, M., Ebi, K. L., Estrada, Y. O., Genova, R. C., Girma, B., Kissel, E. S., Levy, A. N., MacCracken, S., Mastrandrea, P. R., & White, L. L. (2014). Terrestrial and inland water systems. In *Climate change 2014: Impacts, adaptation, and vulnerability. Part A: Global and sectoral aspects. Contribution of working group II to the fifth assessment report of the Intergovernmental Panel on Climate Change* (pp. 271–359). Cambridge University Press.

Sharma, E., Chettri, N., & Oli, K. P. (2010). Mountain biodiversity conservation and management: A paradigm shift in policies and practices in the Hindu Kush-Himalayas. *Ecological Research, 25*, 909–923.

Shen, M., Wang, S., Jiang, N., Sun, J., Cao, R., Ling, X., et al. (2022). Plant phenology changes and drivers on the Qinghai–Tibetan Plateau. *Nature Reviews Earth & Environment, 3*(10), 633–651.

Siegmund, J., Wiedermann, M., Donges, J., & Donner, R. (2016). Impact of temperature and precipitation extremes on the flowering dates of four German wildlife shrub species. *Biogeosciences, 13*, 5541–5555.

Singh, S. P., Singh, V., & Skutsch, M. (2010). Rapid warming in the Himalayas: Ecosystem responses and development options. *Climate and Development, 2*(3), 221–232.

Song, Z., Du, Y., Primack, R. B., Miller-Rushing, A. J., Ye, W., & Huang, Z. (2021). Surprising roles of climate in regulating flowering phenology in a subtropical ecosystem. *Ecography, 44*(9), 1379–1390.

Sparks, T. H., & Carey, P. D. (1995). The response of species to climate over two centuries: An analysis of the Marsham phenological records. *Journal of Ecology, 83*, 321–329.

Steinbauer, M. J., Grytnes, J. A., Jurasinski, G., Kulonen, A., Lenoir, J., Pauli, H., Rixen, C., Winkler, M., Bardy-Durchhalter, M., Barni, E., Bjorkman, A. D., Breiner, F. T., Burg, S., Czortek, P., Dawes, M. A., Delimat, A., Dullinger, S., Erschbamer, B., Felde, V. A., Fernandez-Arberas, O., Fossheim, K. F., Gomez-Garcia, D., Georges, D., Grindrud, E. T., Haider, S., Haugum, S. V., Henriksen, H., Herreros, M. J., Jaroszewicz, B., Jaroszynska, F., Kanka, R., Kapfer, J., Klanderud, K., Kuhn, I., Lamprecht, A., Matteodo, M., di Cella, U. M., Normand, S., Odland, A., Olsen, S. L., Palacio, S., Petey, M., Piscova, V., Sedlakova, B., Steinbauer, K., Stockli, V., Svenning, J. C., Teppa, G., Theurillat, J. P., Vittoz, P., Woodin, S. J., Zimmermann, N. E., & Wipf, S. (2018). Accelerated increase in plant species richness on mountain summits is linked to warming. *Nature, 556*, 231–234.

Stenstrom, A., & Jonsdbttir, I. (2006). Effects of simulated climate change on phenology and life history traits in *Carexbigelowii*. *Nordic Journal of Botany, 24*, 355–371.

Thakur, N., Sharma, V., & Kishore, K. (2016). Leaf senescence: an overview. *Indian Journal of Plant Physiology, 21*, 225–238.

Vitasse, Y., Rebetez, M., Filippa, G., Cremonese, E., Klein, G., & Rixen, C. (2016). 'Hearing' alpine plants growing after snowmelt: Ultrasonic snow sensors provide long-term series of alpine plant phenology. *International Journal of Biometeorology, 61*, 349–361.

Vitasse, Y., Ursenbacher, S., Klein, G., Bohnenstengel, T., Chittaro, Y., Delestrade, A., et al. (2021). Phenological and elevational shifts of plants, animals and fungi under climate change in the European Alps. *Biological Reviews, 96*(5), 1816–1835.

Walther, G., Post, E., Convey, P., Menzel, A., Parmesan, C., Beebee, T., Fromentin, J. M., Hoegh, G., Ove, & Bairlein, F. (2012). Ecological responses to recent climate change. *Nature, 416*, 389–395.

Wang, C., Cao, R., Chen, J., & Tang, Y. (2015). Temperature sensitivity of spring vegetation phenology correlates to within-spring warming speed over the Northern Hemisphere. *Ecological Indicators, 50*, 62–68.

Wenden, B., Campoy, J., Lecourt, J., Ortega, G., Blanke, M., Radičević, S., Schüller, E., Spornberger, A., Christen, D., Magein, H., Giovannini, D., Campillo, T., Carlos, Malchev, S., Peris, J., Meland, M., Stehr, R., Charlot, G., & Quero, J. (2016). A collection of European sweet cherry phenology data for assessing climate change. *Scientific Data, 3*(1), 1–10.

Willis, C. G., Ellwood, E. R., Primack, R. B., Davis, C. C., Pearson, K. D., Gallinat, A. S., Yost, J. M., Nelson, G., Mazer, S. J., Rossington, N. L., Sparks, T. M., & Soltis, P. S. (2017). Old plants, new tricks: Phenological research using herbarium specimens. *Trends in Ecology & Evolution, 32*(7), 531–546.

Winkler, D. E., Butz, R. J., Germino, M. J., Reinhardt, K., & Kueppers, L. M. (2018). Snowmelt timing regulates community composition, phenology, and physiological performance of alpine plants. *Frontiers in Plant Science, 9*, 1140.

Wipf, S., Stoeckli, V., & Bebi, P. (2009). Winter climate change in alpine tundra: Plant responses to changes in snow depth and snowmelt timing. *Climatic Change, 94*, 105–121.

Wolkovich, E. M., Cook, B. I., Allen, J. M., Crimmins, T. M., Bentacourt, J. L., Travers, S. E., Pau, S., Regetz, J., Davies, T. J., Kraft, N. J. B., Ault, T. R., Bolmgren, K., Mazer, S. J., McCabe, G. J., McGill, B. J., Parmesan, C., Salamin, N., Schwartz, M. D., & Cleland, E. E. (2012). Warming experiments underpredict plant phenological responses to climate change. *Nature, 485*, 494–497.

Yu, H., Luedeling, E., & Xu, J. (2010). Winter and spring warming result in delayed spring phenology on the Tibetan Plateau. *PNAS, 107*, 151–156.

Zhang, X., Friedl, M. A., Schaaf, C. B., & Strahler, A. H. (2004). Climate controls on vegetation phenological patterns in northern mid-and high latitudes inferred from MODIS data. *Global Change Biology, 10*, 1–13.

Climate Variability in the Kilum-Ijim Mountain Range of Cameroon: Implications for Rural Livelihoods and Food Security

Suiven John Paul Tume, Nyong Princely Awazi, Henry Bikwibili Tantoh, Tah Kennette Kunsom, Moye Eric Kongnso, Reeves Meli Fokeng, Bailack Kevin Mbuh, and Wirngo Harry Mairomi

Abbreviations

AFOLU	Agriculture, forestry and other land use
AMJ	April, May, June
CA	Conservation agriculture
CCA	Climate change adaptation
CSA	Climate-smart agriculture

S. J. P. Tume (✉) · H. B. Tantoh · B. K. Mbuh
Department of Geography and Planning, The University of Bamenda, Bamenda, Cameroon

N. P. Awazi
Department of Forestry and Wildlife Technology, College of Technology, The University of Bamenda, Bamenda, Cameroon

FOKABS, Ottawa, ON, Canada

FOKABS, Yaounde, Cameroon; https://fokabs.com/

T. K. Kunsom
Society for the Promotion of Initiatives in Sustainable Development and Welfare (SOPISDEW), Bamenda, Cameroon
e-mail: honsumtk@spisdew.org

M. E. Kongnso
Department of Geography and Environmental Planning, University of Dschang, Dschang, Cameroon

R. M. Fokeng
Land Surface Dynamics Department, German Remote Sensing Data Center (DFD), Earth Observation Center (EOC), German Aerospace Center (DLR), Münchener, Germany

W. H. Mairomi
Department of Geography, Higher Teachers' Training College, The University of Bamenda, Bamenda, Cameroon

S. Tripathi et al. (eds.), *Warming Mountains*,
https://doi.org/10.1007/978-3-031-62197-0_6

CV	Coefficient of Variation
EbA	Ecosystem-based Adaptation
GEC	Global Environmental Change
IK	Indigenous knowledge
JAS	July, August, September
JFM	January, February, March
KIMF	Kilum-Ijim Mountain Forest
KIMR	Kilum-Ijim Mountain Range
LEK	Local Environmental Knowledge
MR	Mean rainfall
NbS	Nature-based Solutions
OND	October, November, December
SPI	Standardized Precipitation Index
UNSDGs	United Nations Sustainable Development Goals

1 Introduction

Mountain communities live in fragile environments with limited arable land and low agricultural productivity (Adler et al., 2022). Mountain ranges are vulnerable to environmental-related shocks such as climate change, recurrent natural hazards (slope instability), conflicts over natural resources use, generalized land degradation, limited access to infrastructure and capacity-building opportunities for poverty alleviation, thereby reducing indigenous people's capacity to cope with food insecurity and other socio-environmental shocks (Bhadouria et al., 2023; Tume & Kimengsi, 2021; Tume et al., 2020). According to the FAO and Mountain Partnership (2022), over one billion people live in mountainous areas (15% of the global population), with 648 million people in rural mountainous communities in developing countries, while 346 million of these rural mountain communities are vulnerable to food insecurity. Alder et al., (2022) affirmed that mountains are vital in the context of climate variability and change to meet sustainable development targets. Mountains are located at the juncture of induced local and global warming where significant proportions of rural populations that depend on them are left at the mercy of environmental whims and caprices. The bio-cultural diversity of mountains provides dependable ecosystem services to their inhabitants and downstream communities.

The effects of climate-driven impacts on mountain ecosystems, agriculture and pastoralism are undesirable (Tume et al., 2019). Agricultural ventures are increasingly exposed to agro-meteorological/hydrological droughts, slope instability, changes in the onset of the first rains, timing and availability of water during critical phases of crop growth, pests and pollinators (Adler et al., 2022; FAO and Mountain Partnership, 2022). Other non-climatic stressors, such as land cover/land use dynamics, also play a critical role in the functioning of mountain ecosystems and food security (Kongnso et al., 2021; Mairomi & Tume, 2021).

Local Environmental Knowledge (LEK) is critical in climate change adaptation in highland areas for food security (Kimengsi et al., 2022; Tume et al., 2019). Food security is guaranteed when societies have physical and financial access to enough, safe and healthy food to meet their nutritional needs and food choices for an active and productive life at all times (FAO, 1996). A range of factors affects food production and food security (climate, land-based physical factors and human-related factors). If harmony fails between these interactions, food insecurity sets in. Food insecurity sets in when people are consuming poor quality and inadequate food to meet their nutritional requirements (Jafari & Jafari, 2020). This may be food physical inaccessibility, inadequate socio-economic safety nets, poor food consumption or a blend of these. Food insecurity may be long-lasting, severe, temporary or seasonal as its effects may be on individuals, communities or an entire country (Tume & Nyuyfoni, 2021).

Food security is ascertained when the four pillars of availability, access, utilization and stability are in harmony with the climate and human–environment interactions (Tume et al., 2020). Conversely, climate change for food security is assessed by using actual real-time climatic data, impact and vulnerability assessment, valuation of the adaptation of food systems and mitigation of impacts (Tume, 2021a, b). Food security is a multidimensional problem where climate variability and change are not the only triggers, but some of the significant dimensions. Other important issues such as conflict, population dynamics, land use, land cover change and land deterioration can impact food security (Awazi & Avana, 2020). Governments are trying to achieve the United Nations Sustainable Development Goals (UNSDGs), which are interlinked at the global level. SDG-13 highlights the impact of climate change, which directly impacts poverty (SDG-1), end hunger (SDG-2) and water and sanitation (SDG-6). In line with these SDGs, this chapter sets out to (1) establish climate variability within the KIMR, (2) assess food security in the KIMR and (3) assess rural livelihood options for climate change adaptation.

2 Methodology

The study area is located between latitudes 6°60′ and 6°160′ N and longitude 10°240′ to 10°360′ E. It stretches over 1466.79 km^2. The Kilum-Ijim Mountain Range (KIMR) covers all of Oku and stretches into the Kom Highlands (Fig. 1).

The KIMR has a tropical climate of the Cameroon highland type with about 8 months of the rainy season (mid-March to mid-November) and 4 months of the dry season (mid-November to mid-March). Annual rainfall ranges from 2200 to 3000 mm (Fogwe & Kwei, 2015). A maximum temperature of about 30 °C is recorded at lower altitudes, especially in February, while a minimum temperature of about 25.9 °C is recorded. The climatic conditions also permit all year-round development of the montane forest ecosystem (Kiming et al., 2020).

The vegetation of the KIMR has a direct link with its climate, soils (volcanic), relief and hydrographic networks. The vegetation of this area is also influenced by anthropogenic drivers such as population pressure and land use. This vegetation has about 145 species of Afro-montane forests and grasslands that extend from about 1300 m to 3011 m above sea level. The land under the natural forest including

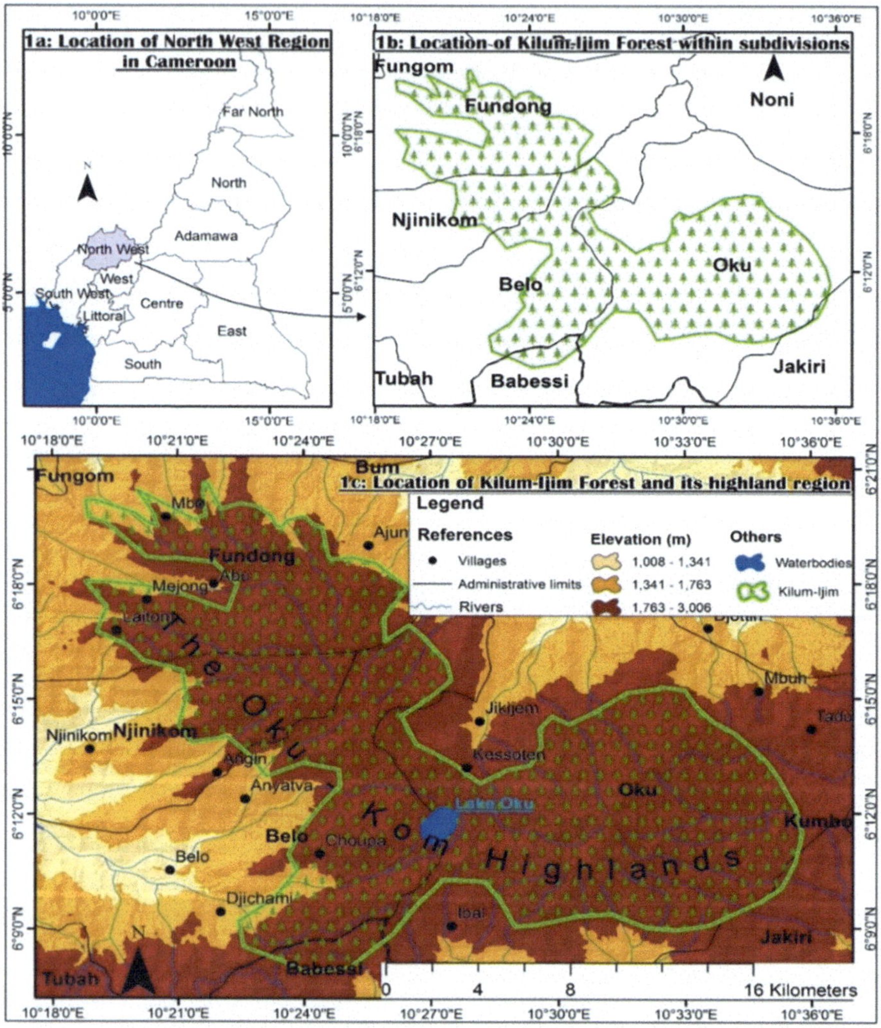

Fig. 1 Location of the Kilum-Ijim Highlands. (Source: Administrative Map of Cameroon (2021))

Table 1 Standardized Precipitation Index classification

SPI value	Category	Probability (%)
>2.00	Extreme wet	2.3
1.5 to 1.99	Severely wet	4.4
1.00 to 1.49	Moderately wet	9.2
0 to 0.99	Mildly wet	34.1
−0.1 to −0.99	Mild dryness	34.1
−1.00 to −1.49	Moderate dryness	9.2
−1.50 to −1.99	Severe dryness	4.4
<−2	Extreme dryness	2.3

Source: World Meteorological Organization (2012)

degraded forest is approximately 6900 ha. The main species include *Agauria salicifolia*, *Nuxia congesta*, *Polyscias fulva* and *Rapanea neurophyla* dominating slopes between 2000 m and 2350 m (Cheek & Onana, 2000). Beyond this altitude, *Phillipia manii*, *Myrica aborea* and *Lasiosiphong laucus* are more prevalent. They are succeeded by zones of colder patches of grassland with strands of wild shrubs and flowers that form the upland prairie zone of vegetation. Between these successions are zones of discontinuous patches of tall thick alpine bamboo forest. This species of bamboo is known as *Arundinaria alpine* (Foncha & Ewule, 2020). The area under the tree and shrub savannah is about 2400 ha and the grass savannah is approximately 1240 ha. This area was converted to a reserve in 1931, but the forest boundaries were demarcated in 1975. In the 1990s, forest laws transferred state ownership and management to local communities. Despite the relatively high rainfall, the area now lacks lush vegetation due to arable farming, grazing and settlement expansion. Livestock rearing (not authorized) is practised by a small proportion of the population.

3 Data Sources and Analysis

Climatic data (rainfall) were obtained for three stations (Oku, Fundong and Belo) from the Regional Meteorological Service for the Northwest Region of Cameroon and the Oku Wildlife Sanctuary from 1986 to 2019 (34 years) from where change was established. The rainfall index used in this study is the Standardized Precipitation Index (SPI), a tool which was developed primarily for defining and monitoring drought (Table 1).

SPI is used in assessing the recurrence of droughts. Statistically, SPI is equivalent to the Z-score (McKee et al., 1993):

$$Z - \text{score} = x - \frac{\mu}{\delta}$$

where *Z*-score expresses the *x* score's distance from the mean (μ) in standard deviation (δ) units.

Standard deviation (σ) and coefficient of variation (CV) were also used to establish rainfall reliability as follows:

$$\sigma = \frac{\sqrt{\Sigma\left(Y - \bar{Y}\right)^2}}{N}; \text{CV} = \sigma \times \frac{100}{\bar{Y}.}$$

where $\bar{Y}$ is the mean and *N* is the sample size.

From these analyses, tables were generated to summarize climatic characteristics. The agro-meteorological droughts for this study were assessed by reconstructing historical occurrences in SPI episodes on a 12-month time scale. To establish rainfall change, the mean monthly rainfall was grouped into three categories (JFM; AMJ; JAS; OND).

Crop data (maize, *Solanum* potato, beans) were obtained from the Northwest Regional Delegation of Agriculture and Rural Development from 1982 to 2019, from where trends were established. Change detection of forest cover was adopted in areal changes in secondary forest cover as the leading indicator of forest cover loss. Forest loss analysis was done using remote sensing data from Landsat images (1979–2023) (Table 2).

The acquired images were classified using the supervised image classification technique.

4 Results

4.1 Climate Variability and Change in the KIMR

The climate of the KIMR is a tropical-montane Afro-montane type, with generally high mean rainfall of more than 160 mm per year. To establish rainfall change, the mean monthly rainfall was grouped into three categories (JFM; AMJ; JAS; OND) (Table 3).

Analysis shows a rainfall reduction of 89.42 mm from 1986 to 2019 for the three stations. The rainfall decrease has been most significant at Belo (−109.12 mm), followed by Fundong (−92.18 mm) and Oku (−66.95 mm). This is significant and has implications for crop production, water management and other components of the man–environment relationship. A glaring trend in rainfall is shown by the Standardized Precipitation Index (Table 7 in Appendix). The trend in Oku is decreasing (Fig. 2).

The SPI occurrences are as follows: 2 moderately wet (1998, 2002); 21 mildly wet incidents (1986, 1987, 1988, 1989, 1990, 1991, 1993, 1994, 1995, 1996, 1999, 2000, 2001, 2003, 2004, 2005, 2006, 2009, 2010, 2015, 2018); 8 mild dryness (1992, 1997, 2008, 2011, 2012, 2013, 2014, 2017); 2 severe dryness (2007, 2016) and 1 extreme dryness (2019). Agro-meteorological droughts set in when SPI values become negative. The rainfall trend in Belo is near stable (Fig. 3).

Table 2 Remotely sensed data used in the study

Sr. no.	Platform	Sensor	Cell size (m)	Acquisition date	Source
1	Landsat 3	MSS	60	1979/12/09	USGS
2	Landsat 5	TM	30	1984/12/31	USGS
3	Landsat 7	ETM+	30	2002/02/12	USGS
4	Landsat 8	OLI/TIRS	30	2015/12/21	USGS
5	Landsat 9	OLI/TIRS	30	2023/01/24	USGS

Source: *TM* Thematic Mapper, *OLI/TIRS* Operational Land Imager/Thermal Infrared Sensor, *NASA* National Aeronautics and Space Administration

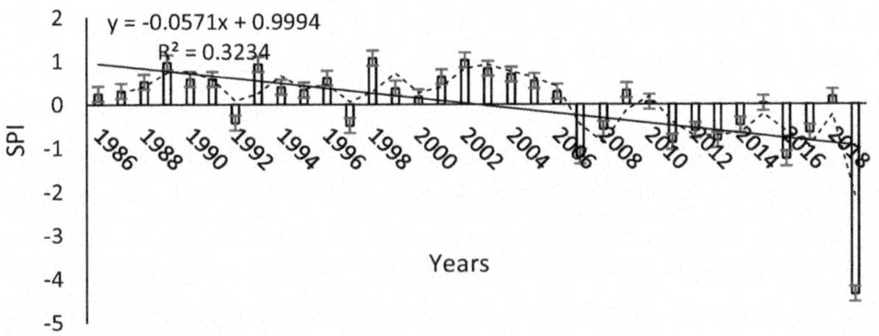

Fig. 2 Standardized Precipitation Index trend for Oku (1986–2019). (Data source: Oku Wildlife Sanctuary (2019))

Table 3 Rainfall change in the Kilum-Ijim Mountain Range

	Oku			Fundong			Belo		
Months	MR (mm)	% Δ	Δ (mm)	MR (mm)	% Δ	Δ (mm)	MR (mm)	% Δ	Δ (mm)
JFM	34.58	20.71	−146.24	53.66	27.92	−164.26	56.05	26.80	−182.32
AMJ	235.18	140.86	−26.09	255.92	133.16	−59.02	280.73	134.24	−74.88
JAS	307.11	183.95	17.00	357.53	186.04	−6.15	389.36	186.19	−22.93
OND	90.94	54.47	−112.48	101.62	52.88	−139.31	110.34	52.76	−156.36
Mean	166.95		−66.95	192.18		−92.18	209.12		−109.12

Source: Meteorological data for Oku, Fundong and Belo (1986–2019)

The SPI manifestations are as follows: 2 severely wet (1997, 2003), 3 moderately wet (1998, 2009, 2013); 11 mildly wet (1999, 2000, 2001, 2002, 2005, 2006, 2008, 2010, 2011, 2012, 2017); 16 mild dryness (1986, 1987, 1988, 1989, 1990, 1991, 1993, 1994, 1995, 1996, 2004, 2014, 2015, 2016, 2018, 2019) and 2 moderate dryness (1992, 2007). Belo had more dry incidents than wet episodes. Fundong is still experiencing an increasing rainfall trend (Fig. 4).

The SPI incidents are as follows: 18 mildly wet (1992, 1993, 1998, 1999, 2002, 2003, 2004, 2005, 2006, 2008, 2009, 2010, 2011, 2012, 2013, 2016, 2017, 2019) and 16 mild dryness (1986, 1987, 1988, 1989, 1990, 1991, 1994, 1995, 1996, 1997, 2000, 2001, 2007, 2014, 2015, 2018). In total, 55.88% of the SPI episodes were wet between 1986 and 2019, while 44.12% were dry (Fig. 5).

SPI trends were complemented by rainfall reliability. Within the tropics, a rainfall coefficient of variation (CV) of less than 20% indicates that rainfall is reliable, while unreliability sets in when the threshold goes beyond 20% (Ayoade, 1998). The situation in the KIMR is shown in Table 4.

Rainfall was still reliable in Oku (19.75%) because of its windward location, while in Fundong and Belo it was unreliable, with 28.73% and 25.92%, respectively. Rainfall unreliability indicates that the area has been experiencing recurrent agro-meteorological–hydrological droughts. This has negative implications for crop production, water resources and other livelihood sources of rural masses.

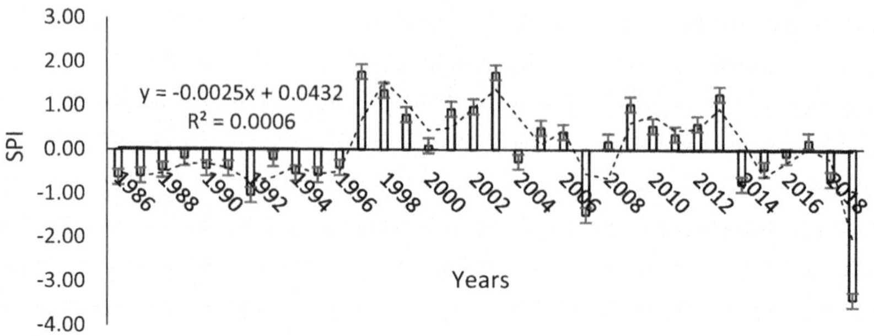

Fig. 3 Standardized Precipitation Index trend for Belo (1986–2019). (Data source: Regional Meteorological Service for the Northwest, Bamenda, Cameroon (2019))

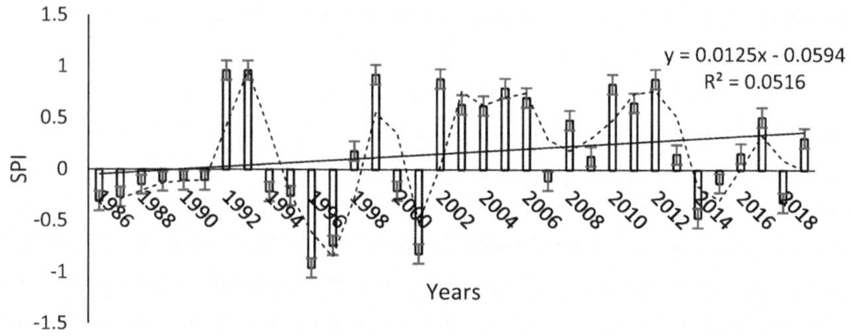

Fig. 4 Standardized Precipitation Index trend for Fundong (1986–2019). (Data source: Regional Meteorological Service for the Northwest, Bamenda, Cameroon (2019))

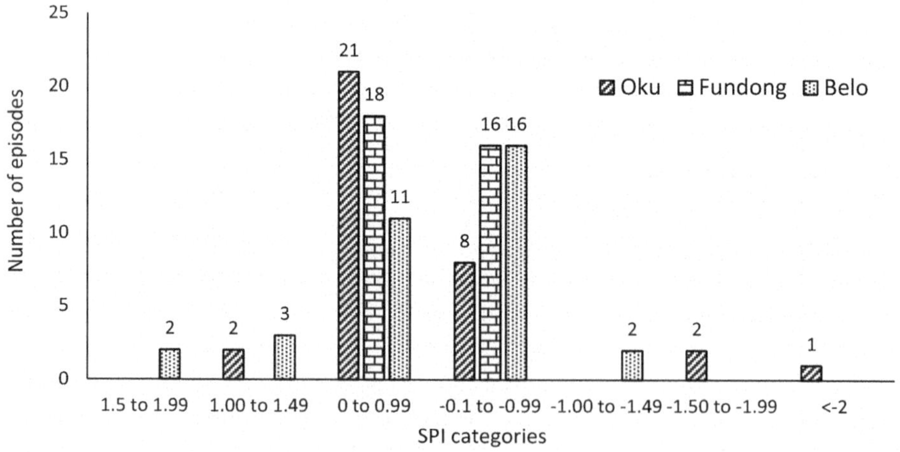

Fig. 5 SPI episodes in the KIMR. (Source: Figs. 2, 3, and 4)

Table 4 Summary of rainfall reliability in the KIMR

Location	MR (mm)	SD	CV (%)	Reliability	Trend
Oku	166.95	32.97	19.75	Reliable	Decreasing
Fundong	192.18	55.22	28.73	Unreliable	Increasing
Belo	209.12	54.21	25.92	Unreliable	Slight decrease
Mean	189.42	47.47	24.80	Unreliable	Decreasing

Source: Calculated from meteorological data for Oku, Fundong and Belo (1986–2019)

4.2 *State of Food Security in the KIMR*

The crops on which the rural populations of the KIMR depend are maize, *Solanum* potato and beans. These crops serve as food and cash crops. Whether the production is abundant or not, the farmers are compelled to sell some to meet other requirements such as hospital bills and payment of children's school fees. Maize, *Solanum* potato and beans production are declining (Figs. 6, 7, and 8).

These declining trends are directly associated with the declining rainfall pattern and the recurrence of agro-meteorological–hydrological droughts. It has been observed that crop failure is mostly associated with dry spells of about 3–4 weeks after the onset of the first rains. Tender crops cannot tolerate such weather conditions and tend to wilt. With these, food insecurity has set in because of weather threats in the food system (Fig. 9).

Food systems involve production, processing, packaging, distributing, retailing and consumption. These dynamics in the Global Environmental Change (GEC) processes are influenced by food system dynamics, drivers, social, environmental and food security fallouts. These drivers provide links between and within physical and human environments (with climate variability and change at the centre) and dictate how food systems function. Outcomes contribute to food security and socio-environmental changes such as deforestation and land degradation in the KIMR. Some of the drivers affect food system outcomes directly through household size and financial capability.

There is interconnectedness between the diverse groups of food system outcomes. Some include regulating ecosystem services and food available from local production, or interconnections between incomes and food access. In addition, food system activities and outcomes result in processes that feed back to environmental and socioeconomic drivers, because food production systems are drivers of '*glocal*' environmental change. Food security is the main result of any given food system, although in many cases food security is hardly achieved (Ericksen et al., 2010). Alternatively, people may be undernourished and face regular hungry seasons (the period between the sowing of crops and the first harvest), scuffles with a multitude of non-communicable diet-related health issues, or income expenses on poor or insufficient diets.

Drivers sustaining food security in a given unit of analysis can be elucidated in four pillars (availability, access, utilization, affordability) and stability (output), each of which has many elements (Tume et al., 2020). Food availability relates to

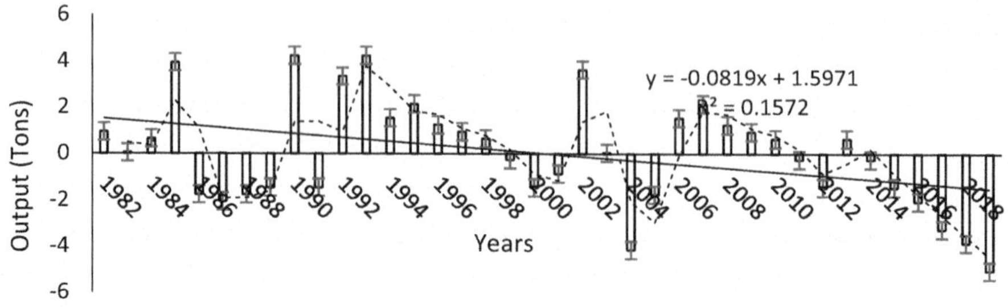

Fig. 6 Maize production trend in the KIMR (1982–2019). (Data source: Regional Delegation of Agriculture and Rural Development, Northwest, Bamenda, Cameroon (2019))

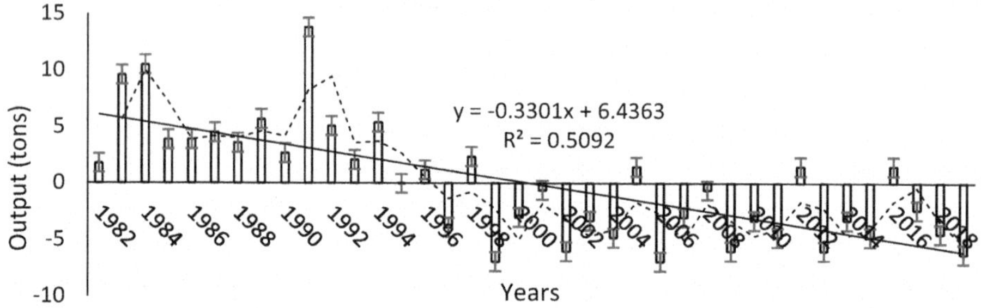

Fig. 7 *Solanum* potato production trend in the KIMR (1982–2019). (Data source: Regional Delegation of Agriculture and Rural Development, Northwest, Bamenda, Cameroon (2019))

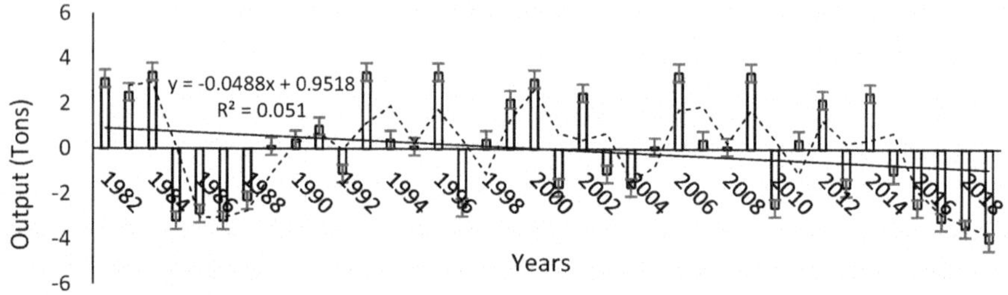

Fig. 8 Beans production trend in the KIMR (1982–2019). (Data source: Regional Delegation of Agriculture and Rural Development, Northwest, Bamenda, Cameroon (2019))

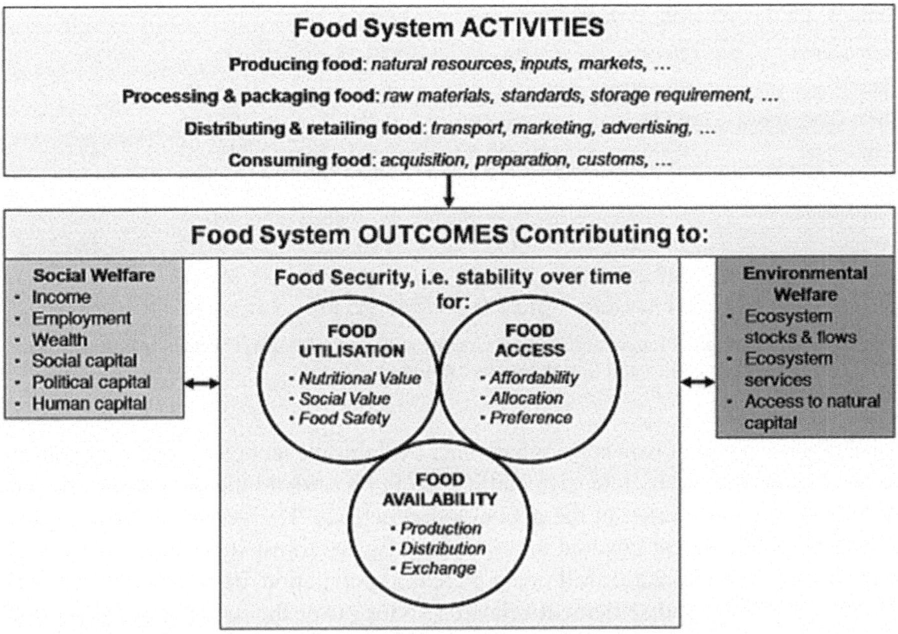

Fig. 9 Components of the food system. (Source: Erickesen et al., (2010))

the amount, type and quality of food a unit has at its disposal to consume at the grassroots level. Food can be available through local production, *distribution* channels *and exchange mechanisms to earn* money, labour and the value of food (cultural acceptance). Access to food relates to the required quality and quantity. It englobes *affordability*, the organization of distribution channels such as markets and government policies, as well as consumers' social *preferences*. Food utilization shows the *nutritional* and *social* values, *the safety* of available food and affordability. Food systems are also influenced by socio-political and environmental drivers to ensure stability (Ericksen et al., 2010).

Environmental outcomes of food systems include the natural resource-base and ecosystem services (KIMR biodiversity hotspot), including welfare outcomes that may arise as people rely on food systems as sources of livelihood. These outcomes include wealth, income and health status of the rural population.

4.3 Land Cover Change in the KIMR (1979–2023)

The KIMR is a dynamic landscape with human pressure on the land, water resources, forests and rangelands. The mainland covers patches of primary forest, secondary forest (from regeneration and conservation), grassland, farmlands, settlement, water sources and bare land (Table 5).

Table 5 Land cover changes (ha)

Land cover	1979	1984	2001	2015	2023
Forest	17,550.6	12,827.52	6143.04	10,134.09	12,665.79
Secondary forest	4278.24	3680.46	10,259.28	5947.65	3153.87
Grassland	9306	10,955.52	10,851.12	7467.84	3474.18
Farmlands	–	3515.13	3590.19	6862.77	10,683.81
Settlement	–	54.45	70.56	266.4	435.69
Water	196.56	195.21	203.22	198.72	186.21
Bare land	216.21	319.32	430.2	670.14	948.06
Total	31,547.61	31,547.61	31,547.61	31,547.61	31,547.61

Source: *TM* Thematic Mapper, *OLI/TIRS* Operational Land Imager/Thermal Infrared Sensor, *NASA* National Aeronautics and Space Administration

The areal extent of bare soils or bare land explains the increased land degradation of the plateau caused by intensive cattle grazing by settled Fulani herders since the 1950s. Forest loss started in the post-colonial periods. The year 1979 showed that with part of the forest covered as secondary forest, forest loss became alarming between the 1980s to 2000s following increased population, increased grazing, fuel wood exploitation and settlement-related activities over the forest area. Forest loss reversal followed the implementation of the Kilum-Ijim Forest Restoration Program in 2015 (state of the forest in 2015 and 2023) (Figs. 10, 11, 12, 13, and 14).

Forest gain is also partly linked to the stability of the forest as a result of the ongoing socio-political crisis, which has led to depopulation of the mountain area, reduced cropping and grazing intensities and less human intervention in this fragile ecosystem (Fig. 15).

Despite the increase in farmlands, crop production has been decreasing. The population has resorted to other livelihood activities such as forest exploitation and grazing to supplement their food supplies, leading to land degradation.

4.4 Sustainable Rural Livelihood Options for Climate Change Adaptation and Food Security in the KIMR

The rural masses of the KIMR have adapted their farming to changing climate and environmental and socio-political conditions to ensure food security (Table 6). Most of these strategies are indigenous, making use of the immediate environment and its resources to sustain rural livelihoods. Some include improved seeds, local seeds, altering planting dates, use of organic manures, mono-cropping, alternative income-generating activities, agroforestry and post-harvest preservation. In the livestock sector, ruminants, particularly goats and sheep, are grazed especially on montane grasslands for livelihood diversification (2000–3000 m above sea level). This cuts across basalt pavement grasslands (2300–2680 m above sea level) with pastures such as *Loudetia simplex*, *Eragrostis volkensii*, *Tristachya* and *Pennisetum*

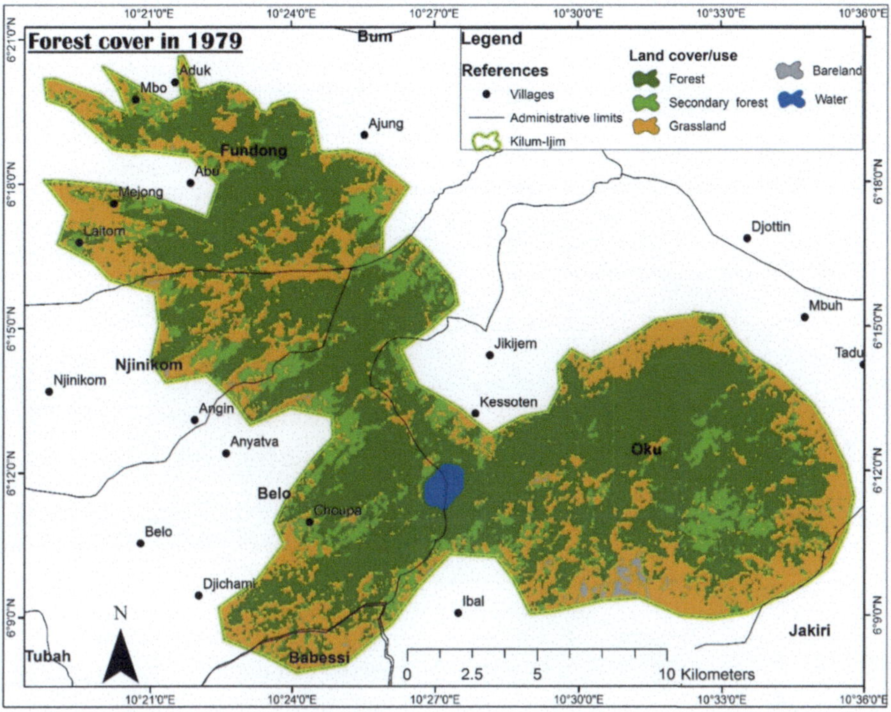

Fig. 10 Forest cover in the KIMR (1979)

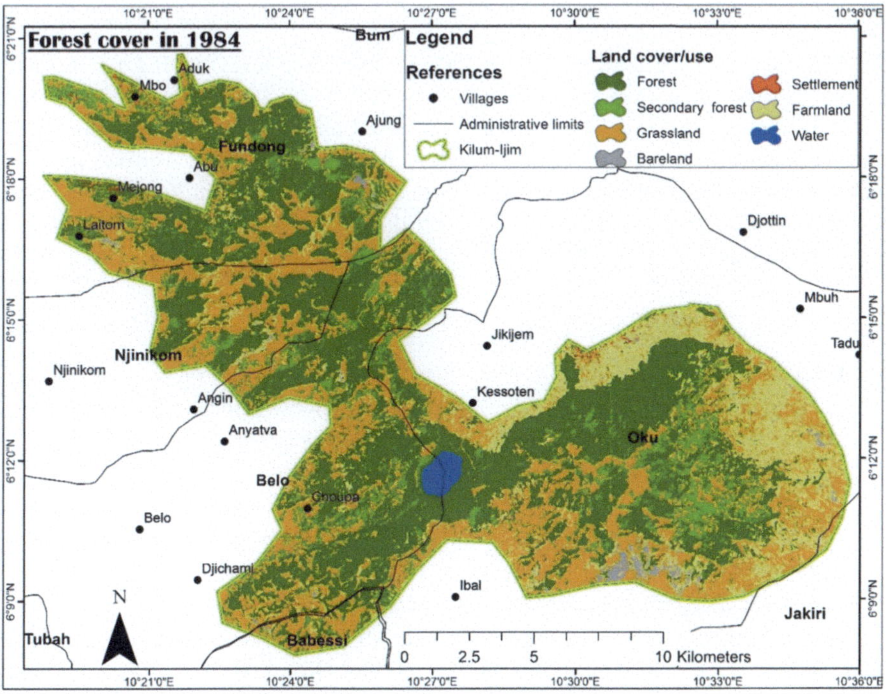

Fig. 11 Forest cover in the KIMR (1984)

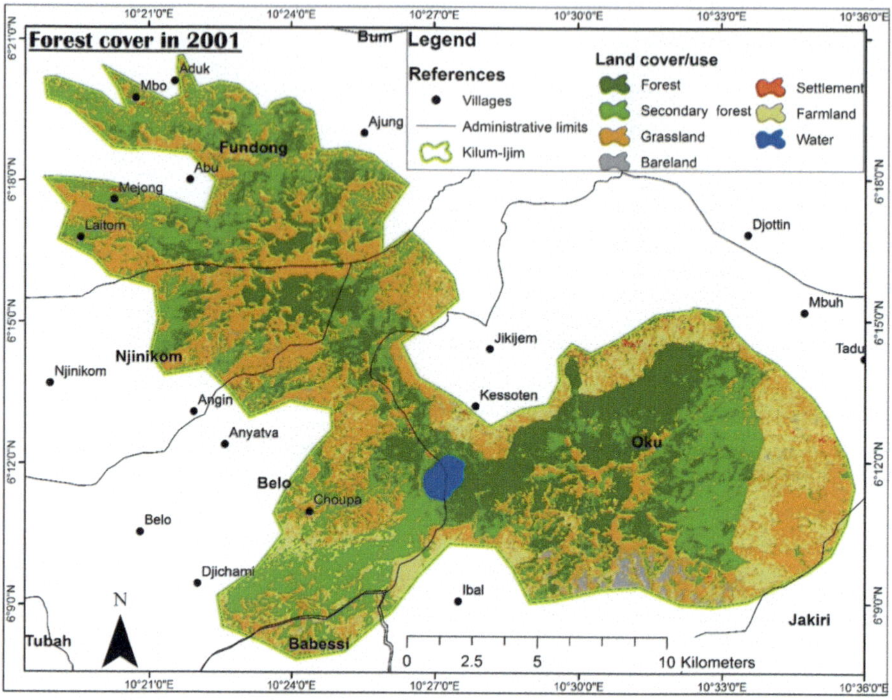

Fig. 12 Forest cover in the KIMR (2001)

clandestinum; rocky grassland (2680–2900 m above sea level) with *Loudetia* and *Eragrostis*; and threatened species such as *Bafutia tenuicaulis*, *Helichrysum camerunenses* and *Crotolaric lamnedae* are also used as grazing land. Summit grasslands and scrub with endemic species such as *Alchemilla fischeri*, *Agrostis mannii*, *Habenaria obovata/microceras* and *Anthospermum aspermuliodes* are also exploited. These areas sustain grazing even during the dry season. Cattle are not permitted in high elevations to limit degradation.

5 Discussion

Pertinent climatic elements in the food security system are dominantly rainfall and temperature related. However, solar radiation, wind, relative humidity and other extreme weather episodes are also important (Tume et al., 2020). Climate change projections through these variables influence food availability, access, utilization and stability, leading to complex impacts on local livelihoods and food security (Niba & Bailack, 2022; Awazi et al., 2021a, b).

The land–climate nexus is complex with multiple biophysical and biogeochemical responses across different spatio-temporal scales (Kongnso et al., 2021). Climate

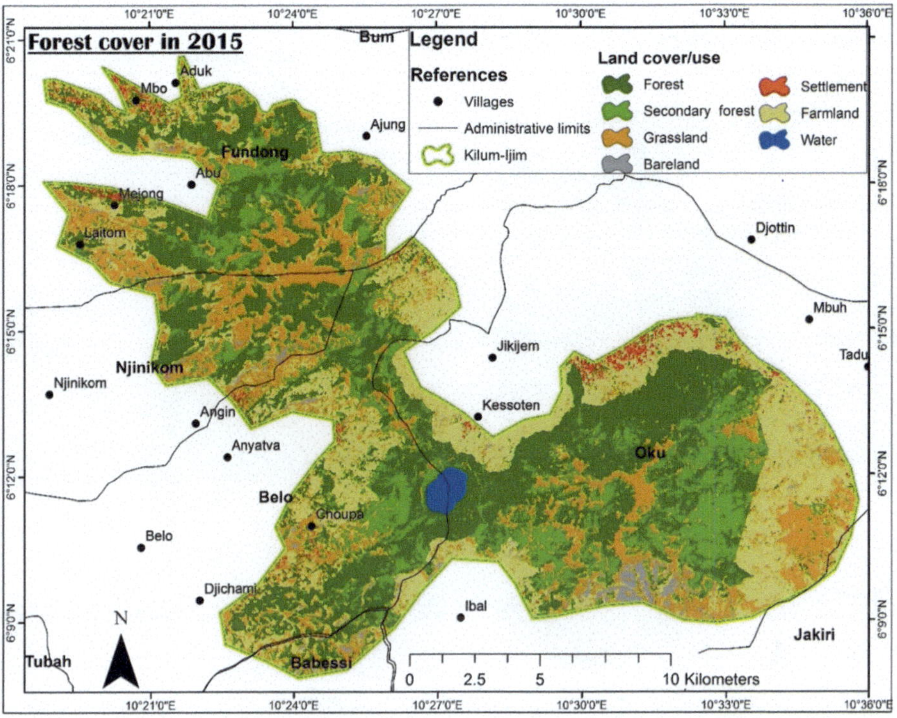

Fig. 13 Forest cover in the KIMR (2015)

drives land cover/use change and land use/cover drives climate through anthropogenic ramifications (Tume & Fogwe, 2018). Agriculture, forestry and other land uses (AFOLU) (Bailack et al., 2023; Awazi et al., 2021a, b) are critical sources of greenhouse gas (GHG) emissions, contributing to anthropogenic emissions of CO_2, CH_4 and N_2O (Jia et al., 2019). Climate change affects food security through increasing temperatures, decreasing precipitation and frequency of extreme events (Tume & Fogwe, 2018). Research that separates climate change from other determinants of crop yields has shown that some crops (maize, *Solanum* potato, beans) in the tropics (Mbow et al., 2019; Jia et al., 2019) have been experiencing declining outputs. Based on indigenous knowledge (IK), climate change is affecting food security in the savannahs and high mountain regions (Tume et al., 2019). The majority of the world's population is fed through the food system and it supports the livelihoods of rural people (Niba and Bailack, 2022). On the other hand, the food system is burdened by non-climate stressors such as rural poverty, population growth, income reduction and demand for animal-sourced products, which weaken food availability, access, utilization and stability (Mbow et al., 2019).

Adaptation of food systems to climate change can be achieved by enabling conditions, policies, markets, institutions and governance (Mbow et al., 2019). For adaptation to work, resilience to extreme events can be accomplished through

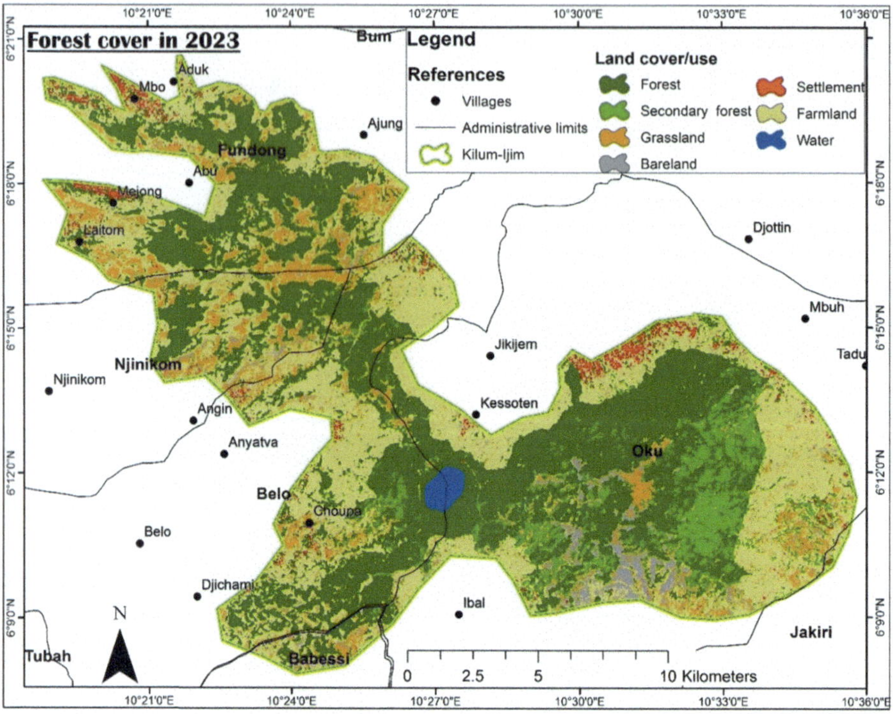

Fig. 14 Forest cover in the KIMR (2023)

risk-sharing and transfer mechanisms such as insurance (Awazi et al., 2019; Azong et al., 2018). Resounding public health policies to improve school feeding programmes, health insurance and sensitization can change demand, lessen healthcare costs and limit GHG emissions. Without inclusive food system responses in wider climate change policies, the mitigation and adaptation potentials and food security will be jeopardized.

Adaptation reduces the impacts of climate shocks on natural and anthropogenic systems (Hallegatte, 2009). It is a combination of behavioural (psychology), technological and structural adjustments. Adaptation varies in terms of:

- adapting agent (societies vs. natural systems; private vs. public),
- purposefulness (active vs. passive; autonomous vs. planned),
- scope (localized vs. regional; short-term vs. long-term) and,
- timing (ex-ante vs. ex-post; anticipatory vs. reactive).

Examples of adaptation actions include altering crop varieties and farming systems, drought-resilient crops, livelihood alterations and land use planning (Bailack et al., 2023; Tume, 2022a, b). Societies through time have adapted to discrepancies in their climate using knowledge amassed through experiences of past climates (IIED, 2015a, b) using local knowledge systems (Carr & Onzere, 2018). The tropics, for example, have historically frequently faced extreme events such as droughts, necessitating adaptation to unreliable rainfall by diversifying livelihood sources and new crop varieties (Awazi et al., 2022; Mairomi et al., 2022; Ericksen et al., 2011).

Climate change adaptation (CCA) is framed within '*Adaptation to What*', '*Who Adapts*', '*How Adaptation Occurs*' and '*How Good is the Adaptation*' (Smith et al., 2009; Smit & Pilifosova, 2001) (Fig. 16). Food crop production adaptation relates to climate-related stimuli, systems to be adapted, types of adaptation and the evaluation of adaptation options using the local ecosystem lens. This involves fast-tracking the robustness of socio-economic activities to reduce climate change vulnerability (Tantoh et al., 2022; Mathews et al., 2014; Kolawole et al., 2014). Adaptation is a set of adjustments within a field of practices, processes or structures of a system that is undergoing actual change (Smit & Pilifosova, 2001). Change can be spontaneous, planned or a response to some anticipated alterations in the normal functioning of local climate and food systems.

'*Adaptation to What*' can be various manifestations of stress. This cuts across crops adapting to annual average precipitation uncertainties, daily precipitation or experiencing hourly precipitation extremes, droughts, crop failures or income loss (Tume & Kongnso, 2022). Relevant climate characteristics that agriculture needs to adapt to are temperature, precipitation and moisture content to meet crop specificities for different stages of growth (Tume & Fogwe, 2018).

'*Who and What to Adapt*' is a system definition of the subject and boundaries. Subjects and boundaries can be micro-ecological zones, farmers, crops and water resources. Farmers adapt to changing agricultural patterns, changing land uses or engaging in other activities, which are income-generating to help balance the income loss from agriculture (Bailack et al., 2023; Kiming et al., 2020). Ecologically, crops adapt by growing new species of tubers (as in the case of *Solanum* potato). Also, new crop species can be introduced into the system through new hybrids that are resilient to climate change (Tume & Tani, 2018). It is important to note that the vulnerability and adaptive capacity of crops will depend on the system characteristics to promote, exhibit and stimulate the occurrence of the nature of adaptation (Tantoh et al., 2022; Awazi et al. 2020a, b).

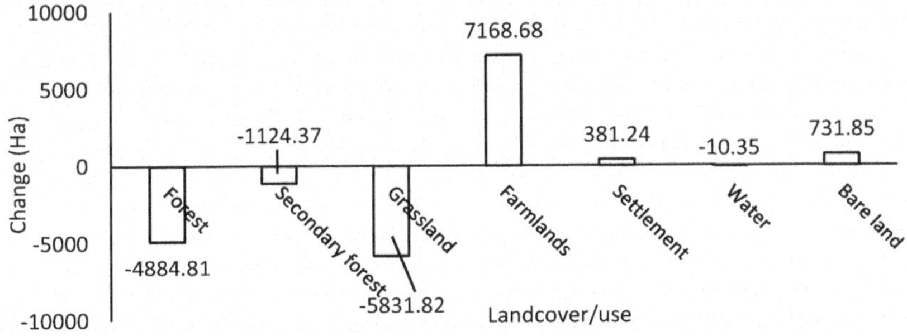

Fig. 15 Relative changes in land cover/use in the KIMR (1979–2023). (Source: Calculated from Table 5)

Table 6 Transformations of peasant agricultural systems in the Kilum-Ijim Mountain Forest (KIMF)

Trends	Farming system	Drivers of change	Outcomes
Changes in farming methods	Traditional methods: keeping seeds, planting local crop varieties	NGO programmes and government policies	Integrated farming (contour farming, planting in rows, crop geometry, using improved seeds, agrochemicals; agroforestry)
Changes in crop varieties cultivated	Main food crops: maize, beans, *Solanum* potato, sweet potatoes, cocoyam; main cash crop cultivated: coffee; cultivation of mostly local varieties; cash crop seeds provided by cooperatives; local seed banks	NGO programmes; climate variability and change; government policies; increasing pests and diseases; new market opportunities; changing farmer preferences; price fluctuations	Food crops: maize, beans, *Solanum* potato, sweet potato; new crops: carrots, onions, green pepper, tomatoes, cabbages; a combination of both local and improved varieties; local seed banks
Increase in local food crop production	The distinction between food crops and cash crops; food crops meant for feeding the family; cash crops sold in local markets and cooperatives	New market opportunities; increased demand for financial resources; price fluctuations	Food crops for consumption and income generation; all crops sold in local and national markets
Use of natural resources to generate income	Natural resources are used as food and medication; the resource base also used as a coping strategy for various shocks and stresses	New market opportunities; increased demand for financial resources	Natural resources exploitation for income generation (stone and sand quarrying, timber exploitation from eucalyptus trees, *apiculture*)
Increase in off-farm activities	Livestock decline; off-farm activities used as a coping strategy to multiple shocks and stresses	Political instability; increased demand for financial resources; increase in off-farm opportunities; declining livestock numbers	Increase in livestock numbers; diverse livelihood strategies; off-farm activities used as a coping strategy to multiple shocks and stresses
Fluctuations in livestock numbers	Integration of mixed farming; livestock used as a socio-economic safety net, accumulating farm assets and generating income	Political instability; land fragmentation; new market opportunities; new crop varieties	Livestock numbers increasing gradually; less communal/marginal land is available for grazing

Source: Adapted from Zetem (2022)

'*How Adaptation Occurs*' depends on the response of food systems to climate stimuli. These are autonomous, spontaneous, passive or natural adaptations (Tume & Maghah, 2021). By autonomous response, crops with high sensitivity to climate stimuli such as *Solanum* potato tend to respond faster and are highly vulnerable to this stimulus, needing much time to adapt. Crops such as maize that are favourably

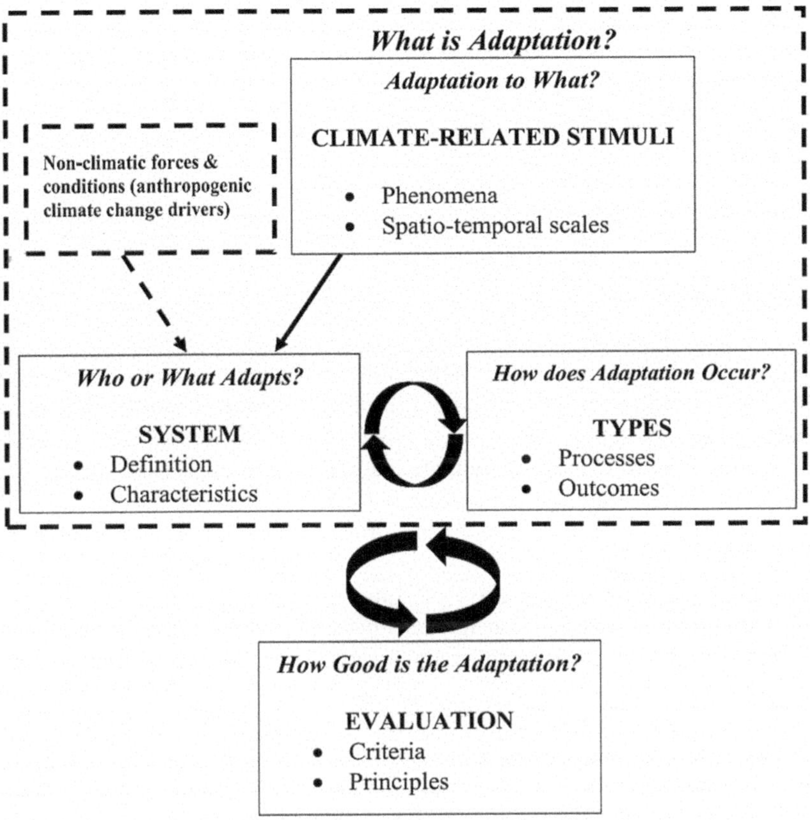

Fig. 16 Climate change adaptation framework in the agricultural sector. (Source: Smit and Pilifosova (2001))

grown under warm conditions are not very sensitive to climate stimuli and, as such, adapt spontaneously.

'*How Good is the Adaptation*' is based on the evaluation that is dependent on several criteria and principles of evaluation. This takes into consideration the use of conceptual models and numerical models (Smit & Pilifosova, 2001). Conceptual models are based on the use of astronomic data from meteorological stations correlating with crop production over a variety of crop species (Kiming et al., 2020). Numeric evaluation is based on comparing the productivity ratio of a given improved species to the farmer.

One option for climate change adaptation is nature-based solutions (NbS) and ecosystem-based adaptation (EbA). These can contribute over a third of the total climate change mitigation needed by 2030 to keep global warming below 2 °C. NbS is inspired by the natural environment and practices natural processes to contribute to ecosystem management (UN-Water, 2018). NbS involves rehabilitating ecosystems and triggering natural processes in man-made ecosystems such as watersheds (Griscom et al., 2019). An important subset of NbS is EbA, which helps people adapt to climate change. EbA is a nature-based approach that uses ecosystem services to help people adapt to the adverse effects of climate change (Tume & Ngwa, 2022; International Institute for Environment and Development-IIED, 2019). It involves landscape restoration (tree planting, slope stabilization by tree planting to prevent landslides and integrated water resources management). EbA uses indigenous knowledge with growing evidence on climate change and has been applied to address the local challenges of climate change impacts (Tume & Kongnso, 2022; Tume, 2021a, b).

6 Conclusion

The Kilum-Ijim Mountain Range is a high-energy tropical montane forest area that has been undergoing anthropogenic and ecological pressures. Rainfall has reduced by nearly 90 mm in the past 34 years, amidst rising temperatures and diminishing water resources, as well as other dynamic natural resources. The foodstuffs output on which the population relies for subsistence and cash has been reducing, leading to the emergence of food insecurity. All the dimensions of food security are directly linked to the climate in the KIMR. Changes in climatic conditions have already affected the production (availability) of some maize, *Solanum potato* and beans. Higher temperatures will have an impact on yields, while rainfall reduction will affect both crop quality and quantity. Climate change is also increasing the prices of major crops and affecting affordability and accessibility. For the most vulnerable rural people, lower agricultural output means lower incomes. Under these conditions, the poorest people, who already use most of their income on food, sacrifice additional income and other assets to meet their nutritional requirements or resort to poor coping strategies. Climate-related risks affect calorie intake (utilization), particularly with the emergence of food insecurity, amidst the socio-political crises that have beset Anglophone Cameroon since 2016. Changing climatic conditions also create a vicious cycle of disease and hunger. Nutrition is likely to be affected by climate change through related impacts on food security, dietary diversity, care practices and health. Overall, climatic variability and change in the KIMR are produced by recurrent extreme weather events that offset community cohesion and national (state) food security strategies, creating fluctuations in the food security pillars.

Short-term adjustments and long-term adaptation to these emerging food security challenges have been through indigenous production systems, with the assistance of NGO programmes and government policies. Some of these strategies include improved seeds, local seeds, altering planting dates, use of organic manures, mono-cropping, alternative income-generating activities, agroforestry and post-harvest preservation. To ensure sustainability, medium- and long-term policies must embrace 'nature-based solutions' and 'ecosystem-based adaptation'. In the face of an uncertain climate, farmers must safeguard food security through agroecological practices, climate-smart agriculture (CSA), conservation agriculture (CA) and modification of production systems. Agroecology delivers knowledge for inclusive natural, social, economic, political and cultural proportions. It shapes resilience through indigenous knowledge systems in synergy with conventional production systems through stakeholder networks. CSA is a pathway towards development and food security (increasing output and incomes, livelihood resilience, ecosystem services productivity and reduction in greenhouse gas emissions). CA, on the other hand, relies on the principle of minimum soil disturbance, permanent soil cover (cover cropping) and crop rotation. CA, CSA, NbS and EbA react positively to the needs of rural farmers facing economic and environmental pressures. Soil and residue management that limit erosion and improve soil quality, by increasing organic matter content, porosity, structural stability, infiltration and water retention capacity, are incorporated together with the diversification of production systems to promote food security and deliver supplementary ecosystem services.

Appendix

Table 7 Standardized Precipitation Index episodes for the Kilum-Ijim Mountain Range (1986–2019)

SPI value	Category	Location/years	Episodes
Fundong			
0 to 0.99	Mildly wet	1992, 1993, 1998, 1999, 2002, 2003, 2004, 2005, 2006, 2008, 2009, 2010, 2011, 2012, 2013, 2016, 2017, 2019	18
−0.1 to −0.99	Mild dryness	1986, 1987, 1988, 1989, 1990, 1991, 1994, 1995, 1996, 1997, 2000, 2001, 2007, 2014, 2015, 2018	16
Belo			
1.5 to 1.99	Severely wet	1997, 2003	2
1.00 to 1.49	Moderately wet	1998, 2009, 2013	3
0 to 0.99	Mildly wet	1999, 2000, 2001, 2002, 2005, 2006, 2008, 2010, 2011, 2012, 2017	11
−0.1 to −0.99	Mild dryness	1986, 1987, 1988, 1989, 1990, 1991, 1993, 1994, 1995, 1996, 2004, 2014, 2015, 2016, 2018, 2019	16
−1.00 to −1.49	Moderate dryness	1992, 2007	2
Oku			
1.00 to 1.49	Moderately wet	1998, 2002	2
0 to 0.99	Mildly wet	1986, 1987, 1988, 1989, 1990, 1991, 1993, 1994, 1995, 1996, 1999, 2000, 2001, 2003, 2004, 2005, 2006, 2009, 2010, 2015, 2018	21
−0.1 to −0.99	Mild dryness	1992, 1997, 2008, 2011, 2012, 2013, 2014, 2017	8
−1.50 to −1.99	Severe dryness	2007, 2016	2
<−2	Extreme dryness	2019	1

References

Adler, C., Wester, P., Bhatt, I., Huggel, C., Insarov, G. E., Morecroft, M. D., Muccione, V., & Prakash, A. (2022). Cross-chapter paper 5: Mountains. In H.-O. Pörtner, D. C. Roberts, M. Tignor, E. S. Poloczanska, K. Mintenbeck, A. Alegría, M. Craig, S. Langsdorf, S. Löschke, V. Möller, A. Okem, & B. Rama (Eds.), *Climate change 2022: Impacts, adaptation and vulnerability* (Contribution of Working Group II to the sixth assessment report of the Intergovernmental Panel on Climate Change) (pp. 2273–2318). Cambridge University Press. https://doi.org/10.1017/9781009325844.022

Awazi, N. P., & Avana, T. M.-L. (2020). Agroforestry as a sustainable means to farmer-grazier conflict mitigation in Cameroon. *Agroforestry Systems, 94*(6), 2147–2165. https://doi.org/10.1007/s10457-020-00537-y(0123456789

Awazi, N. P., Tchamba, N. M., & Avana, T. M. L. (2019). Climate change resiliency choices of small-scale farmers in Cameroon: Determinants and policy implications. *Journal of Environmental Management, 250*, 109560. https://doi.org/10.1016/j.jenvman.2019.109560

Awazi, N. P., Tchamba, M. N., Temgoua, L. F., & Avana, T. M.-L. (2020a). Farmers' adaptive capacity to climate change in Africa: Small-scale farmers in Cameroon. In F. W. Leal, N. Ogugu, L. Adelake, D. Ayal, & I. da Silva (Eds.), *African handbook of climate change adaptation*. Springer. https://doi.org/10.1007/978-3-030-42091-8_9-1

Awazi, N. P., Tchamba, N. M., & Temgoua, L. F. (2020b). Enhancement of resilience to climate variability and change through agroforestry practices in smallholder farming systems in Cameroon. *Agroforestry Systems, 94*(3), 687–705. https://doi.org/10.1007/s10457-019-00435-y

Awazi, N. P., Tchamba, M. N., Temgoua, L. F., Avana, T. M.-L., Shidiki, A. A., Forje, G. W., & Nfornkah, B. N. (2021a). Climate-smart and agro-ecological farming systems of smallholder farmers. In C. M. Galanakis (Ed.), *Environment and climate-smart food production*. Springer Nature Switzerland AG. https://doi.org/10.1007/978-3-030-71571-7_2

Awazi, N. P., Tchamba, N. M., Temgoua, L. F., & Avana, T. M. L. (2021b). Agroforestry as an adaptation option to climate change in Cameroon: Assessing farmers' preferences. *Agricultural Research, 11*, 309–320. https://doi.org/10.1007/s40003-021-00562-y

Awazi, N. P., Quandt, A., & Kimengsi, J. N. (2022). Endogenous livelihood assets and climate change resilience in the Mezam highlands of Cameroon. *GeoJournal, 88*, 2491–2508. https://doi.org/10.1007/s10708-022-10755-9

Ayoade, J.O. (1998). Introduction to Climatology for the Tropics. Fourth edition, Spectrum Books, Ibadan, pp. 1–25.

Azong, M., Kelso, C. J., & Naidoo, K. (2018). Vulnerability and resilience of female farmers in Oku, Cameroon, to climate change. *African Sociological Review, 22*(1), 31–53.

Bailack, K. M., Tume, S. J. P., & Boeyeo, A. C. (2023). Agrarian land management dynamics and crop production: Insights from the Bui Highlands of Cameroon. *Journal of Advance Research in Food, Agriculture and Environmental Science, 9*(2), 1–11. https://doi.org/10.53555/nnfaes.v9i2.1547

Bhadouria, R., Singh, S., Tripathi, S., & Singh, P. (Eds.). (2023). *Understanding soils of mountainous landscapes: Sustainable use of soil ecosystem services and management* (p. 439). Elsevier.

Carr, E. R., & Onzere, S. N. (2018). Really effective (for 15% of the men): Lessons in understanding and addressing user needs in climate services from Mali. *Climate Risk Management, 22*, 82–95.

Cheek, M., & Onana, J. M. (2000). *The plants of Mount Oku and the Ijim Ridge, Cameroon: A conservation checklist* (Vol. 215). Royal Botanical Gardens Kew.

Ericksen, P., Stewart, B., Dixon, J., Barling, D., Loring, P., Anderson, M., & Ingram, J. (2010). The Value of Food System Approach. In J. Ingram, P. Ericksen, & D. Liverman (Eds.), *Food security and global environmental change* (pp. 27–30). Earthscan.

Ericksen, P., Thornton, P., Notenbaert, A., Cramer, L., Jones, P., & Herrero, M. (2011). *Mapping hotspots of climate change and food insecurity in the global tropics* (CGIAR Research

Program on Climate Change, Agriculture and Food Security (CCAFS). CCAFS Report No. 5) (p. 50). CCAFS.

FAO. (1996). *Rome declaration and world food summit plan of action* (pp. 1–5). FAO.

FAO & Mountain Partnership. (2022). *Vulnerability of mountain peoples to food insecurity* (pp. 1–3). FAO.

Fogwe, Z. N., & Kwei, J. (2015). Cameroonian protected Kilum-Ijim forests for the development of Oku Forest Fringe Community. *Journal of Environmental Research and Management, 6*(5), 0293–0303.

Foncha, J. N., & Ewule, D. M. (2020). Community forest management: A strategy for rehabilitation, conservation and livelihood sustainability: The case of Mount Oku, Cameroon. *Journal of Geoscience and Environment Protection, 8*, 1–14. https://doi.org/10.4236/gep.2020.82001

Griscom, B. W., Lomax, G., Kroeger, T., Fargione, J. E., Adams, J., & Almond, L. (2019). We need both natural and energy solutions to stabilize our climate. *Global Change Biology, 25*, 1889–1890.

Hallegatte, S. (2009). Strategies to adapt to an uncertain climate change. *Global Environmental Change, 19*, 240–247.

International Institute for Environment and Development-IIED. (2015a). *Vulnerable communities: Getting their needs and knowledge into climate policy* (IIED Briefing 17328IIED) (pp. 1–4). IIED.

International Institute for Environment and Development-IIED. (2015b). *Vulnerable communities: Climate adaptation that works for the poor* (IIED Briefing 17329IIED) (pp. 1–4). IIED.

International Institute for Environment and Development-IIED. (2019). *Nature-based solutions to climate change adaptation* (pp. 1–4). IIED.

Jafari, M., & Jafari, V. (2020). Climate change and food security: A glance and principles and a strategic road map. In V. R. Squires & M. K. Gaur (Eds.), *Food security and land use change under conditions of climate variability: A multidimensional perspective* (pp. 1–17). Springer. https://doi.org/10.1007/978-3-030-36762-6_6

Jia, G., Shevliakova, E., Artaxo, P., De Noblet-Ducoudré, N., Houghton, R., House, J., Kitajima, K., Lennard, C., Popp, A., Sirin, A., Sukumar, R., & Verchot, L. (2019). Land–climate interactions. In P. R. Shukla, J. Skea, E. Calvo Buendia, V. Masson-Delmotte, H.-O. Pörtner, D. C. Roberts, P. Zhai, R. Slade, S. Connors, R. van Diemen, M. Ferrat, E. Haughey, S. Luz, S. Neogi, M. Pathak, J. Petzold, J. Portugal Pereira, P. Vyas, E. Huntley, K. Kissick, M. Belkacemi, & J. Malley (Eds.), *Climate change and land: An IPCC special report on climate change, desertification, Land Degradation, sustainable land management, food security, and greenhouse gas fluxes in terrestrial ecosystems* (pp. 131–246). IPCC. https://doi.org/10.1017/9781009157988.004

Kimengsi, J. N., Akumbo, C. M., Balgah, R. A., Tingum, E. N., Tume, S. J. P., & Akhere, G. S. (2022). Farm-based climate adaptation dynamics: Insights from the vegetable sector in the western highlands of Cameroon. *Cogent Social Sciences, 8*(2126452), 1–22. https://doi.org/10.1080/23311886.2022.2126452

Kiming, I. N., Kongnso, M. E., & Kwei, J. (2020). Mountain apiculture and environmental dynamics: Impact of climate variability on bee farming in Oku, Cameroon. *International Journal of Environmental, Sustainability, and Social Science, 1*(3), 01–11. https://doi.org/10.38142/ijesss.v1i1.33

Kolawole, D. O., Wolski, P., Ngwenya, B., & Mmopelwa, G. (2014). Ethno-meteorology and scientific weather forecasting: Small farmers and scientists' perspectives on climate variability in the Okavango Delta, Botswana. *Climate Risk Management, 4-5*, 43–58.

Kongnso, M. E., Buba, U. H., & Nfor, J. T. (2021). Implications of climatic stressors on agro-pastoral resources among Mbororo communities along the slopes of Kilum-Ijim Mountain, North West Region, Cameroon. *Frontier in Sustainable Food Systems, 5*, 685071. https://doi.org/10.3389/fsufs.2021.685071

Mairomi, H. W., & Tume, S. J. P. (2021). Standardized precipitation index valuation of seasonal transitions and adaptation of pastoralist to climate variability in rangelands of the Bamenda

Highlands of Cameroon. *Journal of Ecology and Natural Resources, 5*(1), 1–20. https://doi.org/10.23880/jenr-16000229

Mairomi, H. W., Nsahlai, K. L., Ngah, M., Kininla, K. A., Yungsi, D. W., & Tume, S. J. P. (2022). Land use/land cover dynamics in pioneer fronts and implication on livelihoods: Insights from Nkuh and Nkuv on the Bui Plateau. *Journal of Ecology and Natural Resources, 6*(2), 1–16. https://doi.org/10.23880/jenr-16000277

Mathews, T., Mullan, D., Wilby, R. L., Broderick, C., & Murphy, C. (2014). The household economy approach: Managing the impact of climate change on poverty and food security in developing countries. *Climate Risk Management, 4-5*, 59–68.

Mbow, C., Rosenzweig, C., Barioni, L. G., Benton, T. G., Herrero, M., Krishnapillai, M., Liwenga, E., Pradhan, P., Rivera-Ferre, M. G., Sapkota, T., Tubiello, F. N., & Xu, Y. (2019). Food Security. In P. R. Shukla, J. Skea, E. C. Buendia, V. Masson-Delmotte, H.-O. Pörtner, D. C. Roberts, P. Zhai, R. Slade, S. Connors, R. van Diemen, M. Ferrat, E. Haughey, S. Luz, S. Neogi, M. Pathak, J. Petzold, J. P. Pereira, P. Vyas, E. Huntley, K. Kissick, M. Belkacemi, & J. Malley (Eds.), *Climate change and land: An IPCC special report on climate change, desertification, land degradation, sustainable land management, Food security, and greenhouse gas fluxes in terrestrial ecosystems* (pp. 437–550). https://doi.org/10.1017/9781009157988.007

McKee, T. B., Doesken, N. J., & Kleist, J. (1993). The relationship of drought frequency and duration to time scales. In *Proceedings of the eighth conference on applied climatology* (pp. 179–184). American Meteorological Society.

Niba, M. L. F. & Bailack, K. M. (2022). Assessment of Impact of Rainfall Variability and Vegetable Production in Bamenda III Sub-Division, Cameroon. *Asian Journal of Advances in Agricultural Research, 18*(2): pp. 16–28. https://doi.org/10.9734/ajaar/2022/v18i230214

Smit, B., & Pilifosova, O. (2001). Adaptation to climate change in the context of sustainable development and equity. In *Climate change 2001: Impacts, adaptation and vulnerability* (pp. 879–905). Cambridge University Press.

Smith, B., Sheider, H., Oppenhemmer, M., & Yole, M. (2009). Assessing dangerous climate change through update of IPCC. In *Proceedings of natural academy of science*. www.un.org/esal. Accessed 3 Apr 2022.

Tantoh, H. B., Ebhuoma, E. E., & Leonard, L. (2022). Indigenous women's vulnerability to climate change and adaptation strategies in Central Africa: A systematic review. In E. E. Ebhuoma & L. Leonard (Eds.), *Indigenous Knowledge and Climate Governance* (*Sustainable Development Goals Series*) (pp. 53–66). Springer. https://doi.org/10.1007/978-3-030-99411-2_5

Tume, S. J. P. (2021a). Impact of climate change on domestic water accessibility in Bamenda III Sub-Division, North West Region, Cameroon. *Journal of the Cameroon Academy of Sciences, 17*(2), 131–145. https://doi.org/10.4314/jcas.v17i4.3

Tume, S. J. P. (2021b). Rainfall seasonality and standardized precipitation valuation of water resources susceptibility to climate variability on the Bui Plateau, Northwest Region, Cameroon. *Advances in Hydrology & Meteorology, 1*(1), 1–20. AHM.MS.ID.000502.

Tume, S. J. P. (2022a). Standardized precipitation index comparison along the Limbe-Bamenda Axis of the Cameroon Gulf of Guinea. *Advances in Environmental and Engineering Research, 3*(1), 1–13. https://doi.org/10.21926/aeer.2201002

Tume, S. J. P. (2022b). Standardized precipitation index valuation of climate change in Bamenda. *Geology, Earth and Marine Sciences, 4*(1), 1–6. https://doi.org/10.31038/GEMS.2022414

Tume, S. J. P., & Fogwe, Z. N. (2018). Standardised precipitation index valuation of crop production responses to climate variability on the Bui Plateau, Northwest Region of Cameroon. *Journal of Arts and Humanities, 1*(2), 21–38.

Tume, S. J. P., & Kimengsi, J. N. (2021). Indigenous and modern agro-based climate adaptation practices in rural Cameroon. *International Journal of Environmental Studies, 78*(5), 1–10. https://doi.org/10.1080/00207233.2020.1977538

Tume, S. J. P., & Kongnso, M. E. (2022). Standardized precipitation index valuation of the impact of climate variability and change on domestic water accessibility in Bamenda III Municipality, North West Region, Cameroon. *Journal of Science Frontier Research (H), Environment and Earth Science, 22*(8), 39–55.

Tume, S. J. P., & Maghah, J. A. (2021). Challenges to climate change adaptation in the agricultural sector of Bui Division, North West Region, Cameroon. *International Journal of Resource and Environmental Management, 6*(1), 55–68.

Tume, S. J. P., & Ngwa, C. A. (2022). Impact of rainfall variability on maize production in the Ndop Plain, North West Region, Cameroon (1990–2015). *Journal of Ecology and Natural Resources, 6*(1), 1–6. https://doi.org/10.23880/jenr-16000265

Tume, S. J. P., & Nyuyfoni, R. S. (2021). Rainfall anomaly index valuation of agricultural production in Jakiri Sub-Division, Northwest Region, Cameroon. *Modern Concepts & Developments in Agronomy, 9*(4), 931–943. https://doi.org/10.31031/MCDA.2021.09.000717

Tume, S. J. P., & Tani, B. V. (2018). Stakeholders' signature to climate change adaptation in the agrarian sector of Bui Plateau, Northwest Cameroon. *Journal of Environmental Issues and Agriculture in Developing Countries, 10*(3), 140–156.

Tume, S. J. P., Kimengsi, J. N., & Fogwe, Z. N. (2019). Indigenous knowledge and farmer perceptions of climate and ecological changes in the Bamenda Highlands of Cameroon: Insights from the Bui Plateau. *Climate, 7*(138), 1–18. https://doi.org/10.3390/cli7120138

Tume, S. J. P., Zetem, C. C., Nulah, S. M., Ateh, E. N., Mbuh, B. K., Nyuyfoni, S. R., Ahfembombi, L. L., & Kwei, J. (2020). Climate change and food security in the Bamenda Highlands of Cameroon. In V. R. Squires & M. K. Gaur (Eds.), *Food security and land use change under conditions of climate variability: A multidimensional perspective*. Springer. https://doi.org/10.1007/978-3-030-36762-6_6

UN-Water. (2018). *The United Nations World Water Development Report 2018: Nature-based solutions for water* (pp. 1–25). UNESCO.

World Meteorological Organization. (2012). *Standardized precipitation index user guide* (WMO-No. 1090). WMO, Geneva, p. 50.

Zetem, C. C. (2022). *Rainfall variability incidence on agricultural system dynamics in the Kom Highlands, North West Region of Cameroon* [PhD thesis in Geography and Planning, The University of Bamenda, Cameroon], p. 254.

Increasing Warming of Cameroon's Afromontane Landscape: Realities and Implications for Livelihoods and Sustainability

Nyong Princely Awazi, Suiven John Paul Tume, and Henry Bikwibili Tantoh

1 Introduction

Mountains and their diverse ecosystems have become a priority area for achieving the twin goals of biodiversity conservation and livelihood improvement owing to the enormous potential (in terms of natural resources) they provide. In addition, mountains and their ecosystem attract numerous anthropogenic activities including settlements, agriculture, timber harvesting, mining, grazing, as well as stone and sand exploitation. It is within this framework that at a side event at the UN Convention on Biological Diversity (CBD COP15) in Montreal, Canada, in line with the post-2020 Kunming–Montreal Global Biodiversity Framework, the Mountain Futures Action Plan—a product of the global Mountain Futures Initiative was launched (Purvis, 2023). The Mountain Futures Initiative came into effect in 2016 and its goal is to enhance the diversity, stability, and sustainability of ecosystems; support indigenous mountain peoples and provide them with sustainable solutions that aid in the maintenance of diverse mountain landscapes on which they rely; offer mountain peoples sustainable livelihood options; as well as construct and share green mountain solutions. According to Xu Jianchu (2022), "Mountain ecosystems are central to global change, including global warming."

N. P. Awazi (✉)
Department of Forestry and Wildlife Technology, College of Technology, The University of Bamenda, Bamenda, Cameroon

FOKABS, Ottawa, ON, Canada

FOKABS, Yaounde, Cameroon
e-mail: awazinyong@uniba.cm; pnyong@fokabs.com; https://fokabs.com/

S. J. P. Tume · H. B. Tantoh
Department of Geography and Planning, The University of Bamenda, Bamenda, Cameroon
e-mail: suiven.john@uniba.cm

S. Tripathi et al. (eds.), *Warming Mountains*,
https://doi.org/10.1007/978-3-031-62197-0_7

The Mountain Futures Action Plan is a great blueprint for the conservation of mountain ecosystems and sustainable livelihoods as it targets the revitalization of rural mountain communities focusing on the principles of precautionary development, green and low-carbon development, holistic thinking. co-innovation, and secured land rights (CIFOR-ICRAF, 2023). The Plan defines 15 actions divided into four action areas, i.e., future living, scientific exploration, indigenous wisdom, and ecological restoration, which are linked to the CBD's 2021 Kunming Declaration—specifying the commitments of signatories to recognizing the role of conservation in the protection of human health and prioritizing biodiversity protection (CIFOR-ICRAF, 2023).

Purvis (2023), in a study entitled "Working with mountain communities towards a brighter future for all," notes that mountains are very important for global ecosystem services as they are the water towers of the world, providing about 80% of the Earth's freshwater resources. Over half of the world's 8 billion inhabitants rely on this freshwater to survive, while many communities across the world depend largely on clean energy and food derived from mountain ecosystems for livelihood sustenance (Purvis, 2023). Although mountains provide many services and resources such as building materials, carbon storage, medicines, climate regulation, and fibers owing to rich biodiversity, they face multiple threats with the most common being the adverse impacts of climate change, natural disasters, land degradation, and overexploitation (Macchi & ICIMOD, 2010). Mountains, therefore, need protection especially for the sake of indigenous peoples who call mountains home, and who left on their own cannot deal with the increasing environmental changes occurring around them.

Across sub-Saharan Africa in general and Cameroon in particular, the Afromontane ecosystem is a vital biodiversity hotspot that draws the attention of both researchers and policymakers. Cameroon's Afromontane landscape is one of the most endowed biodiversity hotspots in the world with over 50 threatened and endemic flora and fauna species (Martin & Burgess, 2023). Prominent ecoregions that make up this landscape are: the Bamenda Highlands, Bamboutos Mountain, Mount Mbam, the Lebialem Highlands, the Kilum-Ijim Mountain Range, Mount Cameroon, the Bakossi Highlands, Mount Nlonako, Kupe-Muanenguba mountains, the Tchabal-Mbabo Mountain Range. Cameroon's Afromontane landscape is under serious threat as it has shrunk drastically from over 80,000 km^2 some decades ago to roughly 38,000 km^2 today (European Commission, 2023). Among the different drivers of degradation of this landscape is the increasing warming occurring on the landscape, which is triggering wildfires—further exacerbating the degradation phenomenon. Against this backdrop, this chapter assesses the realities of the warming of Cameroon's Afromontane landscape as well as the livelihood and sustainability implications. More specifically, the chapter:

- Assesses the biodiversity of and threats facing Cameroon's Afromontane landscape;
- Examines the realities of warming of Cameroon's Afromontane landscape;

- Assesses the livelihood opportunities and implications of warming of Cameroon's Afromontane landscape on the livelihoods of adjacent communities;
- Examines the sustainability implications of the warming of Cameroon's Afromontane landscape.

2 Cameroon's Afromontane Landscape: Biodiversity and Threats

Cameroon's Afromontane landscape is made up of different ecosystems dominated by highlands forests and savannah grasslands. This Afromontane landscape embodies the Bamenda Highlands (Big Babanki Montane forest, Kilum-Ijim, Bui Highlands, Mezam Highlands, Momo Highlands, Tubah Highlands, Mbi Crater or Mbingo forest), Kimbi-Fungom montane forest, Mount Bamboutos, Mount Mbam, Lebialem Highlands, Mount Cameroon, Mount Nlonako, Mount Mbappit, Kupe-Manengouba mountains, Bakossi Highlands, Tchabal-Mbabo Mountain Range (Fig. 1). Although Cameroon's Afromontane landscape is quite vast, just some parts of this landscape have received enormous attention, particularly Mount Cameroon, Kilum-Ijim, Kimbi-Fungom, Bakossi Highlands, Mount Bamboutos, and Lebialem Highlands.

According to Martin and Burgess (2023), Cameroon's Afromontane landscape is characterized by high levels of endemism—especially primates, birds, and some plant species. One of Africa's most endangered primates, the lowland gorilla, is found in this landscape. The highland forests that are found on this landscape grow along and extinct range of volcanoes characterized by fertile soils derived from volcanic rocks. These forests are of huge biological importance and are endowed

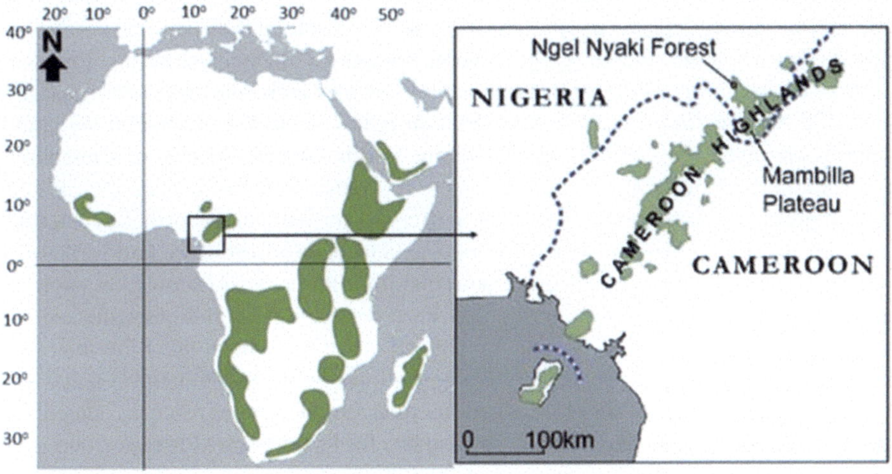

Fig. 1 Map showing Cameroon's Afromontane landscape. (Source: Abiem et al. (2020))

with numerous endemic species. Although tree species diversity in these forests is low, there is a high diversity of nonwoody plant species. Some common endemic and threatened species of tree and nonwoody plants are *Podocarpus latifolius*, *Syzygium guineense*, *Nuxia congesta*, *Rapanea melanophloeos*, *Prunus africana*, *Newtonia camerunensis*, *Chassalia laikomensis*, *Oxyanthus okuensis*, *Ardisia dom*, *Eugenia gilgii*, *Psychotria moseskemei*, *Dombeya ledermannii*, and others (Table 1). A significant amount of endemic flora also occurs in heathlands, moorlands, grasslands, and other non-forested habitats at high elevations such as the case at Mount Oku's summit.

Avian endemism is equally remarkable in Cameroon's Afromontane landscape, with notable examples of endemic avian species including the Bangwa forest warbler, the Bamenda apalis, Bannerman's weaver (*Ploceus bannermani*), which is vulnerable; White-headed vulture (*Trigonoceps occipitalis*), which is critically endangered; Bannerman's Turaco (*Tauraco bannermani*), which is endangered; Banded wattle-eye (*Platysteira laticincta*), which is endangered; Green-breasted bushshrike (*Malaconotus gladiator*), which is vulnerable; Mount Kupe bushshrike (*Chlorophoneus kupeensis*), which is endangered; and the White-throated mountain babbler (*Turdoides gilberti*), which is vulnerable (Ngute et al., 2019). High levels of endemism also occur among amphibians with over 40 species classified as endemic. Different species of endemic mammals, particularly primates, are found within this landscape (especially in the Bakossi Highlands and the Kimbi-Fungom montane forests) such as Eisentraut's striped mouse (*Hybomys eisentrauti*), the Mount Oku hylomyscus (*Hylomyscus grandis*), Mount Oku rat (*Lamottemys okuensis*), Mittendorf's striped grass mouse (*Lemniscomys mittendorfi*), Dieterlen's brush-furred mouse (*Lophuromys dieterleni*), and Eisentraut's brush-furred rat (*L. eisentrauti*); Oku mouse shrew (*Myosorex okuensis*), Rumpi mouse shrew (*M. rumpii*), western vlei rat (*Otomys occidentalis*), Hartwig's soft-furred mouse (*Praomys hartwigi*), and Bioko forest shrew (*Sylvisorex isabellae*); Cross River gorilla (*Gorilla gorilla diehli*), an endemic subspecies of western gorilla, mainland drill (*Mandrillus leucophaeus leucophaeus*), Preuss's red colobus (*Pilocolobus preussi*), chimpanzee (*Pan troglodytes*), and several species of guenon including Preuss's monkey (*Cercopithecus preussi*). Endemic bee species include *Apis mellifera adansonii* and *Apis mellifera monticola*–mountain honey bees. The Afromontane landscape is endowed with a plethora of endemic and threatened species of squamates and amphibians (Table 2).

Cameroon's Afromontane landscape suffers from inadequate formal protection from the state as most of the protection of this landscape is still in the hands of local authorities. This has led to different activities that threaten the ecoregion such as agriculture and pastoralism reducing the formerly vast highlands forests, for example, to just small patches. However, several measures have been taken by the state and other stakeholders to protect this ecoregion from rampant degradation and wanton exploitation. Some of these measures include the creation of reserves, sanctuaries, and national parks. The Nlonako mountain, for example, was transformed into a faunal reserve that provides protection from logging; the loss of forest within the Kilum-Ijim Mountain Range has been halted since 1986 thanks to projects that have

Table 1 Cameroon's Afromontane forest trees

Name of species	Endemic to Cameroon Highlands (Yes/No)	Occurring also below 2000 m alt. (Yes/No)	Species forming 90% of the canopy (Yes/No)
Astropanax abyssinica (A.Rich.) Seem.	No	No	Yes
Astropanax mannii (Hook.f.) Seem.	Yes	No	Yes
Prunus africana (L.) Kalkman	No	No	Yes
Syzygium staudtii (Engl.) Mildbr.	No	No	Yes
Myrsine melanophloeos (L.) Sweet	No	No	Yes
Oldeania alpina (K.Schum.) Stapleton	No	No	Yes
Carapa oreophila Kenfack	Yes	No	Yes
Bersama abyssinica Fresen.	No	No	Yes
Ixora foliosa Hiern	Yes	No	Yes
Clausena anisata (Willd.) Benth	No	Yes	Yes
Nuxia congesta Fresen.	No	No	No
Lasiosiphon glaucus Fresen.	No	No	No
Hypericum revolutum Vahl subsp. *revolutum.*	No	No	No
Maesa lanceolata G.Don	No	No	No
Alsophila dregei (Kunze) R.M.Tryon	No	No	No
Podocarpus latifolius (Thunb.) Mirb.	No	Yes	No
Croton macrostachyus Delile	No	Yes	No
Albizia gummifera (J.F.Gmel) C.A.Smith	No	Yes	No
Cassipourea malosana (Baker) Alston	No	No	No
Brucea antidysenterica J.F.Mill.	No	No	No
Ilex mitis (L.) Radlk.	No	No	No
Neoboutonia mannii Benth. & Hook.f.	No	Yes	No
Olea capensis subsp. *macrocarpa (C.H.Wright) I.Verd.*	No	No	No
Eugenia gilgii Engl. & Brehmer	Yes	No	No

(continued)

Table 1 (continued)

Name of species	Endemic to Cameroon Highlands (Yes/No)	Occurring also below 2000 m alt. (Yes/No)	Species forming 90% of the canopy (Yes/No)
Agarista salicifolia (Lam.) G.Don	No	No	No
Dovyalis cameroonensis Cheek & Ngolan	Yes	No	No
Ternstroemia cameroonensis Cheek	Yes	Yes	No
Deinbollia onanae Cheek	Yes	Yes	No

Source: Adapted from Cheek et al. (2021)

employed the participatory approach (working with local chiefs and community members). Although these measures have yielded some positive outcomes, much damage has already been done as large mammals have almost gone extinct in the Kilum-Ijim Mountain Range, and grazing of goats, cattle, and sheep impedes forest succession (Buchenrieder & Balgah, 2013; Cuni-Sanchez, 2018; Solefack et al., 2018). Besides, cattle herders set wildfires, making the remaining fragments of the highlands forest highly vulnerable.

Studies show that the highlands forest of Cameroon's Afromontane landscape has declined by over 50% since the 1960s mainly due to agricultural activities attracted by the fertile volcanic soils and reliable rainfall (Cheek et al., 2021). These fertile soils have attracted a high population density in this ecoregion and in some parts of Cameroon's Afromontane landscape such as the Bamenda Highlands, the population density goes up to over 300 people per km^2 (Martin & Burgess, 2023). Besides agriculture, there are other threats to this landscape including the collection of fuelwood and construction materials, unsustainable harvesting and use of timber, hunting as well as fires from farms and grazing lands. Uncontrolled exploitation of some tree species such as *Prunus africana* exploited for pharmaceutical use has led to a serious decline of the species' population, making it threatened (Foncha & Ewule, 2020).

As indicated earlier, some parts of Cameroon's Afromontane landscape have received a lot of attention as opposed to others. In the parts that have received attention, studies have uncovered the enormous biodiversity and threats facing these landscapes. Cheek and Onana (2000), for example, noted that over 96.5% of the original vegetation in Oku-Ijim in the Bamenda Highlands has been lost and with it several endemic species, and this has worsened in the past 20 years (Cheek et al., 2021; Martin & Burgess, 2023). In the same vein, 2435 plant species known to be native to the region were documented in Mount Cameroon (Cable & Cheek, 2000). Furthermore, 2412 species, of which 232 are assessed as globally threatened and 82 as strictly endemic, from an area of 2390 km^2 were examined in Kupe-Muanenguba and the Bakossi Highlands (Cheek et al., 2004). This shows the high diversity and endemism in this area. It has also been reported that the high population density within Kilum-Ijim and Kupe Mountain environments is a major threat to the

Table 2 Endemic and threatened species of squamates and amphibians in Cameroon's Afromontane landscape

Group	Endemic and threatened species
Squamates	Pfeffer's Chameleon (*Trioceros pfefferi*)
	Perret's montane Chameleon (*Trioceros perreti*)
	Angel's five-toed skink (*Lacertaspis lepesmei*)
	Pauliana's five-toed skink (*Leptosiaphos pauliani*)
Amphibians	*Wolterstorffina mirei*
	Petropedetes parkeri
	Petropedetes perreti
	Phrynobatrachus cricogaster
	Phrynobatrachus werneri
	Phrynobatrachus species
	Phrynodon species
	Cardioglossa pulchra
	Cardioglossa schioetzi
	Cardioglossa trifasciata
	Cardioglossa venusta
	Astylosternus nganhanus
	Astylosternus perreti
	Astylosternus montanus
	Astylosternus rheophilus
	Leptodactylodon axillaris
	Leptodactylodon bicolor
	Leptodactylodon boulengeri
	Leptodactylodon erythrogaster
	Leptodactylodon mertensi
	Cardioglossa oreas
	Leptodactylodon polyacanthus
	Leptodactylodon perreti
	Afrixalus lacteus
	Phrynobatrachus steindachneri
	Hyperolius ademetzi
	Hyperolius riggenbachi
	Leptopelis nordequatorialis
	Xenopus amieti
	Xenopus species
	Bufo villiersi
	Werneria bambutensis
	Werneria tandyi
	Cardioglossa melanogaster

landscape as different anthropogenic influences including settlements and agro-pastoral activities exist (Ngwa & Fonjong, 2002; BirdLife International, 2003). With roughly 200,000 people living within a day's walk to the Kilim-Ijim Mountain forest, Ngwa and Fonjong (2002) concur that the Kilum-Ijim ecoregion is one of the most vulnerable Afromontane landscapes in Cameroon and Africa. Agro-pastoral activities mainly include tuber crop production and cocoa for the Kupe ecoregion and coffee, staple food crops such as maize, beans, potatoes, and livestock rearing for the Kilum-Ijim ecoregion. Besides agro-pastoral activities, the population in these two ecoregions (Kilum-Ijim and Kupe) depend very much on the highlands forest for different basic needs such as fuelwood, honey, game, building materials, water resources and medicines, as well as spiritual and cultural purposes. It is argued that the priority conservation actions for the Kilum-Ijim ecoregion include supporting alternative livelihood opportunities for local communities dependent on the forests, developing Participatory Forest Management Plans for threatened forests, and establishing more protected areas and increasing their effectiveness (Martin & Burgess, 2023).

3 Warming of Cameroon's Afromontane Landscape: The Reality

Cameroon's Afromontane landscape has been experiencing significant levels of warming in recent times, which is attributable to changing climate. Studies conducted across Cameroon's Afromontane landscape have increasingly shown trends toward higher temperatures, which are unusual in this supposedly cool landscape. Increasing temperatures have largely been reported by studies carried out in the Bamenda Highlands (Fig. 2), Mount Bamboutos, Mount Mbam, Lebialem Highlands, Mount Cameroon, Mount Nlonako, Mount Mbappit, Mount Kupe, Bakossi Highlands, and the Tchabal-Mbabo Mountain Range.

It has been reported that extreme weather events such as high temperatures adversely affect local communities in Oku Subdivision (the Kilum landscape), which is part of the Bamenda Highlands forest region (Azong, 2021). In a bid to look for solutions to respond to these adversities, the local communities resorted to prayers, the use of traditional charms to protect farms, the use of traditional medicines, and conducting traditional rites to appease the gods (Azong, 2021). Similar studies conducted in the Mezam Highlands revealed significant levels of variation in temperature characterized by an increasing trend (Amawa et al., 2015). This was found following an analysis of 10 years of temperature data (2001–2010). The studies of Awazi and colleagues (Awazi, 2018, 2020, 2022a, b; Awazi & Tchamba, 2018; Awazi et al., 2019a, b, 2020a, b, c, 2021a, b, c, 2022a, b; Awazi & Tume, 2023) across different parts of the Western Highlands of Cameroon, particularly the Mezam Highlands, the Momo Highlands, and the Boyo Highlands, found that temperatures have been on an increase since the 1960s. Lézine et al. (2020) in a study

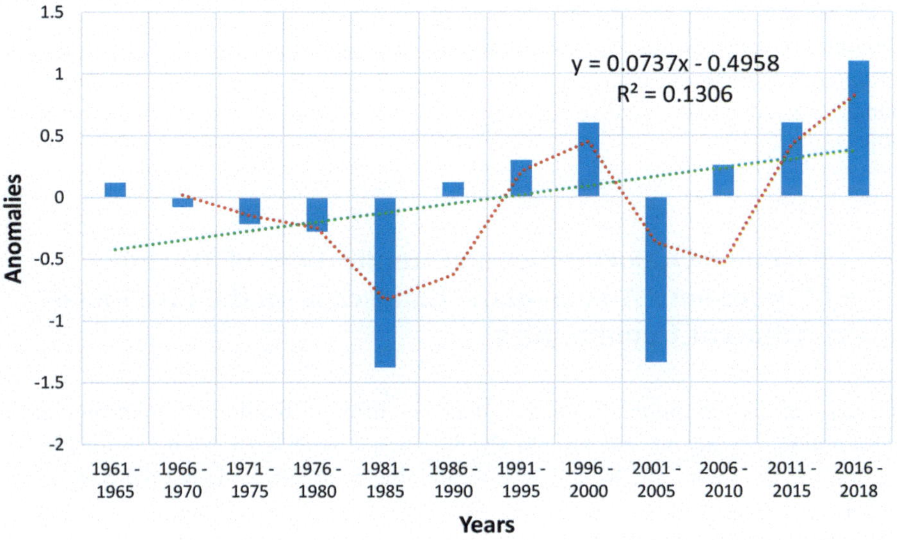

Fig. 2 Temperatures for the Bamenda Highlands from 1961 to 2018 showing an increasingly warming trend. (Source: Awazi (2022a))

within the Mbi Crater of the Bamenda Highlands found that there has been a post-glacial forest change occurred gradually at Mbi, and that the forest-wooded grassland ecotone was highly influenced by climate variability and increasing temperatures, resulting in forest disruption and diversity loss. Ntoko (2020) in a study conducted in and around Mount Cameroon National Park found that the climate in the region has changed over time in terms of perceived temperature increase, affecting the occurrence and availability of forest products. Kiming et al. (2020) reported the impact of climate change on mountain apiculture in the Kilum-Ijim ecoregion that rising temperature adversely affects periods of flowering for bee plants as acknowledged by 97% of bee farmers. There has also been a decreasing rate of hive colonization, with an increasing rate of absconding (34–46%), resulting in a reduction in honey yields. Increasing temperatures have equally exacerbated bushfires and forest degradation while reducing the growing period of some bee plants. With the limited coping strategies of bee farmers, there is the absolute need for better policies to help these farmers adapt to the rising temperatures while maintaining the quality of the Oku honey—a geographic indication and certified product. The study of Kongnso et al. (2021) examining the implications of climatic stressors on agro-pastoral resources among Mbororo Communities along the slopes of Kilum-Ijim Mountain in the Bamenda Highlands of Cameroon uncovered that increasing temperatures and drier conditions are making things difficult for both farmers and herders as pasture and water are becoming scarce, exacerbating the already volatile situation of farmer–herder conflicts. In a study carried out in the Bui Highlands of the northwest region of Cameroon, Yufenyuy and Nguetsop (2020) found that temperatures have increased in recent years around Kumbo, which is the town located at the highest elevation in Cameroon, and this has led to the rapid spread of vectors of malaria and an increase in malaria prevalence. These findings indicate that temperatures are on an increasing trend across Cameroon's Afromontane landscape with many adverse repercussions on communities living in and around this landscape.

4 Livelihood Opportunities and Implications of Warming of Cameroon's Afromontane Landscape on the Livelihoods of Adjacent Communities

Cameroon's Afromontane landscape provides different livelihood opportunities to communities living in and around the landscape. The most common livelihood options for communities living in and around this landscape include food crop farming, market gardening, fruit tree farming, silviculture (mainly Eucalyptus, Podocarpus, cypress, and pines), agroforestry, poultry farming, pig farming, goat herding, cattle and sheep herding, hunting, harvesting of timber, and non-timber forest products, apiculture, fish farming, palm wine tapping, milling of oil palm nuts, as well as off-farm activities such as bike riding.

In the Kilum-Ijim ecoregion, different studies have uncovered diverse livelihood opportunities for adjacent communities as well as the implications of the increasing warming of this ecoregion on the same. Buchenrieder and Balgah (2013) in a study conducted in the Kilum-Ijim ecoregion of the Bamenda Highlands uncovered that wildlife domestication is a sustainable livelihood option for communities around the community forest as bushmeat consumption remains high owing to the absence of alternative sources of protein. Cuni-Sanchez et al. (2019) in a study carried out in the Bui Highlands, particularly around Kumbo and Oku, found that mountains provide diverse ecosystem services to farmers and pastoralists that serve as livelihood options. The most common ecosystem services identified by farmers and pastoralists were wildlife (which serves as meat for farmers in particular), fodder, medicines, as well as timber and non-timber forest products and tourism. There were, however, some differences in perceptions regarding the contribution of these ecosystem services between pastoralists and farmers. Fogwe and Kwei (2015) in a study conducted in the Kilum-Ijim mountain forests in the Bamenda Highlands found that local communities such as those of Oku make use of this ecoregion for livelihood sustenance, particularly in the domains of agriculture and pastoralism. Pastoralism provides products such as meat and milk, while agriculture provides different food through food crops such as maize, potato, vegetables, and beans. The forest-adjacent communities of this ecoregion also benefit from timber and non-timber forest products as well as fresh water. Still, in the Kilum-Ijim Mountain landscape, studies conducted by Kimengsi et al. (2019a, b, 2020) found that medicinal plants extraction, domestication of non-timber forest products, and ecotourism are sustainable livelihood options for the communities. Concerning ecotourism, preference was given to the production and sale of arts and crafts items and the promotion of cultural heritage sites. Non-forest activities were equally common, with migration being the most preferred option. Kimengsi et al. (2020), however, noted that forest-dependent livelihood options are the most cherished by communities in the Kilum-Ijim landscape which has major consequences for conservation. Revising Cameroon's forest legislation and policy, promoting the domestication of non-timber forest products, and establishing mutually agreed forest use quotas for communities are recommended by the authors as measures to mitigate and/or reduce forest-dependent diversification of livelihoods in the Kilum-Ijim landscape. In an ethnomycological study carried out in the Kilum-Ijim Mountain Forest ecoregion, Teke et al. (2018) uncovered that the exploitation of non-timber forest products (NTFPs), particularly mushrooms, was a major livelihood option in the community. These edible mushrooms were used as food and medicines. There were eight edible species of these mushrooms: *Polyporus dictyopus*, *Termitomyces striatus*, *Auricularia polytricha*, *Termitomyces microcarpus*, *Laetiporus sulphureus*, *Termitomyces* sp.1, *Termitomyces* sp.2, and *Polyporus tenuiculus*. Meanwhile, there were nine mushroom species used for medicinal purposes and traditional health care: *Xylaria* sp., *Daldinia concentrica*, *Lentinus squarrosulus*, *Vascellum pretense*, *Polyporus dictyopus*, *Termitomyces microcarpus*, *Ganoderma applanatum*, *Trametes versicolor*, and *Auricularia polytricha*. The mushrooms played an important role as a substitute for meat and a source for protein. One edible species of mushroom, i.e., *Polyporus dictyopus*, was reported for the first time as being edible. Nchamcham (2010) in a study conducted on *Prunus africana* in the Kilum-Ijim mountain landscape found that the exploitation and sale of *Prunus africana* is a major livelihood option for local forest-dependent communities. This is most common in the vicinities of Fundong, Oku, Belo, and Jakiri. Abbot et al. (2001) in a study conducted in the Kilum-Ijim Mountain landscape uncovered that different livelihood options, particularly apiculture, agriculture, and livestock keeping, are common in and around the highlands. In addition, community members are moving toward sustainable livelihood options such as apiculture and wildlife domestication in order to benefit from different ecosystem services including fuelwood, fencing wood, meat, timber, Prunus bark, medicinal plants, poles, honey, and carving wood. In and around Mount Oku, Foncha and Ewule (2020) found that the main livelihood strategies/options of local forest-dependent communities include selective exploitation and sales of *Prunus* spp., tourist guides, agricultural intensification, bee farming, agroforestry, capacity building for skill acquisition, and direct employment. In a study carried out on different Mbororo communities along the slopes of the Kilum-Ijim ecoregion, Kongnso et al. (2021) uncovered that the main livelihood options were agriculture and grazing of livestock, and these livelihood options are being adversely affected by increasingly warmer and drier conditions, driving pasture loss, shortage of water, and decreasing soil fertility. This has exacerbated the already volatile state of farmer–herder conflicts in the ecoregion. Studies conducted around the Bamenda Highlands and Mount Mbam by Gardner (2002) and Cuni-Sanchez (2018) reveal that the forests located adjacent to these communities are very important for the people living around them in terms of livelihood improvement, as they supply water, fuelwood, medicines, honey, and other products and have cultural and spiritual importance. In and around Mount Mbam and Mount Oku for example, the main livelihood options are farming and herding of livestock. Farmers are mainly involved in the growing of maize, beans, and potato, while pastoralists own cattle and goats. The adjacent forests are very important to both farmers and herders as they serve as a source for fuelwood, medicine, honey, building construction materials, wild fruits, and grazing grounds for livestock of pastoralists. For the Oku people in particular, White honey, made from *Nuxia congesta* flowers, is a vital source of income.

In and around Mount Cameroon, studies have shown that different livelihood options abound although increasing warming of the landscape will have major adverse implications on these livelihoods. Ako et al. (2022) in a study assessing climate variability and its impact on the environment and water resources in the Mount Cameroon area found that there is a high level of variability in climate parameters, particularly temperature, with predictions showing that Mount Cameroon will have an average temperature of 27 °C by 2030—which is too high for a Mountainous region. The findings further reveal that these climatic extremes will lead to droughts and floods as well as environmental degradation, which will go a long to adversely impact the livelihoods of local communities living adjacent the mountain. Laird et al. (2007) in a study carried out in and around Mount Cameroon found that the main livelihood option in the community is cocoa cultivation, which is playing an important role to promote biodiversity conservation and local livelihoods as about 50 tree species are normally retained or planted on cocoa farms in order to obtain diverse services including timber and food—most of these tree species are of high conservation value. Although these cocoa farms are not as devastating as commercial plantations, they are putting a lot of pressure on forest reserves, and the use of chemical inputs is environmentally malignant. For a more sustainable livelihood, this study found that cocoa farmers are resorting to the diversification of their livelihood options, especially through the exploitation of non-timber forest products (NTFPs) as well as the extensive use of fallow and home gardens. Njoh and Wanie (2018) in a study in and around the Mount Cameroon National Park found that NTFPs serve as a major livelihood option for local communities with the main NTFPs harvested and sold being *Bush onion* (*Afrosfyrax lepidophyllus*), *Njangsang* (*Ricinodendron heudelotii*), *Bush mango* (*Irvingia gabonensis*), *Bitter kola* (*Garcinia kola*), *Bush pepper* (*Piper guineense*), *Pygeum* (*Prunus africana*), *Eru* (*Gnetum africanum*), and *Plum* (*Dacryodes edulis*). Although most of these NTFPs were naturally abundant in the Mount Cameroon National Park as well as some community forests and farmlands, their supply has been dwindling recently owing to threats posed by deforestation, unsustainable harvesting, overexploitation, unsustainable agriculture and infrastructural development, logging, and above all galloping human population. Tata and Lambi (2014) conducted a study within and around the Mount Cameroon National Park uncovering that the main livelihood option of communities living around the park is agriculture. They, however, noted that agriculture is largely put in place after forest clearance, which is detrimental to conservation. Meanwhile, MINFOF (2014) in its management plan for the Mount

Cameroon National Park emphasizes on the importance of ecotourism as a livelihood option for the forest-dependent communities around the Mount Cameroon National Park. Mount Cameroon has enormous ecotourism potential including wildlife, trekking, scenic, and other attractions.

In and around Mount Bamboutos, the studies of different researchers have uncovered the different livelihood options and the implications of the increasing warming of this ecoregion on these livelihoods. Josiane et al. (2020) in a study on Mount Bamboutos revealed that temperature increased significantly, which has a negative impact on agricultural practices in this ecoregion—thus the need for better adaptation measures and the putting in place of policies that foster the adoption of sustainable adaptation measures by farmers. Ngimdoh et al. (2020) in a study conducted on the Bamboutos mountains found that the main livelihood option was agriculture, which involves crop rotation and mixed farming. The practice of agriculture in and around this ecoregion is, however, plagued by insecurity, farmer–herder conflicts, farmers' reticence to adopt new techniques in the face of environmental changes, expensive farm inputs, low incomes, and poor farm-to-market roads. The authors propose the improvement of farm infrastructures, improvement of land tenure laws, training and recruitment of more agricultural extension officials, and reestablishing agricultural banks as ways to improve the sustainability of agricultural livelihoods on the Bamboutos mountain. Nkembi et al. (2021) on their part, while analyzing smallholder farmer households' situation of food security in and around the Bamboutos ecoregion, found that the major livelihood options were agriculture and livestock keeping. Using the Household Food (In)Security Access Scale (HFIAS) to measure the food insecurity of households in and around this ecoregion, it was found that only 45% of sampled households were food secure, implying that the remaining 55% were food insecure. This shows that the agricultural practices in and around this ecoregion are either not tailored to meet food security needs or are unsustainable. Nkengafac et al. (2021) in a study conducted to examine the role played by agroforestry to food security in the Mount Bamboutos landscape found that the main livelihood option of communities around Mount Bamboutos was agroforestry, with the main agroforestry practices involved the planting of fodder, fertilizer, fruit, and timber trees. Over 50% of farmers adopt agroforestry in order to have better crop yields owing to decreasing soil fertility. However, there are constraints to this adoption including the need for much labor, lack of training, and lack of seeds. The recommendation is for extension agents and other stakeholders to intensify efforts toward educating and training more farmers on the benefits of agroforestry.

Studies by Price et al. (2011) and Oyono et al. (2012) on Cameroon's montane forests in particular and the Afromontane landscape in general reveal that these landscapes serve as a source of livelihood for all forest-dependent communities, particularly through the provision of medicinal and aromatic plants, (eco)tourism and recreation, and sacred and cultural services. It was, however, observed that livelihoods have not improved even though communities living adjacent the Afromontane landscapes have been granted and exercise new community rights, namely management and marketing rights. The authors further uncovered that, despite the resource base staying almost the same, it is largely being threatened by poor local-level institutional arrangements and social and biophysical management strategies, coupled with weak central level regulation and monitoring actions.

Thus, Cameroon's Afromontane landscape offers numerous livelihood opportunities, but the increasing warming of the landscape is adversely affecting the livelihoods of adjacent communities as watersheds are degraded, wildfires are recurrent, pastures are dwindling, and soil infertility rampant, making the livelihoods of farmers, herders/graziers, market gardeners, and harvesters of non-timber forest products in particular extremely difficult.

5 Sustainability Implications of the Increasing Warming of Cameroon's Afromontane Landscape

Increasing warming of Cameroon's Afromontane landscape has major sustainability implications, i.e., in the economic, social, and environmental dimensions. Socially, with the warming landscape, conflicts have become very recurrent between different actors competing for the limited resources of this landscape. The most common conflicts are between farmers and herders, farmers and farmers, herders and herders, as well as interethnic conflicts (Awazi & Avana, 2020; Kongnso et al., 2021; Kimengsi et al., 2024). Moreover, scarcity of resources encourages gender discrimination as the male gender lords over the female gender as far as controlling Afromontane resources is concerned (Kimengsi et al., 2022). Economically, with the warming of the landscape, local communities living adjacent are impoverished as the potential of income generation from the resources of the Afromontane landscape reduces–due to the increasing degradation induced by increasing warming. There is also the need to procure fertilizers to improve soil fertility, buy feed to feed livestock, procure pesticides to control increasing pests and diseases (Ndoh et al., 2016; Tume & Kimengsi, 2021), as well as spend more money on drugs as malaria becomes the norm (Yufenyuy & Nguetsop, 2020), which further impoverishes local communities pushing them to degrade the landscape even further. Environmentally, the excessive warming of the landscape has led to watershed degradation, soil infertility, loss of different species of plants and wildlife, as well as recurrent wildfires that exacerbate degradation, making it difficult for this biodiversity-rich landscape to provide adequate ecosystem services to sustain the numerous local communities living in and around it (Kongnso et al., 2021).

Different studies have been carried out across various ecoregions within Cameroon's Afromontane landscape, revealing that the landscape serves to enhance sustainability. However, the increasing warming of this landscape is a major threat to sustainability. In the Bamenda Highlands in particular and the Western Highlands in general, studies have shown that increasing warming of the landscape has major sustainability implications. Reif et al. (2007) in a study carried out on a mosaic landscape with montane forest remnants in the Bamenda Highlands of Cameroon detected 71 avifauna species with the most abundant of these species being the Northern Double-collared Sunbird *Cynniris reichenowi*, the Oriole Finch *Linurgus olivaceus*, the Common Stonechat *Saxicola torquata*, the Thick-billed Seed-eater *Serinus burtoni*, the Black-crowned Waxbill *Estrilda nonnula*, the Brown-backed *Cisticola Cisticola chubbi*, and the Yellow-breasted Boubou *Laniarius atroflavus*. The authors conclude that these montane forest ecoregions are home to both endemic and non-endemic species and most of which are categorized as threatened. In the same light, Sedláček et al. (2007) in a study conducted within a non-protected montane forest mosaic in the Bamenda Highlands of Cameroon identified 12 species endemic to the Cameroon Mountain landscape and many other bird species that call the Afromontane landscape home—this despite the massive forest fragmentation. The study also recorded a bird species (Great Blue Turaco *Corythaeola cristata*) considered extinct in Mount Oku due to forest clearing as well as the Red Kite *Milvus milvus*, which is the southernmost observation of this Palaearctic migrant in Africa and the first for Cameroon. With increasing temperatures, this bird diversity is threatened as food sources and water becomes scarce and recurrent wildfires make it impossible for birds to breed. Nchamcham (2010) found that Mountain resources are overexploited. Using the Kilum-Ijim Mountain forest as case study, it was found that *Prunus africana* is being overexploited, seriously threatening the species, as some of the stems of the species are dying owing to poor debarking techniques. Higher temperatures coupled with poor debarking techniques will lead to the death of many *Prunus africana* trees, endangering the species across the Afromontane landscape. Foncha and Ewule (2020) in a study conducted around Mount Oku found that local communities need to have more access and ownership of the adjacent montane forest of Mount Oku as this leads to social stability, transparency, and more accountability, which contributes to the conservation of the forest for improved livelihoods. The study recommends putting in place of better capacity-building frameworks to improve livelihoods and conservation as well as implement a system where stakeholders have better conditions for information sharing and learning. Abbot et al. (2001) in a study in the Bamenda Highlands uncovered that the Mountains in Cameroon in general and the Bamenda Highlands in particular play an important role in biodiversity conservation and in the enhancement of environmental sustainability. They noted that the Kilum-Ijim Mountain landscape, for example, is home to some of the world's endemic species such as the Bannerman's turaco, the Banded wattle-eye, and *Platysteira laticincta*. Rising temperatures is therefore a major threat to the conservation of the biodiversity of this ecoregion.

Asanga (2001) in a study carried out on community forestry in the Kilum-Ijim Mountain forest area noted that community forestry is the best way to go as far as the conservation of biodiversity is concerned. They indicate that conventional forest conservation measures have yielded little success largely due to lack of community involvement. They recommend community participation in forest management, building trust and communicating with all concerned parties, managing and resolving conflicts, facilitating joint action, and finding convergence between conservation objectives and those of local resource use as the best methods to promote conservation through community forestry in the local communities of Kilum-Ijim. Gardner (2002) in a study on community forests in the Bamenda Highlands noted that community forestry within the Afromontane landscape of the Bamenda Highlands has developed as a partnership between local communities—mainly interested in the different benefits derived from the forest, and the conservation community—whose major interest is in the conservation of biological diversity. The case of community forestry in the Afromontane landscape of the Bamenda Highlands is peculiar because unlike other forested regions of Cameroon where communities derive enormous benefits including income generation from community forests through timber exploitation, in the Bamenda Highlands it is rather a partnership between local populations and conservation agencies. Examples of these successful partnerships include the Kilum-Ijim Forest Project (KIFP), which was later scaled up to the Bamenda Highlands Forestry Project (BHFP) under the tutelage of the Ministry of Forestry and Wildlife (MINFOF) and the Ministry of Environment, Protection of Nature and Sustainable Development (MINEPDED). The author noted that community forestry within the Bamenda Highlands is key to ensuring the sustainability of the forests, and the success of the community forest initiative is largely due to the genuine interest expressed by the community in managing and conserving their forests. In a study on rare plants on Mount Oku found in the Bamenda Highlands, Maisels et al. (2000) uncovered that the forests within Mount Oku constitute the largest remaining patch of montane forest in West Africa, with the highest altitude. Within this ecoregion lies a small *Sphagnum* community with associated wetland plant species located at the summit of Mount Oku, which is 2900 m. This *Sphagnum* bog is extremely important for conservation as it has one of the highest concentrations of endemic plants in the ecoregion, as it is the source of the highest stream in West Africa. A study on the extirpation of large mammals on the Kilum-Ijim montane forest landscape highlighted the serious adverse consequences of this extirpation on montane forest conservation. Of the 77 mammal species recorded within the landscape, most are small mammals such as rodents, insectivores, and bats. Mega-fauna such as elephant *Loxodonta africana*, buffalo *Syncerus caffer*, and other large mammals have been extirpated. The current large mammal population density is very low, with many species close to extinction at the local level. The long-term consequences of these extinctions are enormous as most of these species are involved in the dispersal of different tree species—especially monkeys and ruminants. Maley and Brenac (1997) in a study on vegetation dynamics, paleoenvironments, and climatic changes in the forests of western Cameroon found that the landscape has been warming tremendously in recent years to the degradation and disappearance of some species and colonization by other species.

In studies conducted around Mount Oku and Mount Mbam, Cuni-Sanchez (2018) found that both forests are important habitats for several endangered species, such as the Bannerman's Turaco (a colorful bird whose feathers are highly valued for cultural purposes), the Oku sardine (a rodent, *Lamottemys okuensis*), and the Lake Oku Clawed Frog (*Xenopus longipes*). The most pressing threats identified are wildfires (triggered by increasing temperatures and prolonged dry spells) and the increasing number of goats and cattle that graze inside the forest during the dry season, damaging tree seedlings. In a study conducted within the surroundings of the Mbi Crater in the Bamenda Highlands of Cameroon, Lézine et al. (2020) found that climate variability particularly increasing warming of the ecoregion is a major cause of the loss of biodiversity. Mbeng and Buba (2017) in a study carried out to understand to analyze the forms of indigenous exploitations, the current mode of community management and its effects on biodiversity conservation in the Santchou reserve found in the Western Highlands region noted that there are widespread conflicts between the administration in general and the indigenous population or local communities. These tensions undermine the sustainability of the reserve, leading to a rapid loss of biodiversity as poaching persists as a means of survival. Limited sources of income worsen the scenario as the community is left with no option than to get into the reserve. In a study highlighting the Savannization of tropical montane cloud forests in the Bamenda Highlands of Cameroon, Ndenecho (2005) uncovers that anthropogenic and pyrogenic factors have contributed to the origin and maintenance of the grasslands found today in the Bamenda Highlands. The study further shows that the complex mosaic of grass savannah, montane woodlands, fallow fields, farms, and tree/shrub savannah found today in the Bamenda Highlands is attributable to poor biological resource management with severe adverse implications on biodiversity. These pyrogenic factors are whipped up by increasing temperatures, making it easier for wildfires to spread across vast swaths of the landscape. The study calls for swift action to rehabilitate, sustainably manage, conserve, and monitor the Afromontane landscape in the region. Thomas et al. (2000) in a study on community conservation in the Kilum-Ijim Afromontane landscape found that the landscape is critical to ensuring the long-term survival of two endemic and threatened species of birds, which are the Bannerman's Turaco *Tauraco bannennani* and Banded Wattle-eye *Platytiera laticincta*, whose habitat is threatened by anthropogenic pressures and wildfires driven by increasing temperatures and drier weather conditions. WHINCONET et al. (2007) in a study on the plant and animal species in the Western Highlands of Cameroon revealed that this landscape is blessed with a diversity of plants, birds, mammals, reptiles, and amphibians. Most of these species are endemic to the region, making it a priority area for conservation programs.

In a study on the floristic composition and functional diversity of plant communities of Mount Oku forests in Cameroon, Solefack et al. (2018) found that humans have altered the functional composition of vegetation by rearing domestic animals in the understory forest near the summit. The results also show that it is sometimes difficult to separate the effect of the altitude from the effect of human presence on the plant composition. Greater efforts of biodiversity conservation in high species-rich areas in Mount Oku should be made. A study carried out on Avifauna in Mount Mbam by Ngute et al. (2019) revealed that 225 bird species have been recorded in the area including three globally threatened species: White-headed vulture *Trigonoceps occipitalis* (critically endangered), Bannerman's Turaco *Tauraco bannermani* (endangered), and Bannerman's Weaver *Ploceus bannermani* (vulnerable), with three that are near threatened (Bateleur *Terathopius ecaudatus*, Crowned Eagle *Stephanoaetus coronatus*, and Bangwa Forest Warbler *Bradypterus bangwaensis*). Twelve species are endemic to the Cameroon Mountains and 26 are restricted to the

Afrotropical Highlands (Afromontane) biome. This area is therefore a birds' paradise and deserves to be conserved and protected especially as anthropogenic activities such as agriculture and grazing are driving its degradation as well as wildfires, which are becoming more recurrent owing to global warming and climate change. Nyugah (2021) in a study examining the conservation status and threat analysis of the endangered Bannerman's Tauraco in the Mount Mbam Forest found that the area is an unprotected important bird area (IBA) home to over 220 bird species including three globally threatened species, which are the critically endangered White-headed vulture (*Trigonoceps occipitalis*), the endangered Bannerman's Turaco (*Tauraco bannermani*), and the vulnerable Bannerman's Weaver (*Ploceus bannermani*). Since the ecoregion is neither legally recognized nor has a protection status, it is in grave danger of losing its highly coveted biodiversity owing to habitat loss driven by forest encroachment for farming, extensive cattle grazing, wildfires and fuelwood/lumbering. Coupled to this is the pressing global threat posed by climate change, particularly increasing warming and drier conditions, which are worsening the already precarious situation of wildfires in the ecoregion. Ewane et al. (2021) in studies conducted in the Mount Bamboutos landscape found that the natural vegetation of the ecoregion made up of grassland, primary forest, and secondary forest has decreased tremendously and replaced by farmlands, tea plantations, eucalyptus plantations, and expanded built-up areas. Agriculture, settlement, and grazing are identified as the key drivers of the land use/land cover changes experienced on Mount Bamboutos and this can be explained by a sharp fall in resources such as fertile soils and water caused by climate change and global warming, which is pushing the population deeper into the mountain environment. Similar findings have been reported by Bhadouria et al. (2023) in the soils of tropical mountain landscapes.

In and around the Mount Cameroon landscape, studies have shown that the landscape is vital for sustainability with increasing warming, posing a major threat to this sustainability. Tata and Lambi (2014) in a study carried out around the Mount Cameroon National Park (MCNP) noted that the park is endowed with a variety of wildlife including elephants, chimpanzees, and monkeys, which need to be protected alongside their habitat, i.e., the park. Common flora species within the park are *Plectranthus insignis*, *Schefflera abyssinica*, *Syzygium staudtii*, *Prunus africana*, *Rapanea melanophlaeos*, *Canthium dunlapii*, *Nuxia congesta*, *S. mannii*, *Hypericum lanceolatum*, *Clausena anisata*, *Ficus* spp., *Agauria salicifolia*, *Pavetta hookeriana*, *Myrica arborea*, *Allophylus bullatus*, *Lasiosiphon glaucus*, *Ilex mitis*, *Mimulopsis solmsii*, *Acanthopale decempedalis*, *Oreacanthus mannii*, and *Brillantaisia* spp. The park has two endemic species of chameleons, an endemic species of squirrel, an endemic species of lizard, three endemic species of butterflies, with 20 of the 28 bird species endemic to mountains. High demand for bushmeat by the surrounding urban markets has led to high rates of poaching, which has contributed to reduce the numbers of wildlife species, particularly mammals—both small and large. Wildfires triggered driven by increasing temperatures and drier weather conditions have degraded vast swaths of the landscape, disturbing the natural habitat of plants and wildlife. Birdlife International has classified the Mount Cameroon landscape as an "endemic bird area" owing to the high level of endemism among the region's avifauna. Besides forest clearance for agriculture, geo-ecological challenges, and illegal wildlife trade, which are major threats to the park, wildfires driven by higher temperatures have recently become the main threat to the sustainability of the park. Cheek et al. (1996) in a study undertaken within the Mount Cameroon Area found that Mount Cameroon is one of Africa's centers of plant diversity where there are 50 endemic species and three genera, with 50 other species categorized as near-endemic. The study notes that pressure from agriculture and logging poses a major threat to the sustainability of the landscape. MINFOF (2014) in its plan for the management of the Mount Cameroon National Park has taken different measures to ensure the sustainability of the park with one of the most important being ecotourism. This is built on the enormous ecotourism potentials of the park including the following: (1) the rich and diverse endemic flora and wildlife species including threatened animals fauna such as forest mountain elephants, chimpanzee, *Francolinus camerunensis*, *Picathartes oreas*, drill, and other endangered apes and birds; (2) the highest peak (4100 m) of West and Central Africa; (3) a highly diverse and beautiful landscapes; (4) diverse cultures and cultural sites (traditional dances, Bimbia old slave village, etc.); (5) picturesque nature sites such as old larva flows, crater lakes, craters, lakes, caves, and waterfalls; (6) availability of some high standard touristic infrastructures (hotels, restaurants, etc.); (7) proximity to the beautiful and unique volcanic sandy coastal beaches at the foot of Mt Cameroon; (8) local tour operators; (9) the presence of a botanical and zoological garden in Limbe; (10) a continuous gradient of vegetation that stretches from the lowland forest at sea level, through submontane and montane forests to alpine savanna at its peak; and (11) proximity to international airport and road network. Ecotourism and other strengths such as the existence of a proper institutional and legal framework for the park are sure fire strategies to ensure the sustainability of the park.

Betti et al. (2011) in a study conducted in the Mount Cameroon Forest found that the sustainability of the mountain's vegetal resources, particularly its stock of *Prunus africana*, is dwindling rapidly owing to overexploitation. However, the highest density of *Prunus* trees continuous to be recorded inside the national park of Mount Cameroon, which is at least 1.5 times higher than the density of trees recorded outside the park that shows the important role played by Mount Cameroon National Park in the conservation of *P. africana*. Lawrence et al. (2000) in a study exploring local values that enhance biodiversity conservation on Mount Cameroon found that decision-making is the key to ensuring the sustainability of the mountain's vegetal and wildlife resources. With decision-making still in the hands of politicians and scientists (top–bottom approach), ensuring sustainability is difficult as communities are most often not involved in the decision-making process. Although there is a general trend toward more participatory approaches, local people most of the time are scarcely consulted when decisions affecting them and conservation priorities are being made. Ntoko and Schmidt (2021) in a study conducted in the Mount Cameroon area on the role played by indigenous knowledge in biodiversity conservation uncovered that the current loss of biodiversity in and around the ecoregion is largely due to the erosion and weakening of indigenous or local knowledge systems. Through totems, sacred sites, libations, traditional ceremonies, and taboos, local institutions have always played a major role toward ensuring the sustainability of forest resources within the Mount Cameroon area. A study on the zonation of vegetation and soils on Mount Cameroon by Proctor et al. (2007) revealed that the ecoregion is very rich in flora diversity, with Schefflera species dominating, and that the main type of soils are andisols. However, the effects of wildfires triggered by increasing warming and drier conditions have degraded the landscape threatening the flora diversity and sustainability. Sedláček et al. (2013) in a study on bird diversity on Mount Cameroon found a diversity of bird species including Waller's Starling *Onychognathus walleri*, Western Mountain Greenbul *Andropadus tephrolaemus*, Northern Double-collared Sunbird *Cinnyris reichenowi*, Yellow-breasted Boubou *Laniarius atroflavus* and Gray Apalis *Apalis cinerea*, and the African Thrush *Turdus pelios*. The foregoing is a clear testament to the fact that the Mount Cameroon Afromontane landscape is very rich in flora and fauna species although highly threatened by wildfires in particular driven by the increasing warming of the landscape.

Besides the Western Highlands (Bamenda Highlands in particular) and the Mount Cameroon Afromontane landscape where numerous studies have been carried out showing the importance of these landscapes for sustainability, studies have also been carried out in other parts of Cameroon's Afromontane landscape—although few. Betti and Ambara (2013) in a study conducted on Tchabal Mbabo and Tchabal Gang Daba Mountain Forests in Cameroon found that a major vegetal product of the ecoregion—*Prunus africana*—is being threatened by a high rate of debarking, which threatens the sustainability of the species in the future. The study noted that the average sustainable mass of pygeum tree in Adamaoua region will be about 69.3 kg of fresh bark per tree but this average is always surpassed as harvesters always seek to make more financial benefits with little concern for the sustainability of the species. Cunningham and Mbekum (1993) in a study carried out on the sustainability of *Prunus africana* harvesting in Cameroon found that felling of trees and tree die-offs are recurrent in sites of high conservation priority despite attempts at ensuring the sustainable harvesting of barks of wild populations. This is largely the case in the Afromontane landscapes. The need for cultivation of the species to ensure and alternative source of supply and practical actions to promote the sustainable debarking of *Prunus africana* are key recommendations of the study. Ingram et al. (2009) in a study on *Prunus africana* in Cameroon uncovered that the Pygeum or Prunus tree largely grows in Afromontane landscapes and is mostly used for medicine, timber, and fuelwood. Debarking is a common practice and the barks are sold locally and abroad for medicinal use—as it serves as a raw material in drugs used to treat prostate problems and other health supplements. Sales of its bark is a major income source for forest-dependent communities in Cameroon's Afromontane landscape. The tree represents a key species in the Afromontane landscape and contributes enormously to biological diversity in the Afromontane ecosystem. It is an endangered species and with the increasing environmental changes occurring within the Afromontane landscape, particularly increasing warming, there is an absolute need for adequate measures geared toward ensuring the sustainability of the species. Cheek et al. (2021) in a study conducted on the Montane forests of the Cameroon

Highlands found that there are over 28 tree species endemic to these highlands with the tree species previously known as *Deinbollia* sp. 2. Uncovered as a new species for science formally characterized and named as *Deinbollia onanae* (Sapindaceae-Litchi clade). The species is assessed as endangered following the IUCN 2012 standard and the main threats to the species are Afromontane forest fragmentation driven by agriculture and wildfires. With the increasing warming of the Afromontane landscape making vital resources such as water, fertile soils, and pasture scarce, the population is rapidly colonizing the landscape, thereby threatening the sustainability of the species. Bussmann (2004) in a study carried out across tropical mountain forests in the world—Cameroon inclusive—found that the impact of anthropogenic influences, e.g., slash and burn, timber extraction, exploitation of forest products, the introduction of exotic species, and wildfires on the one hand, as well as climate change and global warming, on the other hand, are key drivers influencing the sustainability of Afromontane forest ecosystems. In a study on endemic chameleons in the Cameroon Highlands (particularly around Mount Oku, Mount Mbam, and Tchabal Mbabo), Gonwouo et al. (2006) uncovered between May 2003 and December 2005 different species of chameleons including *Chamaeleo montium*, *C. pfefferi*, *C. w. wiedersheimi*, *C. pfefferi*, *C. q. quadricornis*, *C. q. gracilior*, *C. montium*, *C. w. wiedersheimi*, and *C. wiedersheimi perreti*, with the most common subspecies being *C. w. wiedersheimi* and the rarest species being *C. pfefferi.* Larson and Pulhin (2012) in a study on land tenure reforms on forests in general and Afromontane forests in particular in Cameroon, it was uncovered that forest tenure reforms have offered new opportunities for communities to obtain formal rights to forests and forest benefits, but at the same time, a variety of limitations are placed on livelihood options. It identifies three types of regulations, namely rules that delineate conservation areas and impose related limits on use; rules that limit areas available to local communities; and bureaucratic requirements for permits and management plans, which restrict the commercial use and marketing of valuable forest products. These regulations are key to ensuring the sustainability of Afromontane forest resources in Cameroon. From the foregoing, it is seen that Cameroon's Afromontane landscape has a major role to play in ensuring the sustainability of endemic and threatened flora and fauna. However, this sustainability is threatened by the increasing warming of the landscape, which calls for swift actions and appropriate policies to help communities adapt to the environmental changes in a bid to reduce pressure on the Afromontane landscape.

6 Conclusions: Avenues for Further Empirical Research and Recommendations

Cameroon's Afromontane landscape is one of the most endowed biodiversity hotspots in the world, with over 50 threatened and endemic plant and wildlife species. This landscape is, however, under serious threat as it has shrunk drastically from over 80,000 km^2 some decades ago to less than 38,000 km^2 today. Among the different drivers of degradation of this landscape is the increasing warming occurring on the landscape, which is triggering wildfires—further exacerbating the degradation phenomenon. Against this backdrop, this chapter looks at the realities of the warming of Cameroon's Afromontane landscape as well as the livelihood and sustainability implications. Findings indicate that the livelihood options of communities living in and around this landscape are mainly food crop farming, market gardening, fruit tree farming, silviculture (mainly Eucalyptus, Cypress, Pine, Podocarpus), agroforestry, poultry farming, pig farming, goat herding, grazing of cattle and sheep, hunting, harvesting of timber and non-timber forest products, apiculture, and fish farming. Increasing warming of this landscape is adversely affecting the livelihoods of adjacent communities as watersheds are degraded, wildfires are recurrent, pastures are dwindling, and soil infertility is rampant, making the livelihoods of farmers, herders/graziers, market gardeners, and harvesters of non-timber forest products in particular extremely difficult. Increasing warming of Cameroon's Afromontane landscape also has sustainability implications, i.e., in the economic, social, and environmental dimensions.

Socially, with the warming landscape, conflicts have become very recurrent between different actors competing for the limited resources of this landscape. The most common conflicts are between farmers and herders, farmers and farmers, herders and herders, as well as inter-ethnic conflicts. Moreover, scarcity of resources encourages gender discrimination as the male gender lords over the female gender as far as controlling Afromontane resources are concerned. Economically, with the warming of the landscape, local communities living adjacent are impoverished as the potential of income generation from the resources of the Afromontane landscape reduces as degradation induced by increasing warming goes up. The need to buy fertilizers to improve soil fertility, buy feed to feed livestock, buy pesticides to control increasing pests and diseases, as well as spend more money on drugs as malaria becomes the norm, which further impoverishes local communities pushing them to degrade the landscape even more. Environmentally, the excessive warming of the landscape has led to watershed degradation, soil infertility, loss of different species of plants and wildlife, as well as recurrent wildfires exacerbating degradation, making it difficult for this biodiversity-rich landscape to provide adequate ecosystem services to sustain the numerous local communities living in and around it. From the foregoing, it is evident that there is an increasing warming trend in Cameroon's Afromontane landscape, which has major adverse implications for livelihoods and the sustainability of the landscape.

The study uncovers avenues for further empirical research as there are several Afromontane landscapes/ecoregions in Cameroon that require more attention to understand the warming trend and its socioeconomic and environmental implications. This study found that just some of these Afromontane landscapes/ecoregions have received due attention through research and conservation projects, particularly the Bamenda Highlands, Mount Bamboutos, and Mount Cameroon. More empirical research should be conducted on the other Afromontane landscapes such as Mount Nlonako, Tchabal Mbabo, the Bakossi Highlands, and the Lebialem Highlands to better understand the realities of warming and the socioeconomic and environmental implications on the landscape.

It is therefore recommended that adequate measures should be taken to restore the landscape through afforestation and reforestation campaigns, as well as encouraging the adoption of best practices such as agroforestry, which will go a long way to improve livelihoods and ensure sustainability. Nature-based solutions (NbS) and ecosystem-based adaptation (EbA) are highly recommended. NbS can potentially contribute over one-third of the total climate change mitigation needed by 2030 to keep global warming to just below 2 °C. NbS is inspired and supported by nature and use, or imitates natural processes to contribute to the improved management of water (UN-Water, 2018). NbS can involve conserving or rehabilitating natural ecosystems and/or the enhancement or creation of natural processes in modified or artificial ecosystems such as watersheds. Within nature-based solutions, the restoration of watersheds is a key element. This could involve action to better manage approximately 2.5 billion hectares of forest, crop, and grazing land (restoration and avoided degradation), as well as restoring over 230 million hectares of natural cover (Griscom et al., 2019). An important subset of NbS is EbA, which helps people adapt to climate change. EbA is a nature-based approach that uses biodiversity and ecosystem services to help people adapt to the adverse effects of climate change (International Institute for Environment and Development-IIED, 2019). It involves activities such as water catchment restoration by tree planting, slope stabilization by tree planting to prevent landslides, applying integrated water resource management to address water shortages, and managing forests sustainably to prevent erosion and regulate water flow. EbA rests on combining local knowledge with evolving information on climate change and has been applied to address the local challenges of climate change impacts.

Acknowledgments We sincerely acknowledge material support from IDEA WILD through its equipment grant to the corresponding author. Many thanks to Sylvanus Fonbah Tah for liaising with the IDEA WILD team to bring the equipment to Cameroon. Morale, logistics, and financial support from the FOKABS and the African Food Security and Climate Army (AFSACA) is highly appreciated. We also sincerely thank the reviewers and editors of this chapter for their helpful comments and suggestions, which enabled us to improve the quality of the chapter.

Conflict of Interest Authors declare that there are no conflicts of interest.

References

Abbot, J. I. O., Thomas, D. H. L., Gardner, A. A., Neba, S. E., & Khen, M. W. (2001). Understanding the links between conservation and development in the Bamenda Highlands, Cameroon. *World Development, 29*(7), 1115–1136. https://doi.org/10.1016/S0305-750X(01)00033-X

Abiem, I., Arellano, G., Kenfack, D., & Chapman, H. (2020). Afromontane forest diversity and the role of grassland-forest transition in tree species distribution. *Diversity, 12*, 30. https://doi.org/10.3390/d12010030

Ako, A. A., Zita, M. M., & Fongoh, E. J. (2022). Assessment of climate variability and its impact on the environment and water resources in the Mount Cameroon Area. In H. Chenchouni et al. (Eds.), *New prospects in environmental geosciences and hydrogeosciences. CAJG 2019. Advances in Science, Technology & Innovation*. Springer. https://doi.org/10.1007/978-3-030-72543-3_91

Amawa, A. S., Kimengsi, J. N., Sunjo, T. E., & Awambeng, A. E. (2015). The implications of climate variability on market gardening in Santa Sub-Division, North-West Region of Cameroon. *Environment and Natural Resources Research, 5*(2), 14–23.

Asanga, C. A. (2001). Facilitating viable partnerships in community forest management in Cameroon: The case of Kilum-Ijim Mountain Forest Area. In E. Wollenberg, D. Edmunds, L. Buck, J. Fox, & S. Brodt (Eds.), *Social learning in community forests* (pp. 21–44). CIFOR and the East West Center.

Awazi, N. P. (2018). *Adaptation options enhancing farmers' resilience to climate change*. LAP LAMBERT Academic Publishing. 132 p. ISBN-10: 3330027940; ISBN-13: 978-3330027947.

Awazi, N. P. (2020). *Vulnerability attenuation and resilience enhancement to climate variability and change through agroforestry practices in smallholder farming systems in the north-west region of Cameroon* [PhD thesis, Agronomy. Université de Dschang, Cameroun], 147p.

Awazi, N. P. (2022a). Assessing the role of irrigation as an adaptive measure to climate change induced water insecurity: Case study of the market gardening sector in parts of the north-west and west regions of Cameroon. *Frontiers in Water, 4*, 902438. https://doi.org/10.3389/frwa.2022.902438

Awazi, N. P. (2022b). Achieving agricultural sustainability in the midst of climate change through agroforestry: Anecdotal evidence from smallholder farmers and key agricultural stakeholders in rural Cameroon. *Forestist, 73*, 1–3. https://doi.org/10.5152/forestist.2022.22045. https://forestist.org/en/achieving-agricultural-sustainability-in-the-midst-of-climate-change-through-agroforestry-anecdotal-evidence-from-smallholder-farmers-and-key-agricultural-stakeholders-in-rural-cameroon-162616

Awazi, N. P., & Avana, T. M. L. (2020). Agroforestry as a sustainable means to farmer-grazier conflict mitigation in Cameroon. *Agroforestry Systems, 94*(6), 2147–2165. https://doi.org/10.1007/s10457-020-00537-y(0123456789

Awazi, N. P., & Tchamba, N. M. (2018). Determinants of small-scale farmers' adaptation decision to climate variability and change in the North-West Region of Cameroon. *African Journal of Agricultural Research, 13*(12), 534–543.

Awazi, N. P., & Tume, S. J. P. (2023). Impacts, vulnerability, resilience and adaptation to climate variability and change in rural communities: A review of cases in the North West Region of Cameroon. In A. Alam (Ed.), *Disaster risk, resilient agriculture, and livelihood: Methods and applications*. Taylor and Francis Routledge.

Awazi, N. P., Tchamba, N. M., & Avana, T. M. L. (2019a). Climate change resiliency choices of small-scale farmers in Cameroon: Determinants and policy implications. *Journal of Environmental Management, 250*, 109560. https://doi.org/10.1016/j.jenvman.2019.109560

Awazi, N. P., Tchamba, N. M., & Tabi, F. O. (2019b). An assessment of adaptation options enhancing smallholder farmers' resilience to climate variability and change: Case of Mbengwi Central sub-Division, North-West Region of Cameroon. *African Journal of Agricultural Research, 14*(6), 321–334.

Awazi, N. P., Tchamba, N. M., & Temgoua, L. F. (2020a). Enhancement of resilience to climate variability and change through agroforestry practices in smallholder farming systems in Cameroon. *Agroforestry Systems, 94*(3), 687–705. https://doi.org/10.1007/s10457-019-00435-y

Awazi, N. P., Tchamba, N. M., & Temgoua, L. F. (2020b). Climate-smart practices of smallholder farmers in Cameroon faced with climate variability and change: The example of agroforestry. *Agricultural Research, 10*(1), 83–96. https://doi.org/10.1007/s40003-020-00477-0

Awazi, N. P., Tchamba, N. M., Temgoua, L. F., & Avana, T. M. L. (2020c). Appraisal of smallholder farmers' vulnerability to climatic variations and changes in Cameroon. *Scientific African, 10*, e00637. https://doi.org/10.1016/j.sciaf.2020.e00637

Awazi, N. P., Tchamba, M. N., Temgoua, L. F., Avana, T. M.-L., Shidiki, A. A., Forje, G. W., & Nfornkah, B. N. (2021a). Climate-smart and agroecological farming systems of smallholder farmers. In C. M. Galanakis (Ed.), *Environment and climate-smart food production*. Springer Nature Switzerland AG. https://doi.org/10.1007/978-3-030-71571-7_2

Awazi, N. P., Tchamba, N. M., Temgoua, L. F., & Avana, T. M. L. (2021b). Agroforestry as an adaptation option to climate change in Cameroon: Assessing farmers' preferences. *Agricultural Research, 11*, 309–320. https://doi.org/10.1007/s40003-021-00562-y

Awazi, N. P., Temgoua, L. F., Tientcheu-Avana, M.-L., & Tchamba, M. N. (2022a). Reducing vulnerability to climate change through agroforestry: Case study of small-scale farmers in the northwest region of Cameroon. *Forestist, 73*, 11–20. https://doi.org/10.5152/forestist.2022.22031. https://forestist.org/en/reducing-vulnerability-to-climate-change-through-agroforestry-case-study-of-small-scale-farmers-in-the-northwest-region-of-cameroon-162612

Awazi, N. P., Temgoua, L. F., Tientcheu-Avana, M.-L., Chimi, D. C., & Tchamba, M. N. (2022b). Climate change and peasant farmers' adoption of multifunctional agroforestry systems in Cameroon: Determinants and policy ramifications. *Forestist, 73*, 117–130. https://doi.org/10.5152/forestist.2022.22023. https://forestist.org/en/climate-change-and-peasant-farmers-adoption-of-multifunctional-agroforestry-systems-in-cameroon-determinants-and-policy-ramifications-162611

Azong, M. N. (2021). Impact of cultural beliefs on smallholders' response to climate change: The case of Bamenda Highlands, Cameroon. *International Journal of Environmental Studies, 78*(4), 663–678. https://doi.org/10.1080/00207233.2020.1824866

Betti, J. L., & Ambara, J. (2013). Mass of *Prunus africana* stem barks on Tchabal Mbabo and Tchabal Gang Daba Mountain Forests, Cameroon. *African Journal of Environmental Science and Technology, 7*(5), 204–221. https://www.ajol.info/index.php/ajest/article/view/93778

Betti, J. L., Belinga, S. J., & Samba, D. (2011). Stock of *Prunus africana* stems on the mount Cameroon forest. *African Journal of Plant Science, 5*(13), 781–789. https://academicjournals.org/journal/AJPS/article-full-text-pdf/E20A7A311161

Bhadouria, R., Singh, S., Tripathi, S., & Singh, P. (Eds.). (2023). *Understanding soils of mountainous landscapes: Sustainable use of soil ecosystem services and management*. Elsevier.

BirdLife International. (2003). *Mid-term review of the community-based conservation in the Bamenda Highlands Project*. https://www.conservationevidence.com/individual-study/155

Buchenrieder, G., & Balgah, R. A. (2013). Sustaining livelihoods around community forests. What is the potential contribution of wildlife domestication? *The Journal of Modern African Studies, 51*(1), 57–84. https://doi.org/10.1017/S0022278X12000596

Bussmann, R. W. (2004). Regeneration and succession patterns in African, Andean and Pacific Tropical Mountain Forests: The role of natural and anthropogenic disturbance. *Lyonia, 6*(1), 93–111.

Cable, S., & Cheek, M. (2000). *The plants of Mount Cameroon: A conservation checklist* (p. 277p). Royal Botanical Gardens Kew.

Cheek, M., & Onana, J.-M. (2000). *The plants of Mount Oku and the Ijim Ridge, Cameroon: A conservation checklist*. Royal Botanical Gardens Kew. 215p.

Cheek, M., Cable, S., Hepper, F. N., Ndam, N., & Watts, J. (1996). Mapping plant biodiversity on Mount Cameroon. In L. J. G. van der Maesen, X. M. van der Burgt, & J. M. van Medenbach de Rooy (Eds.), *The biodiversity of African plants*. Springer. https://doi.org/10.1007/978-94-009-0285-5_16

Cheek, M., Pollard, B. J., Darbyshire, I., & Onana, J.-M. (2004). *Plants of Kupe, Mwanenguba and the Bakossi Mountains, Cameroon: A conservation checklist*. Royal Botanical Gardens Kew. 508p.

Cheek, M., Onana, J. M., & Chapman, H. M. (2021). The montane trees of the Cameroon Highlands, West Central Africa, with *Deinbollia onanae* sp. nov. (Sapindaceae), a new primate-dispersed, Endangered species. *Peer J, 9*, e11036. https://doi.org/10.7717/peerj.11036

CIFOR-ICRAF (2023). A meeting for the mountains: High-altitude innovation for biodiversity and livelihoods. https://www.ciforicraf.org/news/corporate-news/a-meeting-for-the-mountains-high-altitude-innovation-for-biodiversity-and-livelihoods/

Cuni-Sanchez, A. (2018, July 11). *New insights from the Montane Forests of Western Cameroon*. Mountain Sentinels. https://mountainsentinels.org/new-insights-montane-forests-western-cameroon/

Cuni-Sanchez, A., Ngute, A. S. K., Sonké, B., Sainge, M. N., Burgess, N. D., Klein, J. A., & Marchant, R. (2019). The importance of livelihood strategy and ethnicity in forest ecosystem services' perceptions by local communities in north-western Cameroon. *Ecosystem Services, 40*, 101000. https://doi.org/10.1016/j.ecoser.2019.101000

Cunningam, A. B., & Mbenkum, F. T. (1993, May). *Sustainability of harvesting Prunus africana bark in Cameroon: A medicinal plant in international trade* (People and Plants Working Paper No. 2), 28p. https://www.doc-developpement-durable.org/file/Culture/Plantes-Medicinales-Aromatiques/FICHES_PLANTES/Prunus%20africana/Sustainability%20of%20harvesting%20Prunus%20africana%20bark%20in%20Cameroon.pdf

European Commission. (2023). *Cameroon highlands forests*. DOPA Explorer. https://dopa--explorer.jrc.ec.europa.eu/ecoregion/30103

Ewane, B. E., Asabaimbi, D.-N., Njiaghait, Y. M., & Nkembi, L. (2021). Agricultural expansion and land use land cover changes in the Mount Bamboutos landscape, Western Cameroon: Implications for local land use planning and sustainable development. *International Journal of Environmental Studies, 80*, 186–206. https://doi.org/10.1080/00207233.2021.2006911

Fogwe, Z. N., & Kwei, J. (2015). Cameroonian protected Kilium-Ijim forests for the development of Oku forest fringe community. *Journal of Environmental Research and Management, 6*(5), 0293–0303. https://www.e3journals.org/cms/articles/1450419913_Zephania%20and%20Jude.pdf

Foncha, J. N., & Ewule, D. M. (2020). Community forest management: A strategy for rehabilitation, conservation and livelihood sustainability: The case of Mount Oku, Cameroon. *Journal of Geoscience and Environment Protection, 8*, 1–14. https://doi.org/10.4236/gep.2020.82001

Gardner, A. (2002). *Community forestry in the Bamenda Highlands region of Cameroon: A partnership for conservation*. Second International Workshop on Participatory Forestry in Africa. Defining the Way Forward: Sustainable Livelihoods and Sustainable Forest Management Through Participatory Forestry, pp. 153–160. https://www.fao.org/3/Y4807B/Y4807B12.pdf

Gonwouo, L. N., Lebreton, M., Wild, C., Chirio, L., Ngassam, P., & Tchamba, M. N. (2006). Geographic and ecological distribution of the endemic montane chameleons along the Cameroon mountain range. *Salamandra, 42*(4), 213–230. https://www.mosaic.cm/downloads/Gonwouo_chameleons_cameroon.pdf

Griscom, B. W., Lomax, G., Kroeger, T., Fargione, J. E., Adams, J., & Almond, L. (2019). We need both natural and energy solutions to stabilize our climate. *Global Change Biology, 25*, 1889–1890.

Ingram, V., Awono, A., Schure, J., & Ndam, N. (2009). *National Prunus africana Management Plan Cameroon*. CIFOR and GTZ. 154p. https://www.fao.org/forestry/18283-06774e55e0bca4ec099548d6bd66b707b.pdf

International Institute for Environment and Development-IIED. (2019). *Nature-based solutions to climate change adaptation* (pp. 1–4). IIED.

Josiane, F. K., Tsalefac, M., & Kongnso, M. E. (2020). Perception of climate variability along the slopes of Mount Bamboutos, West Cameroon. *International Journal of Social, Political and Economic Research, 7*(3), 687–707. https://doi.org/10.46291/IJOSPERvol7iss3pp687-707

Kimengsi, J. N., Kechia, M. A., Balgah, R. A., Pretzsch, J., & Kwei, J. (2019a). Households' assets dynamics and ecotourism choices in the Western Highlands of Cameroon. *Sustainability, 11*(7), 1844. https://doi.org/10.3390/su11071844

Kimengsi, J. N., Pretzsch, J., Kechia, M. A., & Ongolo, S. (2019b). Measuring livelihood diversification and forest conservation choices: Insights from rural Cameroon. *Forests, 10*(2), 81. https://doi.org/10.3390/f10020081

Kimengsi, J. N., Mukong, A. K., & Balgah, R. A. (2020). Livelihood diversification and household well-being: Insights and policy implications for forest-based communities in Cameroon. *Society & Natural Resources, 33*, 2020. https://doi.org/10.1080/08941920.2020.1769243

Kimengsi, J. N., Awazi, N. P., Nyugap, A. R., Kwalar, G., & Tamnjong, B. (2024). An analysis of power and conviction around land and linked natural resources in the western highlands of Cameroon. *Journal of Environmental Planning and Management, 67*, 85–107. https://doi.org/10.1080/09640568.2022.2099358

Kiming, I. N., Moye, E. K., & Jude, K. (2020). Mountain apiculture and environmental dynamics: Impact of climate variability on bee farming in OKU, Cameroon. *International Journal of Environmental, Sustainability, and Social Science, 1*(3), 1–11. https://doi.org/10.38142/ijesss.v1i1.33

Kongnso, M. E., Buba, U. H., & Nfor, J. T. (2021). Implications of climatic stressors on agro-pastoral resources among Mbororo communities along the slopes of Kilum-Ijim Mountain, North West Region, Cameroon. *Frontiers in Sustainable Food Systems, 5*, 685071. https://doi.org/10.3389/fsufs.2021.685071

Laird, S. A., Awung, G. L., & Lysinge, R. J. (2007). Cocoa farms in the Mount Cameroon region: Biological and cultural diversity in local livelihoods. *Biodiversity and Conservation, 16*, 2401–2427. https://doi.org/10.1007/s10531-007-9188-0

Larson, A. M., & Pulhin, J. M. (2012). Enhancing forest tenure reforms through more responsive regulations. *Conservation and Society, 10*(2), 103–113. https://www.jstor.org/stable/26393068

Lawrence, A., Ambrose-Oji, B., Lysinge, R., & Tako, C. (2000). Exploring local values for forest biodiversity on Mount Cameroon. *Mountain Research and Development, 20*(2), 112–115. https://doi.org/10.1659/0276-4741(2000)020[0112:ELVFFB]2.0.CO;2

Lézine, A.-M., Izumi, K., & Achoundong, G. (2020). Mbi Crater (Cameroon) illustrates the relations between mountain and lowland forests over the past 15,000 years in western equatorial Africa. *Quaternary International, 657*, 67–76. https://doi.org/10.1016/j.quaint.2020.12.014

Macchi, M., & ICIMOD. (2010). *Mountains of the world—Ecosystem services in a time of global and climate change* (21p). ICIMOD.

Maisels, F. G., Cheek, M., & Wild, C. (2000). Rare plants on Mount Oku summit, Cameroon. *Oryx, 34*(2), 136–140. https://doi.org/10.1046/j.1365-3008.2000.00107.x

Maley, J., & Brenac, P. (1997). Vegetation dynamics, palaeoenvironments and climatic changes in the forests of western Cameroon during the last 28,000 years. *Review of Palaeobotany and Palynology, 99*(1998), 157–187.

Martin, E., & Burgess, N. (2023). Cameroon highlands forests. *One Earth*. https://www.oneearth.org/ecoregions/cameroon-highlands-forests/

Mbeng, E., & Buba, U. H. (2017). *Community livelihoods versus biodiversity conservation: Constraints to sustainability of the Santchou forest reserve in the Western highland region of Cameroon*. 6th Symposium for Research in Protected Areas 2 to 3 November 2017, Salzburg, pp. 417–420. http://www.parcs.at/npa/pdf_public/2018/36388_20180524_114308_115_Mbeng_FINAL_4p_pag.pdf

MINFOF. (2014). *The management plan of the Mount Cameroon National Park and its Peripheral Zone: 2015 to 2019*. MINFOF and Partners, 108p. https://documents1.worldbank.org/curated/en/290571468007177234/pdf/E23260V30REPLA00Box391426B00PUBLIC0.pdf

Nchamcham, C. N. (2010). *Structural distribution of wild Prunus africana populations on the Kilum mountain forest* [Masters of Science Research findings University of Yaounde I, Cameroon], 55p.

Ndenecho, E. N. (2005). Savannization of tropical Montane cloud forests in the Bamenda Highlands, Cameroon. *Journal of the Cameroon Academy of Sciences, 5*(1), 3–10.

Ndoh, I. M., Bitondo, D., & Azibo, B. R. (2016). Climate variability and change in the Bamenda Highlands of The North-West Region of Cameroon: Perceptions, impacts and coping mechanisms. *British Journal of Applied Sciences and Technology, 12*(5), 1–18.

Ngimdoh, M. M., Tchekote, H., & Achamoh, V. N. (2020). Exploratory study on agricultural practices on the Bamboutos Mountains. *International Journal of Innovative Science and Research Technology, 5*(10), 1–17.

Ngute, A. S. K., Hulme, M., Whytock, R. C., Awa, I. I. T., & Mayaka, T. B. (2019). The avifauna of Mount Mbam, Cameroon. *Bulletin of the African Bird Club, 26*(1), 82–91.

Ngwa, N. S. E., & Fonjong, L. N. (2002). Actors, options and the challenges of forest management in anglophone Cameroon. *GeoJournal, 57*, 95–111. https://doi.org/10.1023/A:1026042621101

Njoh, G. E., & Wanie, C. M. (2018). Spatial distribution and species abundance area of Non-Timber Forest Products in the Mount Cameroon National Park and adjoining forest zones. *International Journal of Forest, Animal and Fisheries Research (IJFAF), 2*(4), 113–122.

Nkembi, L., Deh, N. H., Tankou, C. M., & Nkengafac, N. J. (2021). Analysis of small-scale farmers households food security in the Mount Bamboutos ecosystem. *Journal of Food Security, 9*(2), 56–61. https://doi.org/10.12691/jfs-9-2-3

Nkengafac, N. J., Nkembi, L., Forghab, N. E., & Deh, N. A. D. (2021). Agroforestry practices and food security in the mount Bamboutos landscape, Western Highlands of Cameroon. *World Journal of Advanced Research and Reviews, 11*(01), 203–211. https://doi.org/10.30574/wjarr.2021.11.1.0311

Ntoko, V. N. (2020). *Climate change in the Mount Cameroon National Park region: Local perceptions, natural resources and adaptation strategies, the Republic of Cameroon* [PhD thesis, Institute of Geography, University of Augsburg Germany], 301pp. https://opus.bibliothek.uni-augsburg.de/opus4/frontdoor/deliver/index/docId/79479/file/Ntoko_Diss.pdf

Ntoko, V. N., & Schmidt, M. (2021). Indigenous knowledge systems and biodiversity conservation on Mount Cameroon. *Forests, Trees and Livelihoods, 30*(4), 227–241. https://doi.org/10.1080/14728028.2021.1980117

Nyugah, D. (2021). *Conservation status and threat analysis of the endangered Bannerman's Tauraco in the Mount Mbam Forest, Cameroon*. African Bird Club, 14p. https://www.africanbirdclub.org/sites/default/files/2020%20-%20Bannermans%20Turaco%20-%20Cameroon%20-%20Lefor%20Denis.pdf

Oyono, P. R., Biyong, M. B., & Samba, S. K. (2012). Beyond the decade of policy and community euphoria: The state of livelihoods under new local rights to forest in rural Cameroon. *Conservation and Society, 10*(2), 173–181.

Price, M. F., Gratzer, G., Duguma, L. A., Kohler, T., Maselli, D., & Rosalaura, R. (Eds.). (2011). *Mountain forests in a changing world – Realizing values, addressing challenges*. FAO/MPS and SDC. https://boris.unibe.ch/6755/1/FAO_Mountain-Forests-in-a-Changing-World.pdf

Proctor, J., Edwards, I. D., Payton, R. W., & Nagy, L. (2007). Zonation of forest vegetation and soils of Mount Cameroon, West Africa. *Plant Ecology, 192*(2), 251–269. https://www.jstor.org/stable/40212965

Purvis, B. (2023). *Working with mountain Communities towards a brighter future for all*. ICRAF. 2p. https://www.worldagroforestry.org/blog/2023/03/09/working-mountain-communities-towards-brighter-future-all

Reif, J., Sedlácek, O., Horák, D., Riegert, J., Pešata, M., Hrázský, Z., & Janecek, Š. (2007). Habitat preferences of birds in a montane forest mosaic in the Bamenda Highlands, Cameroon. *Ostrich, 78*(1), 31–36.

Sedláček, O., Reif, J., Hořák, D., Riegert, J., Pešat, M., & Klvaňa, P. (2007). The birds of a montane forest mosaic in Big Babanki area, Bamenda Highlands, Cameroon. *Malimbus, 29*, 89–100.

Sedláček, O., Vokurková, J., Ferenc, M., Djomo, E. N., Albrecht, T., & Hořák, D. (2013). A comparison of point counts with a new acoustic sampling method: A case study of a bird community from the montane forests of Mount Cameroon. *Ostrich: Journal of African Ornithology, 86*(3), 1–8. https://doi.org/10.2989/00306525.2015.1049669

Solefack, M. C. M., Fedoung, E. F., & Temgoua, L. F. (2018). Factors determining floristic composition and functional diversity of plant communities of Mount Oku forests, Cameroon. *Journal of Asia-Pacific Biodiversity, 11*(2), 284–293. https://doi.org/10.1016/j.japb.2018.03.005

Tata, E. S., & Lambi, C. M. (2014). Challenges and opportunities of the Mount Cameroon Forest Region as a national park. *Journal of International Wildlife Law & Policy, 17*(4), 197–212. https://doi.org/10.1080/13880292.2014.957036

Teke, N. A., Kinge, T. R., Bechem, E., Nji, T. M., Ndam, L. M., & Mih, A. M. (2018). Ethnomycological study in the Kilum-Ijim mountain forest, Northwest Region, Cameroon.

Journal of Ethnobiology and Ethnomedicine, 14, 25. (2018). https://doi.org/10.1186/s13002-018-0225-8

Thomas, D. H. L., Anders, S., & Penn, N. J. (2000). Conservation in the community: The Kilum-Ijim Forest Project, Cameroon. *Ostrich, 71*(1 & 2), 157–161.

Tume, S. J. P., & Kimengsi, J. N. (2021). Indigenous and modern agro-based climate adaptation practices in rural Cameroon. *International Journal of Environmental Studies, 79*(2), 1–10. https://doi.org/10.1080/00207233.2021.1977538

UN-Water. (2018). *The United Nations World Water Development Report 2018: Nature-based solutions for water* (pp. 1–25). UNESCO.

WHINCONET., Ingram, V., & Nsom, J. A. (2007). Plant and animal guide for the Western Cameroon Highlands. Nature & Fauna, 22, 29–36.

Xu, J. (2022). *Mountain futures program, world agroforestry centre*. https://www.worldagroforestry.org/blog/2023/03/09/working-mountain-communities-towards-brighter-future-all

Yufenyuy, M., & Nguetsop, V. F. (2020). Climate variability and the emergence of Malaria: Case of Kumbo Central Sub-Division, North West Region, Cameroon. *International Journal of Global Sustainability, 4*(1), 104–127.

Impact of Climate Change on Health and Livelihood of People in the Hindu Kush Himalayan Region

Meghnath Dhimal, Luna Thapa, Sailaja Ghimire, Mandira Lamichhane Dhimal, and Ruth Mueller

1 Introduction

The Hindu Kush Himalaya (HKH) region is a vast mountain range extending over 4.2 million square kilometers across eight countries, including Afghanistan, Bhutan, Bangladesh, China, India, Myanmar and Pakistan (Sharma et al., 2019). It comprises the Hindu Kush, the Karakoram, and the westernmost parts of the Himalayas, and is often referred to as the "Third Pole" of the world. With over 14 peaks higher than 8000 meters and hundreds of peaks exceeding 7000 meters, the HKH region is a significant source of freshwater, serving as the origin of ten major rivers and is the source of livelihood for about 240 million people of the world (Ryan, Cooper and Tauer, 2013). This region has a high level of diversity and disparity, partly due to the high climate variations and rugged geography. It is habitat of around 35,000 plant species and 200 animal species.

The HKH region has the largest ice reserves outside of the polar regions, with 18.23% of its landmass covered in snow, making this region highly vulnerable to climate change. Over the different periods of time, temperature differences have been observed. HKH faced warming from 1901 to 1940, cooling from 1940 to 1970, and again warming from 1970 to the present, with an annual increase in the mean surface air temperature at the rate of 0.10 °C per decade from 1901 to 2014 (Xu

M. Dhimal (✉) · L. Thapa · S. Ghimire
Nepal Health Research Council, Kathmandu, Nepal

M. L. Dhimal
Planetary Health Research Centre, Kathmandu, Nepal

R. Mueller
Institute of Tropical Medicine, Antwerp, Belgium
e-mail: rmuller@itg.be

S. Tripathi et al. (eds.), *Warming Mountains*,
https://doi.org/10.1007/978-3-031-62197-0_8

et al., 2019). Warming in the HKH region is at a much higher rate than the global mean (Wester et al., 2019). Changes have been observed in the number of extreme warm and cold days and nights. On the one hand, occurrences of extreme warm days and nights have increased, while the occurrences of extreme cold days and nights have declined (Wester et al., 2020). There have been tremendous impacts on the glacial lakes due to the elevated temperatures. Higher temperatures cause permafrost and glaciers to decline in much of the HKH region. Climate change in the HKH region has been observed not only in the form of rise in temperature but also in regard to the change in precipitation pattern. The HKH region experiences the extreme weather that is consistent with warmer and wetter climate. The eastern (Himalayas and Trans Himalaya) and the northern (Karakoram and Hindu Kush) part of the Indus Basin showed significantly increasing trends in extreme precipitation, whereas the western part presented the downward trend. In the Tibetan Plateau (TP), which occupies the major HKH region, precipitation has slightly increased since 1960s. In general, TP is overall getting warmer and wetter during the recent periods (Qing-Long et al., 2017). It is estimated that climate change will alter distribution of species as well as cryospheric process over the the HKH region in future (ICIMOD, 2019).

2 Consequences of Climate Change for Human Health in the HKH Region

Climate change has been the biggest global health threat of the twenty-first century as it is affecting the environmental and social determinants of health (Watts et al., 2020). Climate change is expected to cause approximately 250,000 additional deaths per year globally, which are anticipated to be resulted from malnutrition, malaria, diarrhea, and heat stress (WHO, 2021). The extreme weather events have brought climate-sensitive health determinants and outcomes to the HKH region. The most climate-sensitive health risks linked with climate change in the HKH region includes respiratory illness, water-borne diseases, vector-borne diseases, malnutrition, and noncommunicable diseases (NCDs) along with mental and psychosocial health (Dhimal et al., 2021a). It is predicted that the HKH region have higher frequencies of climate-induced disasters with differential impacts among countries and are likely to suffer disproportionately (Dhimal et al., 2024). It may disrupt mostly to the vulnerable population such as children, women, and marginalized people living in the poverty region (Goodrich et al., 2019; Dhimal et al., 2024).

The risk of extreme heat exposure and its consequences is gradually increasing with time. The intense heat waves can have significant impacts on health, causing heat stress and stroke. Heat waves in 2016 killed thousands of people in Andra Pradesh, India (Van Oldenborgh et al., 2018). The climate catastrophe increased the likelihood of the heat wave that killed people and broke records in India and Pakistan by about 30 times (World Economic Forum, 2022).

2.1 *Shift in Vector-Borne Diseases Patterns Due to Climate Change in the HKH Region*

The HKH region is endemic to at least six major vector-borne diseases (VBDs) which are chikungunya fever, dengue fever, malaria, lymphatic filariasis (LF), Japanese encephalitis (JE) and visceral leishmaniasis (VL). These diseases are caused by pathogens transmitted to human beings by vectors, mainly by mosquitoes and sandflies (Dhimal et al., 2021b).

Climate change enables the rapid expansion of the distribution of hosts and vectors and the ability of microbes to colonize new hosts (Gerlitz et al., 2017). The HKH region is already facing changes in temperature, wind, precipitation, and other climatic factors. As a result, changes are observed in the transmission patterns of pathogens along with the geographical distribution of the diseases, contributing the presence of different vector-borne diseases in the areas that were non-endemic previously along with the outbreak of diseases. Examples include the expansion of the dengue virus in different geographical range of the HKH. Prior until 2004, Nepal and Bhutan were free of dengue fever. Since 2004, the DENV and its vectors have spread rapidly, putting the majority of the population at risk. During the first dengue outbreak in 2006 in Nepal, 32 cases were reported from nine districts. During the outbreak of dengue in 2022, more than 54,000 cases and 80 deaths were reported from all ecological regions and districts of Nepal, showing rapid spatial and temporal expansion of dengue in Nepal. Malaria and its vector were progressively reported from the HKH region's hills and mountains between 1980 and 2010 (Dhimal et al., 2021b).

A number of autochthonous VL cases have been documented in new places, particularly in hill and mountain regions of Nepal and Bhutan (Dhimal et al., 2021b). *Culex quinquefasciatus*, the primary vector of LF, has already established its range over 2000 m, and LF cases are increasingly being recorded from mountain regions (Dhimal et al., 2014a; Dhimal, 2015a). The risk of ZIKV has also reported in KKH region (Dhimal et al., 2018). Systematic reviews of literature show the positive association of the climate change to the geographical expansion of the diseases and vectors (Dhimal et al., 2021b; Dhimal et al., 2015b).

2.2 *Shift in Water- and Food-Borne Disease Patterns Due to Climate Change in the HKH Region*

Climate change plays a major role in increasing the incidence of diseases through contaminated water, sanitation, and hygiene. Precipitation can influence the transport and spread of infectious agents, particularly through water and sanitation systems (Cissé, 2019). Surface water runoff can bring pollutants and toxins into water sources during periods of heavy rainfall, raising the risk of water-borne diseases, while floods and standing water provide breeding grounds for disease-carrying vectors such as mosquitoes and allow disease agents causing cholera, typhoid, and diarrhea to proliferate (Liang & Messenger, 2018). Prolonged droughts, on the other hand, can reduce water availability and quality, forcing populations to rely on dangerous water sources raising the risk of water-borne infection (World Health Organization, 2023).

Changes in the frequency of extreme events such as hurricanes, as well as ocean–atmosphere oscillations can have an impact on water-borne illness patterns. Flooding can occur if extreme events become more frequent and powerful, polluting water sources and raising the danger of water-borne infections (Seneviratne et al., 2012).

Ample evidences show the association of increase in the ambient temperature with an increase in all-cause and bacterial diarrhea. Every 1 °C increase in maximum temperature was connected with a 0.6% increase in diarrhea cases, whereas a 1 mm increase in rainfall was associated with a 5% increase in Bhutan (Wangdi & Clements, 2017). Similarly in Nepal, every 1 °C increase in average temperature caused 4.4% increase diarrheal diseases among children (Dhimal et al., 2022). However, there are limited studies on climate change and diarrheal diseases in low- and middle-income countries (Dhimal et al., 2023). Table 1 summarizes findings of studies on climate change and infectious diseases in HKH region.

2.3 *Impact of Climate Change on Noncommunicable Diseases, Injuries, and Mental Health in the HKH Region*

Climate change is expected to worsen NCDs such as cardiovascular disease, cancer, respiratory issues, mental health problems, injuries, and malnutrition (Kjellstrom et al., 2010; Friel et al., 2011a; Campbell-Lendrum & Prüss-Ustün, 2019; Nugent & Fottrell, 2019). Mental health will also be affected (Berry et al., 2010; Bourque & Cunsolo Willox, 2014; Hayes et al., 2018; Cianconi et al., 2020).

Research indicates that extreme temperatures, either too hot or too cold, raise the risk of heart- and lung-related deaths due to increased blood pressure, viscosity, and heart rate for heart issues and narrowed airways for lung problems (Mcmichael et al., 2006). Rising temperatures can harm agriculture, especially in poorer tropical regions. If global temperatures increase by over 3 °C, global food production may decrease, leading to more NCDs (Friel et al., 2011a).

Environmental factors and climate change are closely tied to the risk factors for NCDs, posing a significant threat to global health when combined with air pollution (Campbell-Lendrum & Prüss-Ustün, 2019). With worsening global warming, severe heatwaves, droughts, storms, and floods will become more frequent, worsening NCD occurrences (Friel et al., 2011b).

Climate-related temperature increases can also harm air quality by producing more tropospheric ozone. This air pollution, along with extreme temperatures, can heighten the risk of heart disease through various pathways, including dietary changes (Friel et al., 2011b).

Table 1 Inflence of climatic factors on vector-borne and water-borne diseases in the HKH region

Sr no.	Diseases	Causative agent	Climate relation	Health impacts	References
1	Cholera	*Vibrio Cholerae*	Outbreak of disease at the mid-monsoon period, correlated with environment variables as low precipitation and high temperature provide the quicker bacterial replication	Frequent epidemics; high risk individuals from contaminated water, vegetables, and foods	Zahra et al. (2019)
2	Dengue, chikungunya	Mosquito vector (*Aedes aegypti* and *Aedes albopictus*)	Rise in temperature increases the incidence of diseases; shifts the distribution range of disease vectors and diseases transmission from tropical to temperate regions and highlands	Emerging public health threat; higher frequency of dengue infection was reported mostly in post-monsoon season and vector was reported at least 2000 m above mean sea level in Nepal.	Dhimal et al. (2015b)
3	Diarrhea	Rota virus/ *E. coli*	Climate warming such as drought, flood, and rainfall are likely to reduce the quantity of safe water; prevalence of diarrheal diseases mostly in children	Incidences of diarrhea are increasing more in mountain and hill regions compared to Terai region	Dhimal et al. (2022)
4	Malaria	*Plasmodium vivax*/*P. falciparum*	Malaria transmission and distribution are highly influenced by climatic factors in most part of the HKH region, creating complication in eradicating malaria	1 °C increase in temperature leads to increased malaria incidence by 27%	Dhimal et al. (2014b)
5	Visceral leishmaniasis	*Leishmania donovani*	Expanded in non-endemic areas of the HKH region, creating a new threat in elimination	Risk is expected to increase faster with a higher degree of global warming	Dhimal et al. (2021a)
6	Japanese encephalitis	*Flavivirus*/ mosquito vector	Increase in the cases of JE at high Himalaya region		Dhimal et al. (2021b)

(continued)

Table 1 (continued)

Sr no.	Diseases	Causative agent	Climate relation	Health impacts	References
7	Zika virus	Mosquito vector (*Aedes aegypti* and *Aedes albopictus*)	Increasing threat in the Himalaya region due to the climate change; high abundance and wide distribution of human-biting mosquito vectors	Infecting thousands of people and chance of microcephaly to children born from infected women; mainly China, Myanmar, and India have reported the Zika as well as other Asian region	Dhimal et al. (2018)

Apart from NCDs, mountain communities face mental health challenges due to climate-induced disasters such as droughts (Cianconi et al., 2020). Those exposed to these disasters can experience mental health issues such as stress, depression, anxiety, substance misuse, and suicidal thoughts due to factors such as loss of loved ones, resources, social support, and forced relocation (Dhimal et al., 2021a, b, c).

In the HKH region, climate change leads to more frequent and severe natural disasters. These can be grouped into "rapid-onset" hazards such as floods, landslides, glacial lake outburst floods (GLOFs), flash floods, and debris flows, as well as "slow-onset" hazards such as droughts, heat waves, and ecological changes (Eriksson et al., 2008).

The rise in temperatures causes glaciers to melt and forms melt water lakes, heightening the risk of GLOFs. GLOFs not only harm the environment but also cause significant social and economic difficulties for affected communities. Their occurrence is on the rise, with 20 GLOF events recorded in the Himalayan region over the past 70 years, leading to loss of life, property, infrastructure, and farmland (Eriksson et al., 2008).

Flash floods devastate crops, impacting the food and water supply chain, sanitation, and hygiene. Gender differences also play a role, as floods disproportionately affect women. In Nepal, a severe flood in 1993 showed higher mortality rates for girls and women compared to boys and men (Pradhan et al., 2007).

Vulnerable groups, especially the poor and marginalized, suffer more due to their settlement locations. Settlements along riverbanks and floodplains put them at greater risk. Delayed rescue and support operations during and after disasters further impact these groups (Gurung et al., 2019).

The consequences of hazards such as floods, GLOFs, and landslides are lasting, affecting both livelihoods and health. These disasters damage property, agricultural land, and disrupt income sources. Beyond immediate effects such as resource scarcity, psychological stress also negatively affects health (Eriksson et al., 2008).

"Slow-onset" disasters, though less obvious, are equally destructive. Climate change alters hydrological cycles, leading to water stress and related illnesses. Even with economic growth, water-borne diseases persist in regions such as China's

Yunnan province. Changes in ecosystems and biodiversity due to climate change impact diets and nutrition, with evidence of malnutrition tied to the decline of local flora and fauna (Eriksson et al., 2008).

2.4 Implication of Climate Change in the Livelihood of the People in the HKH Region

Mountain people heavily rely on agriculture for their livelihood. Crop-livestock agriculture (CLA), which employs millions of people on farms, has been the source of livelihood for the majority of people residing in the HKH region. Agriculture accounts for 40% of the region's GDP, with more than 200 million smallholder farmers. It also produces the majority of livestock products (75% of milk and 60% of meat). Smallholder farmers' livelihoods in CLA are heavily reliant on natural resources to support the human and animal populations (Wester et al., 2019). Besides this, remittances from migrants, small companies, tourism, wage labor, collection and sale of medicinal plants, and other herbs all help to support livelihoods and food security for the people of this region (Merrey et al., 2018). However, climate change has far-reaching implications for the livelihood of the people residing here. Unprecedented changes in precipitation patterns, hydrological imbalances, increase in temperature, and frequent floods have significant impacts on day-to-day life of the people (Dhimal et al., 2021a, b, c). Increased warming and accelerated melting of snow and glaciers have direct impacts among those living in immediately downstream areas owing to decreased dependability of local water resources and increased incidence of hazards such as glacial lake outburst floods (Molden et al., 2022; Mayewski et al., 2020). The decline of glaciers, snow, and permafrost has negative socioeconomic consequences (such as natural hazards, infrastructure destruction, and changes in the amount, seasonality, and quality of glacier-fed water), as well as negative impacts on tourism and recreation that ultimately threatens the water, food, energy, and livelihood of the people living in this region (Mayewski et al., 2020).

Agricultural patterns are heavily influenced by weather patterns, which dictate the timing of planting and harvesting as well as cultivating of the crops; however, high variability of the rainfall patterns in terms of amount and precipitation altered the agricultural pattern along with disturbing the functioning of the ecosystem (Molden et al., 2022). Some parts of this region are already experiencing water stress. On the other hand, due to the heat stress, farmers have endured skin ailments such as ringworm, measles, and prickly heat (Hussain et al., 2019). According to climate projections, these changes are likely to intensify over time. Climate change led to migration of people. It was reported that the likelihood of migration of household members for work was higher in settlements exposed to rapid onset water hazards than those exposed to a slow onset of water-related hazards (Piguet, 2008; Jakobeit, 2017).

2.5 Gender Dimension of Climate Change in the HKH

Society has differentiated the roles and responsibilities of males and females. Males are supposed to be the breadwinners of the family, while females are obligated to take care of the family members and their children (Leduc et al., 2008). The gender component is thus very important in understanding the effects of climate change on livelihoods. Because of gendered societal norms and the obligation of caring for their houses, children, collecting firewood, water, and agricultural labor, women are more vulnerable to climate change (Goodrich et al., 2017). Gender differentiated health impacts of climate change with more effects on women is reported in a study from Nepal (Lamichhane Dhimal, 2018).

In the HKH region, collection of water, fuel, wood, and fodder are typically the responsibility of women (ICIMOD, 2022). As climate change disrupts the moisture content of the soil, it leads to less production of food items, resulting in food insecurity. Longer periods of drought lead to the depletion of the quality and quantity of natural resources, making it difficult for women to reach water sources, ultimately increasing women's drudgery and affecting the entire family. On the other hand, men who are not able to fulfill their family's needs through their traditional livelihood approach are migrating to other countries and cities in search of better opportunities, leaving the poor people in the community. The poorest males must take low-wage employment and live in deplorable circumstances. Aside from the immediate economic difficulty, they are also under a great deal of mental and emotional stress (Leduc et al., 2008). Because of the outmigration of active male partners of this region, only the elderly, children, and women are left behind, increasing the feminization of the agricultural labor, particularly in the mountain areas of HKH countries such as Nepal, India, Pakistan, and China (ICIMOD, 2022). Moreover, disasters such as floods have destroyed health facilities and increased exposure to communicable diseases due to inadequate infrastructure, poor sanitation, and limited heath care access, and the women face multiple challenges in reproductive health care and nutrition.

3 Climate Action Endeavors and Governmental Strategies for Environmental Sustainability

HKH countries are taking action against climate change by making promises and using tools to adapt and lessen its impact. They are investing in eco-friendly energy systems and nature-centered fixes such as planting trees and restoring ecosystems. Adaptation efforts are focused on nature-based solutions for water management, strong farming methods, sustainable forests, and disaster risk reduction (Rana et al., 2021).

3.1 Climate Change Mitigation and Nationally Determined Contributions (NDCs) of the HKH Region

The Paris Agreement asks every country to create, share, and update their nationally determined contributions (NDCs)—promises about fighting climate change. These aim to keep global warming well below 2 °C, and even better at 1.5 °C, than previous industries. Countries update NDCs every 5 years with bigger goals. NDCs show how nations cut emissions and adapt to climate impacts. All eight HKH countries shared their Intended Nationally Determined Contributions (INDCs) and NDCs, but only Bhutan and Nepal submitted their second NDC to the United Nations Framework Convention on Climate Change (UNFCCC) secretariat. Bangladesh improved its NDC before the Conference of the Parties (COP26) in Glasgow, UK (Rana et al., 2021).

3.1.1 Mitigation Priorities of the HKH Countries

In 2018, about 16,098 Metric tons of carbon dioxide ($MtCO_{2e}$) of greenhouse gases were released in the HKH region, mostly from energy, agriculture, and industry. HKH countries committed to reducing emissions through INDCs/NDCs that are aligned with national strategies. Priorities differ due to economic development stages. Bhutan is carbon negative, aiming for carbon neutrality by 2050. China targets peak emissions by 2030 and carbon neutrality by 2060. Nepal aims for net-zero by 2050. Like globally, energy, agriculture, and industry are the leading sources of emissions in the HKH region, necessitating immediate action. Energy and land-use sectors are key, focusing on diversifying energy sources and increasing renewable, while also enhancing carbon sinks through afforestation and restoration (Rana et al., 2021).

3.2 Government Initiatives

Afghanistan

In Afghanistan, efforts to address climate change fall into two categories: mitigating its effects and adapting to the changes it brings. While reducing greenhouse gases might not be a top priority due to the country's limited emissions, preparing for climate impacts is critical. Afghanistan faces significant risks from climate change, making changes in practices such as farming, water management, and energy crucial for safeguarding the nation (Mayar, 2022).

Afghanistan is incorporating climate change into national programs, such as water and natural resource management, aiming to prepare for climate challenges. National Environmental Protection Agency (NEPA) has launched various climate-related projects to mitigate climate impact such as the Agromet Project and LDCF initiatives. The country has a network of institutions handling climate and disaster risks. Afghanistan's National Adaptation Programme of Action (NAPA) focuses on adapting to climate change impacts, creating connections with environmental agreements and integrating climate concerns into planning. Afghanistan's NDC outlines key actions to enhance adaptive capacity, integrate climate change into planning, promote economic growth, strengthen technical capabilities, and raise awareness about climate impacts and adaptation strategies (NEPA, 2017).

Bangladesh

Since the 1970s, Bangladesh has invested in flood management, coastal protection, cyclone shelters, and more. These efforts have improved economic progress and reduced poverty. The government collaborates with ministries, agencies, and society to address the complexity of climate change. Key ministries include Environment and Forests, Disaster Management, and Meteorological Department. Bangladesh integrates climate change into its development strategy and aims for low-carbon growth (MoEF, 2008).

Initiatives such as the Bangladesh Climate Change Strategy and Action Plan (BCCSAP) and NDC prioritize adaptation, mitigation, and international support. The NAPA focuses on immediate adaptation needs, identifying vulnerabilities, proposing measures, and seeking global assistance (MoEF, 2008). Moreover, Bangladesh Delta Plan 2100 also aimed to address long-term challenges of climate change and rising sea levels, outlines strategies for integrated water resource management and sustainable development in the delta region (GED, 2018).

Bhutan

Bhutan's NAPA assesses its climate vulnerabilities in its unique mountainous setting. Despite political stability and development progress, Bhutan faces risks due to fragile ecosystems, reliance on monsoons, and GLOFs. To protect its largely agrarian population, adaptation measures include reducing dangerous lakes, hazard mapping, and managing hydropower infrastructure (United Nations Development Nations and Global Environment Fund, 2006).

NDC Mitigation Components (RGB, 2021):

Bhutan focuses on various sectors for mitigation:

- *Forest conservation*: Bhutan's comprehensive REDD+ framework emphasizes forest conservation to balance climate goals, biodiversity, and livelihoods.
- *Low emission agriculture*: Bhutan aims to cut agriculture's 14.5% emissions by shifting to organic fertilizers and enhancing agricultural practices.
- *Sustainable industries*: Mitigation in industries targets reductions by improving energy efficiency, using renewable inputs, and transforming manufacturing processes.
- *Green transport*: Bhutan's transport strategy promotes mass transit, electric vehicles, greener freight, and better infrastructure.
- *Waste management*: Bhutan commits to achieving "Zero Waste" by 2030 through the National Waste Management Strategy and circular economy principles.
- *Sustainable hydropower*: Bhutan's hydropower policy guides low-emission energy development, supporting carbon neutrality and economic growth.

- *Renewable energy diversification*: Bhutan's alternative energy program aims to diversify the energy mix, integrating mini hydro, solar, wind, and waste-to-energy technologies.
- *Green hydrogen roadmap*: Bhutan explores hydrogen generated from renewables for energy storage and fuel replacement, aligning with sustainable hydropower policies.
- *Energy efficiency and conservation*: Bhutan adopts energy efficiency policies, targeting reduction potential and cost savings in various sectors.
- *Cooperative mechanisms*: Bhutan engages in cooperative strategies under the Paris Agreement to enhance mitigation while ensuring sustainable growth.

Bhutan's efforts include the NDC, Technology Needs Assessment, Disaster Management Act, and Climate Change Policy, which aim to achieve sustainable development and climate resilience (World Bank Group and Asian Development Bank, 2021). Similarly, Bhutan is committed to maintaining carbon neutrality. Its constitution mandates maintaining forest cover, which absorbs more carbon than the country emits. Bhutan's efforts in protecting its forests contribute to its carbon neutrality goal (GVI, 2022).

China

China aims to peak CO_2 emissions by 2030 and achieve carbon neutrality by 2060. They target reducing CO_2 emissions per GDP unit by over 65% from 2005, increasing nonfossil fuel energy consumption to about 25%, and raising forest stock volume by 6 billion cubic meters from 2005. They also aim for a wind and solar power capacity of over 1.2 billion kilowatts by 2030 (MFAPRC, 2021).

Some of the implementation measures include:

- *Green transition*: Integrating carbon goals into socioeconomic planning, emphasizing green infrastructure and sustainable layouts.
- *Emission reduction plan*: Outlining strategies for energy transition, efficient industries, eco-friendly transport, and circular economy practices.
- *Energy revolution*: Prioritizing energy conservation, nonfossil fuels, and limiting coal projects to improve energy efficiency.
- *Dual energy control*: Focusing on energy management, sector-specific consumption reduction, and carbon intensity control.
- *Sector-specific actions*: Promoting green development in industries, urban–rural construction, transportation, and agriculture (MFAPRC, 2021).

Climate mitigation:

China's efforts encompass technology advancement, renewable energy transition, tree planting, population management, regulations enforcement, institutional enhancement, capacity building, and public engagement (NDRC, 2007).

Recent policy enhancements:

China's post-2021 policies include stronger legislation, green finance promotion, the world's largest carbon market launch, enhanced greenhouse gas monitoring, technological innovation focus, professional training, and public engagement initiatives (MEE PRC, 2019).

China's ecological civilization initiative:

China has launched various policies and programs to combat climate change, including the "Ecological Civilization" initiative that promotes green development and low-carbon growth. It aims to reduce carbon intensity and transition to clean energy. China's "Ecological Civilization" concept redefines development for twenty-first-century challenges, focusing on political factors, cultural factors, and balanced human–nature relationships. It emphasizes pollution reduction, circular and low-carbon economies, green development, aligned with finance, law, and innovation. This concept, beyond sustainable development, aims to be fully in place by 2035. Challenges include aligning it with economic reforms, yet linking opportunities with national policies can make it thrive, benefiting global goals such as UN Sustainable Development Goals and climate change efforts. Ecological civilization values ecological services and resource protection for transformative impact (Hanson, 2019).

China's national carbon market:

China launched the world's largest carbon market in 2021, covering more than 2200 power companies and aiming to reduce emissions from the power sector (Nogrady, 2021).

India

The Indian government launched the National Action Plan on Climate Change (NAPCC) in 2008, outlining eight national missions aimed at mitigating greenhouse gas emissions and promoting sustainable development (*National Action Plan on Climate Change (NAPCC*), 2021). India has been actively promoting renewable energy sources such as solar, wind, and hydropower. The International Solar Alliance (ISA) was launched by India in partnership with France to accelerate solar energy adoption globally. India's ambitious target of achieving 175 GW of renewable energy capacity by 2022 is a testament to its commitment (Anand, 2023).

India's solar initiatives:

The Jawaharlal Nehru National Solar Mission aims to achieve 100 GW of solar capacity by 2022 (SECI, 2008). The Kurnool Ultra Mega Solar Park in Andhra Pradesh (Anjali Jaiswal, 2017) and the Kamuthi Solar Power Project in Tamil Nadu (EJOLT, 2019) are among the world's largest solar installations.

Myanmar

The government formulated and adopted the Myanmar Climate Change Policy (MCCP) to provide long-term guidance for climate change actions, both adaptation and mitigation. It serves as a roadmap for promoting sustainable and low-carbon development across sectors. Aligned with the MCCP, the Myanmar Climate Change Strategy (MCCS) outlines actionable strategies and plans to address climate-related risks and opportunities over the next 13 years. It focuses on inclusive development, resource efficiency, integrated policy alignment, and result-oriented goals (Thein, 2019).

Nepal

Nepal has taken key steps to address climate-related health risks:

Climate action plans:

Initiatives such as the National Adaptation Plan of Actions to Climate Change (NAPA) 2010 and Climate Change Policy updated in 2019 emphasize health resilience and national adaptation plan (2021–2050).

Health policies integration:

National Health Policy and Strategy includes climate considerations, strengthening readiness for climate-related health hazards.

Adaptive strategies:

Health National Adaptation Plan (H-NAP) 2017–2021 focuses on healthcare resilience through improved infrastructure and public health systems (Ministry of Health and Population, 2018). Furthermore, Nepal is updating its H-NAP.

Cleaner practices:

Nepal's NDC targets cleaner cooking with electric stoves, upgraded cookstoves, and waste management advancements.

National Adaptation Plan:

Nepal's plan prioritizes climate-resilient health, water, and sanitation sectors, emphasizing capacity building, WASH services, disease surveillance, urban health, and water conservation (MoFE, 2018).

The WWF-led Hariyo Ban project:

Combining community action and landscape conservation, it empowers regions to adapt and prosper. From forest restoration to cardamom farming, this initiative shows promise in securing water sources and rebuilding ecosystems across Nepal's diverse landscapes. It has empowered communities to safeguard their environment while improving livelihoods (Seiff, 2017).

Pakistan

The Government of Pakistan has taken significant steps to tackle climate change. It launched the "Pakistan Climate Change Policy" in 2012, emphasizing adaptation and mitigation strategies. The "Green Pakistan Program" aims to increase forest cover, combat deforestation, and enhance ecosystem resilience. The "Clean Green Pakistan Initiative" targets urban areas for improved waste management and environmental sustainability (Ministry of Climate Change – Pakistan, 2019).

3.3 Local Initiatives

Bangladesh's floating gardens: In flood-prone areas, locals have developed floating gardens to grow crops during inundation, ensuring food security even during floods (Sundar, 2020).

Pakistan's Billion Tree Tsunami: Pakistan launched a reforestation campaign, "Billion Tree Tsunami," aimed at planting one billion trees by 2023. The initiative contributes to carbon sequestration, biodiversity conservation, and local employment (Hess, 2021).

Pakistan's solar energy adoption: The "Sindh Solar Energy Project" installed solar panels on schools and health facilities in remote areas, providing clean energy access and reducing reliance on fossil fuels.

India's Dharnai solar microgrid: The village of Dharnai in Bihar became the first Indian village to be fully solar-powered, showcasing the potential of off-grid solar solutions for rural electrification (Mathai, 2022).

3.4 International Initiatives

International Centre for Integrated Mountain Development (ICIMOD): ICIMOD works in the HKH region to foster regional cooperation and implement climate change adaptation and mitigation projects. It facilitates knowledge sharing and research among member countries (ICIMOD, 2021).

3.5 Success Stories

Bangladesh cyclone shelters: The construction of over 2000 cyclone shelters has significantly reduced fatalities during tropical cyclones. These shelters provide safe havens for communities in vulnerable areas (Haque et al., 2012).

Bangladesh's climate-resilient crops: The development and adoption of flood-tolerant rice varieties have helped farmers maintain food production despite regular flooding (Bairagi et al., 2021).

China's renewable energy expansion: China has made significant strides in renewable energy deployment. It is now the world's largest producer of solar panels and wind turbines, reducing its reliance on coal and decreasing emissions. China has surged ahead in renewable energy, recently accelerating wind power capacity. In the past 2 years, China added more wind capacity than in the previous 7 years, generating 46% more wind power than all of Europe in 2022, securing its global green energy leadership. With a record 34.2% of electricity from clean sources, China's commitment to greening its energy system stands firm. Wind's steady generation complements solar's fluctuations. China's operational wind capacity hit 278,353 MW, mainly modern, efficient installations. This boom reshapes local energy mixes, bolsters industries, and contributes to China's ongoing green transition (Maguire, 2023).

China's Shenzhen's electric bus fleet: Shenzhen has successfully converted its entire bus fleet to electric buses, reducing emissions and improving air quality in the city (Berlin et al., 2020).

Pakistan's reforestation efforts: The Billion Tree Tsunami project in Pakistan has successfully surpassed its tree planting targets, contributing to increased forest cover and carbon sequestration (Root, 2021).

Myanmar's mangrove restoration in Ayeyarwady Delta: Local communities have worked together to restore degraded mangrove forests, which act as natural barriers against storm surges and protect coastal communities from climate-related hazards (Ichikawa, 2012).

Sikkim's organic farming: The state of Sikkim achieved 100% organic farming, reducing the use of chemical fertilizers and pesticides, which has positive environmental and health impacts (Parikh, 2023).

4 Conclusion

Despite having minimal share of emission and contribution in climate change and the disasters, people from the HKH region are facing several climate-related adversities in addition to existing environmental and socioeconomic issues. The spectrum of the problem is not only limited to extreme weather events such as flood, heatstroke, and landslide but many communicable and noncommunicable diseases are also associated with climate disasters. Climate change has adversely affected the health, economic, and social life of mountain people along with exaggerating social inequality. Crucial steps are now needed to protect the climate and to protect the health and livelihood of people from harmful impacts due to climate change. In order to stop the climate change, it is very essential to dramatically cut down the amount of fossil fuels we use and promote greenery. The natural environment is excellent at removing pollutants, because photosynthesizing plants absorb carbon dioxide as they grow and store it in soils; hence, afforestation is required. Finally, developing a climate-resilient and low carbon health system including climate-resilient society is of utmost importance for protecting the health of the planet and people.

References

Anand, S. (2023). India, France discuss ways to accelerate global energy transition | Mint. *MINT*. Available at: https://www.livemint.com/news/india/india-france-discuss-ways-to-accelerate-global-energy-transition-11686731071650.html. Accessed 13 Aug 2023.

Bairagi, S., et al. (2021). Flood-tolerant rice improves climate resilience, profitability, and household consumption in Bangladesh. *Food Policy, 105*, 102183. https://doi.org/10.1016/J.FOODPOL.2021.102183

Berlin, A., Zhang, X., & Chen, Y. (2020). *Case study: Electric buses in Shenzhen, China* (p. 9). EV Outlook 2020 International Energy Agency. Available at: https://iea.blob.core.windows.net/assets/db408b53-276c-47d6-8b05-52e53b1208e1/e-bus-case-study-Shenzhen.pdf

Berry, H. L., Bowen, K., & Kjellstrom, T. (2010). Climate change and mental health: A causal pathways framework. *International Journal of Public Health, 55*(2), 123–132. https://doi.org/10.1007/S00038-009-0112-0

Bourque, F., & Cunsolo Willox, A. (2014). Climate change: The next challenge for public mental health? *International Review of Psychiatry, 26*(4), 415–422. https://doi.org/10.3109/09540261.2014.925851

Campbell-Lendrum, D., & Prüss-Ustün, A. (2019). Climate change, air pollution and noncommunicable diseases. *Bulletin of the World Health Organization, 97*(2), 160–161. https://doi.org/10.2471/BLT.18.224295

Cianconi, P., Betrò, S., & Janiri, L. (2020). The impact of climate change on mental health: A systematic descriptive review. *Frontiers in Psychiatry, 11*, 74. https://doi.org/10.3389/FPSYT.2020.00074

Cissé, G. (2019). Food-borne and water-borne diseases under climate change in low- and middle-income countries: Further efforts needed for reducing environmental health exposure risks. *Acta Tropica, 194*, 181–188.

Dhimal, M. L. (2018). Gender dimensions of health impact of climate change in Nepal. In *Schriften zur interdisziplinären Frauen- und Geschlechterforschung. | Series on Interdisciplinary Women's and Gender Studies* (1st ed.). Nomos.

Dhimal, M., Dahal, S., Dhimal, M.L., Mishra, S., Karki, K.B., Aryal, K., Haque, U., Md Iqbal Kabir, Guin, P et al., (2018). Threats of Zika virus transmission for Asia and its Hindu-Kush Himalayan region. *Infectious diseases of poverty, [online] 7(1).* https://doi.org/10.1186/s40249-018-0426-3

Dhimal, M., Gautam, I., Kreß, A., Müller, R., & Kuch, U. (2014a). Spatio-temporal distribution of dengue and lymphatic filariasis vectors along an altitudinal transect in Central Nepal. *PLoS neglected tropical diseases, 8(7), e3035.*

Dhimal, M., O'Hara, R.B., Karki, R., Thakur, G.D., Kuch, U., & Ahrens, B. (2014b). Spatio-temporal distribution of malaria and its association with climatic factors and vector-control interventions in two high-risk districts of Nepal. *Malaria journal, [online] 13(1).* https://doi.org/10.1186/1475-2875-13-457.

Dhimal, M., Ahrens, B., Kuch, U. (2015a). Climate change and spatiotemporal distributions of vector-borne diseases in Nepal – A systematic synthesis of literature. *PloS one, [online] 10(6), pp.e0129869–e0129869.*

Dhimal, M., Gautam, I., Joshi, H. D., O'Hara, R. B., Ahrens, B., & Kuch, U. (2015b). Risk factors for the presence of chikungunya and dengue vectors (Aedes aegypti and Aedes albopictus), their altitudinal distribution and climatic determinants of their abundance in central Nepal. *PLoS neglected tropical diseases, 9(3), e0003545.*

Dhimal, M., Kramer, I. M., Phuyal, P., Budhathoki, S. S., Hartke, J., Ahrens, B., Kuch, U., Groneberg, D. A., Nepal, S., et al. (2021a). Climate change and its association with the expansion of vectors and vector-borne diseases in the Hindu Kush Himalayan region: A systematic synthesis of the literature. *Advances in Climate Change Research, 12*(3), 421–429. https://doi.org/10.1016/j.accre.2021.05.003

Dhimal, M., Bhandari, D., Dhimal, M. L., et al. (2021b). Impact of climate change on health and well-being of people in Hindu Kush Himalayan Region: A narrative review. *Frontiers in Physiology, 12*, 651189. https://doi.org/10.3389/fphys.2021.651189

Dhimal, M., Kramer, I. M., Phuyal, P., Budhathoki, S. S., Hartke, J., Ahrens, B., Kuch, U., Groneberg, D. A., Cun-rui, H., et al. (2021c). Climate change and its association with the expansion of vectors and vector-borne diseases in the Hindu Kush Himalayan region: A systematic synthesis of the literature. *Advances in Climate Change Research, 12*(3), 421–429. https://doi.org/10.1016/j.accre.2021.05.003

Dhimal, M., Bhandari, D., Karki, K.B., Shrestha, SL., Khanal, M., Shrestha, RRP., Dahal, S., Bista, B., Ebi, KL., Cissé G, et al. (2022). Effects of climatic factors on Diarrheal diseases among children below 5 years of age at national and subnational levels in Nepal: An ecological study. *International Journal of Environmental Research and Public Health. 19(10):6138.* https://doi.org/10.3390/ijerph1910613

Dhimal, M., & Bhandari, D. (2023). Climate change and imperatives to ascertain causes of infectious diarrhoea in low-income and middle-income countries. *The Lancet Global Health, 11(3), e308–e309.*

Dhimal, M., Bhandari, D., & Lamichhane Dhimal, M. (2024). Climate change and human health: vulnerability, impact and adaptation in Hindu Kush Himalayan Region. In Climate Change and Human Health Scenarios: International Case Studies (pp. 159–169). Cham: Springer Nature Switzerland.

EJOLT. (2019). *Kamuthi Solar Power Project, Tamil Nadu, India | EJAtlas* [Online]. Environmental Justice Atlas. Available at: https://ejatlas.org/conflict/kamuthi-solar-power-project-648-mw-tamil-nadu-india. Accessed 3 Sept 2023.

Eriksson, M., et al. (2008). *How does climate change affect human health in the Hindu Kush-Himalaya region.* Conference Paper – Regional Health Forum [Preprint]. Available at: https://www.uncclearn.org/wp-content/uploads/library/who44.pdf#page=19. Accessed 14 Aug 2023.

Friel, S., et al. (2011a). Climate change, noncommunicable diseases, and development: The relationships and common policy opportunities. *Annual Review of Public Health, 32*, 133–147. https://doi.org/10.1146/annurev-publhealth-071910-140612

Friel, S., et al. (2011b). Climate change, noncommunicable diseases, and development: The relationships and common policy opportunities. *Annual Review of Public Health, 32*, 133–147. https://doi.org/10.1146/ANNUREV-PUBLHEALTH-071910-140612

General Economics Division (GED) Bangladesh Planning Commission. (2018). Bangladesh Delta Plan 2100. *Ministry of Planning Government of the People's Republic of Bangladesh, 55*, 1–42. Available at: https://plandiv.gov.bd/sites/default/files/files/plandiv.portal.gov.bd/files/2b2db593_2ebd_482e_a0e3_56a7bfe300c8/BDP%202100%20Abridged%20Version%20English.pdf

Gerlitz, J., et al. (2017). The multidimensional livelihood vulnerability index—An instrument to measure livelihood vulnerability to change in the Hindu Kush Himalayas. *Climate and Development, 9*(2), 124–140. https://doi.org/10.1080/17565529.2016.1145099

Goodrich, C. G., Mehta, M., & Bisht, S. (2017). *Status of gender, vulnerabilities and adaptation to climate change in the Hindu Kush Himalaya: Impacts and implications for livelihoods, and sustainable mountain development.* ICIMOD. Available at: https://lib.icimod.org/record/32618

Goodrich, C.G., Prakash, A., Udas, P.B. (2019). Gendered vulnerability and adaptation in Hindu-Kush Himalayas: Environment Development | HimalDoc. [online] Icimod.org. Available at: https://www.lib.icimod.org/record/34431 [Accessed 7 Jul. 2024].

Gurung, C., Bhushan, P., & Larrington-spencer, H. (2019). Conceptualizing gendered vulnerability to climate change in the Hindu Kush Himalaya: Contextual conditions and drivers of change. *Environmental Development, 31*(November 2018), 9–18. https://doi.org/10.1016/j.envdev.2018.11.003

GVI. (2022). *How Bhutan became a carbon-negative country | GVI.* Available at: https://www.gvi.co.uk/blog/bhutan-carbon-negative-country-world/. Accessed 13 Aug 2023.

Hanson, A. (2019). *Ecological civilization: Values, action, and future needs* (ADB East Asia Working Paper Series No. 21). Available at: https://www.adb.org/publications/ecological-civilization-values-action-future-needs

Haque, U., et al. (2012). Reduced death rates from cyclones in Bangladesh: What more needs to be done? *Bulletin of the World Health Organization, 90*(2), 150–156. https://doi.org/10.2471/BLT.11.088302

Hayes, K., et al. (2018). Climate change and mental health: Risks, impacts and priority actions. *International Journal of Mental Health Systems, 12*(1), 28. https://doi.org/10.1186/S13033-018-0210-6

Hess, L. (2021). Pakistan is on its way to planting a 10 Billion Tree Tsunami. Here's how. *Landscape News*. Available at: https://news.globallandscapesforum.org/52934/pakistan-is-on-its-way-to-planting-10-billion-tree-tsunami-heres-how/. Accessed 13 Aug 2023.

Hussain, A., et al. (2019). Climate change perspective in mountain area: Impacts and adaptations in Naltar Valley, western Himalaya, Pakistan. *Fresenius Environmental Bulletin, 28*(9), 6683–6691. Available at: https://www.researchgate.net/publication/335260249_CLIMATE_CHANGE_PERSPECTIVE_IN_MOUNTAIN_AREA_IMPACTS_AND_ADAPTATIONS_IN_NALTAR_VALLEY_WESTERN_HIMALAYA_PAKISTAN

Ichikawa, K. (2012). *Socio-ecological production landscapes in Asia*. United Nations University Institute of Advanced Studies [Preprint]. Available at: https://satoyama-initiative.org/wp-content/uploads/2018/01/SEPL_in_Asia_report_2nd_Printing.web_.pdf

International Centre for Integrated Mountain Development [ICIMOD]. (2019). *Summary of the Hindu Kush Himalaya Assessment Report*. Kathmandu, Nepal. Available at: https://lib.icimod.org/record/34450

International Centre for Integrated Mountain Development [ICIMOD]. (2021). *ICIMOD—International Centre for Integrated Mountain Development* [Online]. Available at: https://www.icimod.org/. Accessed 3 Sept 2023.

International Centre for Integrated Mountain Development [ICIMOD]. (2022). *State of gender equality and climate change in South Asia and the Hindu Kush Himalaya*. Available at: www.icimod.org/himaldoc

Jaiswal, A. (2017). The World's Largest Solar Park – Kurnool, India. *NRDC*. Available at: https://www.nrdc.org/bio/anjali-jaiswal/worlds-largest-solar-park-kurnool-india. Accessed 13 Aug 2023.

Jakobeit, C. (2017). Climate change, migration, and displacement: The underestimated disaster. *Climate Change*, p. 40.

Kjellstrom, T., et al. (2010). Public health impact of global heating due to climate change: Potential effects on chronic non-communicable diseases. *International Journal of Public Health, 55*(2), 97–103. https://doi.org/10.1007/S00038-009-0090-2

Leduc, B., Shrestha, A., & Bhattarai, B. (2008). *Case study: Gender and climate change in the Hindu Kush Himalayas of Nepal*. ICIMOD [Preprint]. Available at: https://www.researchgate.net/publication/254908887_Case_Study_Gender_and_Climate_Change_in_the_Hindu_Kush_Himalayas_of_Nepal. Accessed 1 Sept 2023.

Liang, S. Y., & Messenger, N. (2018). Infectious diseases after hydrologic disasters. *Emergency Medicine Clinics of North America, 36*(4), 835–851. https://doi.org/10.1016/j.emc.2018.07.002

Maguire, G. (2023). Column: China widens renewable energy supply lead with wind power push | Reuters. *Reuters*. Available at: https://www.reuters.com/markets/commodities/china-widens-renewable-energy-supply-lead-with-wind-power-push-2023-03-01/. Accessed 13 Aug 2023.

Mathai, M. V. (2022). *Lessons from Dharnai, "India's First Fully Solar Powered Village": A case study*. Available at: https://www.researchgate.net/publication/360724868_Lessons_from_Dharnai_India's_First_Fully_Solar_Powered_Village_A_Case_Study. Accessed 13 Aug 2023.

Mayar, M. A. (2022). *The climate change crisis in Afghanistan: The catastrophe worsens – What hope for action? – Afghanistan Analysts Network – English*. Available at: https://www.afghanistan-analysts.org/en/reports/economy-development-environment/the-climate-change-crisis-in-afghanistan-the-catastrophe-worsens-what-hope-for-action/. Accessed 10 Aug 2023.

Mayewski, P. A., et al. (2020). Reflection climate change in the Hindu Kush Himalayas: Basis and gaps. *One Earth, 3*(5), 551–555. https://doi.org/10.1016/j.oneear.2020.10.007

Mcmichael, A. J., Woodruff, R. E., & Hales, S. (2006). Climate change and human health: Present and future risks. *The Lancet, 367*(9513), 859–869. https://doi.org/10.1016/S0140-6736(06)68079-3

MEE PRC. (2019). *China's policies and actions for addressing climate change—The Progress Report 2011*. Available at: http://english.mee.gov.cn/News_service/news_release/201812/P020181203536441502157.pdf

Merrey, D. J., et al. (2018). Evolving high altitude livelihoods and climate change: a study from Rasuwa District, Nepal. *Food Security, 10*(4), 1055–1071. https://doi.org/10.1007/s12571-018-0827-y

MFAPRC. (2021). China Nationally Determined Contributions Update. *UNFCCC*, pp. 1–62. Available at: https://transition-china.org/mobilityposts/chinas-updated-nationally-determined-contributions-ndcs/

Ministry of Climate Change – Pakistan. (2019). *Pakistan's Sixth National Report to the United Nations Convention on Biological Diversity Ministry of Climate Change-Pakistan*. Available at: https://www.cbd.int/doc/nr/nr-06/pk-nr-06-en.pdf

Ministry of Health and Population. (2018). *Climate change health adaptation strategies and action plans of Nepal (2017–2021)*. Available at: https://www4.unfccc.int/sites/NAPC/Documents/Parties/NepalHNAP.pdf

MoEF. (2008). *Climate change strategy and climate change strategy and, strategy*. Available at: https://policy.asiapacificenergy.org/sites/default/files/Bangladesh Climate Change Strategy and Action Plan_reduced_0.pdf

MoFE. (2018). *Nepal's National Adaptation Plan (NAP) process: Reflecting on lessons learned and the way forward*, p. 54. Available at: http://napglobalnetwork.org/wp-content/uploads/2018/07/napgn-en-2018-nepal-nap-process.pdf

Molden, D. J., Shrestha, A. B., & Immerzeel, W. W. (2022). *The great glacier and snow-dependent rivers of Asia and climate change: Heading for troubled waters*. Springer. https://doi.org/10.1007/978-981-16-5493-0

National Action Plan on Climate Change (NAPCC). (2021). Available at: https://static.pib.gov.in/WriteReadData/specificdocs/documents/2021/dec/doc202112101.pdf

NDRC. (2007, June). *China's National Climate Change Programme – The National Development and Reform Commission (NDRC) – The Peoples Republic of China*, p. 63. Available at: http://en.ndrc.gov.cn/newsrelease/200706/P020070604561191006823.pdf

NEPA. (2017, December). *Second national communication under the United Nations Framework Convention on Climate Change (UNFCCC)*. National Environmental Protection Agency, pp. 1–111.

Nogrady, B. (2021). China launches world's largest carbon market: But is it ambitious enough? *Nature, 595*(7869), 637. https://doi.org/10.1038/D41586-021-01989-7

Nugent, R., & Fottrell, E. (2019). Non-communicable diseases and climate change: Linked global emergencies. *The Lancet, 394*(10199), 622–623. https://doi.org/10.1016/S0140-6736(19)31762-3

Parikh, A. (2023). Sikkim: Making of the first 100% organic state of the world: paper presented at 14th International... [online] ResearchGate. Available at: https://www.researchgate.net/publication/369753410_Sikkim_Making_of_the_first_100_organic_state_of_the_world_paper_presented_at_14th_International_Conference_on_Environmental_and_Rural_Development_5_March_2023 [Accessed 7 Jul. 2024].

Phuyal, P., et al. (2020). Spatiotemporal distribution of dengue and chikungunya in the Hindu Kush Himalayan region: A systematic review. *International Journal of Environmental Research and Public Health, 17*(18), 1–18. https://doi.org/10.3390/ijerph17186656

Piguet, É. (2008). Migration and climate change. *Futuribles: Analyse et Prospective, 341*, 112p. https://doi.org/10.1051/futur:200834131

Pradhan, E. K., et al. (2007). Risk of flood-related mortality in Nepal. *Disasters, 31*(1), 57–70. https://doi.org/10.1111/J.1467-7717.2007.00340.X

Pradhananga, D. S., Panthi, J., & Bhattarai, D. (2015). International conference on climate change innovation and resilience for sustainable livelihood. *Academia*, pp. 12–14. Available at: https://www.academia.edu/41354791/Conference_proceedngs_climdev

Qing-Long, Y., et al. (2017). An overview of studies of observed climate change in the Hindu Kush Himalayan (HKH) region. *Advances in Climate Change Research, 8*, 141–147. https://doi.org/10.1016/j.accre.2017.04.001

Rana, P. J. B., Kaur, N., & Sharma, D. (2021). *Climate priorities and action in the Hindu Kush Himalaya background*. Available at: www.climate-laws.org

RGB. (2021). *Royal Government of Bhutan: Kingdom of Bhutan second nationally determined contribution*. Available at: https://lib.icimod.org/record/35579

Root, T. (2021). Pakistan's Ten Billion Trees Tsunami, in photos – Washington Post. *The Washington Post*. Available at: https://www.washingtonpost.com/climate-solutions/interactive/2021/pakistan-tree-billion-tsunami-photos/. Accessed 13 Aug 2023.

Ryan, Cooper and Tauer. (2013). *Mountain studies understanding and managing mountains for people and nature, paper knowledge. Toward a media history of documents*.

SECI. (2008). *Jawaharlal Nehru National Solar Mission towards building SOLAR INDIA*, pp. 1–15. Available at: https://www.seci.co.in/upload/static/files/mission_document_JNNSM(1).pdf

Seiff, A. (2017). Adapting to climate change in Nepal. *Magazine Articles | WWF, World Wildlife Magazine*. Available at: https://www.worldwildlife.org/magazine/issues/spring-2017/articles/adapting-to-climate-change-in-nepal. Accessed 13 Aug 2023.

Seneviratne, S. I., et al. (2012). Changes in climate extremes and their impacts on the natural physical environment. In *Managing the risks of extreme events and disasters to advance climate change adaptation: Special report of the Intergovernmental Panel on Climate Change* (Vol. 9781107025, pp. 109–230). https://doi.org/10.1017/CBO9781139177245.006

Sharma, E., et al. (2019). Introduction to the Hindu Kush Himalaya Assessment. In *The Hindu Kush Himalaya Assessment: Mountains, climate change, sustainability and people* (pp. 1–16). https://doi.org/10.1007/978-3-319-92288-1_1

Sundar, K. (2020). The remarkable floating gardens of Bangladesh – BBC Future. *BBC*. Available at: https://www.bbc.com/future/article/20200910-the-remarkable-floating-gardens-of-bangladesh. Accessed 14 Aug 2023.

Thein, U. H. M. (2019). *Myanmar climate change strategy (2018–2030)*. Available at: https://myanmar.un.org/sites/default/files/2019-11/MyanmarClimateChangeStrategy_2019.pdf

United Nations Development Nations and Global Environment Fund. (2006). Bhutan National Adaptation Programme of Action. *Program*, p. 95. Available at: https://unfccc.int/topics/resilience/workstreams/national-adaptation-programmes-of-action/napas-received

United Nations Framework Convention on Climate Change (UNFCCC), "India Updated First Nationally Determined Contribution (NDC)," August 2022, https://unfccc.int/sites/default/files/NDC/2022-08/India%20Updated%20First%20Nationally%20Determined%20Contrib.pdf

Van Oldenborgh, G. J., et al. (2018). Extreme heat in India and anthropogenic climate change. *Natural Hazards and Earth System Sciences, 18*(1), 365–381. https://doi.org/10.5194/nhess-18-365-2018

Wangdi, K., & Clements, A. C. A. (2017). Spatial and temporal patterns of diarrhoea in Bhutan 2003–2013. *BMC Infectious Diseases, 17*(1), 1–9. https://doi.org/10.1186/s12879-017-2611-6

Watts, N., et al. (2020). Review The 2020 report of The Lancet Countdown on health and climate change: Responding to converging crises. *The Lancet, 397*(10269), 129–170. https://doi.org/10.1016/S0140-6736(20)32290-X

Wester, P., et al. (2019). *The Hindu Kush Himalaya Assessment mountains, climate change, sustainability and people*. Springer International Publishing [Preprint]. Available at: https://lib.icimod.org/record/34383

Wester, P., et al. (2020). The Hindu Kush Himalaya Call to Action: Sustaining mountain environments and improving livelihoods. *Mountain Research and Development, 40*(1), 6–10.

WHO. (2021). *Climate change and health*. Available at: https://www.who.int/news-room/fact-sheets/detail/climate-change-and-health. Accessed 1 Sept 2023.

World Bank Group and Asian Development Bank. (2021). *Bhutan climate risk country profile*. Available at: www.worldbank.org. Accessed 1 Sept 2023.

World Economic Forum. (2022). *Deadly heat wave in India and Pakistan was 30x more likely due to climate change, scientists say*. World Economic Forum. Available at: https://www.weforum.org/agenda/2022/06/deadly-heat-wave-in-india-and-pakistan-was-30x-more-likely-due-to-climate-change-scientists-say/. Accessed 1 Sept 2023.

World Health Organization. (2023). *Drought*. Available at: https://www.who.int/health-topics/drought?gclid=Cj0KCQjwl8anBhCFARIsAKbbpyR1KH5qB2b36HRrujVyjsSosY-a1w0MDmuK68iNOn7_UZKod7UXv_MaAhFuEALw_wcB#tab=tab_1. Accesse 1 Sept 2023.

Xu, J., et al. (2019). *Sustaining biodiversity and ecosystem services in the Hindu Kush Himalaya*. Springer International Publishing. https://doi.org/10.1007/978-3-319-92288-1

Zahra, A., Mohammadi, H., Kermani, M., Alireza B., Gholami, M. (2019). The effect of climate change on cholera disease: The road ahead using artificial neural network. *PloS one, [online] 14(11), pp.e0224813–e0224813.* https://doi.org/10.1371/journal.pone.0224813

Part III
Monitoring, Assessing and Managing Disasters in Mountains Under Changing Climate Scenario

Minimizing Carbon Footprint Through the Traditional Approach *Tar Span* in the Mountain Valleys of Himachal Pradesh

Twinkle Thakur and Kesar Chand

1 Introduction

The gravity goods *tar span* (wire ropeway) is a rope-based transport system; it is the oldest mode of transportation system (Thapa Magar, 2016). Even in ancient times, the concept of a rope-based transportation method was used. Since the early 250 B.C., this technology was used to move goods and people (Gyawali et al., 2004). For thousands of years ago, ropes made from animal skins or plant fibres were utilized as essential tools for solving transportation challenges. Their versatile applications in bridging rivers and valleys testify to the resourcefulness and adaptability of early civilizations, laying the groundwork for the evolution of transportation systems we rely on today. Such ancient ropeways were found in various countries including China, Japan, Brazil, New Zealand, and India, showcasing the global presence and historical significance of early rope-based transportation systems (Hoffmann & Zrnić, 2012). *Ropeways* were first developed in Europe, specifically in Germany, Austria, and Switzerland, and then subsequently in Italy and France (Gyawali et al., 2004). The modern ropeway initially appeared in the nineteenth century. In 1862, the first urban cableway was established in Lyon, France. LEITNER built the first passenger ropeway in Central Europe, the "Kohlernbahn," in Bozen, Italy, in 1908 (Hoffmann & Zrnić, 2012). The widespread usage of tramways in steep terrain during World War I between Italy and Austria prepared the way for the development and adaptation of this technology into a reliable mode of transportation for goods (DoLIDAR, 2010). In various European and Asian countries of the world, particularly in hilly countries, *tar span* was used for the transportation of timbers and minerals (Gyawali et al., 2004). Tar span is an alternative and

T. Thakur · K. Chand (✉)
G.B. Pant National Institute of Himalayan Environment, Himachal Regional Centre, Mohal-Kullu, Himachal Pradesh, India

S. Tripathi et al. (eds.), *Warming Mountains*,
https://doi.org/10.1007/978-3-031-62197-0_9

compatible mode of transportation in hilly terrains (Baral, 2012). Mountainous regions often pose significant challenges for conventional road due to steep slopes, rugged terrain, and environmental obstacles. Wire ropeways are designed to overcome these obstacles by spanning across valleys and mountain ranges, providing a direct and efficient transportation solution. Also building roads in mountainous areas can be prohibitively expensive and environmentally disruptive, whereas wire ropeways require fewer infrastructure investments, as they mainly rely on support towers and cable systems. This makes them a more cost-effective and sustainable transportation option, for mountain regions. Wire ropeways have a lower environmental impact compared to traditional transportation options. They require less land clearance, reducing habitat fragmentation and preserving natural ecosystems. Furthermore, cable cars produce minimal noise pollution and have lower carbon emissions, especially if powered by renewable energy sources. The implementation of an aerial cable-car system in Colombia has had a positive impact on geographic accessibility and CO2 emission reduction, enhancing transportation options and promoting environmental sustainability (Sarache & Jiménez-Riaño, 2022). In the past decade, numerous significant cities worldwide have implemented aerial cable-car systems, including Portland (Oregon), Roosevelt Island (NY, USA), Koblenz (Germany), Innsbruck (Austria), Hong Kong (China), Lagos (Nigeria), Constantine (Algeria), Medellin (Colombia), Caracas (Venezuela), Rio de Janeiro (Brazil), and La Paz (Bolivia), demonstrating the global trend towards adopting this mode of transportation for improved mobility and carbon emission reduction (Alshalalfah et al., 2012; Garsous et al., 2019). This sustainable transportation mode is still used in the Himalayan mountainous states (Thapa Magar, 2016).

2 Tar Span in Himalayan State

Ropeways have proved beneficial in solving transportation obstacles in mountainous and isolated areas in India. It contributed to economic development by linking remote areas with markets and roadheads, reducing transportation costs and post-harvest losses. Additionally, they have promoted tourism, offering visitors breathtaking views and convenient access to popular destinations in India (Economic Times, 2023). For the delivery of products, gravity ropeways are in use in various places in Indian Himalayan states such as Himachal Pradesh (Bilaspur, Chamba, Kullu, Shimla, Sirmaur, Kangra, and Hamirpur), Uttarakhand (Haridwar, Mussoorie, Nainital, etc.), Sikkim (Gangtok, Namchi), and Jammu and Kashmir (Gulmarg, Jammu, Srinagar, and Vaishno Devi) (Yadav & Khura, 2015). Himachal Pradesh is predominantly a mountainous state located in Northwest India. The state has highly dissected mountain ranges interspersed with deep gorges and valleys. Tar span technology is used in hilly/mountainous districts of Himachal Pradesh. Out of 12 districts, Kullu district is located in the eastern part of central Himachal Pradesh. Mostly it is used in the Kullu district where most of the people live in the hilly regions. It is located on the banks of the Beas River. Out of the total population of the Kullu district, 9.5% of people live in urban areas while 90.5% live in rural areas. This traditional approach is used mostly in the village as a mode of transportation. Some of the villages in the rural region have no transportation facilities; people who live in these villages depend on the *tar span* to transport their harvested products to markets. The Kullu district in Himachal Pradesh is characterized by a vast expanse of hills and mountains which dominate and encompass the majority of the district's geographical area. In such geographical conditions, the steep and hilly terrain areas, it is difficult for rural communities to carry their goods from higher regions to markets and vice versa. Lack of access to markets due to a lack of proper and reliable means of transportation has been a major obstacle to the commercialization as well as livelihood opportunities of hilly and mountain agriculture (Hine & Ellis, 2001). This technology is most commonly used in the village of Kullu district such as Lug Valley, Tirthan Valley, Parvati Valley, and Upper Valley. It is one of the oldest/traditional transportation technologies which have been used from the ages to transport goods in the villages of Kullu Valley.

3 Working of Tar Span

Tar span runs entirely on gravity, without the use of fossil fuels, electricity, or any other form of external power (Fig. 1a) (DoLIDAR, 2010). It is comprised of two trolleys that move along a support wire. The trolleys are also connected in the centre by a control cable that travels in the typical flywheel system (Fig. 1c). It works in such a way that one trolley loaded at the up-station with goods is pulled downwards, and the second trolley at the bottom station is pulled up by using a control cable (Practical Action Nepal, 1996) (Fig. 1b). The products coming down from the top station are three times heavier the weight of the products coming up from the bottom station, and when loaded, their centre of gravity must be perfectly balanced (Thapa Magar, 2016). A flywheel containing a bracket (Fig. 1d) is used as a brake to slow down the landing speed of the trolley at the bottom station. Around the wheel, brake strips (Fig. 1d) are connected to a fixed support at one end and to the brake handle (Fig. 1e) at the other end. Applying force on the handle, the breaks are applied to the rim of the wheel. The wire rope is used to communicate between the top and bottom stations; the operator hits the wire rope with the help of a stick to send a signal through the wire. Table 1 shows the cost of the tar span when compared to the diesel span.

The cost of the span depends on the quantity of the product, and the quantity of the products depends on the local topography and angle of inclination of the two stations, the distance between the two stations, and local raw material. This table shows that the cost of the gravity span is less when compared to the diesel span. The cost of the goods used to construct the diesel span is Rs. 5,77,693.00 in Lug Valley. Another miscellaneous cost such as cement, stone, sand, and the daily wage of labours to construct the diesel span is approximately Rs. 4,20,459.00.

A.

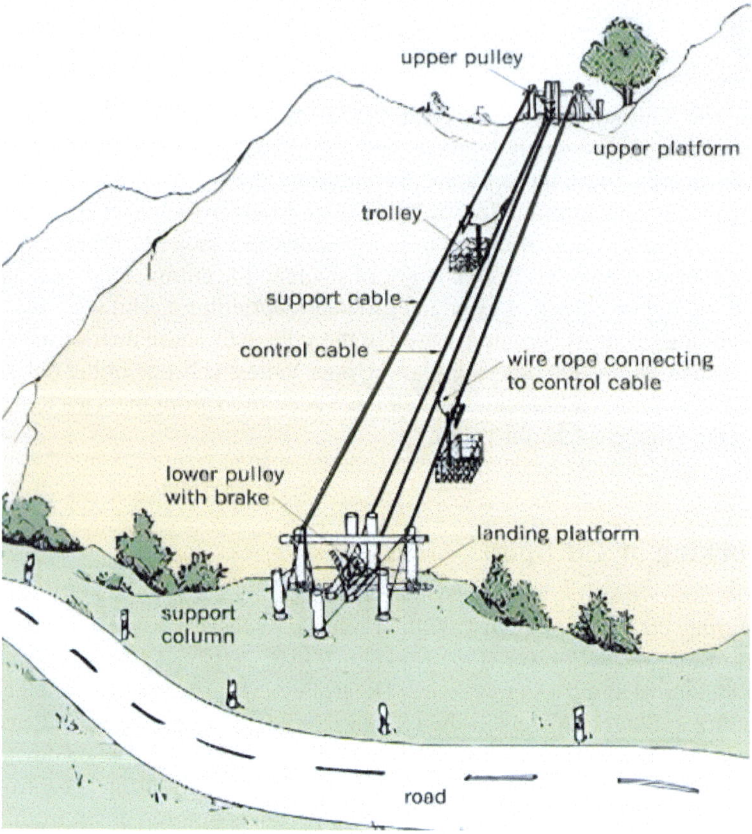

A. Working of ropeways

Fig. 1 Mechanism of working of ropeway (**a**–**e**). (Source: DoLIDAR, 2010)

4 Benefits and Limitation of Tar Span

Tar span technology has a high potential in Kullu's hills and mountainous region to strengthen rural communities' access to the nearby market. It is a highly useful and cost-effective mode of transportation. Its construction cost is lesser as compared to other transportation. It is very time-saving because it is the quickest way to transport in hilly and mountain regions (Thapa Magar, 2016). It saves local people's energy and time and reduces hard work. Hence, it is energy-efficient. This technology is environment-friendly because it is a non-motorized mode of transportation and does not cause air pollution by emitting CO_2 or any other gases (Laxman, 2009). Despite the fact that gravity goods ropeways have various advantages for the local community, we cannot ignore their negative impact. Several environmental impacts such as loss of agricultural land, deforestation, soil erosion, water resources, landslides, etc. are some of the demerits (Kandlikar & Sagar, 1999; Mahato, 2014; Sharma et al., 2009; Kumar et al., 2018). About 70% of the population depend on the agricultural

Fig. 8.1 (continued)

sector, and it is the most important part of the economy in rural areas (Kumar et al., 2018; Sabin et al., 2020; Negi et al., 2021; Singh et al., 2021). Therefore, ropeways facilitate the efficient transportation of agricultural produce, reduce post-harvest losses, enhance market access, and stimulate rural development. These combined benefits contribute to the growth and prosperity of the rural agricultural sector, ultimately boosting the overall economy. If proper mitigation steps are taken during construction, the impact on the environment and ecological system will be negligible (Sarkar, 2011). Due to its low occupancy, it has a minimum impact on the surrounding environment when compared to other transportation projects such as roads and motorable bridges and diesel engines.

For its operation and maintenance, the community does not need to spend money on fuel or a highly skilled workforce (Singh, 2010). It is a very easy and simple technology that can be operated or maintained by the local communities. The daily wage is given to the local labour to maintain the tar span if any damage occurs if any damage occurs. Tar span is designed or constructed by local engineers so this does not require any experts. Accepting that construction of roads to connect the hilly/mountainous regions causes extreme damage to the environment, cost-intensive and extreme events such as landslides, rockfalls, and slope failures occurring during the rainy season and the construction of tar span would give environment-friendly rural

Table 1 Detail of goods, cost for diesel engine span, and gravity span

Sr. No.	Diesel span					Gravity span			
	Descriptions of goods	Qty.	Rate	Per	Amount	Descriptions of goods	Rate	Per	Amount
1.	Wire rope 8 mm	3000 mtrs	28.00/-	mtrs	84,000.00/-	Wire rope 8 mm	28.00/-	mtrs	84,000.00/-
2.	Wire rope 12 mm	3000 mtrs	46.00/-	mtrs	1,38,000.00/-	Wire rope 12 mm	46.00/-	mtrs	1,38,000.00/-
3.	Upper wheel	2 pcs.	10,670.00/-	pcs.	21,340.00/-	Wheels	8000.00/-	pcs.	16,000.00/-
4.	Brake strip	2 pcs.	1550.0/-	pcs.	3100.00/-	Brake strip	1550.0/-	pcs.	3100.00/-
5.	Ultra-Span set	2 pcs.	39,500.00/-	pcs.	79,000.00/-	Carrier bearing type (Garrari)	1100.00/-	pcs.	8800.00/-
6.	Diesel engine 10 HP	2 pcs.	31,5000.00/-	pcs.	63,000.00/-	Chain pulley block (5-ton capacity)	12,500.00/-	pcs.	12,500.00/-
7.	Carrier bearing type (*Garrari*)	8 pcs.	1100.00/-	pcs.	8800.00/-	U-clamps (12 mm)	45.00/-	pcs.	1080.00/-
8.	Chain pulley block (5-ton capacity)	1 pc.	12,500.00/-	pcs.	12,500.00/-	Nut bolts	35.00/-	pcs.	1120.00/-
9.	Transmission belting	15 mtrs	492.00/-	mtrs	7380.00/-	Trolley	6500.00/-	pcs.	26,000.00/-
10.	U-clamps (12 mm)	24 pcs.	45.00/-	pcs.	1080.00/-				
11.	Nut bolts	32 pcs.	35.00/-	pcs.	1120.00/-				
12.	Anchoring girders	590.000 kg	75.00/-	kg	44,250.00/-				
13.	Trolley	4 pcs.	6500.00/-	pcs.	26,000.00/-				
				Total	**5,77,693.00/-**	**Total**			**2,90,600.00/-**

Source: Mahila Kalyan Mandal, Kullu, H.P

transportation option that would not require any input of external energy (Adhikari, 2009).

Despite its numerous advantages, *tar span* (wire ropeway) also has some limitations: it cannot be used on a gentle slope (Thapa Magar, 2016). It needs a slope of at least 15° to run smoothly, and the highest limit can reach as high as 40°; according to current experiences, it is best if it is built in between 20° and 30° slopes. There is an obligatory limitation of *tar span* that they only transport the harvested products or goods but not the people (DoLIDAR, 2010). Operators are the local people, and they must be trained properly in order to make a balance between two trolleys and must be extremely cautious of the weight of the load at both stations when operated. The operators must be very careful of their safety as well as the safety of other people, particularly children, to personal damage (Singh, 2010).

- *Livelihood Opportunities* (Operations and Maintenance): Ropeway systems require skilled personnel for their operation, maintenance, and safety management. This includes ropeway operators, mechanics, engineers, and technicians. These roles provide direct employment and require specialized training, contributing to skill development and livelihood enhancement.
- Tourism and Hospitality: Ropeways often attract tourists, as they offer scenic views and access to popular tourist destinations. This generates opportunities for hospitality services, such as hotels, restaurants, tour guides, and souvenir shops, creating indirect employment and boosting local tourism-related businesses.
- Agriculture and Horticulture: Ropeways can support agriculture and horticulture sectors by facilitating the transportation of agricultural produce from remote areas to markets. This enables farmers to expand their reach and gain better access to buyers, leading to increased income and livelihood improvement for agricultural communities.
- Handicrafts and Local Industries: Ropeway systems can promote the growth of local industries and handicrafts by facilitating the transportation of raw materials, finished products, and traditional handicrafts from rural or hilly areas to markets. This fosters entrepreneurship, preserves cultural heritage, and generates income for local artisans.

5 Government Initiative for Promotion of Wire Ropeways

5.1 National Bank for Agriculture and Rural Development (NABARD) Initiative

In communities in the Lug Valley in the Kullu district, carrying their goods to market was quite challenging. They had to walk for around 2–3 hours on a steep foot carrying agricultural produce and goods on their back to sell the products in the nearest market. Moreover, they have to employ animals (horses, mules) to transport heavy materials like concrete to construct houses. To transport agricultural and horticultural products such as garlic, apple, pears, tomatoes, and cabbage, they hire labourers. According to local people, a handsome amount of money is needed to be paid as a daily wage to the labourers (Rs 70 or more than 70 per turn by labour) and animal owners (Rs.100 per animal per turn), which is earned by them through selling agricultural or horticultural products, thus leaving a very small amount of profit. The less facility for transportation isolated the farmers from the market and increased the travel time which degraded the quality of the products. It is also used in some parts of the Kullu district to shift local people/sick people from far-flung villages to the roadside where transportation facilities are not available in an emergency.

Government organization *NABARD* in collaboration with Mahila Kalyan Mandel, Kullu has taken initiative in Lug Valley to help the local communities. Two diesel spans have been installed in Lug Valley villages (Fig. 2a, b) from *Palalang* to *Mathala* villages for local communities to carry their goods from higher-altitude regions to the market and roadside. The local inhabitants of *Palalang* village have their crop fields in the remote area of *Mathala* village in which they cultivate agricultural and horticultural crops. As the area is remote, the farmers have to carry their crops and products on foot to the road. A total of 81 families *of Mathala, Palalang, Kharsi,* and *Bhajling* are beneficiaries of this initiative. They have mentioned that this can help them to carry their goods and other products from the field to the market and from the market to their houses.

Therefore, transporting products with diesel span saves money and time when compared to manpower. They have mentioned that the travel time has reduced from 2 to 3 hours to 10 to 15 minutes.

Natural and anthropogenic calamities occurred in the Kullu district including landslides, earthquakes, flashfloods, etc. which caused the loss of human life as well as animals. During disasters like floods and landslides in hilly terrain, local communities faced the problem of connectivity because small bridges and constructed roads were washed out by floods and landslides. So, this technology may be useful during natural disasters to maintain communication with each other and to conveniently shift them from a disaster affected area to a safe area.

There are some limitations of diesel span, such as, during winter, diesel engines may not start due to cold temperatures and loss of fluidity. After the installation of the diesel span in the villages, a committee was formed to operate and maintain this diesel span.

Fig. 2 (**a**) Diesel engine (**b**) Installation of diesel span in Lug Valley. (Source: Mahila Kalyan Mandal, Kullu. https://socialprimes.com/ngo/370b10e1-01f1-4e73-ba0a-58942e7e49d1)

Local beneficiaries have mentioned that they have to pay a handsome amount of money to transport their goods, but after installation of the diesel span, they are paying Rs. 30–40 per turn. It is benefiting the majority of the local people, but it lacks some environmental concerns like its engine (Photo: a) which is operated by fossil fuel and emits a huge amount of gaseous pollution (Reşitoğlu et al., 2015; Dobrzyńska et al., 2020).

6 Conclusion

In conclusion, gravity goods ropeways have proven to be an excellent transport system with the dual advantage of enhancing transportation efficiency and reducing carbon footprint. These innovative systems leverage the force of gravity to transport goods, providing a sustainable and environment-friendly alternative to conventional transportation methods. Gravity goods ropeways offer several benefits, including the ability to overcome challenging terrains and provide connectivity to remote areas. They significantly reduce the reliance on road infrastructure, which not only helps to relief congestion but also minimizes the environmental impact associated with road construction and maintenance.

Furthermore, by utilizing gravitational force and minimal energy consumption, gravity goods ropeways have a significantly lower carbon footprint compared to traditional transport modes, such as trucks or other motorized vehicles. They contribute to reducing greenhouse gas emissions, improving air quality, and mitigating climate change impacts.

Additionally, gravity goods ropeways promote efficient logistics and supply chain management, enabling the timely and cost-effective transportation of goods. This system has been successfully implemented in various regions, including mountainous areas and rural communities, where it has facilitated the movement of agricultural produce, construction materials, and other goods, thereby enhancing local economies and livelihoods.

7 Recommendation

Based on the numerous advantages of gravity goods ropeways, here are some recommendations for their implementation and further development:

1. Many countries have used solar energy ropeways/cable cars to carry goods and passengers up to the top of the mountain. So we can also apply this solar power in our small tar span technology to transfer the products and people that it contributes to carbon neutrality. If this renewable energy is used, it can be multiple in other hilly terrain areas which also contributes to sustainable approaches. Therefore, the government needs to encourage solar energy/power ropeways and

community-operated traditional tar spans which are used in every season in the mountainous region (Zhang & Fujimori, 2020; Shaheen & Lipman, 2007). This kind of effort can be a contributor to global mitigators which is mentioned in IPCC AR6 report Climate Change and Mitigation Report (IPCC, 2022).

2. Research and Feasibility Studies: Conduct thorough research and feasibility studies to identify suitable locations and assess the potential benefits and challenges of implementing gravity goods ropeways. This includes analysing terrain conditions, traffic patterns, market demand, and environmental impacts.
3. Collaboration and Partnerships: Foster collaborations and partnerships between government bodies, private sector entities, and local communities to ensure effective planning, funding, and operation of gravity goods ropeway projects. Engage stakeholders at all stages, from design and construction to maintenance and operation.
4. Capacity Building and Training: Provide training and capacity-building programs for local communities, ropeway operators, and maintenance personnel. This will ensure the sustainable and safe operation of gravity goods ropeways, including proper maintenance, emergency response protocols, and adherence to safety standards.

These types of small-small steps and sustainable practices help to reduce the GHG emission and the impact of climate change in hilly mountainous regions, so this type of technology should be supported and encouraged (IPCC AR6, Climate Change, and Mitigation).

References

Adhikari, D. P. (2009). Mainstreaming livelihood centred approach to disaster risk reduction: Nepal component (mid-term evaluation report).

Alshalalfah, B., Shalaby, A., Dale, S., & Othman, F. M. Y. (2012). Aerial ropeway transportation systems in the urban environment: State of the art. *Journal of Transportation Engineering, 138*(3), 253–262.

Baral, S. D. (2012). Gravity ropeway: Could be a reliable source of transport the goods and services in hills and mountainous region of Nepal. In *6th international symposium "networks for mobility"*, Stuttgart, Germany.

Department of Local Infrastructure Development and Agricultural Roads & Practical Action Nepal. (2010). *Technical guidelines for gravity goods ropeway*. Ministry of Local Development, Nepal Printing Office.

Dobrzyńska, E., Szewczyńska, M., Pośniak, M., Szczotka, A., Puchałka, B., & Woodburn, J. (2020). Exhaust emissions from diesel engines fueled by different blends with the addition of nanomodifiers and hydrotreated vegetable oil HVO. *Environmental Pollution, 259*, 113772.

Economic Times – Ropeways to improve connectivity in hilly, remote areas. (2023).

Garsous, G., Suárez-Alemán, A., & Serebrisky, T. (2019). Cable cars in urban transport: Travel time savings from La Paz-El Alto (Bolivia). *Transport Policy, 75*, 171–182.

Gyawali, D., Dixit, A., & Upadhya, M. (2004). *Ropeways in Nepal: Context, constraints, and co-evolution*. Nepal Water Conservation Foundation with Kathmandu Electric Vehicle Alliance.

Hine, J. L., & Ellis, S. D. (2001). *Agricultural marketing and access to transport services*. Transport Research Laboratory.

Hoffmann, K., & Zrnić, N. (2012). *A contribution on the history of ropeways* (pp. 381–394). Springer.

IPCC. (2022). Summary for policymakers [H.-O. Pörtner, D.C. Roberts, E.S. Poloczanska, K. Mintenbeck, M. Tignor, A. Alegría, M. Craig, S. Langsdorf, S. Löschke, V. Möller, A. Okem (Eds.)]. In *Climate change 2022: Impacts, adaptation, and vulnerability. Contribution of Working Group II to the Sixth Assessment Report of the Intergovernmental Panel on Climate Change* [H.-O. Pörtner, D.C. Roberts, M. Tignor, E.S. Poloczanska, K. Mintenbeck, A. Alegría, M. Craig, S. Langsdorf, S. Löschke, V. Möller, A. Okem, B. Rama (Eds.)]. Cambridge University Press. In Press.

Kandlikar, M., & Sagar, A. (1999). Climate change research and analysis in India: An integrated assessment of a South–North divide. *Global Environmental Change, 9*(2), 119–138.

Kumar, P., Tokas, J., Kumar, N., Lal, M., & Singal, H. R. (2018). Climate change consequences and its impact on agriculture and food security. *International Journal of Chemical Studies, 6*(6), 124–133.

Laxman, K. C. (2009). Gravity goods ropeway an alternative sustainable solution for rural transportation without hampering to the natural environment and climate: A case study from Janagaun village. *IOP Conference Series. Earth and Environmental Science, 6*(20). https://doi.org/10.1088/1755-1307/6/20/202019. IOP Publishing.

Mahato, A. (2014). Climate change and its impact on agriculture. *International Journal of Scientific and Research Publications, 4*(4), 1–6.

Negi, V. S., Thakur, S., Dhyani, R., Bhatt, I. D., & Rawal, R. S. (2021). Climate change observations of indigenous communities in the Indian Himalaya. *Weather, Climate, and Society, 13*(2), 245–257.

Reşitoğlu, İ. A., Altinişik, K., & Keskin, A. (2015). The pollutant emissions from diesel-engine vehicles and exhaust after treatment systems. *Clean Technologies and Environmental Policy, 17*(1), 15–27.

Ropeways in Nepal: Conceptual analysis of context and viability, interdisciplinary analysis, practical action Nepal. (1996).

Sabin, T. P., Krishnan, R., Vellore, R., Priya, P., Borgaonkar, H. P., Singh, B. B., & Sagar, A. (2020). Climate change over the Himalayas. In *Assessment of climate change over the Indian region* (pp. 207–222). Springer.

Sarache, W., & Jiménez-Riaño, E. (2022). The impact of a new aerial cable-car project on accessibility and CO2 emissions considering socioeconomic stratum. A case study in Colombia. *Journal of Cleaner Production, 340*, 130802.

Sarkar, J. (2011). Conditional nod to the Girnar Ropeway project. *Current Science, 100*(7), 965.

Shaheen, S. A., & Lipman, T. E. (2007). Reducing greenhouse emissions and fuel consumption: Sustainable approaches for surface transportation. *IATSS Research, 31*(1), 6–20.

Sharma, E., Chettri, N., Tse-ring, K., Shrestha, A. B., Jing, F., Mool, P., & Eriksson, M. (2009). *Climate change impacts and vulnerability in the Eastern Himalayas*. ICIMOD.

Singh, R. B. (2010). Gravity goods ropeway: A complementary means of rural transportation. In *Transport in mountains, an international workshop.*

Singh, P., Vaidya, M. K., & Pathania, K. (2021). *Economic impact of climate change on agriculture: Present, past and future.*

Thapa Magar, R. (2016). Gravity goods ropeways: A sustainable solution for rural transportation in hilly and mountainous regions of Nepal.

www.ucost.in/update/rnd/gravity-based-ropeway.html

www.ucost.in/update/rnd/gravity-based-ropeway.html.himachal

Yadav, S. N., & Khura, T. K. (2015). Techno-economic feasibility study of low cost gravity ropeway for carrying agricultural produce in hilly terrain. *Agricultural Engineering Today, 39*(4), 1–8.

Zhang, R., & Fujimori, S. (2020). The role of transport electrification in global climate change mitigation scenarios. *Environmental Research Letters, 15*(3), 034019.

Impact of Climate Change and Mitigation Plans for Building Climate-Resilient Village Ecosystem: Challenges and Adaptive Strategies

Rishav Rawal and Harshit Pant Jugran

1 Introduction

Climate change, a worldwide crisis, is characterised by persistently changing weather patterns that have been observed for a long time. It could also be used to describe a shift in the weather around longer-term average conditions or a fluctuation in the timing of the weather around longer-term average conditions (i.e. more or fewer extreme weather events) (Houghton et al., 1991). CC is currently posing a threat to human sustainability and way of life. Since the beginning of the Earth's evolution, the climate has changed at its own pace, but recently, it has accelerated due to unintentional anthropogenic perturbations (Jugran Pant & Chaudhary, 2021). Climate change is undeniable and has intensified since the 1950s (IPCC, 2013). Changing rainfall patterns, drought, flooding and the geographical dispersion of pests and diseases are just a few examples of how climate change affects agricultural output both directly and indirectly (IPCC, 2013).

The Himalayan region is vulnerable to the climatic changes and has faced more atmospheric temperature warming than the global average with the effects of climate change becoming a more intensifying issue (Xu et al., 2009, Seidler and Bawa, 2016) because of two primary reasons: (i) alarming trends are stronger there, and (ii) the impacts are amplified by the abrupt shift in altitude over short distances

R. Rawal (✉)
Department of History, D.S.B. Campus, Kumaun University, Nainital, Uttarakhand, India

G.B. Pant National Institute of Himalayan Environment, (GBPNIHE), Kosi-Katarmal, Almora, Uttarakhand, India

H. P. Jugran
Centre for Socio-Economic Development (CSED) G.B. Pant National Institute of Himalayan Environment, (GBPNIHE), Kosi-Katarmal, Almora, Uttarakhand, India

S. Tripathi et al. (eds.), *Warming Mountains*,
https://doi.org/10.1007/978-3-031-62197-0_10

(Negi, 2021). Agriculture, cattle, forestry, tourism, fisheries and human health are all affected by the effects of climate change on the socioeconomic circumstances of Himalayan residents (Sharma et al., 2009). The Himalayas have demonstrated a warming trend and a persistently changing weather pattern, according to several studies (Xu et al., 2009, Seidler and Bawa, 2016) that have evaluated the influence of climate change at local climate levels. Rapid warming over the Himalayan region's high peaks and valleys has a significant impact on climate-dependent industries including agriculture and water resources, supporting a strong and reliable view for the region's climate in the future (Shrestha et al., 2015; Sharma et al., 2019).

A climate-resilient village is the one that has the potential to reduce the effects of climatic variability with the utilisation of existing natural resources such as land, soil, water and genetic resources integrated with improved technological practices. Social and ecological systems of the resilience of agricultural systems are important, and the knowledge on the vulnerability of agricultural systems is essential for the deployment of technologies that can enhance resilience. Social systems of resilience consist of farmers, regions and communities, the level of which depends both on the property and techniques the farmers can share and the welfare given by institutions and governments. Climate-resilient villages consist a well-managed community development planning in addition to a synchronised plan with proper participation and sustainable utilisation of natural resources. In these villages, climate risks are easily addressed by increasing the adaptive capacity helping in the formation of a resilient ecosystem helping in harnessing the benefits of the surroundings even under harsh climatic conditions.

Climate-resilient villages aim to integrate local knowledge and system with the promotion of scientific innovation for a holistic approach for climate change adaptation in the local level that can be integrated at the community level through community participation.

2 Challenges for Climate-Resilient Village Ecosystem

Rural households frequently rely significantly on naturally occurring resources such as fuelwood and wild herbs as well as climate-sensitive activities such as arable farming and livestock husbandry. Climate-sensitive resources include local water supplies and agricultural land (Shackleton and Gumbo, 2010). The rural life in the state of Uttarakhand is highly dependent on farming and resources obtained from the forest, but due to variability in climate, these resources' support had decrease (Rawat et al., 1997).

Loss of livelihood capital, changing agro-livestock conditions and the emergence of alien invasive species are all consequences of the region's changing climatic conditions (Negi, 2021). The region has also experienced unpredictable rainfall patterns in the past, which had a negative impact on agro-horticultural operations. The agriculture of the region is harshly affected by climate change as most of native crops are not getting a favourable climate for their growth. The climate change scenario had led the community to migrate permanently from their native homes in search for better opportunities (Rawat et al., 1997).

Climate change may restrict the availability of local natural resources, reducing the livelihood generation and resource conservation possibilities for rural households that rely on them for trade or consumption.

Migration is also a major concern in the context of rural regions. In the face of climate-related environmental change, such as the decline of productive agricultural land, rural residents may be forced to migrate in search for work (Hunter and Sutton, 2004). Migration is a tactic of survival that people used to take when available resources get exhausted, and it becomes tough for natives to survive in these villages leading towards abandoned villages. In mountain villages across the Himalayas, outmigration is a major phenomenon that is happening in a large proportion since the past few decades, and the youth has a major contribution in this phenomenon (Sati and Sati, 2020). However, the reason behind this is quite clear; i.e. the rural areas are not able to provide the facilities and opportunities in response to contemporary needs and demands; thus, the local inhabitants had to leave home in search for new opportunities and avenues.

In Uttarakhand, outmigration has severe consequences, mostly as a result of the ongoing migration. Due to an exodus of residents, some settlements are completely empty and referred to as "ghost villages". Several communities in almost every district are depopulated (Uttrakhand Migration Commission Report, 2023). It has been noted that there are many villages with no residents in the districts where out-migration is strong. According to various studies (Sati and Sati, 2020), it is noticed that Almora followed by Tehri and Pauri is amongst the districts with the highest number of out-migrants in the past decades.

Although agriculture, which is the main source of food and rural income in the area, is focusing on subsistence based farming, there is a significant lack of arable land and low crop productivity. A significant section of the adult male population must leave the mountain region due to the limitations of the subsistence economy in order to find a living and better job opportunities and education (Tiwari & Joshi, 2015). Mountain communities have evolved adaptations to shifting environmental conditions over millennia (Aase et al., 2013; UNEP, 2004; ICIMOD, 2010). Male youth mobility is a significant adaptation strategy to the limitations of subsistence economies and shifting environmental circumstances throughout the world's alpine regions (Sherpa, 2007). Labour migration has unintentionally created a divide in mountain civilisations where women are burdened with more obligations while they are also playing a significant role in improving economic conditions and ensuring food security through remittance (Sherpa, 2007; Hoermann & Kollmair, 2007; Kollmair et al., 2006). Therefore, there is a need to have a solid and sustainable basic framework for controlling this kind of population drain from mountains, and it is only possible by making a resilient and sustainable ecosystem in these rural areas. In this study, we investigated challenges and strategies for achieving a climate-resilient village ecosystem through a baseline survey for the projection of expected changes.

The present study aims to analyse the challenges that natives of villages in Takula block faces due to changing climate and try to lay down some mitigation plans through which they can become resilient enough to survive in these conditions. The objectives of the present study are as follows:

1. To identify the challenges being faced by the native community in the wake of climate change
2. To suggest mitigation studies based on local perception and adopted adaptive measures

The methodology followed is based on questionnaire-based survey and observation-based group discussion following Lise (2000) and Jugran Pant and Chaudhary (2021) and suggest the remedial measures for solutions at the local level.

3 Description of the Study Site

The present study focused on rural regions of Almora district in the state of Uttarakhand. For this purpose, secondary data has been obtained from the survey conducted at Bhetuli village by Jugran Pant and Chaudhary (2021). Bhetuli village is situated at 29° 40′ 42″ N 79° 42′ 45″ E, at an altitude ranging between 1700 and 1800 m asl, and comes under the Takula developmental block of Almora district. According to Jugan Pant and Chaudhary (2021), this village has 151 households with a total population of 713 (male, 379; female, 334), which has increased from the 2011 census. The study aims to lay down a better implementation plan for the establishment of climate-resilient ecosystem villages on the basis of people's perception of Bhetuli village (Table 1).

4 Survey Method

The survey was done with the help of a semi-structured questionnaire through the views and perceptions of people in regard to climate change and its impacts. The questionnaire consists of varied questions for the assessment of the perception of

Table 1 General description and demographic details of studied van panchayat (VP)

VP name	Developmental block (district)	Year of formation	Area (ha.)	Dominant forest	Density (t/ha.)	TBA (total basal area) (m^2/ha.)	Dependent population (2011)	Dependent population (2019)
Bhetuli	Takula (Almora)	1978	20.0	*Pinus roxburghii*	1000	57.3	660	713

the community regarding climate change and its impact in natural resources. In the study conducted by Jugran Pant and Chaudhary (2021), the questionnaire consisted of 19 different questions (Annexure I) for the assessment of the perception of the community regarding climate change and its impacts on agriculture and related natural resources. Three options were provided to the respondents: "Yes" (if yes, then what), "No" (if no, then why) and "Don't know" (Dalal et al., 2018) for the clearance of observational studies (Box 1). The survey comprised PRA exercises for sensitisation of village communities and group discussions followed by questionnaire-based survey and observational assessment (Fig. 1). The survey also conducted PRA exercises for sensitisation of village communities and group discussions followed with questionnaire-based survey and observational assessment. The total survey size is 50% of the respondents of the village population.

Box 1 Sample questions regarding climate change impacts on agricultural resources

No.	Questions on climate change impact on the community resources
1	Are you familiar with climate change? If you answered "yes", what effects are you now seeing on your natural resources?
2	Decline in water resources from forest
3	Decline in biodiversity of the forest
4	Change in forest species composition
5	Do you notice changes in forest species (decline/increase)?
6	Do you notice changes in traditional agricultural practice?
7	What are the techniques used in crop farming?
8	What kind of modification you have done for your crop sustenance?
9	What kind of modification you have done for your water resource sustenance?
10	What kind of modification you have done for your forest resource sustenance?

Modified from Pant and Chaudhary (2021)

Observation is extensively used in this study as a major tool for gaining information regarding the concerned subject. It is one of those techniques that is used in community research or in social science. In this process, we observed the community without participating in or getting engaged in the day-to-day activities of communities. In this process, our observations are recorded directly on paper which is both quantitative and qualitative on the basis of our required research. We also had used a process of making mental notes of what we observe and can make jotted notes about the local perception regarding climate change and related issues. These jotted notes had helped us in writing a set of full field notes for a better analysis of the topic. This work is conducted by making noting patterns in the data and by drawing conclusions about various facts of the community in relation to climatic challenges and its mitigation. It is also to be noted that while performing research, various

Fig. 1 Questionnaire-based survey and group discussion in the target village

methodological concerns are also kept in mind such as accessibility of observational opportunities and appropriation of recording unobtrusive observations.

In this work, we have also used ethnographical approach, popularly known as participant observation (PO). The ethnographical approach is amongst the most used methods in the community research domain. With the help of this technique, we had participated in community activities by taking prior permission from those being observed in these activities. An advantage of ethnography is that we are able to observe aspects by being in the natural settings of the community. We had also taken care of the methodological issues that are also taken full care in this study. Ethically, as a researcher, we had provided information about our identity and affiliation, an account or description of our work on how our findings will be analysed and disseminated and the risks and benefits that accrue to those being observed. Both the methodological and ethical concerns are anticipated and addressed before the study is engaged. In this process, we are able to gain a wide range of perspectives regarding climatic problems and issues of the villagers and their resilient adaptation in real-life situations.

Table 2 Occupational details and landholding percentage of stakeholders in the study area

Village	Occupational structure			Landholding percentage (%)		
	Primary sector (%)	Secondary sector (%)	Tertiary sector (%)	Semi-marginal (below 0.5 ha)	Marginal (0.5–1 ha)	Small (1–2 ha)
Bhetuli	54	10	36	97.35	1.32	1.32

5 Results

The findings demonstrated that the local community of the hamlet is strongly dependent on climate-sensitive industries, such as agriculture and forestry, and their interconnections. Jugran Pant and Chaudhry (2021) claim that this condition has the potential to lead to a fall in food and nutritional security as well as a reduction in the systems that support livelihoods.

Forest vegetation in the study zone is dominant with *Pinus roxburghii* Sarg., locally known as chir pine which is a fire-prone species and possesses a risk of unintended cases of forest fires in the study region. In Bhetuli, high sapling density was recorded and the region is abundant with a very high number of *Pinus roxburghii* saplings (866).

According to the research by Jugran Pant and Chaudhary (2021), *Pinus roxburghii* is the dominant tree species in the examined van panchayats, and the higher regeneration status of this species is a strong indicator of its higher regeneration potential and its capacity to colonise in unfavourable conditions.

In Bhetuli village, 54% of the working population is under the primary sector including cultivators and agricultural labours, 10% is under the secondary sector which includes self-employed people (i.e. shop keepers), and 36% are involved in the tertiary sector which includes people in government and private jobs. The stakeholders of the study village have marginal landholdings, with 97.35% of people with semi-marginal landholding size (i.e. below 0.5 ha) and 1.32% of people with small (1–2 ha) landholding size indicating that almost all the village people have below semi-marginal landholding size agricultural land (Table 2).

The population of the village is greatly dependent on agriculture and rainwater seems to be the only major source of irrigation, but due to inconsistent rainfall and changing weather patterns, people are forced to decrease the cultivation of wheat and rice crops. It is also predicted that if this condition of rising temperature and declining rainfall continues, then the cultivation of crops such as wheat will become impossible. The major crops cultivated in the agricultural area are presented in Table 3.

Local people have also perceived that in the past decade, it was observed that invasive species had paved their way and are also getting spread. According to locals, introduction of alien plant is due to changing climatic conditions, and these invasive plant species are also the cause of various ailments such as respiratory problems, skin problems, fever, and headaches.

Table 3 Major crops cultivated in the study area

Crops	Common name	Botanical name
Cereals	Wheat	*Triticum aestivum*
	Rice	*Oryza sativa*
	Maize	*Zea mays*
	Millets	*Panicum miliaceum*
Pulses	Horse gram	*Macrotyloma uniflorum*
	Black gram	*Vigna mungo* L.
	Lentils	*Lens esculenta*
	Chickpea	*Cicer arietinum*
	Soyabean	*Glycine max*
Other vegetables	Chilli	*Capsicum frutescens* L.
	Capsicum	*Capsicum annuum*
	Potato	*Solanum tuberosum*
	Tomato	*Solanum lycopersicum*
	Onion	*Allium cepa*

6 Adaptive Measures

Well-managed natural resources, improved food security, the development of social and human capital and reinforced institutional frameworks are necessary for adaptation (Adger et al., 2003). Moreover, multilevel interventions on climate change and variability are required, taking into account the participation of a wide variety of stakeholder (Mongi et al., 2010).

In regard to the above context, it is clear that the village ecosystem in the mountain region of Himalayas is in great threat, making villages vulnerable to thrive. According to various studies, it was assessed that the inhabitants residing in the remotest of the areas are those who are least able to adapt. Therefore, it alarms for the need for adaptive strategies for making these rural settlements resistant to the changing climate. The IPCC Third Assessment Report states that adaptation has the ability to lessen the negative effects of climate change while enhancing positive effects (IPCC, 2007). There is a need for the development of a knowledge base with a basic framework with a concept for climate change adaptation enabling communities for harnessing the benefits through the proper and better implementation of the model.

Designing effective climate change adaptation strategies necessitates a consistent understanding of the context of vulnerability and local ground realities, which is impossible without taking into account the indigenous knowledge and experiences of local resident communities that have helped them cope with extreme weather and environmental change over centuries (Negi et al., 2017). This information gained by the locals will help in the development of strategically planned activities or outline based on the survival tactics that was continued since centuries by these local inhabitants as the locals created their adaptation tactics to deal with climate change in reaction to changes in the weather and climate using their prior knowledge and indigenous wisdom (Negi et al., 2017). Also, it is believed that if local communities are adequately trained to use their indigenous knowledge, adaption techniques based on their knowledge and capabilities will result in a far better adaptation by the community (Jugran Pant & Chaudhry, 2021). Formal adaptation and mitigation measures would become more regionally nuanced and, therefore, more effective at managing the region's increased weather unpredictability due to climate change if local perspective and traditional knowledge were to be incorporated (Maikhuri et al., 2019). The actions and attitudes related to hazards are greatly influenced by the local context since they are ingrained via daily experience (Roy et al., 2012).

On the basis of their traditional knowledge and learning experiences, the locals of mountainous terrain are experimenting with novel techniques to combat climate change. These communities are more concerned about how they will combat and survive in the face of climate change than they are about how much change is taking place. Agricultural communities observed climatic conditions as part of a socio-ecological system, and this traditional knowledge would in fact support scientific understanding of climate change rather than replacing it (Saxena and Rao, 2009).

Hence, mountain agricultural initiatives must be integrated with the region's ongoing poverty eradication programmes, food and livelihood improvement plans, rural job programmes, ecological restoration efforts and climate change adaptation methods. Especially for the impoverished and underprivileged in mountain communities, tourism would greatly help to secure viable alternative livelihood options by leveraging both the socio-cultural and ecological qualities of the mountain region. Yet, tourism needs to be connected with regional agricultural and food systems in order to become an ecologically friendly, commercially viable and pro-poor livelihood adaptation strategy (Tiwari & Joshi, 2015).

Designing effective climate change adaptation strategies necessitates a consistent understanding of the context of vulnerability and local ground realities, which is impossible without taking into account the indigenous knowledge and experiences of local resident communities that have helped them cope with extreme weather and environmental change over centuries (Negi et al., 2017). The above statements clarify the fact that resilience strategies for these villages in the mountain region can be formed on the basis of adaptive strategies that had been continued since long past and is managed successfully by the communities. In this process, traditional knowledge system (TKS) plays a major role for framing the outline of climate-resilient village ecosystem.

The current study also demonstrates that there is a significant impact of climate change on the community's livelihood opportunities, which has a major perception of its impact on their daily activities as well as their vigour for sustainable utilisation of natural resources in their vicinity, as the abrupt weather conditions and climatic anomalies have impacted the socio-economical and livelihood generation sectors, with the prominent one being agriculture. It is important to realise that environmental changes affect rural households directly and indirectly in addition to occasionally acting as a driving force behind migration due to livelihood. The options for rural households who rely on natural resources for consumption or to improve their standard of living are limited as a result of climate change's reduction in the availability of local natural resources.

7 Discussion

In the present scenario, it became really important for these rural areas to have sound adaptive measures against irregular climatic phenomena so that people living here can become resilient to climate change. Resilience can be viewed of as a strategy to bridge the conceptual gap between mitigation strategies for climate change and adaptation strategies, on the one hand, and enhancing the capacity of human and non-human systems to endure and respond to changes in the Earth's climate, on the other hand.

According to the World Resources Institute (WRI), resilience could be built by the following measures:

- Building the long-term capacity of the central ministries of finance, economy and planning and agriculture of developing countries to comprehend and manage climate threats
- Promoting a paradigm change in adaptation so that local actors and communities have more authority and resources to develop climate resilience
- Promoting transformative ways to adaptation while increasing investments in digital climate advisory services, demand-driven approaches to research and development and resilient agricultural solutions.
- Encouraging governments, funding agencies and practitioners to build and reinforce pledges, investments and partnerships in order to support the scaling-up of nature-based solutions for adaptation.

Therefore, the present work highlights the effects of climate change on society and the economy of communities residing in lower altitude regions of the Himalayas or at lesser Himalayas. This work on the basis of native perception also describes various kinds of problems that rural communities face due to climate change and its devastating effects that can be a threat to their livelihood as well as to their survival. This work mainly tries to put down a detailed account on the mitigation strategies and measures that can be adopted by the residents of these regions to shield themselves against these continued changing weather patterns so that these villages will be developed into climate-resilient ecosystems.

8 Conclusion

The conditions of communities living in the Himalayan region are indicative of climate change, and we can see its effects in a variety of areas such as agriculture, cattle and other sectors. However, there is little precise information and research on how CC affects human well-being in the Himalayas (Negi & Mukherjee 2020).

Because mountain communities are heavily reliant on the forest or other natural resources, the effects of climate change there are most severe.

Natural resource governance that is collaborative and community-based strengthens institutional capacity by supporting local organisations with institutional and policy support, imparting necessary training and providing operational resources. Collaboration and institutional frameworks promote "social capital", which enables locals to acquire commodities that are otherwise impossible to obtain through individual efforts (Johnson, 2004). Dynamic social structures in rural and mountainous areas enable locals to implement adaptive solutions by facilitating access to resources, encouraging mutual trust and supporting decision-making (Belay & Fekadu, 2021; Postigo, 2021). Effective CBA has been seen to result from proactive local involvement in the development process (Forsyth, 2013).

In this study, we had taken the perception of the community in relation to climate change. According to the natives of the studied region, there is the significant reduction in agricultural outputs due to inappropriate weather distribution in the past few decades. In the study area, mostly people are negatively affected by climatic variations, and according to them, unpredicted weather scenarios and increasing crop pests were amongst the greatest challenges. The availability of livelihood options is less, which makes communities highly dependent on natural resources.

In response to climate change, only a few adaptation strategies were observed, which raise a question as to whether these plans are sufficient enough to stand against future changing scenarios or not. The lack of facilities, opportunities and challenges to live in tough conditions had forced the people of the region to seek new alternatives, and in this process, the villages will continue to be depopulated and soon no human population will remain in the mountain village ecosystems.

The present study shows that the majority of people living in the most remote areas and belonging to the lowest economic strata are the ones who are least able to adapt to changing circumstances. It is noted that implementing adaptation strategies requires the availability of resources (e.g. the potential to migrate or to engage in off-farm activities). However, because they have more land and more livestock, which diversifies their income sources and enables them to build safety nets against problems, the wealthier and better-educated population that lives close to market centres tends to be the most robust to adopt towards changing environmental conditions.

There is a need for enhancement in adaptive capacity to mitigate climatic adversity. There are various measures and practices that can help in increasing resilience against changing climate. Climatic disturbances could cause drought, flood, uneven weather patterns, etc., which could cause immense amounts of damage. Therefore, there is a need to develop resilient communities that could resist the negative impacts of climate change. According to the resource endowments of the farmers, the notion of climate-resilient villages (CRVs) entails applying these resilient practices at a scale that covers the entire village in a saturation mode with one or more interventions for enhancing resilience to the production systems. A village comprises adjacent farms that are well-integrated into the surrounding landscape and also have residences (Rao et al., 2016, 2020). Coordination at the farm and landscape scales can achieve significant synergies for agricultural output and climate adaptation and mitigation, as well as other livelihood and environmental objectives, through climate-resilient agricultural landscapes/villages (Scherr et al., 2012). A climate-resilient village (CRV) is required at the local level to stabilise the way of life and increase the available sources of income by diversifying the opportunities in the context of extreme climatic events such as earthquakes.

In recent years, there is a need to create a well-managed environment that can adapt to climate change. Ecosystems and environment have a great impact in human development in each and every context whether it is social, cultural, or economic. Because of their capacity to control climate and natural hazards, ecosystems are essential natural capital in the context of climate change and natural catastrophes. By continuing to offer protective functions and other ecosystem services that are at risk of climate change, well-managed ecosystems can aid in disaster risk reduction and climate change adaptation (Uy & Shaw, 2013). The management of land, water and living resources in an integrated manner for the promotion and sustainable conservation and sustainable use in an equitable way is important for risk reduction and climate change adaptation in the context of disaster risk. This can result in reduced losses and safer and more resilient communities in a proactive way.

In order to "find significant linkages amongst different dimensions and different forms of hierarchy (physical systems; biological, social, and organisational hierarchies)", an ecosystem method might be utilised (Johnson et al., 1999).

To cope up with changing climate, community needs to be more resilient, and for this purpose, communities can align their activities on the basis of climatic events or by inventing unique practices to cope up with continued varying climatic conditions, for the same purpose of forming an eco-sensitive environment and constructing more climate-resistant infrastructure. There is a need for keeping intact with long passing indigenous knowledge practices (IKP) to understand the ways in which natives had tackled climatic problems since centuries in the region.

Appropriate management and utilisation of natural resources for biodiversity conservation along with revival and integration of traditional knowledge system (TKS) in the present time for learning resisting techniques for survival against variation in climatic conditions, bringing out scope to lay down the management strategies for sustainable conservation and for better adjustment along the changing climatic scenario. In addition to the above-mentioned plans, there is a need to highlight alternative livelihood opportunities, so that the residing communities can survive under climate change scenario and preventing the natural resources.

Acknowledgement The authors highly appreciate the research possibilities provided by Prof. (Dr.) Sunil Nautiyal, Director of the GBPNIHE. The authors thankfully acknowledge Professor Sanjay Ghildiyal of the history department at Kumaun University in Nainital for his invaluable contributions to the manuscript. The work's first phase-out was credited to Dr. G.C.S. Negi, a former senior scientist of GBPNIHE, for which the authors are highly grateful. The authors would like to dedicate this work in honour of Dr. Ranbeer Singh Rawal, former director of the GBPNIHE, for his tremendous encouragement and assistance during the project's early stages and his never-ending motivation.

References

Aase, T. H., Chapagain, P. S., & Tiwari, P. C. (2013). Innovation as an expression of adaptive capacity to change in Himalayan farming. *Mountain Research and Development, 33*(1), 4–10.

Adger, W. N., Huq, S., Brown, K., Conway, D., & Hulme, M. (2003). Adaptation to climate change in the developing world. *Progress in Development Studies, 3*(3), 179–195.

Belay, D., & Fekadu, G. (2021). Influence of social capital in adopting climate change adaptation strategies: Empirical evidence from rural areas of Ambo district in Ethiopia. *Climate and Development, 13*(10), 857–868. https://doi.org/10.1080/17565529.2020.1862741

Dalal, N., Joshi, A., Soragi, B., Chaudhary, S., Sharma, S., Naidu, S., & Kazmi, Y. (2018). People's Perception to Climate Change in Remote Himalayan Mountains and Rainfall Variability in the Kailash Sacred Landscape-India. J. Climatol. Weather Forecast, 6.

Forsyth, T. (2013). Community-based adaptation: A review of past and future challenges. *WIREs Climate Change, 4*(5), 439–446. https://doi.org/10.1002/wcc.231

Hoermann, B., & Kollmair, M. (2007). Labour migration and remittances in the Hindu Kush-Himalayan region, Status Report, International Centre for Integrated Mountain Development (ICIMOD), Kathmandu, Nepal.

Houghton, R. A., Lefkowitz, D. S., & Skole, D. L. (1991). Changes in the landscape of Latin America between 1850 and 1985 I. Progressive loss of forests. *Forest Ecology and Management, 38*(3–4), 143–172.

Hunter, L. M., & Sutton, J. (2004). Examining the association between hazardous waste facilities and rural "Brain drain". *Rural Sociology, 69*(2), 197–212.

ICIMOD. (2010). Mountains of the World – Ecosystem services in a time of global and climate change: Seizing opportunities – Meeting challenges, framework paper prepared for the Mountain Initiative of the Government of Nepal by ICIMOD and the Government of Nepal, Ministry of Environment.

IPCC (Intergovernmental Panel on Climate Change). (2013). *Summary for policymakers. Climate change. The physical science basis.* Cambridge University Press.

IPCC (2007). Contribution of working group I to the fourth assessment report of the intergovernmental panel on climate change. In: Solomon S, Qin D, Manning M, Chen Z, Marquis M, et al., editors. Cambridge Univ Press, Cambridge, UK.

Johnson, C. (2004). Uncommon ground: The 'poverty of history' in common property discourse. *Development and Change, 35*(3), 407–434. https://doi.org/10.1111/j.1467-7660.2004.00359.x

Johnson, N. C., Malk, A. J., Szaro, R. C., & Sexton, W. T. (Eds.). (1999). *Ecological stewardship: A common reference for ecosystem management.* Elsevier Science.

Jugran Pant, H., & Chaudhary, N. (2021). A case study towards understanding the impact of changing climatic pattern on village resources in Indian Himalayan Region (IHR). *Journal of Uttarakhand Academy of Administration, Nainital, 2*(3), 25–38.

Kollmair, M., Manandhar, S., Subedi, B., & Thieme, S. (2006). New figures for old stories: Migration and remittance in Nepal. *Migration Letters, 3*, 151–160.

Lise, W. (2000). Factors influencing people's participation in forest management in India. *Ecological Economics, 34*(3), 379–392.

Maikhuri, R. K., Rawat, L. S., Maletha, A., Phondani, P. C., Semwal, R. L., Bahuguna, Y. M., & Bisht, T. S. (2019). Community response and adaptation to climate change in Central Himalaya, Uttarakhand, India. Tropical Ecosystems: Structure, Functions and Challenges in the Face of Global Change, 213–231.

Mongi, H., Majule, A. E., & Lyimo, J. G. (2010). Vulnerability and adaptation of rain fed agriculture to climate change and variability in semi-arid Tanzania. *African Journal of Environmental Science and Technology, 4*(6), 371–381.

Negi, V. S., Maikhuri, R. K., Pharswan, D., Thakur, S., & Dhyani, P. P. (2017). Climate change impact in the Western Himalaya: people's perception and adaptive strategies. *Journal of Mountain Science, 14*(2), 403–416.

Negi, G. C. S., & Mukherjee, S. (2020). Climate change impacts in the Himalayan Mountain ecosystems. Encyclopedia of the World's biomes. Elsevier, 349–354.

Negi, V. S., Thakur, S., Dhyani, R., Bhatt, I. D., & Rawal, R. S. (2021). Climate change observations of indigenous communities in the Indian Himalaya. Weather, *Climate, and Society, 13*(2), 245–257.

Postigo, J. C. (2021). The role of social institutions in indigenous Andean Pastoralists' adaptation to climate-related water hazards. *Climate and Development, 13*(9), 780–791. https://doi.org/10.1080/17565529.2020.1850409

Rama Rao, C. A., Raju, B. M. K., Subba Rao, A. V. M., Rao, K. V., Rao, V. U. M., Kausalya Ramachandran, Venkateswarlu, B., et al. (2016). A district level assessment of vulnerability of Indian agriculture to climate change. *Current Science,* 1939–1946.

Rao, C. S., Prasad, J. V. N. S., Choudhari, S. K., & Singh, A. K. (2020). Mainstreaming climate resilient villages in national programmes towards sustainability of agriculture and environment in India. *Climate Change and Environmental Sustainability, 8*(2), 116–133. https://www.wri.org/equitable-development/climate-resilience

Rawat, D. S., Joshi, M., Sharma, S., Rikhari, H. C., & Palni, L. M. S. (1997). Sustainable development and management of rural ecosystems in the Central Himalaya: a case study from Haigad watershed. *The International Journal of Sustainable Development & World Ecology, 4*(3), 214–225.

Roy, P. N. S., Mondal, S. K., & Joshi, M. (2012). Seismic hazards assessment of Kumaun Himalaya and adjacent region. *Natural Hazards, 64*, 283–297.

Sati, V. P., & Sati, V. P. (2020). Population, Social and Economic Change. *Himalaya on the Threshold of Change*, 113–137.

Saxena, K. G., & Rao, K. S. (2009, April). Climate change, biodiversity and livelihoods in Indian Himalaya: the elements of uncertainty and diverse perceptions. In Proceedings of National Seminar on Climate Change: Data Requirement and Availability 16–17 April 2009 (p. 104).

Scherr, S. J., Shames, S., & Friedman, R. (2012). From climate-smart agriculture to climate-smart landscapes. *Agriculture & Food Security, 1*, 1–15.

Seidler, R., & Bawa, K. S. (2016). India faces a long and winding path to green climate solutions. *Proceedings of the National Academy of Sciences, 113*(44), 12337–12340.

Shackleton, S., & Gumbo, D. (2010). Contribution of non-wood forest products to livelihoods and poverty alleviation. In The dry forests and woodlands of Africa (pp. 63–91). Routledge.

Sharma, E., Chettri, N., Tse-ring, K., Shrestha, A. B., Jing, F., Mool, P., & Eriksson, M. (2009). Climate change impacts and vulnerability in the Eastern Himalayas, ICIMOD, Kathmandu.

Sharma, E., et al. (2019). Introduction to the Hindu Kush Himalaya assessment. In P. Wester, A. Mishra, A. Mukherji, & A. B. Shrestha (Eds.), *The Hindu Kush Himalaya Assessment-Mountains, climate change, sustainability and people*. Springer-Nature.

Sherpa, D. (2007). New vulnerabilities for mountain women: A different light on the Greater Himalaya, Status Report, International Centre for Integrated Mountain Development (ICIMOD), Kathmandu, Nepal.

Shrestha, A. B., Agrawal, N. K., Alfthan, B., Bajracharya, S. R., Marechal, J., & Van Oort, B. (2015). *The Himalayan climate and water atlas: Impact of climate change on water resources in Bve of Asia's major river basins*. GRID-Arendal and CICERO, ICIMOD.

Tiwari, P. C., & Joshi, B. (2015). Climate change and rural out-migration in Himalaya. *Change and Adaptation in Socio-Ecological Systems, 2*(1). https://doi.org/10.1515/cass-2015-0002

UNEP. (2004). *The fall of water*. United Nations Environment Programme, GRID-Arendal. http://www.grida.no/_documents/himalreport_scr.pdf

Uttrakhand Migration Commission Report. (2023). pp 1–97.
Uy, N., & Shaw, R. (2013). Ecosystem resilience and community values: Implications to ecosystem-based adaptation. *Journal of Disaster Research, 8*(1), 201–202.
Xu, J., Grumbine, R. E., Shrestha, A., Eriksson, M., Yang, X., Wang, Y. U. N., & Wilkes, A. (2009). The melting Himalayas: cascading effects of climate change on water, biodiversity, and livelihoods. *Conservation Biology, 23*(3), 520–530.

Managing Disasters in Mountains: Challenges in the Era of Global Warming

Laila Shahzad, Maryam Ali, Faiza Sharif, and Arshad Ali Shedayi

Abbreviations

DRM	Disaster risk management
DRR	Disaster risk reduction
DTM	Digital terrain models
GLOFs	Glacial lake outburst floods
ICS	Incident command system
MSMRMF	Mountain-specific risk management framework
SDGs	Sustainable Development Goals

1 Introduction

Mountains have a significant role in providing water, energy, many kinds of agricultural products, minerals, and recreational values. Mountains are a source of tourism and having many sporting and aesthetic values. They are an important part of the

L. Shahzad (✉)
Institute for Risk and Disaster Reduction, University College London, London, UK

Sustainable Development Study Center, Faculty of Mathematical and Physical Sciences, Government College University, Lahore, Pakistan
e-mail: lailashahzad@gcu.edu.pk

M. Ali · F. Sharif
Sustainable Development Study Center, Faculty of Mathematical and Physical Sciences, Government College University, Lahore, Pakistan

A. A. Shedayi
Karakorum International University, Gilgit, Pakistan

S. Tripathi et al. (eds.), *Warming Mountains*,
https://doi.org/10.1007/978-3-031-62197-0_11

global ecosystem as they serve as a home to biological diversity and a number of endangered species (Alcántara-Ayala & Moreno, 2016). Orographic features contain approximately 25% of continental surfaces, and approximately 26% of the world's population is present in the mountain ranges or within the mountains (Klein et al., 2019). In addition, 40% of the population of the world resides in the watersheds of rivers originating from different mountain ranges. Environmental degradation is happening in many mountain regions. Anthropogenic activities such as land use practices have changed the shape of environment by destroying it (Beniston et al., 1997).

However, the increased human pressure such as climate change will increase the resilience of ecosystems. Different disasters such as landslides, debris flow, ice and snow avalanches, and floods are happening due to climate change. Others are ice volume loss, permafrost degradation, and increased precipitation (Casagli et al., 2023). Many of the glaciered mountain ranges are present on the earth plate boundaries which increase their chances of experiencing climate change, earthquakes, and volcanic eruptions. It has been stated that at the lower latitude, mountainous regions are more sensitive to climate change than other land surfaces (Messerli & Ives, 1997). Different studies and climate models have shown that high elevations such as mountains have been warming at a higher rate than the low elevation surfaces in the past years (Xue et al., 2022).

Since 1999, Tibetan glaciers are melting very fast, with approximately 0.77 km^2 per surface losses recorded (Cullen et al., 2006). IPCC states that 40–50% of mountain glacier mass will disappear in 2100 worldwide. Mountains are the unique areas for the assessment of climate change and its associated impacts, since climate changes rapidly with an elevation over relatively short horizontal distances, as in the case of vegetation and hydrology (Whiteman, 2000). Moreover, plant species occurring in the mountains are often endemic, because species remain isolated at high surfaces as compared to the low vegetation species (IPCC, 2007).

In socioeconomic terms, mountain landscapes provide opportunities for recreation and tourism. Besides this, tourists at a larger number impose environmental stress on mountain resources (Godde et al., 2000). The burgeoning population growth and enhanced industrialization are one of the major players involved in changing environment and climate. Economic growth and population growth are the two main factors that have been imposed by humans on the natural environment (Liu et al., 2009). Different resources such as energy, industrial goods, commodities, and freshwater and agricultural products are consumed by people and overpopulation increased pressure on these resources. As a result, environmental degradation takes place and the overall structure of the environment is lost (Qin et al., 2009). Due to the lack of effectiveness of the risk management system, the mountains are highly vulnerable to disasters (Moos et al., 2018). Population growth, unsustainable practices, and urbanization are considered disturbing factors to minimize the risks (UNISDR, 2015). In the recent era, the management of disasters has become a major challenge for many stakeholders. Anticipation of the future risks of disasters and reduction of exposure and vulnerability to risk factors are included in the disaster risk reduction (DRM) (Vaidya et al., 2019). The multi-hazard-based risk assessment can be used to facilitate different policies in the effective management of risk control (UNISDR, 2015). Past approaches only focus on hazard risks associated with climate change in mountains and do not give the cumulative results of the risk assessment. Bayesian networks, agent-based models, system dynamic models, event and fault trees, and hybrid models are five recent modeling approaches that were used in the multi-risk assessment and climate change adaptation in different mountain areas (Alcantara-Ayala et al., 2015). To investigate the high-mountain processes, digital terrain models (DTMs) are widely used. To see surface displacements in the glaciers and landslides, the multi-temporal data is basically used. A combination of DTMs and the multi-temporal data will provide an easy assessment method for the different potential risks. In the modeling of the hazards and risk assessment, the graphical information system (GIS) is also used (Sekhri et al., 2020).

Coordinated and joint-based operating centers for flood alert and response can also be used for the early detection of different natural hazards (Sorenson, 2000). Early warning systems are also used in which the observations of field, observations of the satellites, different software, data management and storage, and inclusion of the different assessment methods and alert systems were used. The areas in which field data is limiting the remotely sensed data is basically used (Basher, 2006).

This chapter will present the major challenges that were faced by the mountains globally due to the changing climate and their possible solutions and approaches to manage disasters in mountain regions.

2 Challenges due to Global Warming and Climate Change

2.1 Change in the Hydrological Cycle

The warming of the climate can increase or enhance the hydrological cycle (IPCC, 1996). Due to this, the rate of evaporation increases and the liquid to solid precipitation rate also changes, thereby affecting the soil moisture and reserves of the groundwater and also changing the periods of floods and droughts (Gong et al., 2019). Water is present in ample quantity and is associated with the global hydrological cycle. Due to urbanization, different industrial activities, and population growth, there is in increase in the usage of water (Vaidya et al., 2019). During the past years, it has been seen that water resources are polluted and there is also shortage of water. This scenario enhances pressure to the water availability for the generations of the twenty-first century (Shiklomanov, 2001). Snow and the ice are the major components of the hydrological cycle because both are sources of water, but due to climate change, there is an increase in temperature. As a result, glaciers that are sensitive to the increase in temperature begin to melt and the whole process of the precipitation changes (Haeberli & Beniston, 1998). Seasonal-based snow pack changes its shift patterns due to climate change (Liu et al., 2018). With the increased warming in the future due to climate change, the areas that experience snowfall now face precipitation as the temperature increases (Adhikari et al., 2018). To determine the amount of snow which is affected by climate change a snow model is widely used. In the southern part of the French Alps, there is a reduction in the amount of snow and this data is provided by the sensitivity studies (Prasad et al., 2016). The annual cycle of the vegetation of the mountains and seasonal runoff are affected because of the reduced snow cover. The temperature in the winter seasons is exceeding the freezing point, and there is more chances of the early melting of the snow (Manandhar et al., 2015).

2.2 Disturbance in Natural Ecosystems

The plant life present in the mountains is affected by various environmental factors such as temperature, snow, wind, precipitation, and water availability. Plants exhibit some morphological and physiological adaptations through which they respond to climatic changes (Gaire et al., 2015). When climate change induced disturbances exceeding from the adaptive capacity of environment, then plants' growth and their diversity are also affected (Xu and Wang, 2019). When global warming happens, the weather becomes much drier and the temperature of the atmosphere increases, and this phenomenon affects the growth of the plants due to evapotranspiration. Mt. Kinabalu in the Malaysian part of Borneo is the example in which the plant species die due to the shortage of the water. The snow cover that is present in the mountains is characterized by low temperature and precipitation (Davis, 1989). This snow cover also provides protection and water (in spring) to the forest and the plants that are present in the mountains. But due to the global warming, the temperature increases and the snow begins to melt and the plant and vegetation present in the mountains were affected (Zhao et al., 2018). To determine the impacts of climate change on the biomass of the forests and the composition of species, different modeling studies are used. Forest gap models are basically used to determine the impacts of climate change. VEMAP, a continental-scale vegetation response study of the United States, shows that due to the increase in the concentration of the CO_2, some alpine and subalpine regions move toward higher elevations and thus their numbers decrease (González-Arqueros et al., 2018). Gap model applied to British Columbian mountains have also shown that some upward-moving forests disappeared because of the decrease in their potential habitats. The reasons are the lack of cooling and their greater sensitivity toward floods and droughts conditions (Terzi et al., 2019). We tried to highlight the broader challenges of global warming from natural hazards in Fig. 1.

2.3 Risk of Natural Hazards

Mountain's regions are more vulnerable to different natural hazards and risks (Shahzad et al., 2019a). These risks and hazards can cause destruction and death at any time. Landslides and falling of rocks that are basically enhanced and triggered

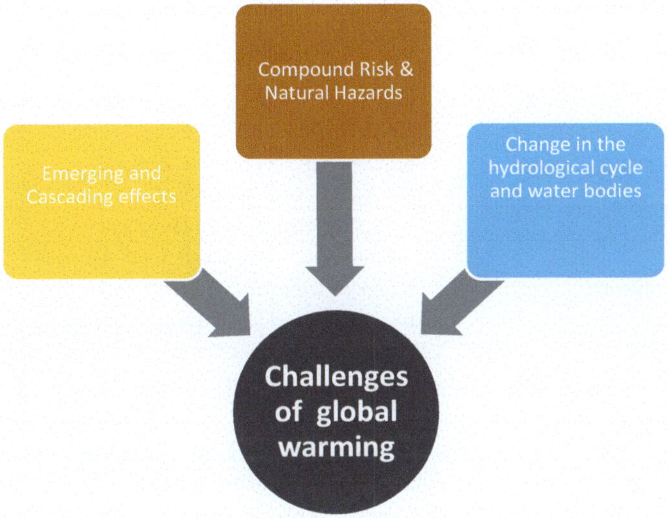

Fig. 1 Challenges due to global warming and climate change

by earthquakes can cause of road and infrastructure destructions (Hewitt, 1997). Mountains are basically present in the tectonically active zone, and they are more susceptible to natural hazards such as earthquakes as compared to other areas. Eighty-eight percent of the area in the Andes is susceptible to earthquakes (Mohanty et al., 2019). Melting of the glaciers can also trigger earthquakes (Sauber & Molnia, 2004; Doser et al., 2007). It is seen that the reduction in the snow cover of the southeast Alaska can be linked or responsible for the seismicity there. It is also seen that ice loss in the active volcanic areas can be responsible for the upward movement of the magma (Sigmundsson et al., 2010).

Climate change causes alterations in the monsoon pattern in the mountain regions of South and Southeast Asia, which may increase the frequency of storms and affect vegetation and rainfall patterns (Shahzad et al., 2019b). When vegetation cover decreases, then there is an increase of erosion. Rock fall and debris flow that occur may pose risk to the livelihoods and infrastructure (Shahzad et al., 2019c; Bavay et al., 2009).

Due to global warming, the areas that are already prone to fire are becoming more sensitive due to the rising temperatures. For example, in recent years, fire incidents have been reported frequently in California or the Blue Mountains in Australia (Papathoma-Köhle et al., 2019). Global warming also increases the risk of insect-prone diseases in the mountains which were previously not present (Beniston, 2006). The hazards that are induced by the changing climate and their affected areas are listed in Table 1.

Floods are other hazards that can also cause infrastructure destruction and deaths of the people. There are multiple factors that may cause floods. Poor infrastructure and maintenance of dams, earthquake, melting of snow and ice, and piping of the

Table 1 Hazards due to the changing climate and their affected areas

Hazards due to the changing climate	Affected areas
Earthquakes (seismic hazard)	South America, Southeast Asia, North and Central America
Snow avalanches	Northern Hemisphere, New Zealand
Floods	Asia, Africa, North America
Droughts	Africa, Caucasus, Eastern Himalaya
Landslides	Southeast Asia, Central and South America
Glacier flow outburst	All regions with valley glaciers
Forest fires	Africa, North and Central America, Eurasia
Insect-borne diseases	Africa, Southern Asia, Central and South America

Source: Ch. Marty based on IPCC (2007), and UNEP-WCMC Mountain Watch (2002)

fine sediments are the causative factors of floods. Due to glacier melting, there is outburst of water in the rivers which may cause flooding in mountain areas, for example, in the Andes (Clague & Evans, 2000). Researchers also found that due to climate change, there is degradation of the permafrost in the European Alps (Hegglin & Huggel, 2008). The global warming that happens in the twentieth century shows that there is an increase of 10 m of the thawing steep mountain slopes, and this will move further deeper as the temperature of the environment increases (Imperiale & Vanclay, 2016). Rockslides, rock falling, permafrost degradation, and debris flow are related to the increase in climate change, i.e., the rate of the increase and decrease of precipitation (Sidle & Ochiai, 2006).

2.4 *Major Consequences of Global Warming and Their Impacts on Human Health*

Human health has always been a subject of interest from multiple perspectives, because the burden of diseases affects countries economically as well as demographically. Consequences of the global warming and their impacts on the human health are described in Fig. 2.

2.5 *Impacts on Socioeconomic Development*

Mountains are important because they provide many resources to the local people. Global warming and climate change-induced disasters have very negative impacts on the infrastructure of the people. Roads are broken due to landslides and rock falling and the local community has been affected. People residing in the hilly areas are more prone to disasters (Beniston, 2002).

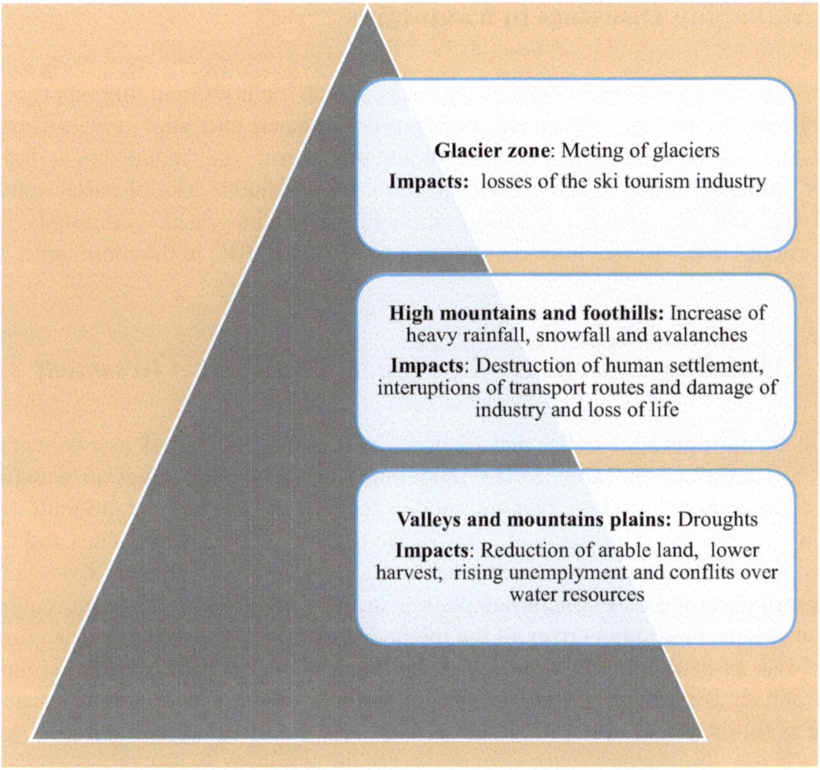

Fig. 2 Consequences of global warming and their harmful impacts

Other important economic factor related to mountains is tourism. It accounts 10% of the world's net financial output. In the developing countries, tourism plays an important role in the income of the people. Patterns of tourism changed due to the addition of new activities (Shahzad et al., 2021). Different resorts and infrastructures were created on the natural remote areas to attract many tourists (Mishra et al., 2019). These infrastructures are mostly located or created in the natural landscapes that impose negative impacts on the sensitive ecosystems and their services that they support and provide. Tourism can both enhance the economic stability of the people of the mountains and create negative impacts to the environment (Papathoma-Köhle et al., 2015). Global warming affects tourism and hence the economic status of the mountain areas in many different ways. In a study, it is shown that if the temperature increases by 2–3 °C by 2050, resorts that were present at below 1200–1500 m above sea level will be badly affected (Xenarios et al., 2019). Floods also affect the tourism badly which is caused by global warming. High mountains also have aesthetic values and their snowy look gives peace to the eyes of people. But because of global warming, the snow cover decreases with the passing of time. The glaciers on Mount Kilimanjaro have almost lost 80% of its ice cover in the last 100 years and will completely lose its snow cover by 2060.

3 Managing Disasters in Mountains

There are different ways to manage disasters, which includes modeling approaches for climate multi-risk assessment, building coordinated and joint centers for risk alert and response, establishment of mountain-specific risk management framework, utilizing remotely sensed data in hazard assessments, use of early warning systems, and designing mitigation strategies and integrated community and ecosystem-based approach for disaster risk reduction (DRR) in the mountains.

3.1 Modeling Approaches for Climate Multi-risk Assessment

There are multiple approaches that are used in the climate multi-risk assessment and they are described in Table 2. The modeling approaches are Bayesian networks, agent-based model, system dynamic models (stock and flow), event and fault trees, and hybrid models (Gallina et al., 2016). We discussed these approaches and their different factors such as uncertainty management, temporal dynamics, spatial analysis, cross-sectoral assessment, adaptation strategy integration, and data required. These approaches do not cover all the methodologies that were used for the climate multi-risk assessment. However, these approaches cover a wide range of complex environmental problems. The three basic features of these modeling approaches were as follows:

- These are integrated environmental modeling techniques in which the qualitative and the quantitative information are present (Mallampalli et al., 2016).
- These may be used in climate change-related uncertainty analysis (Maani, 2013)
- These can be used in the critical infrastructures through the interdependency and cascading effect analysis (Ouyang, 2014).

3.2 Building Coordinated and Joint Centers for Risk Alert and Response

A well-coordinated effort is required for better management of disaster hazards in the mountain areas. Management of flood hazards in the mountain areas is considered the most common theme in the management of all other disaster hazards. For this, the incident command system (ICS) is needed in the emergency management center and can coordinate with the other management centers of a particular region (Malmstadt et al., 2009).

The ICS is an effective and efficient system that can work by integrating a combination of different equipment, facilities, procedures, and protocols and different operational communications that can work within the common organizational

Table 2 Modeling approaches for multi-risk assessment

Modeling approaches	Uncertainty management	Temporal dynamics	Spatial analysis	Cross-sectoral assessment	Adaptation strategy integration	Data required	References
Bayesian networks	Conditional probabilities are considered among the variables	Time step approach is implemented	The model requires integration with GIS software	Socioeconomic parameters are integrated	Assessments of adaptation policies are integrated	Both the limited and excessive data can be dealt with in this model. It contains the qualitative information	Doll and Romero-Lankao (2017)
Agent-based model	Monte Carlo simulations are implemented	Simulated system dynamics	Grid maps are used in this system	Socioeconomic parameters are present in this model	Social adaptation behaviors are monitored	A large number of qualitative data are required	Eid et al. (2017)
System dynamic models (stock and flow)	Monte Carlo simulations are implemented	Changes over time are discussed	The model requires integration with the GIS software	Physical, social, and economic processes are described through this model	Adaptation strategies are included in this model	Qualitative information is required for this model and this model improves system thinking and a large amount of data are required	Duran-Encalada et al. (2017)
Event and fault tree model	Probability of the occurrence of event is considered	In this dynamic, fault tree approach is used	This model requires integration with the GIS software	Socioeconomic parameters are considered	Uncertainty-based assessment of adaptive policies are carried out through this model	For probabilistic assessment, the qualitative data is used, but solutions are also provided in the case of limited data	Pope and Gimblett (2015)
Hybrid models	Monte Carlo simulations are implemented	Exchanging information is carried out	Exchange of information at different spatial levels	Suitable connections between the models are required	Adaptation strategies can be used in the design of these models	High amount of data is required	Bertone et al. (2016)

Table 3 Five essential management functions of the incident command system (ICS)

Incident command	The overall responsibility for the management of the incidents by establishing the strategies, objectives, and procedures of the incident commander
Operations	Different search operations were performed to reach the objectives of the incidents
Planning	The action planning process is supported by the tracking of resources, the collection of the information, and the documentation maintenance
Logistics	In which different resources and services are provided for the achievement of the objectives of the incidents
Finance and administration	The cost related to the all incidents are measured and monitored. The cost analysis is performed

structure. It was developed in the 1970, when there was an outburst of the fire in the California, in which the detailed analysis showed that the reason of the fire was not the lack of resources but the lack of effective management system (FEMA, 2015). This ICS is also integrated at different government levels, i.e., federal, state, local institutions, and agencies, primarily for the better management of disasters. Five major functions are used for the foundation of the ICS, namely, incident command, operations, planning, logistics, and finance and administration. This system works according to the complexity of the incidents. The complex incidents require a larger number of supervisory layers as compared to the smaller incidents (McGinnis, 2014). The five basic major functions of the ICS are mentioned in Table 3.

3.3 Establishment of Mountain-Specific Risk Management Framework (MSMRMF) and Coping Capacity

The framework aims to reduce the different risks by utilizing the internal strength of the system. It utilizes different adaptation techniques to reduce the risk at the maximum level. This system has risk- and vulnerability-based measures, used for the mountainous regions. Different types of indicators are used in this framework (Gill & Malamud, 2016). The overall process of this framework is shown in Fig. 3a, b.

Different physical, social, environmental, and mental factors are considered for the selection of the indicators for this framework. These indicators have different roles and functions for this framework. MSMRMF is a multidisciplinary approach that has basically two main components. The first component is related to the risk as it drives all information related to the risk named “risk information.” The second component of this framework is related to the mitigation of the risk called “risk discourse” (IHCAP, 2018).

Risk information deals with the exposure and vulnerability to climatic conditions and also design methodologies for the assessment of the risk of the mountain areas. The second component deals with the risk mitigation and the coping capacity. The coping capacity is linked to the internal and external strength of the system (Eitner,

2015). The internal strength was gained by the natural or the man-made resources, and the coping capacity is gained by the participation of friends, colleagues, and stakeholders. The external strength is achieved by the availability and the

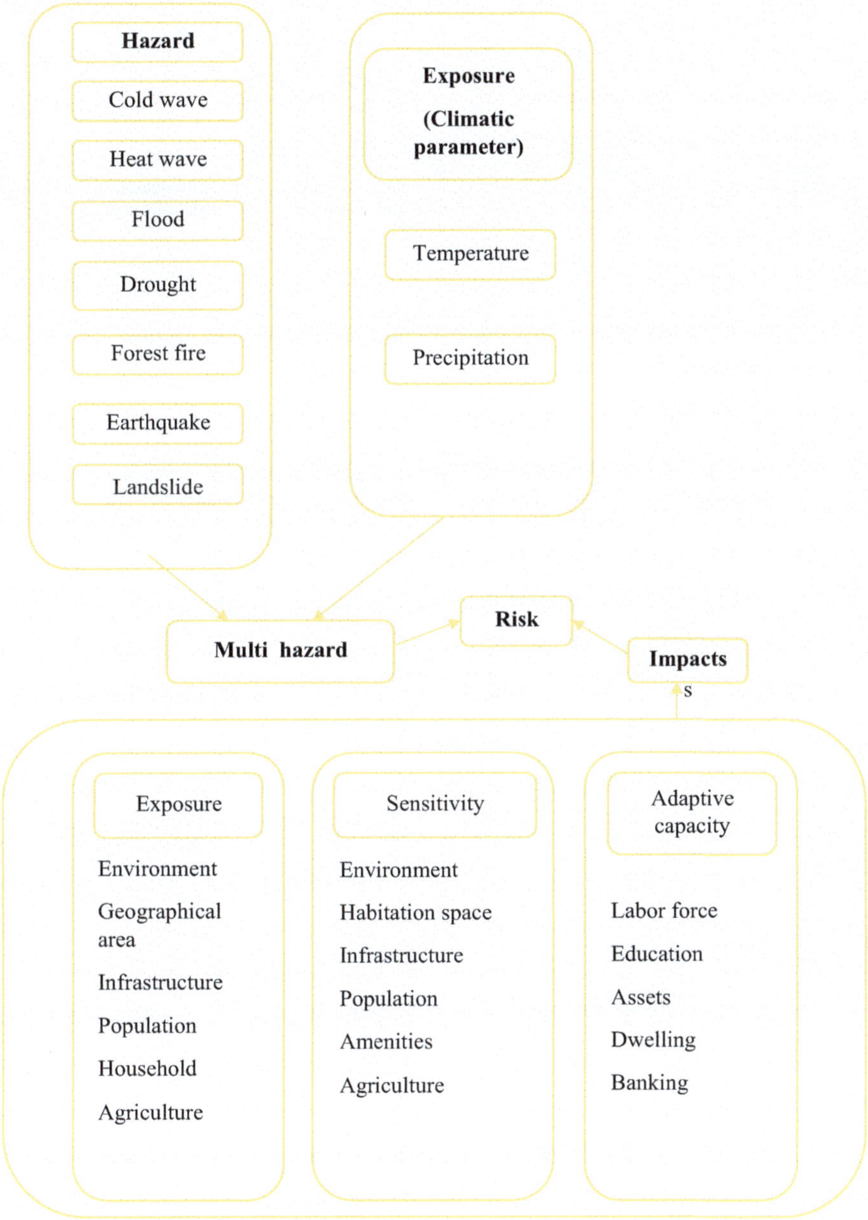

Fig. 3 (**a**) Risk information. (**b**) Risk discourse

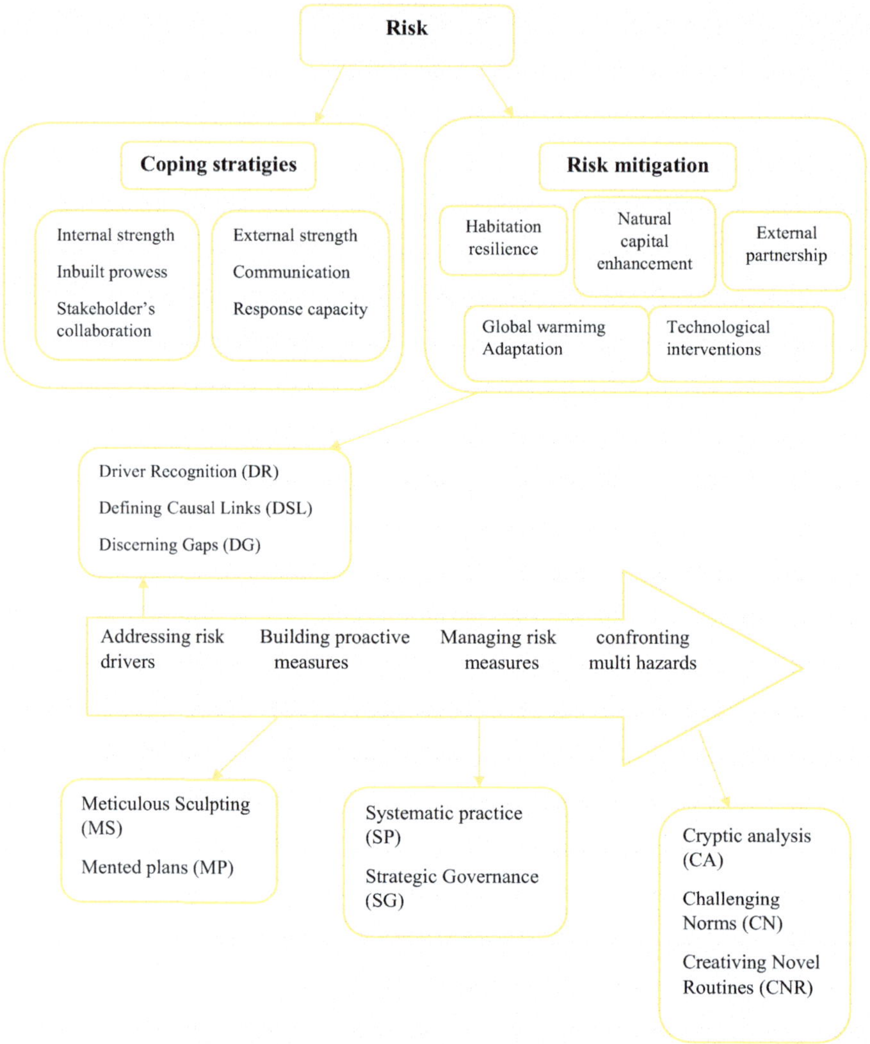

Fig. 3 (continued)

accessibility of the communication. Risk mitigation reduces the severity of the hazards (Scheuer et al., 2011).

3.4 *Utilizing Remotely Sensed Data in Hazard Assessments*

Remotely sensed data can be used in places where the field data is limited. It can play a role in the critical management of floods. This data requires expertise that is usually not present in the rural and mountain areas. This capacity issue is resolved by HAZUS-MH, a GIS-based methodology developed by FEMA (Luke et al., 2015). It can estimate the potential, social, economic, and physical losses very effectively. This information can be used for the mitigation, adaptation, response, and preparation for the natural hazards. Through this, there can be evaluation of the vulnerable populations and the design alternatives for the activation of the different floodways (Pandey et al., 2018).

Besides this, different remote sensing devices such as charged-coupled devices, film photography, radiometers, and infrared active sensors are used for glacier- and permafrost-related hazard assessments in the mountain areas. Different hazard-related areas (present at high altitudes) are present in remote regions, so due to political or physical reasons, it is difficult to obtain information from these areas (Angelopoulou et al., 2019). Different air and spaceborne remote sensing methods are used for hazard assessments in the mountain regions (Xiang et al., 2019). Different image and change detection techniques are also used for the measurement of the different risks (Abdulridha et al., 2019). Digital terrain models (DTMs), synthetic aperture radar, and laser scanning are used in the risk determination of the high mountain areas (Negassa et al., 2020). Large changes in the terrain can be detected by the DTMs. To detect the surface displacements on glaciers, permafrost, and landslides, the multi-temporal data can be used. Geographical information system (GIS) is also used in the risk and hazard assessments of the mountain areas (Szpakowski & Jensen, 2019).

3.5 Use of Early Warning Systems

Advance warning of any hazard can reduce losses of lives and property damages. Because this system can provide early warning signs, the people will have time move to safer locations. Early warning systems have been developed years ago (Sorenson, 2000). In these different systems, we are working on field observations that are related to the depths of the rivers and the precipitation rates and satellite-based observations related to the precipitation and the wind speed. Different software are utilized for weather prediction and understanding hydrological cycles and data storage and management. Besides, hazard assessments and alert systems are also included. There are three types of early warning systems, e.g., by type of hazard, by the level at which it is operated, and by classification as a single or multi-hazard system (Cesarec et al., 2020). In recent years, space technologies have been used as early warning systems, namely, satellite telecommunications and Earth observation. Satellite telecommunications are used in two ways. First, they transmit remotely sensed data for observations, and second, this data is analyzed, which can be used to predict different catastrophic events. Earth observation is used to track the path of cyclones, to assess the severity of cyclones, and to determine the presence of hot spots (Perera et al., 2020). Technological challenges are the limitations of these systems because these are linked to the accuracy and the advance warning of the any system. Perception of the people toward these systems is very important.

These systems focus on both the characteristics and the social vulnerabilities of the people toward disasters. These systems help the people to relocate themselves to the safer locations (Sufri et al., 2020).

3.6 Integrated Community and Ecosystem-Based Approach to DRR in Mountains

In 2015, an earthquake shook Nepal in which a huge destruction of the infrastructure occurred. Many people died due to this disaster. Because of that earthquake, it was realized that there should be an effective disaster risk reduction (DRR) program in mountain areas which can reduce the severity of disasters (Daly et al., 2017). The United Nations (UN) policy frameworks declared that there should be a DRR program to reduce environmental and socioeconomic costs caused by disasters worldwide. Ecosystem-based disaster risk reduction (DRR) is focusing on the restoration, conservation, and the sustainable management of the ecosystem as the major component of the DRR. By integrating the disaster risk reduction (DRR) with the ecosystem, addressing the vulnerabilities of mountain disasters and the ecosystem becomes easy (Sherpa, 2015). However, there are some challenges that DRR faces. It is stated that mountains have many resources but still poor. All the policies that are implemented on the mountains are made by outsiders. Mountains are remote areas but vulnerable to global warming. Demographic shifts are present in mountains and mountains require fine-based scale data but the data is limited. To address these challenges, the ecosystem-based disaster risk reduction (DRR) (Gallen et al., 2017) basically proposed four principles:

- Governance- and institution-based arrangements that can easily fill all the needs of the people
- Encouragement of the communities that are resilient with the capacity building and the sustainable development
- Relief that is basically focused on three things, namely, inclusion, equity, and well-being
- Integrated-based knowledge, learning, and innovation (Mishra et al., 2017)

4 Recommendations

It is recommended that the public should be involved in the effective management of natural disasters. The government should educate and provide awareness to the people on how they can reduce disaster vulnerability. Disaster mitigation planning and disaster response can be used for this purpose (Blancher et al., 2018).

The government should use climate change-detecting tools and models to address different risks and adaptation procedures. Effective monitoring and reporting procedures should be used for proper mitigation of disasters.

Improve the modeling technology for the early and effective management of any kind of risk in mountains areas. The government should increase funding for the effective management of disasters by utilizing modern technologies (Kato & Huang, 2021).

It is recommended that all the stakeholders will work together and identify multiple hazards in mountain areas. All policy makers need to build the capacity of the adaptations and learning programs.

There should be utilization of latest early warning systems and remote sensing technology for attaining the correct information regarding the hazards and risks in mountain areas (Semernya et al., 2017).

5 Conclusion

Mountains are very important as they provide many ecological services, but due to the global warming and climate change, mountains are prone to many disasters such as earthquakes, floods, landslides, forest fire, glacier melting, and permafrost. Mountains faced many challenges due to the global warming and climate change. The major challenges are change in the hydrological system, disturbance in the natural ecosystem, risk of natural hazards, and impacts of global warming on the socioeconomic development. To tackle these challenges, many management strategies can be used. These are modeling approaches for climate multi-risk assessment, coordinated and joint centers for risk alert and response, mountain specific-risk management framework, remotely sensed data in hazard assessments, early warning systems, mitigation strategies, and integrated community- and ecosystem-based approach for disaster risk reduction (DRR) in mountains. In conclusion, there is a need for effective management of disasters in mountains areas as well for reduction of climate change impacts by utilizing the green technology.

References

Abdulridha, J., Batuman, O., & Ampatzidis, Y. (2019). UAV-based remote sensing technique to detect citrus canker disease utilizing hyperspectral imaging and machine learning. *Remote Sensing, 11*(11), 1373. https://doi.org/10.3390/rs11111373

Adhikari, S., Baral, H., & Nitschke, C. (2018). Adaptation to climate change in Panchase Mountain ecological regions of Nepal. *Environments, 5*(3), 42. https://doi.org/10.3390/environments5030042

Alcántara-Ayala, I., & Moreno, A. R. (2016). Landslide risk perception and communication for disaster risk management in mountain areas of developing countries: A Mexican foretaste. *Journal of Mountain Science, 13*, 2079–2093. https://doi.org/10.1007/s11629-015-3823-0

Alcantara-Ayala, I., Altan, O., & Baker, D. (2015). Disaster risks research and assessment to promote risk reduction and management. In A. Ismail-Zadeh, & S. Cutter (Eds.), ICSU-ISSC Ad Hoc Group on Disaster Risk Assessment. http://www.icsu.org/science-for-policy/disasterrisk/documents/DRRsynthesisPaper

Angelopoulou, T., Tziolas, N., Balafoutis, A., Zalidis, G., & Bochtis, D. (2019). Remote sensing techniques for soil organic carbon estimation: A review. *Remote Sensing, 11*(6), 676. https://doi.org/10.3390/rs11060676

Basher, R. (2006). Global early warning systems for natural hazards: Systematic and people-centred. *Philosophical Transactions of the Royal Society, 364*, 2167–2182. https://doi.org/10.1098/rsta.2006.1819

Bavay, M., Lehning, M., Jonas, T., & Lowe, H. (2009). Simulations of future snow cover and discharge in Alpine headwater catchments. *Hydrological Processes, 23*(1), 95–108. https://doi.org/10.1002/hyp.7195

Beniston, M. (Ed.). (2002). *Climatic change. Implications for the hydrological cycle and for water management. Advances in global change research* (503 p). Kluwer Academic Publishers.

Beniston, M. (2006). August 2005 intense rainfall event in Switzerland: Not necessarily an analog for strong convective events in a greenhouse climate. *Geophysical Research Letters, 33*, L05701. https://doi.org/10.1029/2005GL025573

Beniston, M., Diaz, H. F., & Bradley, R. S. (1997). Climatic change at high elevation sites; A review. *Climatic Change, 36*, 233–251. https://doi.org/10.1023/A:1005380714349

Bertone, E., Sahin, O., Richards, R., Mengersen, K., & Roikod, A. (2016). Managing water quality related health risks from extreme events: A coupled Bayesian network and system dynamics modelling approach. Final Rep. – WaterRA Proj: 1071-12.

Blancher, M., Albasini, F., Elsensohn, F., Zafren, K., Hölzl, N., McLaughlin, K., & Paal, P. (2018). Management of multi-casualty incidents in mountain rescue: Evidence-based guidelines of the International Commission for Mountain Emergency Medicine (ICAR MEDCOM). *High Altitude Medicine & Biology, 19*(2), 131–140. https://doi.org/10.1089/ham.2017.0143

Casagli, N., Intrieri, E., Tofani, V., Gigli, G., & Raspini, F. (2023). Landslide detection, monitoring and prediction with remote-sensing techniques. *Nature Reviews Earth & Environment, 4*(1), 51–64. https://doi.org/10.1038/s43017-022-00373-x

Cesarec, I., Mikac, R., & Spevec, D. (2020). The concept of human security as a basis for the application of big data concept in establishment of early warning system for crisis management in the Republic of Croatia. *Croatian International Relations Review, 26*(86), 72–95. https://doi.org/10.37173/cirr.26.86.3

Clague, J. J., & Evans, S. G. (2000). A review of catastrophic drainage of moraine-dammed lakes in British Columbia. *Quaternary Science Reviews, 19*, 1763–1783. https://doi.org/10.1016/S0277-3791(00)00090-1

Cullen, N. J., Moelg, T., Kaser, G., Hussein, K., Steffen, K., & Hardy, D. R. (2006). Kilimanjaro glaciers; Recent areal extent from satellite data and new interpretation of observed 20th century retreat rates. *Geophysical Research Letters, 33*, L16502. https://doi.org/10.1029/2006GL027084

Daly, P., Ninglekhu, S., Hollenbach, P., Duyne Barenstein, J., & Nguyen, D. (2017). Situating local stakeholders within national disaster governance structures. *Environment and Urbanization, 29*(2), 403–424. https://doi.org/10.1177/0956247817721403

Davis, M. B. (1989). Lags in vegetation response to greenhouse warming. *Climatic Change, 15*, 75–82. https://doi.org/10.1007/BF00138846

Doll, P., & Romero-Lankao, P. (2017). How to embrace uncertainty in participatory climate change risk management – A roadmap. *Earth's Future*. https://doi.org/10.1002/2016EF000411

Doser, D. I., Wiest, K. R., & Sauber, J. (2007). Seismicity of the Bering Glacier region and its relation to tectonic and glacial processes. *Tectonophysics, 439*, 119–127. https://ui.adsabs.harvard.edu/link_gateway/2007Tectp.439.119D/doi:10.1016/j.tecto.2007.04.005

Duran-Encalada, J. A., Paucar-Caceres, A., Bandala, E. R., & Wright, G. H. (2017). The impact of global climate change on water quantity and quality: A system dynamics approach to the US? Mexican transborder region. *European Journal of Operational Research, 256*, 567–581. https://doi.org/10.1016/j.ejor.2016.06.016

Eid, M. S., Asce, S. M., El-adaway, I. H., & Asce, M. (2017). Integrating the social vulnerability of host communities and the objective functions of associated stakeholders during disaster recovery processes using agent-based modeling, 1–15. https://doi.org/10.1061/(ASCE)CP.1943-5487.0000680

Eitner, M. A. B. (2015). Ecological vulnerability indicators. *Ecological Indicators, 60*, 329–334. https://doi.org/10.1016/j.ecolind.2015.07.001

FEMA. (2015). *National flood insurance program community rating system: A local official's guide to saving lives, preventing property damage, reducing the cost of flood insurance–FEMA 573*. Federal Emergency Management Agency. http://www.fema.gov/media-library-data/20130726-1708-25045-7720/99032_nfip_small_brochure.pdf

Gaire, S., Castro Delgado, R., & Arcos González, P. (2015). Disaster risk profile and existing legal framework of Nepal: Floods and landslides. *Risk Management and Healthcare Policy*, 139–149. https://doi.org/10.2147/RMHP.S90238

Gallen, S. F., Clark, M. K., Godt, J. W., Roback, K., & Niemi, N. A. (2017). Application and evaluation of a rapid response earthquake-triggered landslide model to the 25 April 2015 Mw 7.8 Gorkha earthquake, Nepal. *Tectonophysics, 714*, 173–187. https://doi.org/10.1016/j.tecto.2016.10.031

Gallina, V., Torresan, S., Critto, A., Sperotto, A., Glade, T., & Marcomini, A. (2016). A review of multi-risk methodologies for natural hazards: Consequences and challenges for a climate change impact assessment. *Journal of Environmental Management, 168*, 123–132. https://doi.org/10.1016/j.jenvman.2015.11.011

Gill, J. C., & Malamud, B. D. (2016). Hazard interactions and interaction networks (cascades) within multi-hazard methodologies. *Earth System Dynamics, 7*, 659–679.

Godde, P., Price, M. F., & Zimmermann, F. M. (2000). *Tourism and development in mountain regions*. CABI Publishing.

Gong, J., Xie, Y., Cao, E., Huang, Q., & Li, H. (2019). Integration of InVEST-habitat quality model with landscape pattern indexes to assess mountain plant biodiversity change: A case study of Bailongjiang watershed in Gansu Province. *Journal of Geographical Sciences, 29*, 1193–1210. https://doi.org/10.1007/s11442-019-1653-7

González-Arqueros, M. L., Mendoza, M. E., Bocco, G., & Castillo, B. S. (2018). Flood susceptibility in rural settlements in remote zones: The case of a mountainous basin in the Sierra-Costa region of Michoacán, Mexico. *Journal of Environmental Management, 223*, 685–693. https://doi.org/10.1016/j.jenvman.2018.06.075

Haeberli, W., & Beniston, M. (1998). Climate change and its impacts on glaciers and permafrost in the Alps. *Ambio, 27*, 258–265.

Hegglin, E., & Huggel, C. (2008). An integrated assessment of vulnerability to glacial hazards. *Mountain Research and Development, 28*, 299–309. https://doi.org/10.1659/mrd.0976

Hewitt, K. (1997). Risks and disasters in mountain lands. In B. Messerli & J. D. Ives (Eds.), *Mountains of the world: A global priority* (pp. 371–408). Parthenon Publishing.

IHCAP. (2018). Indian Himalayan Climate Adaptation Programme – Climate vulnerability assessment for the Indian Himalayan Region using a common framework. Project, Capacity Building on Climate Change Vulnerability Assessment in the States of Indian Himalayan Region, India.

Imperiale, A. J., & Vanclay, F. (2016). Experiencing local community resilience in action: Learning from post-disaster communities. *Journal of Rural Studies, 47*, 204–219. https://doi.org/10.1016/j.jrurstud.2016.08.002

IPCC. (1996). *Climate change. The IPCC second assessment report*. Cambridge University Press. Volumes I (Science), II (Impacts) and III (Socio-economic implications).

IPCC. (2007). Climate change 2007: Impacts, adaptation and vulnerability. In M. L. Parry, O. F. Canziani, J. P. Palutikof, P. J. Van der Linden, & C. E. Hanson (Eds.), *Contribution of*

Working Group II to the fourth assessment report of the Intergovernmental Panel on Climate Change (976 p). Cambridge University Press.

Kato, S., & Huang, W. (2021). Land use management recommendations for reducing the risk of downstream flooding based on a land use change analysis and the concept of ecosystem-based disaster risk reduction. *Journal of Environmental Management, 287*, 112341. https://doi.org/10.1016/j.jenvman.2021.112341

Klein, J. A., Tucker, C. M., Steger, C. E., Nolin, A., Reid, R., Hopping, K. A., & Yager, K. (2019). An integrated community and ecosystem-based approach to disaster risk reduction in mountain systems. *Environmental Science & Policy, 94*, 143–152. https://doi.org/10.1016/j.envsci.2018.12.034

Liu, X., Cheng, Z., Yan, L., & Yin, Z. (2009). Elevation dependency of recent and future minimum surface air temperature trends in the Tibetan Plateau and its surroundings. *Global and Planetary Change, 68*, 164–174. https://doi.org/10.1016/j.gloplacha.2009.03.017

Liu, D., Li, Y., Shen, X., Xie, Y., & Zhang, Y. (2018). Flood risk perception of rural households in western mountainous regions of Henan Province, China. *International Journal of Disaster Risk Reduction, 27*, 155–160. https://doi.org/10.1016/J.IJDRR.2017.09.051

Luke, A., Kaplan, B., Neal, J., Lant, J., Sanders, B., Bates, P., & Alsdork, D. (2015). Hydraulic modeling of the 2011 New Madrid Floodway activation: A case study on floodway activation controls. *Natural Hazards, 77*(3), 1863–1887. https://doi.org/10.1007/s11069-015-1680-3

Maani, K. (2013). Decision-making for climate change adaptation: A systems thinking approach.

Mallampalli, V. R., Mavrommati, G., Thompson, J., Duveneck, M., Meyer, S., LigmannZielinska, A., Druschke, C. G., Hychka, K., Kenney, M. A., Kok, K., & Borsuk, M. E. (2016). Methods for translating narrative scenarios into quantitative assessments of land use change. *Environmental Modelling & Software, 82*, 7–20. https://doi.org/10.1016/j.envsoft.2016.04.011

Malmstadt, J., Scheitlin, K., & Elsner, J. (2009). Florida hurricanes and damage costs. *Southeastern Geographer, 49*(2), 108–131.

Manandhar, S., Pratoomchai, W., Ono, K., Kazama, S., & Komori, D. (2015). Local people's perceptions of climate change and related hazards in mountainous areas of northern Thailand. *International Journal of Disaster Risk Reduction, 11*, 47–59. https://doi.org/10.1016/j.ijdrr.2014.11.002

McGinnis, B. P. (2014). The long slog: What lies ahead for the financing of our nation's inland waterways infrastructure. *Water Resources Impact, 16*(1), 19–20.

Messerli, B., & Ives, J. D. (1997). *Mountains of the world: A global priority*. Parthenon Publishing Group. https://doi.org/10.1002/(SICI)1099145X(200003/04)11:2%3C197::AIDLDR390%3E3.0.CO;2-U.

Mishra, A., Ghate, R., Maharjan, A., Gurung, J., Pathak, G., & Upraity, A. (2017). Building ex ante resilience of disaster-exposed mountain communities: Drawing insights from the Nepal earthquake recovery. *International Journal of Disaster Risk Reduction, 22*, 167–178. https://doi.org/10.1016/j.ijdrr.2017.03.008

Mishra, A., Appadurai, A. N., Choudhury, D., Regmi, B. R., Kelkar, U., Alam, M., & Sharma, U. (2019). Adaptation to climate change in the Hindu Kush Himalaya: Stronger action urgently needed. In *The Hindu Kush Himalaya assessment: Mountains, climate change, sustainability and people* (pp. 457–490). Springer.

Mohanty, A., Hussain, M., Mishra, M., Kattel, D. B., & Pal, I. (2019). Exploring community resilience and early warning solution for flash floods, debris flow and landslides in conflict prone villages of Badakhshan, Afghanistan. *International Journal of Disaster Risk Reduction, 33*, 5–15. https://ui.adsabs.harvard.edu/link_gateway/2019IJDRR..33....5M/doi:10.1016/j.ijdrr.2018.07.012

Moos, C., Bebi, P., Schwarz, M., Stoffel, M., Sudmeier-Rieux, K., & Dorren, L. (2018). Ecosystem-based disaster risk reduction in mountains. *Earth-Science Reviews, 177*, 497–513. https://doi.org/10.1016/j.ecoleng.2021.106485

Negassa, M. D., Mallie, D. T., & Gemeda, D. O. (2020). Forest cover change detection using Geographic Information Systems and remote sensing techniques: A spatio-temporal study on Komto Protected forest priority area, East Wollega Zone, Ethiopia. *Environmental Systems Research, 9*, 1–14. https://doi.org/10.1186/s40068-020-0163-z

Ouyang, M. (2014). Review on modeling and simulation of interdependent critical infrastructure systems. *Reliability Engineering and System Safety, 121*, 43–60. https://doi.org/10.1016/j.ress.2013.06.040

Pandey, R., Alatalo, J. M., Thapliyal, K., Chauhan, S., Archie, K. M., Gupta, A. K., Jha, S. K., & Kumar, M. (2018). Climate change vulnerability in urban slum communities: Investigating household adaptation and decision-making capacity in the Indian Himalaya. *Ecological Indicators, 90*, 379–391. https://doi.org/10.1016/j.ecolind.2018.03.031

Papathoma-Köhle, M., Zischg, A., Fuchs, S., Glade, T., & Keiler, M. (2015). Loss estimation for landslides in mountain areas–An integrated toolbox for vulnerability assessment and damage documentation. *Environmental Modelling & Software, 63*, 156–169. https://doi.org/10.1016/j.envsoft.2014.10.003

Papathoma-Köhle, M., Cristofari, G., Wenk, M., & Fuchs, S. (2019). The importance of indicator weights for vulnerability indices and implications for decision making in disaster management. *International Journal of Disaster Risk Reduction, 36*, 101103. https://doi.org/10.1016/j.ijdrr.2019.101103

Perera, D., Agnihotri, J., Seidou, O., & Djalante, R. (2020). Identifying societal challenges in flood early warning systems. *International Journal of Disaster Risk Reduction, 51*, 101794. https://doi.org/10.1016/j.ijdrr.2020.101794

Pope, A. J., & Gimblett, R. (2015). Linking Bayesian and agent-based models to simulate complex social-ecological systems in semi-arid regions. *Frontiers in Environmental Science, 3*, 1–9. https://doi.org/10.3389/fenvs.2015.00055

Prasad, A. S., Pandey, B. W., Leimgruber, W., & Kunwar, R. M. (2016). Mountain hazard susceptibility and livelihood security in the upper catchment area of the river Beas, Kullu Valley, Himachal Pradesh, India. *Geoenvironmental Disasters, 3*(1), 1–17. https://doi.org/10.1186/s40677-016-0037-x

Qin, J., Yang, K., Liang, S., & Guo, X. (2009). The altitudinal dependence of recent rapid warming over the Tibetan Plateau. *Climatic Change, 97*, 321–327. https://doi.org/10.1007/s10584-009-9733-9

Sauber, J. M., & Molnia, B. F. (2004). Glacier ice mass fluctuations and fault instability in tectonically active Southern Alaska. *Global and Planetary Change, 42*, 279–293. https://doi.org/10.1016/j.gloplacha.2003.11.012

Scheuer, S., Haase, D., & Meyer, V. (2011). Exploring multicriteria flood vulnerability by integrating economic, social and ecological dimensions of flood risk and coping capacity: From a starting point view towards an end point view of vulnerability. *Natural Hazards, 58*, 731–751. https://doi.org/10.1007/s11069-010-9666-7

Sekhri, S., Kumar, P., Fürst, C., & Pandey, R. (2020). Mountain specific multi-hazard risk management framework (MSMRMF): Assessment and mitigation of multi-hazard and climate change risk in the Indian Himalayan Region. *Ecological Indicators, 118*, 106700. https://doi.org/10.1016/j.ecolind.2020.106700

Semernya, L., Ramola, A., Alfthan, B., & Giacovelli, C. (2017). Waste management outlook for mountain regions: Sources and solutions. *Waste Management & Research, 35*(9), 935–939. https://doi.org/10.1177/0734242X17709910

Shahzad, L., Tahir, A., Sharif, F., Khan, W. A., Farooq, A., Abbas, A., & Saqib, Z. A. (2019a). Vulnerability, wellbeing and livelihood adaptation under changing environmental conditions: A case from mountainous region of Pakistan. *Environmental Science and Pollution Research*. https://doi.org/10.1007/s11356-019-05880-x

Shahzad, L., Tahir, A., Sharif, F., Haq, I. U., & Mukhtar, H. (2019b). Assessing the impacts of changing climate on forest ecosystem services and livelihood of Balakot mountainous communities. *Pakistan Journal of Botany, 51*(4), 1405–1414.

Shahzad, L., Tahir, A., Sharif, F., Hayyat, M. U., Ghani, N., Farhan, M., & Dogar, S. S. (2019c). Does livelihood vulnerability index justify the socio-economic status of mountainous community? A case study of post-earthquake ecological adaptation of Balakot population. *Applied Ecology and Environmental Research, 17*(3), 6605–6624.

Shahzad, L., Tahir, A., Dogar, M., & Saeed, S. (2021). A metric-based assessment of climate and tourism in major cities of Pakistan. *Environment Development and Sustainability*. https://doi.org/10.1007/s10668-021-01230-5

Sherpa, L. N. (2015, August 3). Forced smiles. *The Kathmandu Post*. http://kathmandupost.ekantipur.com/news/2015-08-03/forced-smiles.html

Shiklomanov, I. A. (2001). *World water resources at the beginning of the 21st century*. UNESCO Publications.

Sidle, R. C., & Ochiai, H. (2006). Landslides, processes, prediction and land use. *American Geophysical Union, Water Resources Monograph, 18*, 312 p.

Sigmundsson, F., Hreinsdottir, S., Hooper, A., Arnadottir, T., Pedersen, R., Roberts, M. J., Oskarsson, N., Auriac, A., Decriem, J., Einarsson, P., Geirsson, H., Hensch, M., Ofeigsson, B. G., Sturkell, E., Sveinbjornsson, H., & Feigl, K. L. (2010). Intrusion triggering of the 2010 Eyjafljallajokul explosive eruption. *Nature, 468*, 426–430. https://doi.org/10.1038/nature09558

Sorenson, J. (2000). Hazard warning systems: Review of 20 years of progress. *Natural Hazards Review, 1*(2), 119–125. https://doi.org/10.1061/(ASCE)1527-6988(2000)1:2(119)

Sufri, S., Dwirahmadi, F., Phung, D., & Rutherford, S. (2020). A systematic review of community engagement (CE) in disaster early warning systems (EWSs). *Progress in Disaster Science, 5*, 100058. https://doi.org/10.1016/j.pdisas.2019.100058

Szpakowski, D. M., & Jensen, J. L. (2019). A review of the applications of remote sensing in fire ecology. *Remote Sensing, 11*(22), 2638. https://digital.library.txstate.edu/handle/10877/13831

Terzi, S., Torresan, S., Schneiderbauer, S., Critto, A., Zebisch, M., & Marcomini, A. (2019). Multi-risk assessment in mountain regions: A review of modelling approaches for climate change adaptation. *Journal of Environmental Management, 232*, 759–771. https://doi.org/10.1016/j.jenvman.2018.11.100

UNEP-WCMC. (2002). *Mountain watch. Environmental change & sustainable development in mountains*. UNEP-WCMC.

UNISDR. (2015). *Global assessment report making development sustainable: The future of disaster risk management*, Geneva, Switzerland. 9789211320428.

Vaidya, R. A., Shrestha, M. S., Nasab, N., Gurung, D. R., Kozo, N., Pradhan, N. S., & Wasson, R. J. (2019). Disaster risk reduction and building resilience in the Hindu Kush Himalaya. In *The Hindu Kush Himalaya assessment: Mountains, climate change, sustainability and people* (pp. 389–419). https://doi.org/10.1007/978-3-319-92288-1_11

Whiteman, D. (2000). *Mountain meteorology* (355 p). Oxford University Press.

Xenarios, S., Gafurov, A., Schmidt-Vogt, D., Sehring, J., Manandhar, S., Hergarten, C., & Foggin, M. (2019). Climate change and adaptation of mountain societies in Central Asia: Uncertainties, knowledge gaps, and data constraints. *Regional Environmental Change, 19*, 1339–1352. https://doi.org/10.1007/s10113-018-1384-9

Xiang, T. Z., Xia, G. S., & Zhang, L. (2019). Mini-unmanned aerial vehicle-based remote sensing: Techniques, applications, and prospects. *IEEE Geoscience and Remote Sensing Magazine, 7*(3), 29–63. https://doi.org/10.1109/MGRS.2019.2918840

Xu, D., & Wang, X. (2019). Vulnerability of rural household livelihood to climate variability and adaptive strategies in landslide-threatened western mountainous regions of the Three Gorges Reservoir Area, China. *Climate and Development, 11*(6), 469–484. https://doi.org/10.1007%2Fs11069-020-04460-4.

Xue, K., Cao, S., Liu, Y., Xu, D., & Liu, S. (2022). Disaster-risk communication, perceptions and relocation decisions of rural residents in a multi-disaster environment: Evidence from Sichuan, China. *Habitat International, 127*, 102646. https://doi.org/10.1016/j.habitatint.2022.102646

Zhao, G., Pang, B., Xu, Z., Yue, J., & Tu, T. (2018). Mapping flood susceptibility in mountainous areas on a national scale in China. *Science of the Total Environment, 615*, 1133–1142. https://doi.org/10.1016/j.scitotenv.2017.10.037

Ground-Based and Geospatial Measurement of Aerosol and Black Carbon to Assess the Warming Patterns in a Suburban Location in Himachal Pradesh, India

Isha Thakur, Renu Lata, Abhishek Gupta, and Jagdish Chandra Kuniyal

1 Introduction

Warming in mountains is one of the major concerns across the globe. In many mountainous areas of the world, higher warming rates have been seen as elevation has increased or elevation-dependent warming (EDW) tendencies (Pepin et al., 2015, 2022; Rangwala & Miller, 2012). Continuous warming at high elevations may have adverse effects on biodiversity, mountain ecosystems, hydrological regimes, and cryospheric resources (Rangwala & Miller, 2012; Beniston, 2003). Studies of the Tibetan Plateau have shown that aerosols directly absorb and scatter solar radiation which indirectly influence cloud characteristics, suggesting that aerosols play a significant role in the climate system and weather of a region. This is true when surface-dwelling absorbing aerosols such as black carbon (BC) settle into the snow. In fact, it has been seen that aerosols significantly affect how quickly snow and glaciers melt across the Tibetan Plateau (Kang et al., 2019). Anthropogenic aerosols directly scatter and absorb solar radiation which impacts the Earth's radiation budget, which is estimated to be comparable in global extent (Houghton et al., 1996). A number of atmospheric impacts are caused by aerosols. They have a significant role in affecting visibility and formation of clouds, fog, etc. in the lower troposphere and in the production of electrical and chemical effects in the upper region. Further, several studies have also shown that air pollutants, such as black and brown carbon and aerosols, that

I. Thakur · R. Lata (✉) · A. Gupta
G.B. Pant National Institute of Himalayan Environment, Himachal Regional Centre, Mohal- Kullu, Himachal Pradesh, India

J. C. Kuniyal
G.B. Pant National Institute of Himalayan Environment, Kosi-Katarmal, Almora, Uttrakhand, India

S. Tripathi et al. (eds.), *Warming Mountains*,
https://doi.org/10.1007/978-3-031-62197-0_12

absorb light and dust lead to the warming of the atmosphere (Bond et al., 2013; Ramanathan et al., 2007). These particles are regarded as light-absorbing contaminants once they have been added to glacier ice or snow (Light Absorbing Impurities). They can considerably increase the melting of glaciers and snowpack by lowering their surface albedos (Doherty et al., 2010; Flanner et al., 2007; Qian et al., 2014, 2015; Xu et al., 2009). Currently, the study related to air pollutants and climate change including the temperature variations is very limited in Himachal Pradesh. To understand the warming patterns in the Indian Himalayan region, it is essential to understand the behavior of aerosol as well as black carbon in the area. In the present study, a multiwavelength radiometer was used to monitor AOD at Mohal, Himachal Pradesh. The objective of the present study is to understand the diurnal and seasonal pattern of AOD and black carbon in the region as any change in their concentration can alter the dynamics of the atmosphere. This study will be helpful in understanding the pattern of AOD and BC which can also affect the warming patterns in the mountains. The aerosol optical depth was monitored using a multiwavelength radiometer, while black carbon was monitored using Aethalometer, the raw data was further analyzed, and the satellite data was used to predict aerosol concentration over the entire Kullu valley. The present study will help researchers to understand the behavior of aerosol and black carbon and how these can affect the warming patterns of the Himalayan mountain range in India.

2 Study Area

The study area was Mohal which is located in the Kullu district of Himachal Pradesh. Kullu district is located between the Pir Panjal and the Dhaula Dhar mountain ranges at 31.91° North and 77.12° East, at an elevation of 1154 meters (Fig. 1) The Beas River, with its origins in Rohtang Pass, flows through the U-shaped Kullu Valley. The Kullu Valley starts at Largi (957 meters amsl) in the lower Beas Basin which expands up to Rohtang peak (4038 m amsl) in the upper Beas Valley which is extended up to 80 km length and 2 km in width (Kuniyal et al., 2009). The Kullu Valley's surrounding mountains rise to a height of between 3000 and 5000 meters (Sharma et al., 2009). The number of tourist has increased drastically over the past years, in the Kullu district (Sharma et al., 2009) and shows a continuous increase in number every year. The number of domestic and international tourists visiting Himachal Pradesh's various destinations has surged dramatically (Kumar & Sharma, 2022). Most of the Kullu Valley's population lives in rural and semi-urban areas.

3 Materials and Methods

Aerosol optical depth (AOD) and black carbon were observed both at the ground level and using satellite; further satellite data was also used to assess the change in temperature.

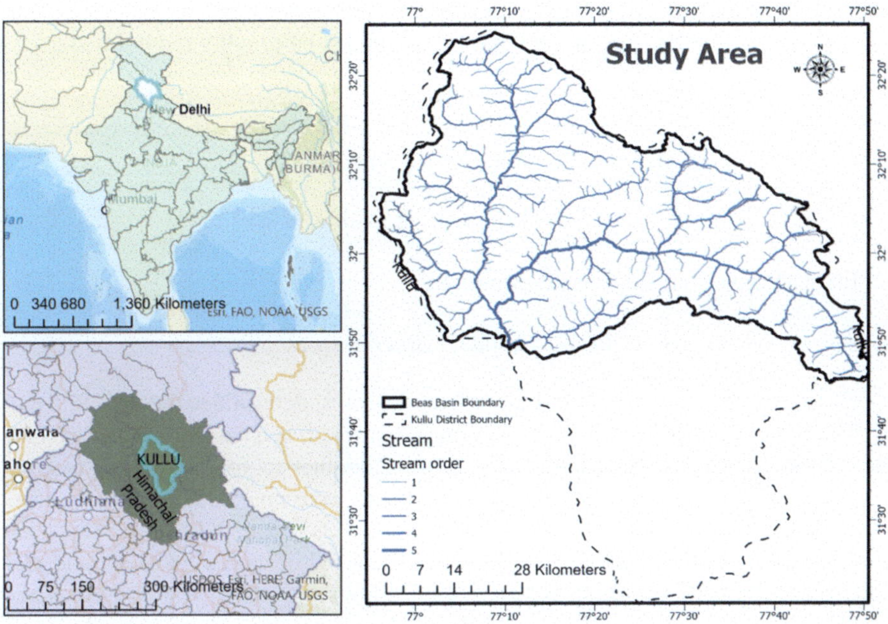

Fig. 1 Study area

3.1 Ground-Based Measurement of AOD

A multiwavelength radiometer (MWR) was used to monitor the aerosol optical depth at Mohal, Kullu. This radiometer has been designed by the Indian Space Research Organisation (ISRO), Space Physics Laboratory. This radiometer works on the principle that the filter wheel of the radiometer makes continuous spectral extinction measurements of ground-reaching, the directly transmitted solar flux at ten narrow spectral bands. Ten filters having wavelengths of 380, 400, 450, 500, 600, 650, 750, 850, 935, and 1025 nm are present in MWR. An automated filter wheel works to pass through solar radiation from UV to NIR. It operates under the principle that aerosols reduce solar energy through the process of scattering and absorption. The change in solar insolation of the Earth's surface can be calculated using the Lambert-Beer law (Shaw et al., 1973; Moorthy et al., 1998) by using the following formula (Moorthy et al., 1999):

$$\boldsymbol{Fo} = \boldsymbol{FO\lambda}\ (\boldsymbol{ror}\ )2\mathrm{e} - \boldsymbol{m\tau\lambda}$$

where $F\lambda$ stands for the Earth's surface solar flux and $Fo\lambda$ stands for top of atmosphere (TOA) *ro*, and *r* stands for mean instantaneous of the Sun to the Earth's distance, whereas *m* signifies relative air mass, signifying increased optical path with the rise in zenith angle. The optical depth varies with wavelength. Taking into account curvature effects m along with refraction can be roughly represented as

$$m = \sec \chi$$

Incident flux is converted into voltage by MWR as:

$$\boldsymbol{Vo} = \boldsymbol{VO\lambda}\ (\boldsymbol{ror}\ )2\mathrm{e} - \boldsymbol{m\tau\lambda}$$

Taking Ln on both sides of the equation:

$$\boldsymbol{lnV\lambda} = \boldsymbol{lnVo\lambda} + \boldsymbol{2lnr0r} - \boldsymbol{m\tau\lambda}$$

The graph of ln $V\lambda$ with m (which is Langley plot) determines the optical depth $\tau\lambda$. The SPL menu-driven program generates one set of AOD at ten wavelengths for the morning section of the measurement day and another set for the evening section. The information was gathered on days when there were no clouds in the proximity of the solar disk and is divided into three categories: clear days (clear sky from morning until evening), hazy days (the sky is clear but obscured by haze), and partially clear days (the morning or afternoon sections that the sky are clear). Since ground-based monitoring cannot be performed on rainy days, they alone are insufficient to assess the regional distribution of aerosols. A plot of ln$E\lambda$ v/s m is therefore a Langley plot that produces a slope of $-\tau\lambda$ with a straight line (Moorthy et al., 1999). Thus, AOD$\tau p\lambda$ is estimated for each observation day using the following equation (Ångström, 1961):

$$\tau \boldsymbol{p}\lambda = \tau\lambda - \left(\tau \boldsymbol{r}\lambda + \tau\alpha\lambda\right)$$

This method follows the Lambert-Beer law in estimating columnar optical depth using Angstrom relation which relates the dependence of AOD on wavelength as total as follows (Ångström, 1961):

$$\tau \boldsymbol{p}\lambda = \boldsymbol{\beta}\,\lambda -$$

3.2 Ground-Based Measurement of Black Carbon

Black carbon aerosols were measured continuously in real-time using an Aethalometer's Model AE-31-ER manufactured by Magee Scientific, USA. A quartz fiber filter tape used in an Aethalometer's operation attenuates a light beam that is being used to measure the amount of aerosol particles in the surrounding air (Raju et al., 2011; Hansen et al., 1984). An average flow rate of 2.9 L m^{-1} is used to operate the instrument 24 hours a day, 5 minutes every cycle. By tracking the shift in transmittance of a quartz filter tape, the BC mass concentration may be calculated (Hansen et al., 1984):

$$\text{ATN} = 100^{*} \ln\left(I_0 / I\right)$$

where I_0 is intensity of the light transmitted through the original filter/blank portion, I is the intensity of the light transmitted through the portion of the filter on which BC is collected, and 100 corresponds to an aerosol spot that is quite dark gray.

3.3 *Satellite-Based Measurement*

Satellite-based measurement was done using MODIS for aerosol and MEERA-2 for black carbon. The data was further analyzed using the RS-GIS technique.

3.4 *Fixed Box Model and Positive Matrix Factorization*

3.4.1 Fixed Box Model

A box model is a receptor model that works on the principle of conservation of mass. In this model, pollutants are assumed to be released into the receptor, which then triggers chemical and physical reactions. Simple meteorology is used as input to the model. Pollutant emissions are allowed both inside and outside the box. The air mass is regarded well-mixed and have uniform concentrations. The input of simple meteorology and detailed chemical reaction schemes and detailed aerosol dynamics treatment are advantages of the fixed box model. A fixed box model, however, simulates the formation of pollutants inside the box after inputs of the initial conditions but does not reveal the local concentrations of the pollutants (Yannawar et al., 2014). For a given pollutant in all of the boxes, one differential equation can summarize the entire mass balance equations. The following formula is used to determine air pollution emissions:

$$C = Co + \left\{(q)(L)\right\} / \left\{(u)(H)\right\} \tag{1}$$

where:

Co = concentration of pollutants entering the town
q = expressed as flux (g/s.m^{-2})
W = width of region
L = length of region
H = height of region
u = velocity of air

3.4.2 Positive Matrix Factorization (PMF) Model

Positive matrix factorization is a receptor model that works on mathematical concepts. This model utilizes mathematical methods based on the composition or unique fingerprints this model calculates based on the contribution of sources to samples. The species profile of each source with the mass g contributed by each factor and the number of factors p aim to resolve each discrete sample (Eq. 2). The receptor model shows the chemical mass balance (CMB) between source profiles. Species concentrations are measured as:

$$xij = \sum_{k=1}^{p} gikfkj + eij \tag{2}$$

For source apportionment in the area, the PMF model is applied, and input concentration and uncertainty were calculated using the following:

$$\text{Unc} = 5/6^{*}\text{MDL}$$

where MDL is the method detection limit for specific species.

4 Results and Discussions

4.1 AOD at Mohal, Kullu, Himachal Pradesh

AOD values at shorter wavelengths were higher, indicating increase in the anthropogenic activities (Fig. 2). The mean AOD at Mohal (1154 m AMSL) in 2021 was found to be 0.31 ± 0.03 nm and 0.33 ± 0.02 nm in the year 2021 and 2022, respectively, the while mean AOD_{500nm}, i.e., mean value of AOD at 500 nm (2021 and 2022), was found to be 0.28 at Mohal, Kullu, Himachal Pradesh.

4.2 Seasonal AOD Value

Analyzing the seasonal AOD, it showed the highest mean value of 0.52 ± 0.07 nm in summer followed by 0.33 ± 0.02 nm in monsoon, 0.24 ± 0.02 nm in autumn, and 0.23 ± 0.04 nm in winter (the lowest) in the year 2021. Similarly, in the year 2022, the mean AOD was highest in summer, i.e., 0.49 ± 0.03, followed by 0.30 ± 0.02 nm in monsoon, 0.29 ± 0.02 nm in autumn and 0.25 ± 0.03 nm in winter (the lowest) (Table 1).

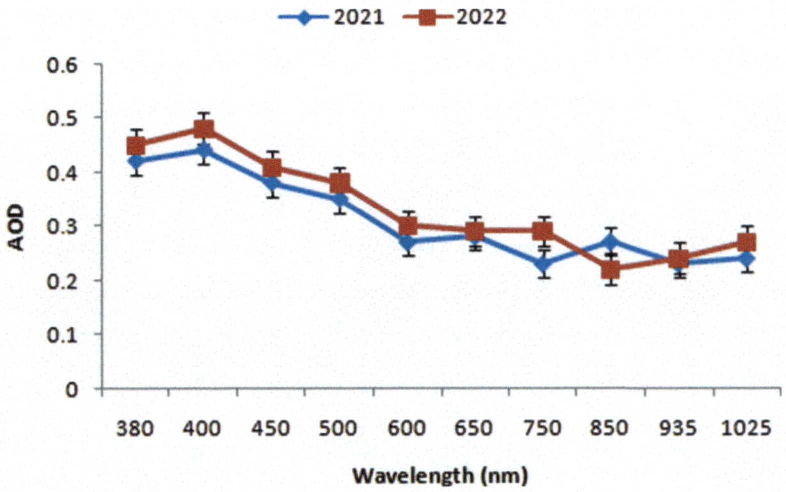

Fig. 2 Aerosol optical depth (AOD) at different wavelengths at Mohal, Kullu (in the year 2021 and 2022)

Table 1 Seasonal mean AOD value at Mohal, Kullu, Himachal Pradesh

Season	Mean AOD value at 500 nm (2021)	Mean AOD value at 500 nm (2022)
Summer	0.52 ± 0.07	0.49 ± 0.03
Monsoon	0.33 ± 0.02	0.30 ± 0.02
Autumn	0.24 ± 0.02	0.29 ± 0.02
Winter	0.23 ± 0.04	0.25 ± 0.03

4.3 Satellite-Based Measurement of AOD

AOD satellite data was obtained from MODIS satellite using the Google Earth Engine which was further analyzed by using geospatial technique. High AOD values were seen during the summer season in both years, while the AOD values were low in the winter season, although the AOD values are quite close in autumn and monsoon season. Seasonal trend of AOD for the entire Kullu Valley was quite similar to the ground-level measurement (Fig. 3).

4.4 Seasonal Variability of Black Carbon (BC) at Mohal, Kullu

The ground-level observation of black carbon showed the highest concentration in winter, followed by summer, autumn, and the lowest in monsoon. Black carbon showed the highest concentration of 6657.4 ± 162.74 ngm^{-3} in winter, followed by 4161.11 ± 160.149 ngm^{-3} in autumn, 3025.65 ± 495.72 ngm^{-3} in summer, and the lowest of 1670.96 ± 256.66 ngm^{-3} in monsoon (Figs. 4 and 5).

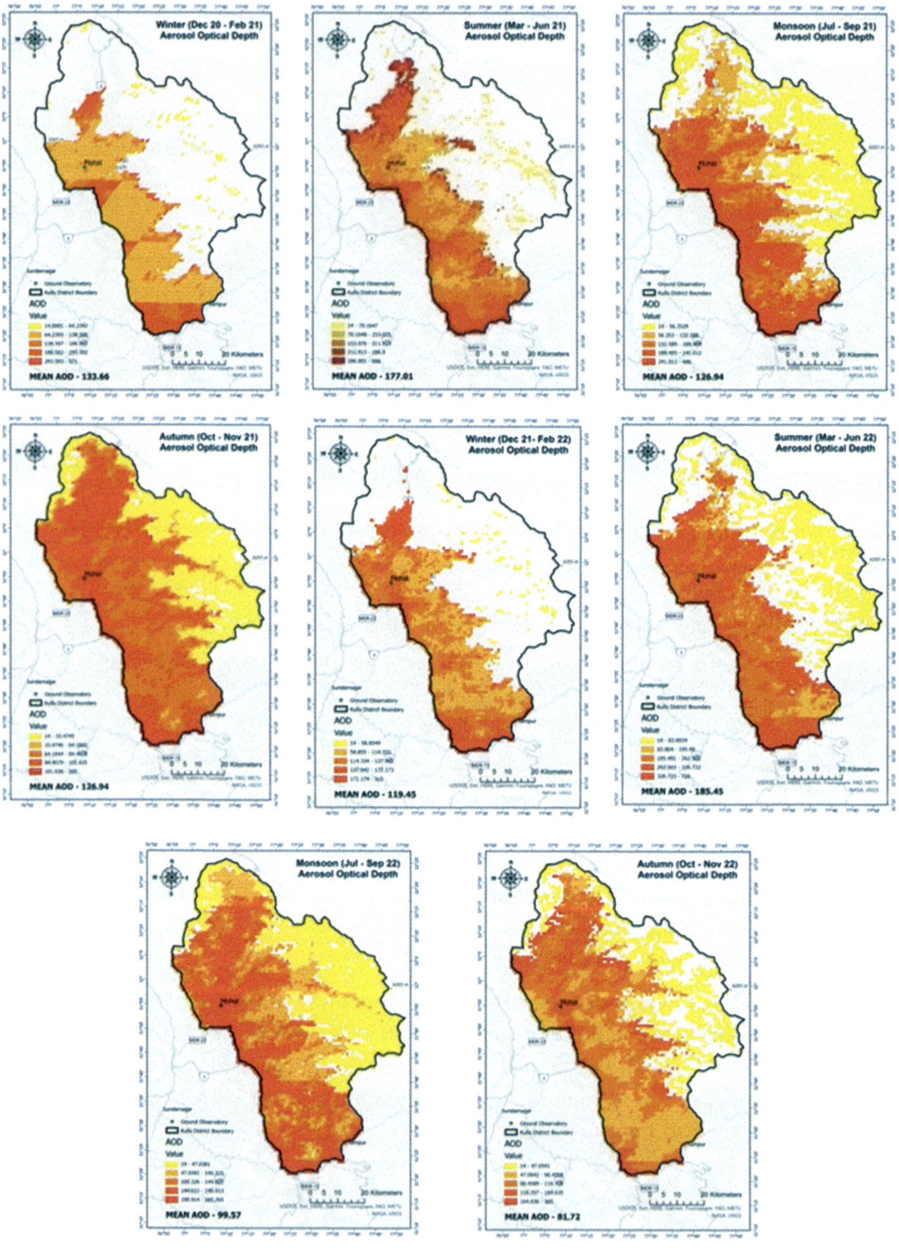

Fig. 3 Seasonal AOD at the Kullu Valley in the year 2021 and 2022 using MODIS

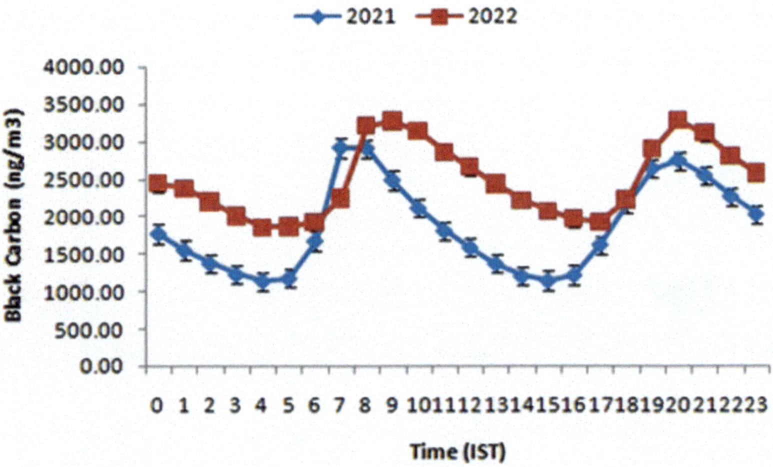

Fig. 4 Diurnal concentration of black carbon at Mohal, Kullu, in 2021 and 2022

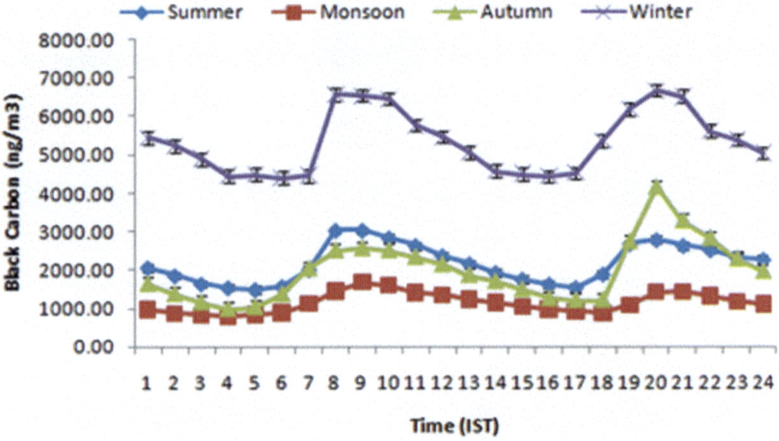

Fig. 5 Seasonal diurnal variability of black carbon at Mohal, Kullu, Himachal Pradesh

4.5 Seasonal Variability of Black Carbon Using Satellite Data

The seasonal variability of black carbon was determined using MEERA-2 satellite data. Black carbon surface mass concentration was determined. The satellite-based results were quite similar to that of ground-based results for black carbon mass concentration for the entire Kullu Valley. Black carbon mass concentration was highest in winter followed by summer and low in autumn and monsoon season (Fig. 6).

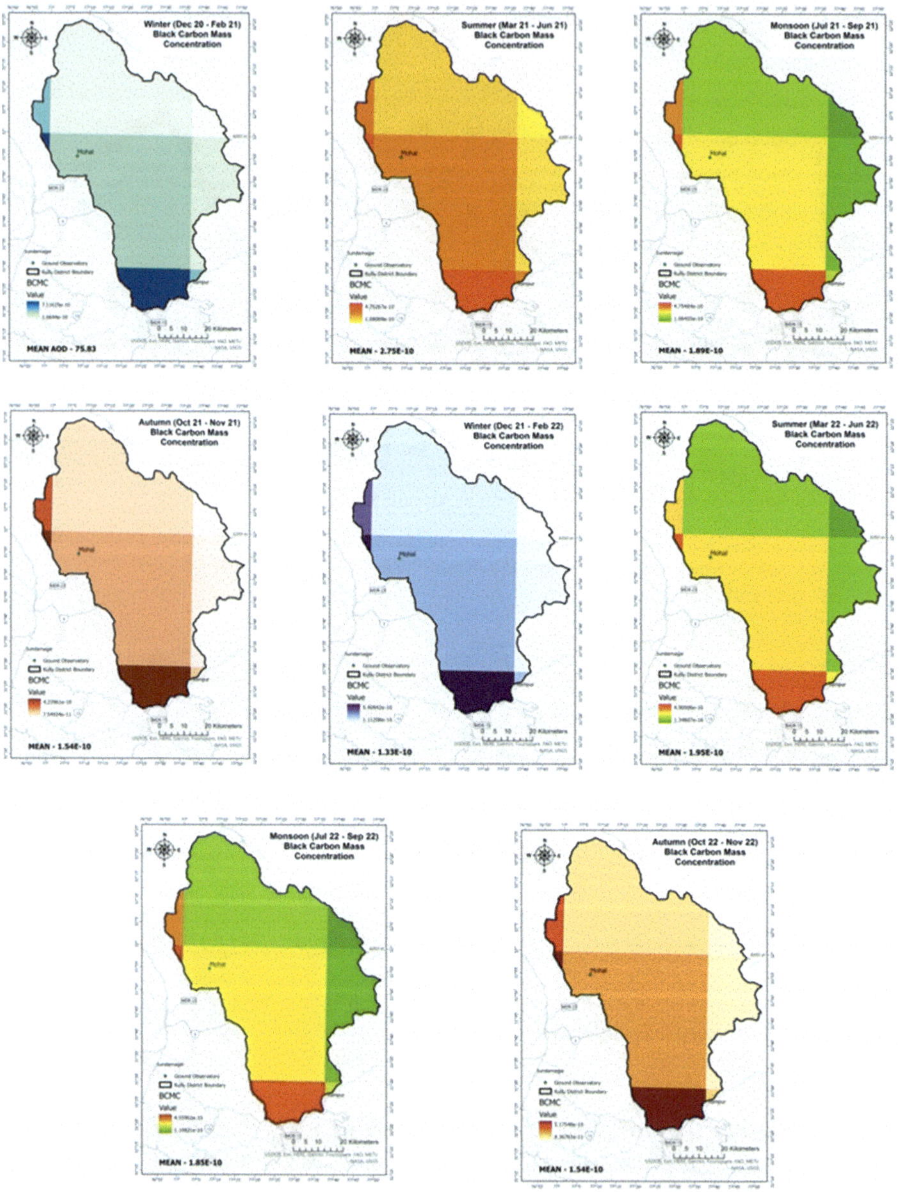

Fig. 6 Seasonal black carbon mass concentration at Mohal Kullu by using MEERA-2

4.6 Fixed Box Model and Positive Matrix Factorization

Fixed box model for black carbon showed that the measured concentration for black carbon was and 0.23 μg/m^3, while the predicted concentration was 0.50 μg/m^3 with an absolute error of 0.28. The use of positive matrix factorization model revealed that the predicted value of black carbon was quite close to the measured value (Fig. 7). The source apportionment of black carbon using positive matrix factorization biomass burning and vehicles are the two contributors of black carbon (BC) in the region. Biomass burning contributed 64.5%, while vehicles contributed 35.4% (Fig. 8).

4.7 HYSPLIT Model

The HYSPLIT model was run to calculate backward trajectories to determine long-range transport of pollutants. Seven-day backward trajectories were run on episodes of high pollution every month. Air mass movement during winter (December–February) and autumn (Oct–Nov) months are from west of the experimental site. During summer (March–June), it is from north-west and south-west, and in rainy season (July–September), it is from south-east or south-west. During rainy season, the movement of air mass trajectories originates over the Arabian Sea and Bay of Bengal. Air originating from marine air was cleaner in comparison to the air originating from land which shows that apart from local sources, long-range transport

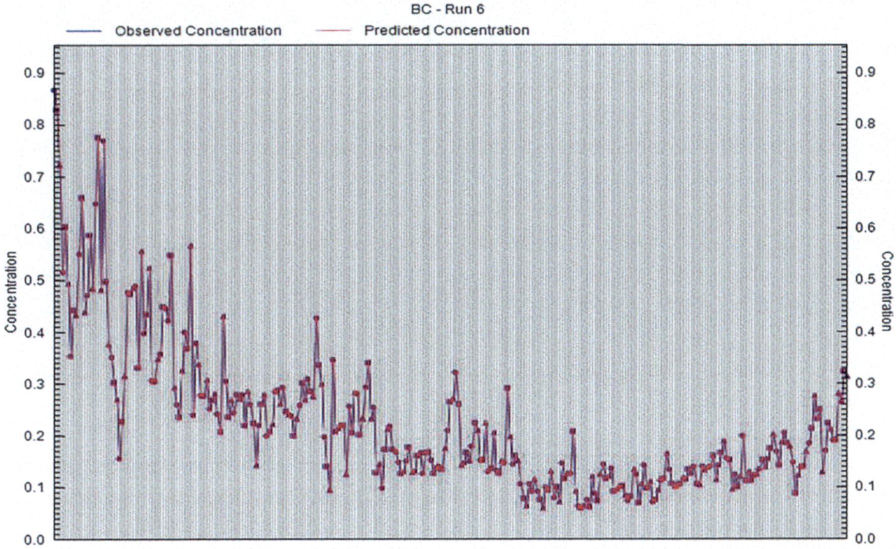

Fig. 7 Observed and predicted value of black carbon at Mohal Kullu using the PMF model

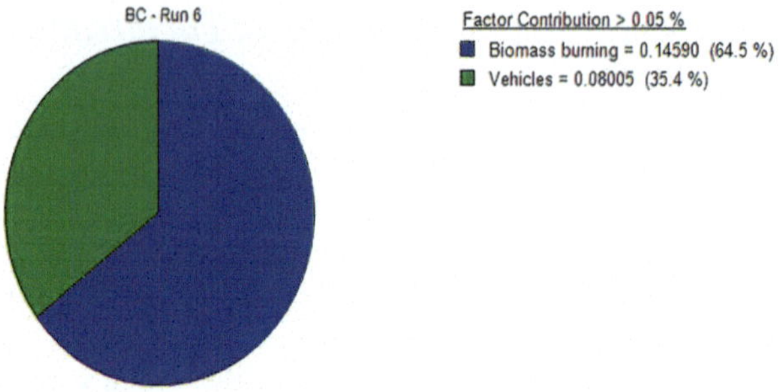

Fig. 8 Source apportionment of black carbon at Mohal, Kullu, Himachal Pradesh

also plays a significant role in contributing to air pollution in the region particularly from the northwestern region during winter season in the months from December to February and in summer season in the months from March to late June (Fig. 9i–xxiv). Marine air masses are relatively clean compared to continental air, which suggests that there is a long-range transport of pollutants from the northwestern region, especially during the winter season, which also contributes to regional air pollution along with local pollutants.

Discussion

The mean AOD of 0.52 ± 0.07 and 0.49 ± 0.03 was highest in summer 2021 and 2022, respectively. Similar results were seen using MEERA-2 satellite. The result also showed that AOD is higher in shorter wavelength and gradually decreases toward longer wavelength. It is important to understand the pattern and seasonal variability of aerosol as the thermal state of the atmosphere can alter due to changes in tropospheric aerosols, which in turn affects the dynamic structure of the atmosphere and results in climate change (Huang et al., 2007) which can eventually lead to warming in the mountains. Black carbon mass concentration was highest in winter in both satellite and ground-level observations. Black carbon absorbs sunlight in all its wavelengths and transfers its warmth to the atmosphere. With roughly a million times the heat-trapping power of CO_2 (Bachmann, 2009), air currents can carry black carbon over great distances. Black carbon also absorbs heat and increases melting when it precipitates on snow or ice because it prevents snow surfaces from reflecting sunlight back into space. This shows how the patterns of BC and increased concentration of black carbon can contribute to melting of snow and contribute to atmospheric warming in the mountains. Regular monitoring and understanding the pattern of AOD and black carbon can help to determine any change in the concentration of pollutants which can affect the dynamic of atmosphere and can lead to warming of mountains in that region. Further regular observation along with meteorological parameters can be helpful in policy framework and suggesting mitigating measures to lower the value of these and other pollutants which are indirectly linked to warming of mountains.

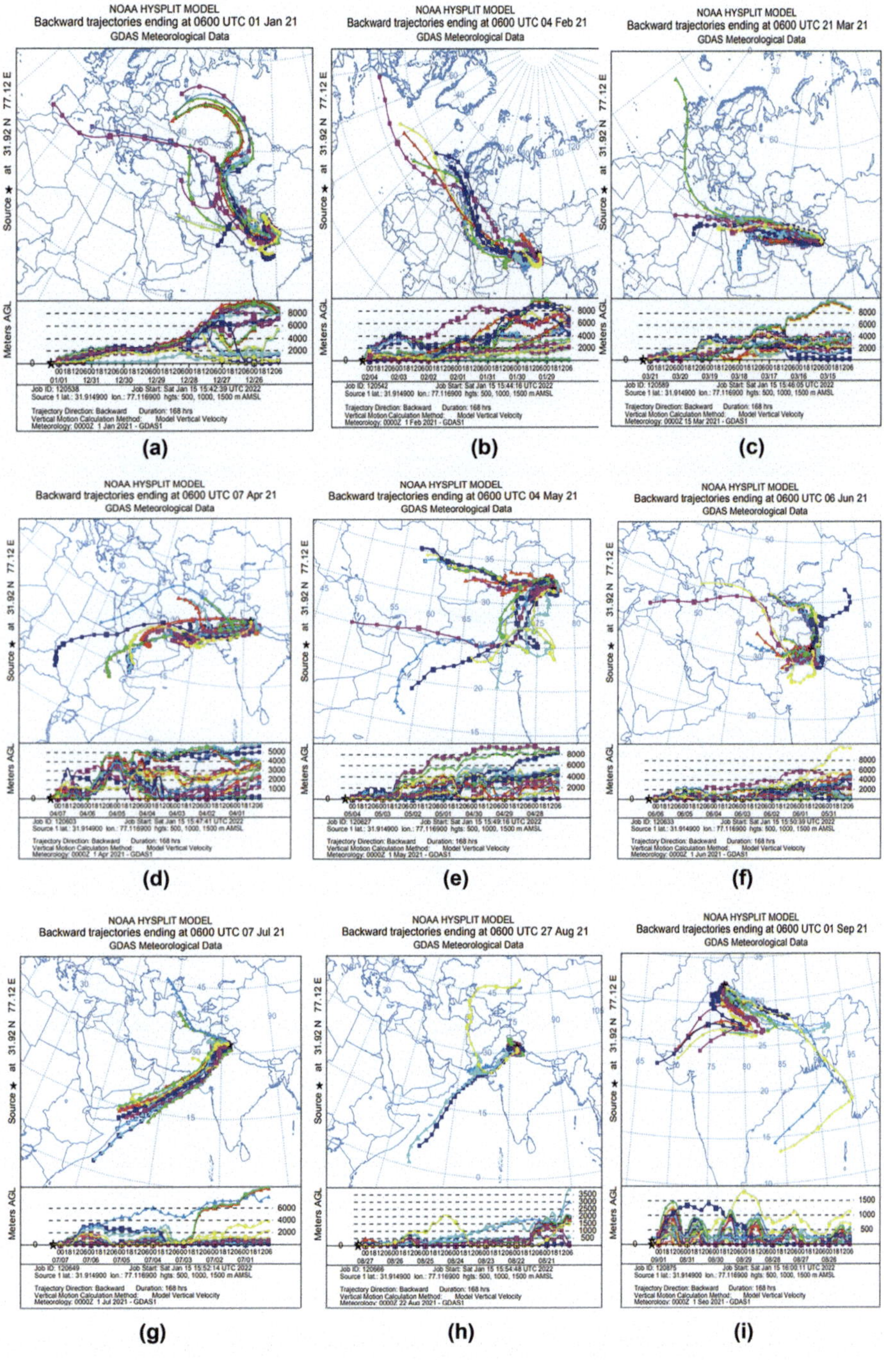

Fig. 9 (**a**–**x**) Monthly backward trajectory for identification of transportation of air parcel during the year 2021–2022

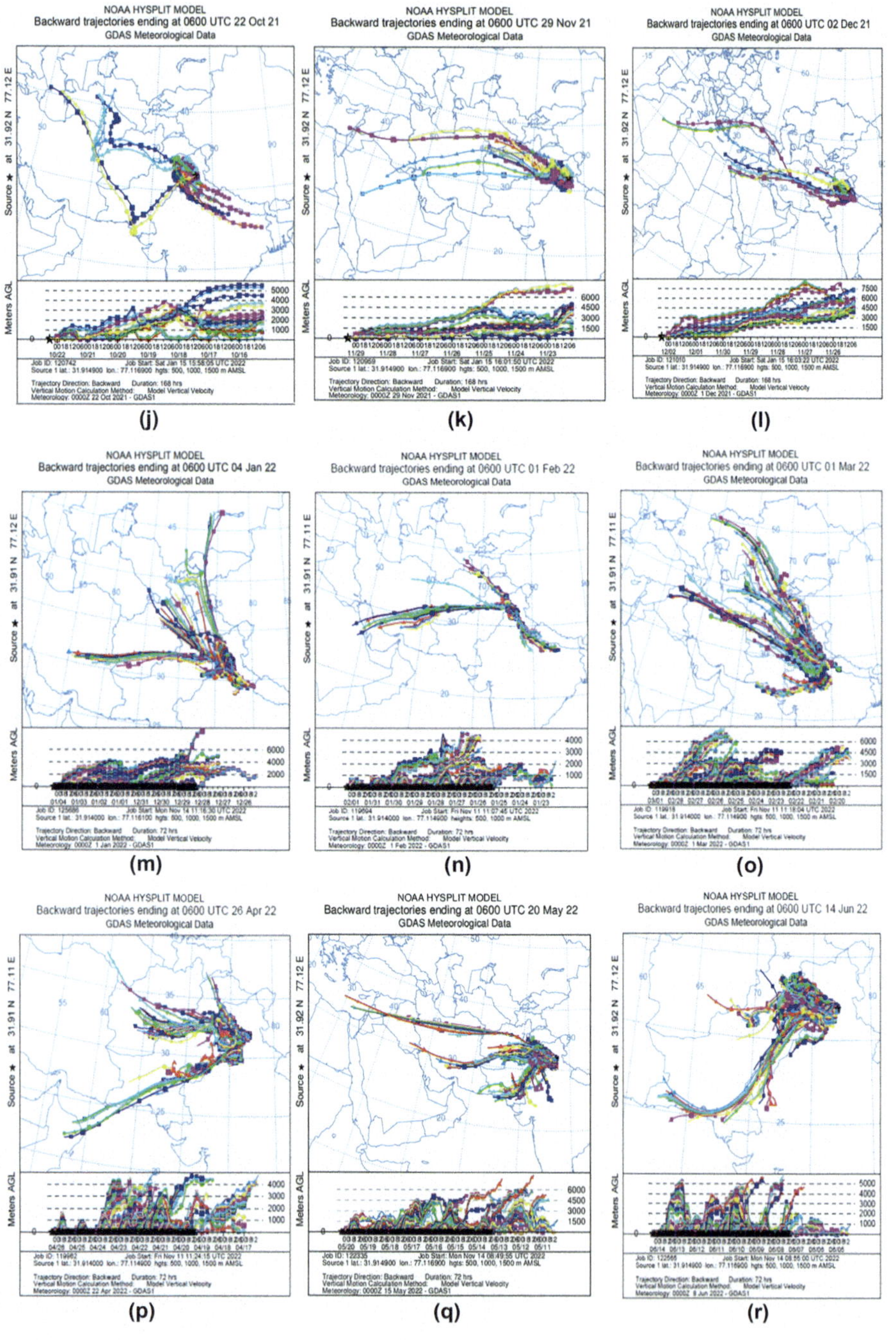

Fig. 9 (continued)

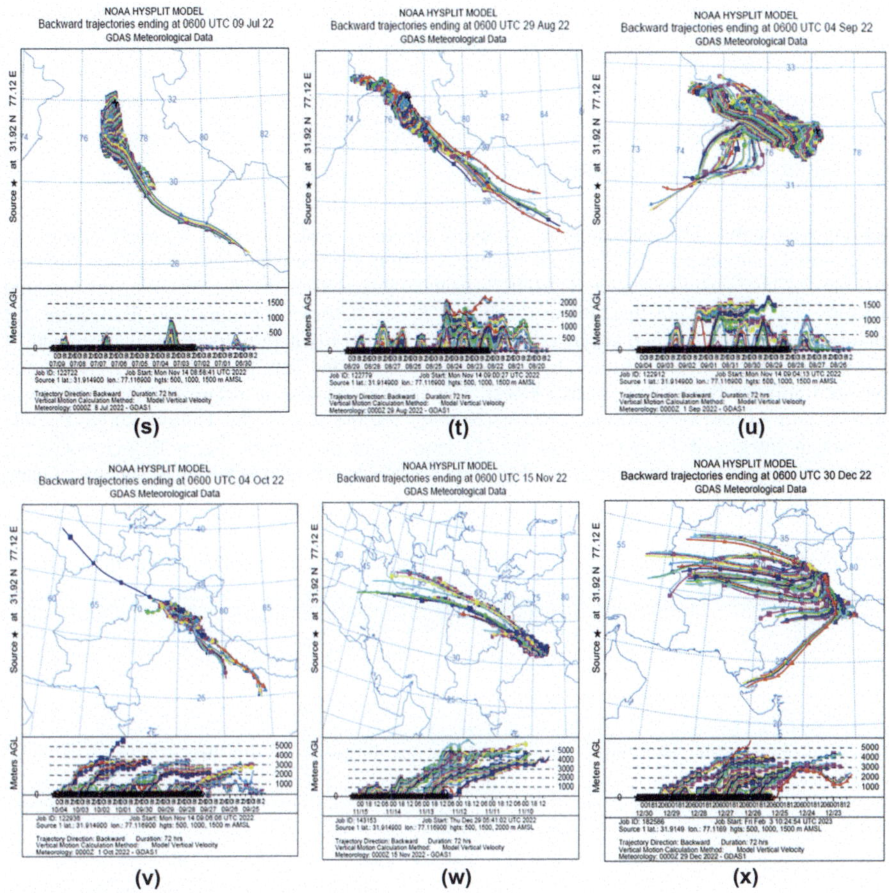

Fig. 9 (continued)

5 Conclusions

To understand the warming patterns in the mountain region, it is essential to understand the various factors such as air pollutants, aerosols, black carbon, and other atmospheric conditions that contribute to warming of mountains. Aerosols and black carbon play key role in maintaining the Earth's heat budget. In Kullu Valley, values of AOD are higher at shorter wavelengths, indicating an increase in the anthropogenic activities. Seasonal variation of aerosol showed high concentration in summer, while black carbon showed highest concentration in winters. Moreover, the fixed box model for black carbon showed that the predicted value was quite close to the observed value, while the source apportionment of black carbon using positive matrix factorization showed that biomass burning and vehicles are the two contributors of black carbon (BC) in the region. Long-range transport of pollutants using the HYSPLIT model showed that apart from local sources, long-range transport of pollutants from the north-west region also contributes to air pollution in the region, especially during the winter months from December to February and summer months from March to late June.

Acknowledgments The authors sincerely thank the director of G.B. Pant National Institute of Himalayan Environment (NIHE), Kosi-Katarmal, Almora, Uttarakhandand G.B. Pant National Institute of Himalayan Environment (NIHE) Himachal Regional Center, for providing the essential facilities. The authors are thankful to the Indian Space Research Organization, Bangalore, for providing necessary financial support to conduct the current research work under two projects of ISRO-GBP which include Atmospheric Chemistry-Transport and Modelling (AT-CTM) and Aerosol Radiative Forces over India (ARFI). The authors also like to acknowledge the National Oceanic and Atmospheric Administration (NOAA) for the use of the Hybrid Single-Particle Lagrangian Integrated Trajectory (HYSPLIT) model and the Google Earth Engine for using MERRA-2 and MODIS data and also to EPA for the PMF model used in the source apportionment of pollutants.

Author's Contribution All authors contributed equally as the main contributors to this work.

References

Ångström, A. (1961). Techniques of determining the turbidity of the atmosphere. *Tellus, 13*, 214–223.

Bachmann, J. (2009). *Black carbon: A science/policy primer*. Pew Center on Global Climate Change Black 45.

Beniston, M. (2003). Climatic change in mountain regions: A review of possible impacts. In H. F. Diaz (Ed.), *Climate variability and change in high elevation regions: Past, present and future* (pp. 5–31). Springer. https://doi.org/10.1007/978-94-015-1252-7_2

Bond, T. C., Doherty, S. J., Fahey, D. W., Forster, P. M., Berntsen, T., DeAngelo, B. J., et al. (2013). Bounding the role of black carbon in the climate system: A scientific assessment. *Journal of Geophysical Research: Atmospheres, 118*(11), 5380–5552. https://doi.org/10.1002/jgrd.50171

Doherty, S. J., Warren, S. G., Grenfell, T. C., Clarke, A. D., & Brandt, R. E. (2010). Light-absorbing impurities in Arctic snow. *Atmospheric Chemistry and Physics, 10*(23), 11647–11680. https://doi.org/10.5194/acp-10-11647-2010

Flanner, M. G., Zender, C. S., Randerson, J. T., & Rasch, P. J. (2007). Present-day climate forcing and response from black carbon in snow. *Journal of Geophysical Research, 112*(D11), D11. https://doi.org/10.1029/2006JD008003

Hansen, A. D. A., Rosen, H., & Novakov, T. (1984). The aethalometer: An instrument for the real-time measurements of optical absorption by aerosol particles. *Science of the Total Environment, 36*, 191–196. https://doi.org/10.1016/0048-9697(84)90265-1

Houghton, J. T., Filho, L. G. M., Bruce, J., Lee, H., Callander, B. A., Haites, E., et al. (1996). Radiative forcing of climate change. In *Climatechange* (p. 231). Cambridge University Press.

Huang, J., Minnis, P., Yi, Y., Tang, Q., Wang, X., Hu, Y., Liu, Z., Ayers, K., Trepte, C., & Winker, D. (2007). Summer dust aerosols detected from CALIPSO over the Tibetan Plateau. *Geophysical Research Letters, 34*, L18805. https://doi.org/10.1029/2007GL029938

Kang, S., Zhang, Q., Qian, Y., Ji, Z., Li, C., et al. (2019). Linking atmospheric pollution to cryospheric change in the Third Pole region: Current progress and future prospects. *National Science Review, 6*, 796–809. https://doi.org/10.1093/nsr/nwz031

Kumar, V., & Sharma, A. (2022). Economic impacts of tourism in himachalPradesh. *International Journal of Multidisciplinary Education Research, 11*(3(5)), ISSN:2277-7881.

Kuniyal, J. C., Thakur, A., Thakur, H. K., Sharma, S., Pant, P., Rawat, P. S., & Moorthy, K. K. (2009). Aerosol optical depths at Mohal-Kullu in the northwestern Indian Himalayan high altitude station during ICARB. *Journal of Earth System Science, 118*(1), 41–48. https://doi.org/10.1007/s12040-009-0004-y

Moorthy, K. K., Satheesh, S. K., & Murthy, B. V. K. (1998). Characteristics of spectral optical depths and size distributions of aerosols over tropical oceanic regions. *Journal of Atmospheric and Solar-Terrestrial Physics, 60*(10), 981–992. https://doi.org/10.1016/S1364-6826(98)00044-3

Moorthy, K. K., Niranjan, K., Narasimhamurthy, B., Agashe, V. V., &Murthy, B. V. K. (1999). *Aerosol cli-matology over India, ISRO-GBPMWR network and data base* [Scientific report], SR-03-99.

Pepin, N. C., Bradley, R. S., Diaz, H. F., Baraer, M., Caceres, E. B., Forsythe, N., et al. (2015). Elevation-dependent warming in mountain regions of the world. *Nature Climate Change, 5*(5), 424–430. https://doi.org/10.1038/nclimate2563

Pepin, N. C., Arnone, E., Gobiet, A., Haslinger, K., Kotlarski, S., Notarnicola, C., et al. (2022). Climate changes and their elevational patterns in the mountains of the world. *Reviews of Geophysics, 60*(1), RG000730. https://doi.org/10.1029/2020RG000730

Qian, Y., Wang, H., Zhang, R., Flanner, M. G., & Rasch, P. J. (2014). A sensitivity study on modeling black carbon in snow and its radiative forcing over the Arctic and Northern China. *Environmental Research Letters, 9*(6), 064001. https://doi.org/10.1088/1748-9326/9/6/064001

Qian, Y., Yasunari, T. J., Doherty, S. J., Flanner, M. G., Lau, W. K. M., Ming, J., et al. (2015). Light-absorbing particles in snow and ice: Measurement and modeling of climatic and hydrological impact. *Advances in Atmospheric Sciences, 32*(1), 64–91. https://doi.org/10.1007/s00376-014-0010-0

Raju, M. P., Safai, P. D., Rao, P. S. P., Devara, P. C. S., & Budhavant, K. B. (2011). Seasonal characteristics of black carbon aerosols over a high altitude station in southwest India. *Atmospheric Research, 100*(1), 103–110. https://doi.org/10.1016/j.atmosres.2011.01.006

Ramanathan, V., Ramana, M. V., Roberts, G., Kim, D., Corrigan, C., Chung, C., & Winker, D. (2007). Warming trends in Asia amplified by brown cloud solar absorption. *Nature, 448*(7153), 575–578. https://doi.org/10.1038/nature06019

Rangwala, I., & Miller, J. R. (2012). Climate change in mountains, a review of elevation-dependent warming and its possible causes. *Climatic Change, 114*(3–4), 527–547. https://doi.org/10.1007/s10584-012-0419-3

Sharma, N. L., Kuniyal, J. C., Singh, M., Negi, A. K., Singh, K., & Sharma, P. (2009). Number concentration characteristics of ultrafine aerosols (atmospheric nanoparticles/Aitkennuclei) during2008 over western Himalayan region, Kullu-Manali, India. *Indian Journal of Radio and Space Physics, 38*, 326–337.

Shaw, G. E., Reagan, J. A., & Herman, B. M. (1973). Investigations of atmospheric extinction using direct solar radiation measurement made with a Multiple Wavelength Radiometer. *Journal of Applied Meteorology, 12*(2), 374–380. https://doi.org/10.1175/1520-0450(1973)012<0374:IOAEUD>2.0.CO;2

Xu, B., Cao, J., Hansen, J., Yao, T., Joswia, D. R., Wang, N., et al. (2009). Black soot and the survival of Tibetan glaciers. *Proceedings of the National Academy of Sciences of the United States of America, 106*(52), 22114–22118. https://doi.org/10.1073/pnas.0910444106

Yannawar, D. V., Bhosle, A., & Yannawar, S. (2014). Prediction of air pollution concentration using a fixed box model. *Research, 6*, 89–92.

Spatiotemporal Assessment of Climate Change-Led Ecological Vulnerability Through DPSIR in Indian Himalayan Region

Rajiv Pandey, Lakshmikant Tiwari, Diksha Verma, Divya Mehta, Ali Jamshed, Arun Jyoti Nath, and Ranjeet Kumar

Highlights

- Ecologically, 19.97% of the North-Eastern Himalayan Region (NEHR) of India was highly vulnerable.
- Temporally, the ecological vulnerability of the region was increased during the last two decades.
- Land-use and land-cover changes (LULC), forest fire, and temperature rise were major drivers of ecological vulnerability in the region.
- The development of human habitation agglomeration in the region was largely at the cost of ecological degradation.

R. Pandey (✉)
Division of Forestry Statistics, Indian Council of Forestry Research & Education, Dehradun, India

L. Tiwari · D. Verma · D. Mehta
Indian Council of Forestry Research and Education, Dehradun, India

A. Jamshed
University of Stuttgart, Stuttgart, Germany
e-mail: ali.jamshed@ireus.uni-stuttgart.de

A. J. Nath
Assam University, Silchar, Assam, India

R. Kumar
Himalayan Forest Research Institute, Shimla, India

S. Tripathi et al. (eds.), *Warming Mountains*,
https://doi.org/10.1007/978-3-031-62197-0_13

1 Introduction

A recent IPCC report has highlighted multiple risks to ecological systems associated with climate change (IPCC, 2022a, b). Climate change impacts mount pressure on social systems such as water security, food production, health, well-being, and settlements, and ecological systems such as evolutionary processes, geographic ranges, and phenology of plant and animal species are already affected by climate change (IPCC, 2021; 2022a, b; Wu, 2021; Li et al., 2022). Moreover, levels of biological organization are affected differentially by climate change-induced stresses (de Lange et al., 2010). Therefore, assessment of the vulnerability of ecological systems is crucial to assist for providing a basis to implement an ecosystem-based approach for climate change mitigation and adaptation (Guo & Wen, 2020; Li et al., 2021).

Vulnerability is defined as a propensity and predisposition to be adversely affected (IPCC, 2014) and corresponds to the characteristics of an entity or system to respond to stressor over time and space (Jamshed et al., 2020; Birkmann et al., 2013). Ecological vulnerability assessment (EVA) accounts for the early detection of potentially damaging changes in the ecosystem and thus safeguards the sustainability of the ecosystem (Weibhuhn et al., 2018) and accounts for the existing interlink ages between the coupled human and ecological systems (Gupta et al., 2019; Kumar et al., 2019). EVA identifies vulnerable areas along with optimal measures for their future management (IPCC, 2014). Thus, EVA is a vital tool that can guide effective policy-making and aid decision-makers in understanding multiple impacts of natural and factitious elements on the ecosystem (He et al., 2018).

Several studies have determined vulnerability using different approaches, e.g., index-based system, threshold approach, or process trend (Cutter & Finch, 2008; de Lange et al., 2010). Index-based approaches are the most common approaches to assess social vulnerability (Cutter & Finch, 2008; Depietri et al., 2013; Cutter et al., 2003), economic vulnerability (Willroth et al., 2011), physical vulnerability (Guillard-Gonçalves & Zêzere, 2018), and ecological vulnerability (Hu et al., 2021; He et al., 2018; Dossou et al., 2021; de Lange et al., 2010). In India, EVA has been conducted for fragile river basins (Rehman et al., 2021), forest ecosystems (Kumar et al., 2019), and agroecological systems (Pandey et al., 2017). However, the ecological vulnerability in mountainous regions, particularly Hindu Kush Himalayas (HKH) at a larger scale, has not been paid sufficient attention in India.

The mountainous regions are more susceptible to climate change and anthropogenic pressures due to high environmental fragility, fragility, marginal landholding, and high dependency on natural systems (Gerlitz et al., 2016; Adler et al., 2022). Mountains support half of humankind for their water requirements and one-fourth of the world's terrestrial biodiversity. Globally, 22% of the globe is covered with mountain regions that provide home to around one billion people, and one-fourth of this population is residing in Hindu Kush Himalayas (HKH) (FAO, 2015; Sharma et al., 2019). However, HKH is undergoing a rapid change driven by climatic and non-climatic stressors (Wester et al., 2019) with serious implications for biodiversity and socioecological system loss due to changes in temperature and precipitation (Chettri & Sharma, 2016; Wester et al., 2019).

Climate change in the North-Eastern Himalayan Region (NEHR), a part of HKH, is detrimental to the rural population due to intricate linkages between livelihood and natural resources (Singha, 2018). NEHR has a heavy annual rainfall ranging from 20,000 to 30,000 mm; however, extreme rainfall events have been observed because of climate change in recent periods (Zahan et al., 2021; Borah, 2022). Climate change is apparent with the reduction in the number of rainy days across the season and an increase in annual maximum and minimum temperatures of 0.01 and 0.04 °C, respectively, during 1991–2015 (Saikia & Hazarika, 2020), leading to increases in the incidence of climate-induced natural disasters (Roy et al., 2021; Zahan et al., 2021). In future climatic scenarios, the forests of NEHR are climate change hotspots (FSI, 2021), and the region is highly vulnerable to various challenges due to the settings such as geo-ecological fragility, strategic location, transboundary river basin, and inherent socioeconomic instability (Roy et al., 2021). The region has the most diverse types of forest ecosystems and includes tropical forest ecosystems in the floodplains to subtropical, temperate, and alpine forest ecosystems in the high mountains (FSI, 2021). The riverine ecosystems are scattered along the Brahmaputra and Barak rivers and their tributaries (NHC, 2006). The agroecosystem diversity in the north-east region ranges from the ecosystems under a variety of shifting cultivation and terrace land cultivation practices in the mountains to settled and intensive cultivation practices in the valleys and plain lands (Deka & Barua, 2015). *Jhum* (shifting) cultivation system is a prime livelihood option for most of the population and has high adverse impacts on land productivity. Most of the population of the region depends on natural resources for their survival and practices subsistence farming on undulating topography with low cropping intensity (Roy et al., 2021).

The impacts of climate change on socio-ecological systems are well understood and documented; however, regional-level assessment is lacking (Sharma et al., 2009; IPCC, 2022a, b). Therefore, the study hypothesized that ecological vulnerability of the NHER varies at a spatiotemporal scale. The specific objective of the study was to evaluate the ecological vulnerability at a spatiotemporal scale for the NEHR and evaluation of the change dynamics contributing to the agglomeration of human habitation in the urban and rural divide vis-à-vis ecological system dynamics. The present study attempted to identify the vulnerable ecosystems of the region, thereby facilitating sustainable management of the systems besides providing an improved understanding about the dynamics of the socio-ecological system. The investigation would facilitate to devise policy perspectives for tailor-made measures for mitigating the ecological vulnerability of the NEHR.

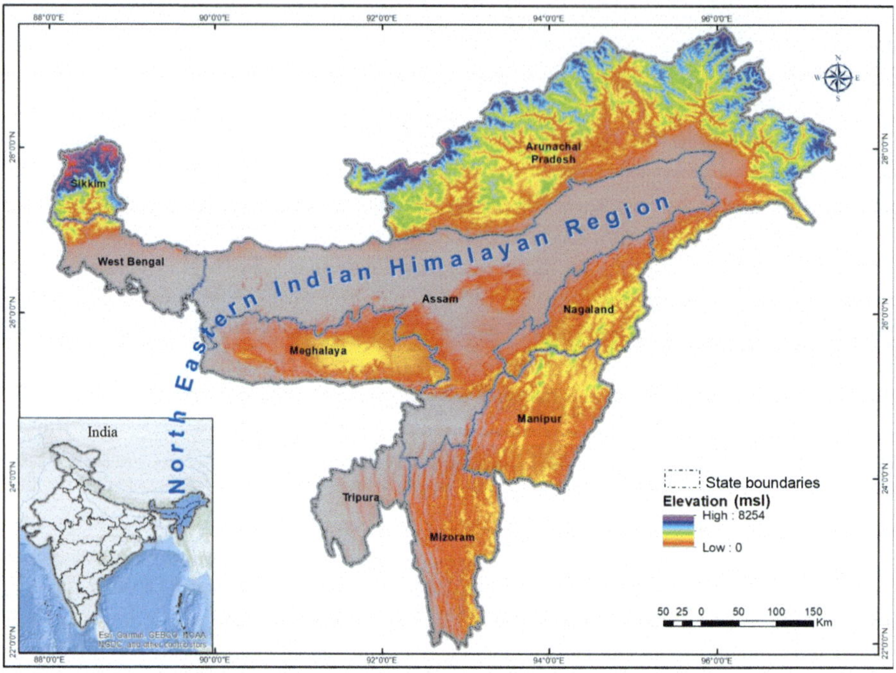

Fig. 1 Location map of North-Eastern Himalayan Region (NEHR)

2 Materials and Methods

2.1 Study Area

The NEHR comprises nine Indian states, viz., Arunachal Pradesh, Assam, Manipur, Meghalaya, Mizoram, Nagaland, Sikkim, Tripura, and the three districts of West Bengal (Darjeeling, Koch Bihar and Jalpaiguri), covering a total area of 2.74 lac km^2, i.e., 8.32% of India's geographical area, and spreading between 21.58°N to 29.30°N and 87.09°E to 97.30°E (Fig. 1). NEHR has a complex geography and includes the Shillong Plateau, North-Eastern Hill, and Brahmaputra Valley. The region receives ≥1000 mm of annual rainfall ranging between 400 and 3800 mm and the temperature varies from 1 to 40 °C (FSI, 2020). With a record of 12,000 mm annual rainfall, Meghalaya receives the highest rainfall in the northeastern region. The region consists of undulating topography with elevation ranging from 6 to 8254 meters above mean sea level (msl). Soil erosion, soil degradation, and soil acidity are major issues in the region due to increased *jhum* cultivation, deforestation, high rainfall, and ecological complexity (De & Singh, 2017). NEHR fosters two biodiversity hotspots: Himalayas and Indo-Burma and its loss is irreplaceable due to the presence of huge endemic flora.

2.2 *Theoretical Settings: Ecological Vulnerability Assessment Through DPSIR Framework*

2.2.1 DPSIR

In the present study, the vulnerability was assessed through the driver, pressure, state, impact, and response (DPSIR) framework, which is a chain of five causal links (factors), i.e., starting with *driving forces* through *pressures* to *states* and *impacts* on the system, eventually leading to *responses* (Kristensen, 2004). DPSIR is an indicator-based bottom-up approach developed for socio-ecological environmental management and is the extended version of the sensitivity-resilience-pressure framework (Bradley & Yee, 2015). The DPSIR framework was applied by identifying ecological indicators and categorizing the indicators into five major (i.e. the five-link factors) and their associated sub-components based on inductive and deductive reasoning (Table 1). A total of 27 selected indicators for all the five main components based on empirical knowledge, previous literature, and discussion with the experts was ascertained for developing a composite index for the assessment of EV (Li et al., 2018) (Fig. 2).

2.2.2 Data Sources and Pre-processing

Satellite data and secondary records of NEHR were collected for the years 2000, 2010, and 2020 (Table 1). Meteorological and satellite data extracted with heterogeneous attributes was transformed into uniform cohesive data (Adhikari et al., 2021). Data layer resolution and format was harmonized through resampling at 100 m grid (Eken & Sayar, 2021).

2.2.3 Data Standardization

All indicators were subjected to standardization to make them comparable depending on their relationship with vulnerability (Pandey & Jha, 2012):

$$X_i' = \frac{X_{ij} - X_{i_{\min}}}{X_{i_{\max}} - X_{i_{\min}}} \text{ for positive indicator} \tag{1}$$

$$X_i' = 1 - \left(\frac{X_{ij} - X_{i_{\min}}}{X_{i_{\max}} - X_{i_{\min}}} \right) \text{ for negative indicator} \tag{2}$$

where X_i' represents the standardized value of the ith indicator varying from 0 to 1, X_{ij} represents the actual value of *i*th indicator for the *j*th spatio-temporal case, $X_{i_{\min}}$

Table 1 Description and source of ecological vulnerability assessment

Major component	Sub-component	Indicator (relationship)	Description (source)	Data source (resolution)
Driver—forces deriving pressure on the environment				
Climate	Precipitation	Mean annual precipitation (+)	Mean annual precipitation (mm) (Gerlitz et al., 2016)	CHELSA (1 km)
		Rainfall pattern (+)	Daily rainfall intensity (mm) (Pandey et al., 2015)	Layer interpolated using IMD daily rainfall data (0.25°)
		Wet period (+)	Annual rainfall exceeding 95th of daily rain intensity (Darabi et al., 2019)	
		Dry period (+)	Longest count of days with less than 1 mm rainfall (Darabi et al., 2019)	Climatic research unit (0.25°)
	Temperature	Mean annual temperature (+)	Mean annual temperature (°C) (Kumar et al., 2021)	CHELSA (1 km)
		Hot period (+)	The average count of days for maximum temperature ≥35 °C	Layer interpolated using IMD daily rainfall data (0.25°)
		Cold period (+)	Average count of sequential ≥6 days exceeding tenth of daily minimum temperature	
Socio-economic	Socio-economic	Occupation (+)	Occupation (Darabi et al., 2019)	Layer interpolated using Census of India and SEDAC by NASA (1 km)
		Purchasing power parity (−)	Purchasing power (Proxy for GDP)	
		Population density (+)	Population per unit area (Li et al., 2020)	CIESIN by Google Earth Engine (1 km)
Pressure—stressors due to the driver on the natural environment				
Distress	Anthropogenic disturbance	Distance to road (+)	The proximity of forest to the road (Li et al., 2020)	SEDAC by NASA (1 km)
	Disaster	Forest fire index (+)	Estimation based on the frequency and area of forest fires (Darabi et al., 2019)	FIRMS by NASA (1 km)
	Moisture index	Air moisture (−)	Mean annual near-surface relative humidity (%)	CHELSA (1 km)
		Soil moisture (−)	Water content at 2–5 cm soil depth (% of total saturation)	ISRIC (250 m)

(continued)

Table 1 (continued)

Major component	Sub-component	Indicator (relationship)	Description (source)	Data source (resolution)
State—condition of the environment				
Topographic	Topography	Altitude (+)	Elevation (m in amsl) (Wu, 2021)	NASADEM by Google Earth Engine (30 m)
		Slope (+)	Degree of steepness of land (Li et al., 2020)	
		Curvature of slope (+)	Rate of change of slope	
		Hill shade (+)	Emphasizing hills or mountains	
Land-use	Land-use dynamics	LULC (+)	Change in LULC (Darabi et al., 2019)	VEGA 2000 dataset (1 km), NRSC 2010 dataset (56 m), and Sentinel 2020 dataset (10 m)
		Agriculture (+)	Cultivable land (Darabi et al., 2019)	
		Pasture land (+)	Grazing land (Darabi et al., 2019)	
		Built up area (+)	Built-up in forest periphery (Darabi et al., 2019)	
Impact—effect on ecosystem				
Landscape characteristic	Soil composition	Soil erosion (+)	Loss of top layer of soil (Darabi et al., 2019)	NRSC land degradation dataset (1 km)
	Landscape	Landscape fragmentation (+)	Fragmented forest patch (Li et al., 2020)	NRSC LULC datasets (1 km)
Response—action of the anthropogenic system to resolve environmental changes				
Vegetation	Vegetation health	NDVI (−)	Greenness in the ecological patch (Kang et al., 2018)	Landsat by NASA (30 m)
	Vegetation structure	Forest density (−)	The proportion of area covered by forest cover (Darabi et al., 2019)	FSI 2019 dataset (56 m)
		Forest type (+)	Forest type (Darabi et al., 2019)	

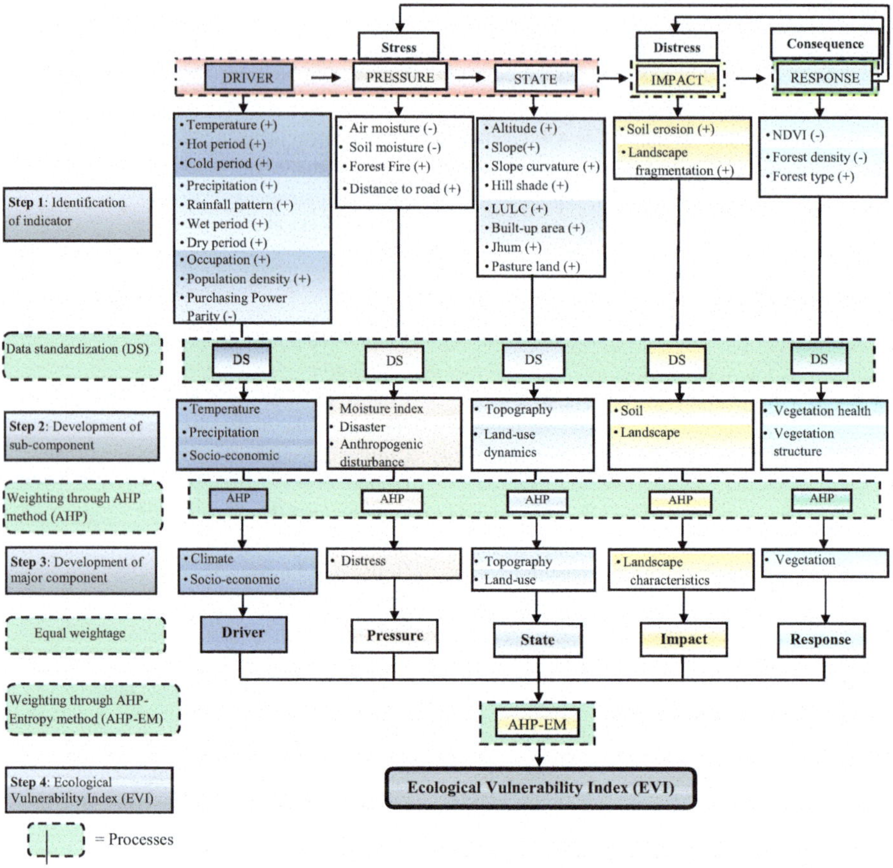

Fig. 2 Ecological Vulnerability Assessment through DPSIR framework

represents the minimum value, and $X_{i_{\max}}$ represents the maximum value for the ith indicator.

2.2.4 Weighing Through AHP-EM

The analytical hierarchy process-entropy method (AHP-EM) was employed for assigning the weights for each sub-component (Zhang et al., 2014). Consistency of decisions was measured through the consistency ratio (CR) (Saaty, 1977; Jiang et al., 2021). The estimated ecological vulnerability index (EVI) was classified into five vulnerability classes, viz., "very high," "high," "moderate," "low," and "very low," based on geometrical interval classification.

2.2.5 Spatial Autocorrelation Analysis

Spatial autocorrelation assesses the relevancy of the spatial distribution of the dataset (Li et al., 2020) and explains spatial cluster pattern. Moran's I index was applied for assessing autocorrelation. The index ranges from −1 to +1 with I > 0 representing positive spatial correlation, I < 0 reflecting negative spatial correlation, and I = 0 as a random pattern between the grids (Ding et al., 2018).

2.2.6 Logistic Regression Analysis

Logistic regression was applied to identify the influential indicators contributing to EVI for the year 2020 (Yang et al., 2020):

$$Y_i = \ln\left(\frac{P_i}{1-P_i}\right) = \beta_0 + \beta_1 i_1 + \beta_2 i_2 + \ldots + \beta_n \tag{4}$$

where Y_i is the EVI, iP_i is the probability, β_0 is the intercept, and β_n is the regression coefficient indicating the effect of the nth indicator.

3 Results

The spatiotemporal distribution of EVA using the DPSIR framework for the three periods across the NEHR was estimated based on all identified indicators and respective sub-components (Table 2). The results showed that the ecological vulnerability was increased in 2000–2010 and a marginal decrease of 0.03% in 2010–2020, with the diverse trend for various major components of the DPSIR. Spatially, the maximum vulnerable area was in the state of Assam, Meghalaya, Mizoram, and Manipur, while Arunachal Pradesh was the least vulnerable state (Table 2).

3.1 Spatial Distribution of Areas for Each Component of DPSIR in NEHR

Each major component and vulnerability was classified into five classes from very low to very high for the years 2000, 2010, and 2020. The diverse range of changes in areas for major components and vulnerability was observed during the time. In general, a high proportion of the area falls under the two classes, i.e., high and very high classes. However, during the successive periods, varied patterns of changes were observed across the major components, i.e., for lower classes, i.e., very low and low, a decreasing trend was observed from 2000 to 2010 and an increasing trend

Table 2 Environmental vulnerabilityVulnerabilities and major component of DPSIR at various spatiotemporal scales in NEHR

State	Year	Major component					EVI
		Driver	Pressure	State	Impact	Response	
Arunachal Pradesh	2000	0.232± 0.055	0.132± 0.065	0.310± 0.132	0.139± 0.239	0.502± 0.235	0.209± 0.067
	2010	0.244 ± 0.052	0.130± 0.061	0.304± 0.118	0.108± 0.146	0.492± 0.241	0.203± 0.044
	2020	0.264 ± 0.048	0.121± 0.060	0.344± 0.108	0.132± 0.151	0.521± 0.221	0.226± 0.050
Assam	2000	0.399±0.041	0.144± 0.063	0.177± 0.087	0.437± 0.253	0.263± 0.269	0.283± 0.056
	2010	0.392± 0.067	0.142± 0.054	0.110± 0.070	0.157± 0.218	0.254± 0.264	0.231± 0.037
	2020	0.406± 0.057	0.136± 0.057	0.136± 0.082	0.294± 0.174	0.325± 0.275	0.274± 0.039
Manipur	2000	0.260± 0.054	0.182± 0.061	0.277± 0.134	0.201± 0.273	0.549± 0.228	0.247± 0.079
	2010	0.272± 0.068	0.170± 0.053	0.242± 0.088	0.150± 0.178	0.520± 0.246	0.224± 0.044
	2020	0.303± 0.050	0.169± 0.060	0.280± 0.095	0.184± 0.164	0.565± 0.228	0.251± 0.043
Meghalaya	2000	0.314± 0.046	0.187± 0.061	0.192± 0.111	0.140± 0.244	0.514± 0.260	0.236± 0.064
	2010	0.402± 0.046	0.175± 0.055	0.196± 0.113	0.170± 0.197	0.523± 0.260	0.251± 0.047
	2020	0.354± 0.053	0.170± 0.057	0.230± 0.099	0.215± 0.171	0.579± 0.216	0.265± 0.043
Mizoram	2000	0.295± 0.045	0.233± 0.101	0.204± 0.089	0.037± 0.127	0.672± 0.172	0.235± 0.047
	2010	0.292± 0.054	0.210± 0.086	0.215± 0.092	0.150± 0.177	0.680± 0.200	0.250± 0.049
	2020	0.332± 0.046	0.216± 0.086	0.265± 0.090	0.142± 0.134	0.679± 0.157	0.264± 0.045
Nagaland	2000	0.265± 0.071	0.186± 0.055	0.324± 0.159	0.169± 0.250	0.636± 0.216	0.267± 0.083
	2010	0.262± 0.066	0.175± 0.051	0.306± 0.134	0.195± 0.198	0.629± 0.224	0.259± 0.052
	2020	0.314± 0.062	0.170± 0.054	0.342± 0.136	0.168± 0.152	0.659± 0.180	0.272± 0.053
Sikkim	2000	0.228± 0.065	0.146± 0.053	0.398± 0.120	0.254± 0.277	0.315± 0.260	0.257± 0.071
	2010	0.351± 0.062	0.146± 0.051	0.425± 0.117	0.132± 0.200	0.253± 0.224	0.268± 0.060
	2020	0.263± 0.050	0.140± 0.051	0.463± 0.094	0.253± 0.143	0.322± 0.258	0.299± 0.058
Tripura	2000	0.367± 0.026	0.201± 0.083	0.130 ± 0.124	0.159± 0.245	0.507± 0.291	0.260± 0.071
	2010	0.436± 0.059	0.184± 0.059	0.152± 0.110	0.246± 0.221	0.446± 0.292	0.293± 0.053
	2020	0.455± 0.047	0.189± 0.073	0.155± 0.100	0.247± 0.144	0.573± 0.234	0.293± 0.046

Table 2 (continued)

State	Year	Major component Driver	Pressure	State	Impact	Response	EVI
West Bengal	2000	0.367± 0.053	0.151± 0.040	0.189± 0.123	0.385± 0.261	0.287± 0.275	0.287± 0.056
	2010	0.476± 0.066	0.152± 0.038	0.143± 0.141	0.115± 0.199	0.234± 0.231	0.253± 0.046
	2020	0.438± 0.072	0.145± 0.038	0.169± 0.151	0.274± 0.134	0.373± 0.273	0.297± 0.046
NEHR	2000	0.290 ± 0.071	0.159 ± 0.073	0.242 ± 0.136	0.239 ± 0.281	0.445 ± 0.284	0.248 ± 0.071
	2010	0.327 ± 0.097	0.153 ± 0.063	0.218 ± 0.136	0.148 ± 0.192	0.431 ± 0.287	0.231 ± 0.050
	2020	0.339 ± 0.084	0.148 ± 0.067	0.251 ± 0.139	0.207 ± 0.172	0.483 ± 0.266	0.258 ± 0.051

was observed from 2010 to 2020; however, the reverse trend was observed for higher classes, i.e., high and very high classes (Table 3).

Overall, across the NEHR, the very high vulnerable area was increased in the last two decades to the tune of 1.59% in 2000 to 19.97% in 2020; however, decrease in areas was observed for lower classes, i.e., very low and low classes, i.e., 55.78% in 2000 to 40% in 2010 and 2020 (Table 3). The changes in areas under various classes across the states were diverse as in Arunachal Pradesh, Mizoram, and Nagaland, and the vulnerability was increased till 2010 and slight decrease thereafter up to 2020. The combined area under low and very low categories was decreased in subsequent decades across all states except Arunachal Pradesh and Nagaland, where it decreased drastically in the first decade and subsequently increased. Proportionally, the geographic area under the very low and low classes of vulnerability was maximum in Assam for all the 3 years followed by Manipur and West Bengal. The area under very high and high categories of vulnerability was maximum in Arunachal Pradesh followed by Assam, Meghalaya, Mizoram, Manipur, and Nagaland with an increasing trend with time. The varied trend with time was observed for changes in areas under the high and very high categories in Arunachal Pradesh, Mizoram, and Nagaland; however, these all three states were having more area under vulnerable categories since 2000 (Fig. 3).

The area under low, moderate, and high vulnerable regions were 86,193.75 km^2, 55,276.77 km^2, and 61,208.01 km^2 for 2000; 54,650.68 km^2, 54,669.72 km^2, and 54,645.63 km^2 for 2010; and 54,693.80 km^2, 54,710.76 km^2, and 54,620.16 km^2 for 2020. The total area under very low vulnerable class in the NE region was found to decline from 66,213.0 km^2 in 2000 to almost equal area with 54,634.1 km^2 by 2020, and very high vulnerable class increased from 4343.9 km^2 in 2000–2010 to 54,576.6 km^2 by 2020. The loss of 4% of very low vulnerable area can lead to the increased sensitivity of ecological components toward driving factors (Fig. 4a–c).

Table 3 Distribution of areas (%) under various categories for the components of DPSIR in NEHR

Major component	Year	Very low	Low	Moderate	High	Very high	Total
Driver	2000	0.24	10.00	28.10	40.85	20.80	100.00
	2010	0.01	2.12	7.01	50.00	40.86	100.00
	2020	0.02	12.24	13.07	45.06	29.61	100.00
Pressure	2000	11.15	20.68	33.68	22.08	12.42	100.00
	2010	4.69	16.71	31.86	30.79	15.96	100.00
	2020	26.40	33.75	27.65	11.39	0.82	100.00
State	2000	22.03	9.00	10.68	36.36	21.92	100.00
	2010	11.93	19.67	28.67	29.45	10.29	100.00
	2020	30.10	14.85	5.51	24.48	25.06	100.00
Impact	2000	55.83	2.81	4.81	34.80	1.74	100.00
	2010	25.94	47.73	2.48	12.51	11.33	100.00
	2020	45.80	12.19	8.51	31.92	1.58	100.00
Response	2000	34.28	13.65	20.06	26.94	5.08	100.00
	2010	33.29	10.98	15.00	29.81	10.92	100.00
	2020	25.70	13.95	13.86	32.12	14.37	100.00
EVI	2000	24.23	31.55	20.23	22.40	1.59	100.00
	2010	20.00	20.00	20.01	20.00	19.99	100.00
	2020	20.00	20.02	20.02	19.99	19.97	100.00

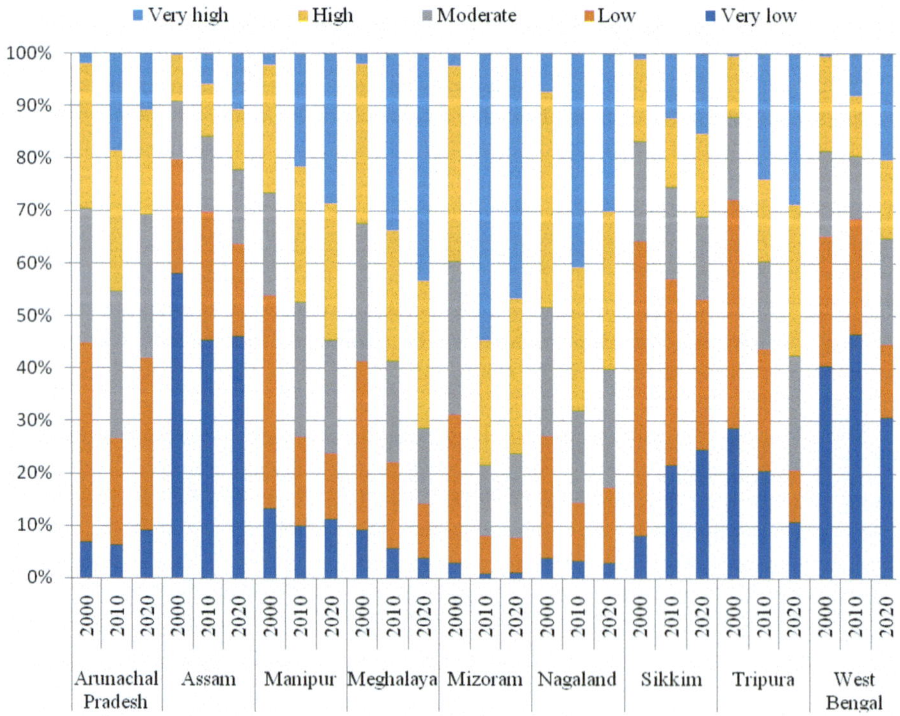

Fig. 3 Spatiotemporal distribution of area (in %) prone to ecological vulnerability under various categories in NE region

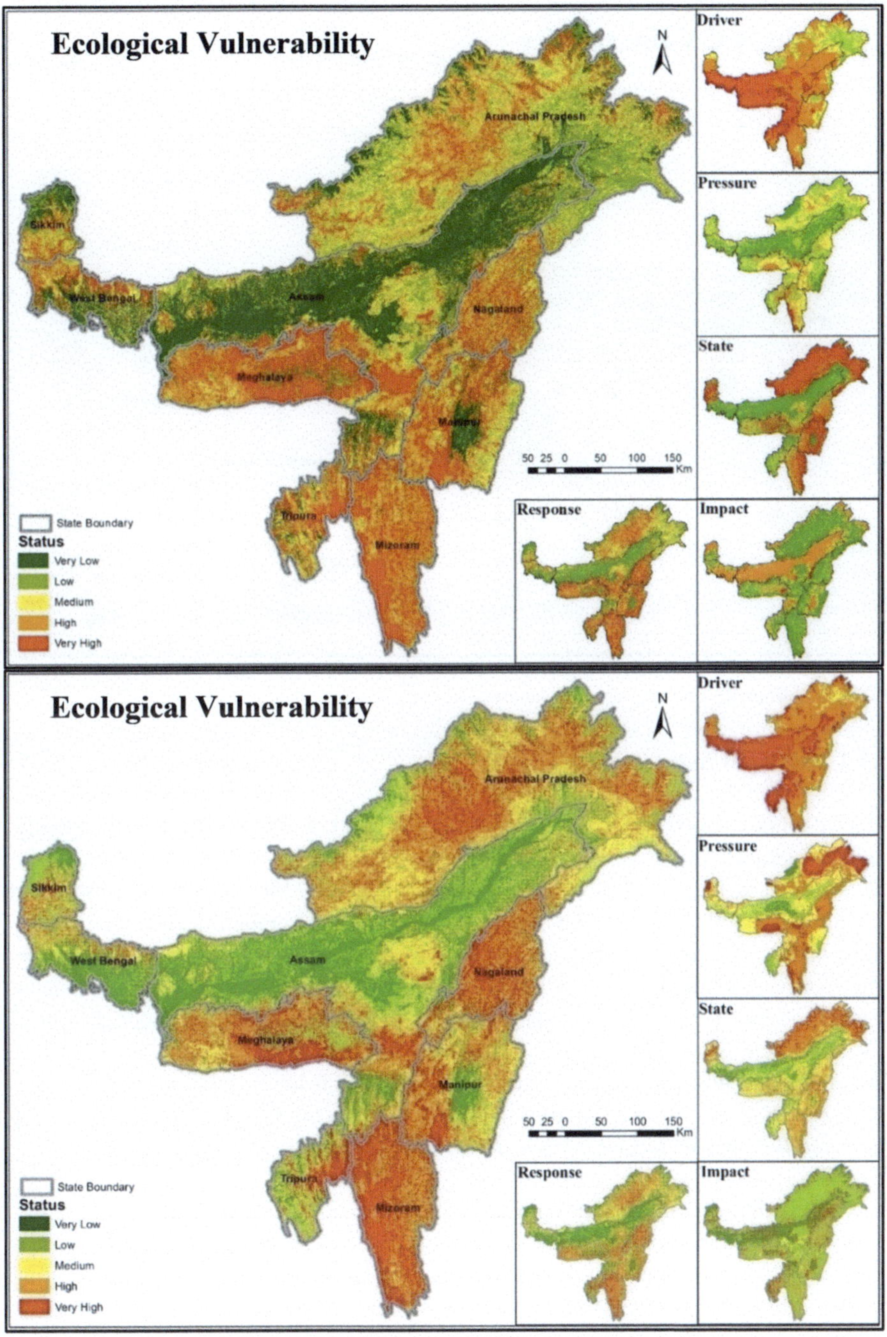

Fig. 4 Ecological vulnerability of NEHR for years (**a**) 2000, (**b**) 2010, and (**c**) 2020

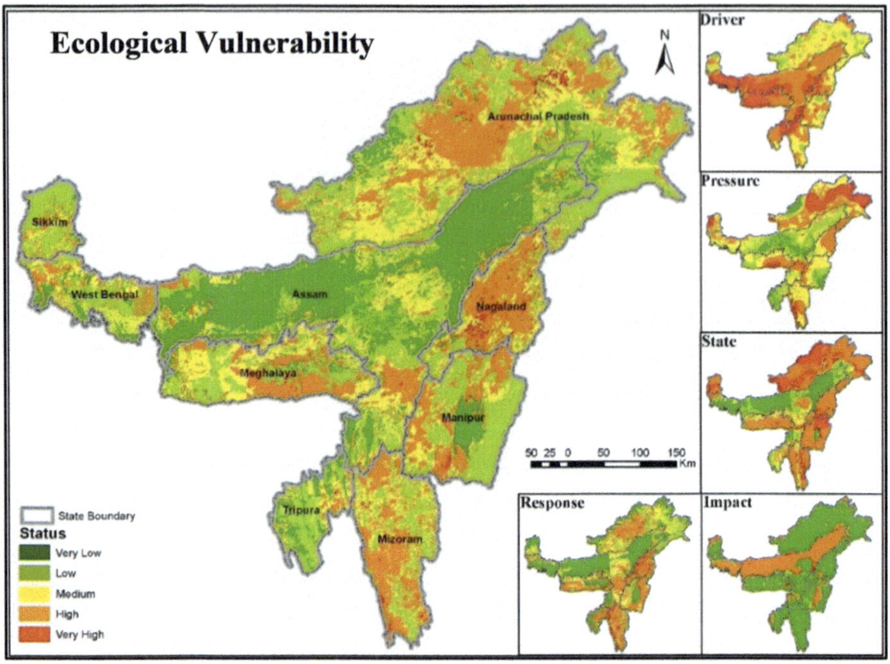

Fig. 4 (continued)

3.1.1 Hotspot Assessment

Moran's index was 0.124, 0.108, and 0.074 for 2000, 2010, and 2020, respectively. A considerable increase in clustering of "high-high" vulnerability areas was observed in Arunachal Pradesh, Mizoram, Nagaland, Meghalaya, Manipur, and Tripura from 2000 to 2010 (Fig. 5a). The "high-high" aggregations of vulnerability were observed mainly in Mizoram, Meghalaya, central Arunachal Pradesh, west Manipur, east Tripura, and southern patches of Assam in 2020. Assam, Arunachal Pradesh, Sikkim, and West Bengal noted an increase in "low-low" aggregation of EV. The central and south Mizoram, west Meghalaya, part of west Manipur, Nagaland, and north Arunachal Pradesh showed "high-low" and "low-high" aggregates of EV (Fig. 5a). The analysis showed that aggregations of vulnerable areas were significantly increased in most of Assam, Meghalaya, Mizoram, and Arunachal Pradesh.

The distribution of hotspot (aggregation of high vulnerable patches) was significantly increased in Mizoram, East Tripura, and west of Manipur and Meghalaya and was under the moderate to high vulnerability class in 2000. Cold spots (aggregation of low vulnerable patches) were generally decreased, while hotspots were increasing in succeeding decades. Hotspot areas remain undisturbed in the central portion of the Arunachal Pradesh, whereas cold spot areas were dissolving significantly. Assam and West Bengal observed maximum changes in cold spot distribution

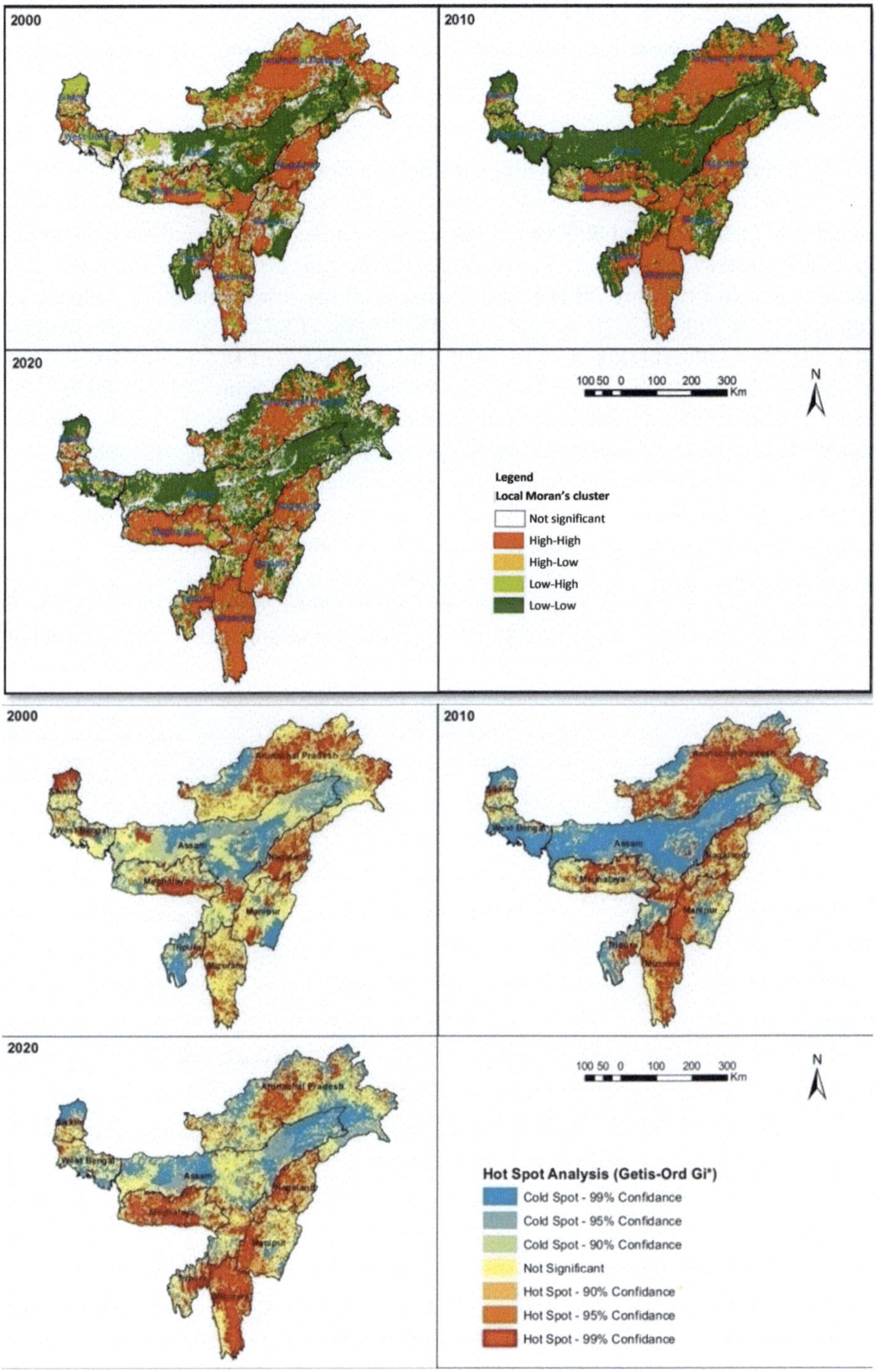

Fig. 5 Local Moran's index. (**a**) Moran's cluster analysis and (**b**) hotspot analysis

among nine states from 2000 to 2020. Mizoram, Meghalaya, and Arunachal Pradesh confronted the increase in hotspot areas from 2000 to 2020, indicating a decrease in ecosystem resilience to time-scale changes (Fig. 5).

3.1.2 Change Trajectory of Vulnerability Classes

Around 1471.50 km^2 and 895.74 km^2 areas were shifted to high and very high vulnerability classes (Fig. 6). A change of 28,263.22 km^2 (10.35%) of the total geographic area of the entire NEHR was shifted from the low to moderate vulnerable category from 2000 to 2010. A total of 37,467.95 km^2 (13.72%) of area was changed to a very high vulnerability class in 2010. The conversion of 17,272 km^2 (6.32%) of moderate class to high vulnerability class was observed from 2010 to 2020. The analysis also observed a decrease from very high to very low (6.40%), low (4.80%), moderate (2.90%), and high (6.00%) categories in 2010–2020 (Fig. 6; Table 4). The increase in very high vulnerability areas is a matter of concern as resource consumption and land utilization are already constrained and distinctly building pressure on the ecosystem's potential to counter disturbances and challenges.

Assam and Arunachal Pradesh noted a maximum shift of vulnerable areas to very high category. In the change observed from 2000 to 2010, 29.97% of very low vulnerability class in Assam and 33.01% of low vulnerability class in Arunachal

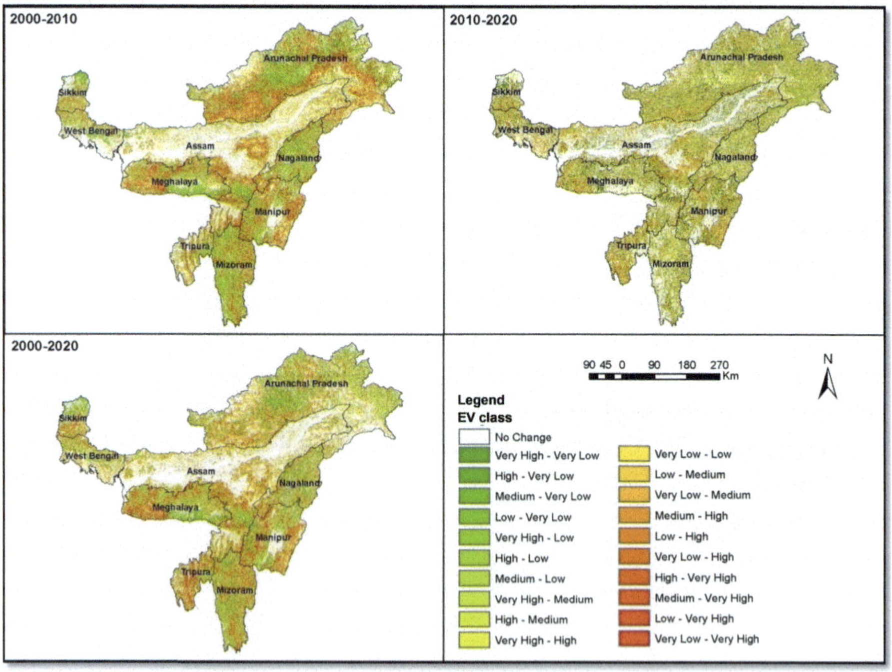

Fig. 6 Decadal change of ecological vulnerability classes in NEHR

Table 4 EVI change pattern for the year 2000–2010 and 2010–2020

Zone	Very low	Low	Moderate	High	Very high
Change in ecological vulnerable zone in the year 2000–2010					
Very low	43,638.99	17,886.86	2796.59	1471.15	895.74
Low	8935.29	26,548.62	28,263.22	1214.94	16,684.22
Moderate	8943.61	6938.47	831.16	14,984.54	19,624.33
High	24.37	4973.08	14,571.11	10,269.95	263.66
Very high	17,888.59	29,580.12	464.1	1083.7	2687.69
Change in ecological vulnerable zone in the year 2010–2020					
Very low	39,049.44	8165.31	4025.59	1923.2	1424.77
Low	13,231.01	19,234.87	9967.2	1286.2	6566.98
Moderate	5605.51	16,166.04	645.44	17,272.06	12,201.25
High	357.95	7889.30	7722.79	15,527.84	3189.93
Very high	17,495.35	13,075.36	7899.92	16,421.87	26,735.09

Pradesh shifted to very high zone (Fig. 7). In Arunachal Pradesh, 42.84% of moderate vulnerable area was shifted to very high vulnerability area and 38.30% of high vulnerability area was shifted to very high vulnerability area. Approximately 569.90 km^2 (40.00%) of very low vulnerable and 1981.10 km^2 (30.17%) of low vulnerable zone were shifted into very high vulnerable area in the Assam state. In Arunachal Pradesh, 3563 km^2 (29.20%) of moderately vulnerable area and 1653.80 km^2 (51.84%) of high vulnerability area were converted into very high vulnerability area (Figs. 6 and 7; Table 4).

3.1.3 Logistic Regression Analysis

Logistic regression was used to derive the contextual parameters resulting in vulnerability having BIC 33286.7. The results indicated that LUCC, forest fire, hot period, soil erosion, forest fragmentation, and cold period were positively influencing vulnerability (Table 5).

4 Discussion

NEHR is characterized by rich biodiversity, large forest cover, and undulating landscape (NRSC, 2019) and is influenced by extreme events such as intense precipitation, high temperature, and topography as steep slopes, high rate of soil erosion, and frequent flash flood-prone basins (Tsering et al., 2010). In NEHR, Mizoram has the largest area under forests (84.53%) followed by Arunachal Pradesh (79.00%), Meghalaya (76.00%), Manipur (74.30%), Nagaland (73.00%), Tripura (73.64%), Sikkim (47.08%), Assam (36.09), and West Bengal (Darjeeling = 75.1%; Jalpaiguri = 45.88%, and Koch Bihar = 10.30%) (FSI, 2021). The forests of the region are affected by climate change, i.e., rainfall and temperature; however, forest fires and other climatic hazards had a greater impact on the ecosystems (Dolan

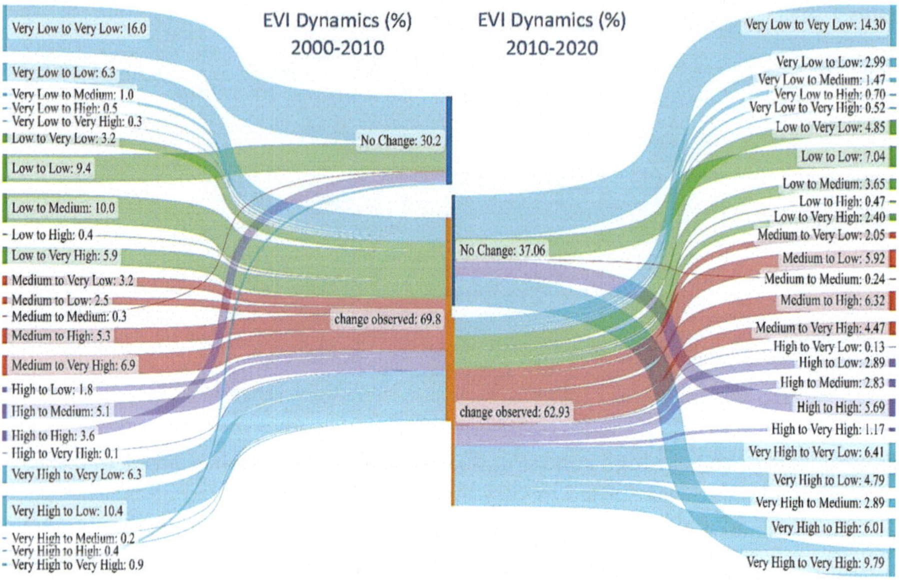

Fig. 7 Sankey diagram for ecological vulnerability dynamics (%)

Table 5 Estimated logistic regression slope (β_n), standard deviation (σ), and odds ratio

Indicator	β_n	σ	Odds ratio	Indicator	β_n	σ	Odds ratio
LULC	1.11	0.05*	3.02	Occupation	0.33	0.01*	1.39
Forest fire	0.87	0.03*	2.40	Distance to road	0.31	0.01*	1.36
Hot period	0.86	0.02*	2.37	Forest density	0.29	0.01*	1.34
Soil erosion	0.75	0.08*	2.11	Dry period	0.29	0.01*	1.33
Forest fragmentation	0.73	0.12*	2.06	Slope	0.28	0.01*	1.33
Cold period	0.67	0.03*	1.96	Moisture	0.26	0.02*	1.29
Population density	0.59	0.07*	1.81	Average annual precipitation	2.52	0.04*	1.24
Forest type	0.47	0.02*	1.60	Land cover	0.16	0.04*	1.17
Mining	0.45	0.03*	1.57	Curvature of slope	0.15	0.03*	1.16
PPP	0.39	0.02*	1.48	Rainfall pattern	0.12	0.03*	1.13
Altitude	0.38	0.02*	1.46	Wet period	0.02	0.02*	1.02
NDVI	0.33	0.01*	1.39				

"*" is used for p-value <0.001

et al., 2017) with further exacerbation of the impacts by extreme events (Chettri et al., 2018).

A consistent rise in air temperature, precipitation, and extreme climatic events is observed in the Himalayas (Mir et al., 2021) including increasing incidences of extreme monsoon events and wet days in the north region of the Himalayas and in the Hindu Kush Himalayas, respectively (Bhardwaj et al., 2021). Moreover, changes in cold days and hot days are also observed in the Himalayas (Manzoor & Ahmad Ahanqger, 2022). The changes in climate, i.e., long-term temperature and precipitation patterns in the Himalayan region, result in hydrometeorological disasters such as flash flood, cloudburst, and landslides (Kanwar & Kuniyal, 2022) and climate extremes (Shekhar et al., 2017; Bhardwaj et al., 2021), consequently leading to changing river patterns, water supply regimes, and livelihood securities for the communities (Eriksson et al., 2009).

Anthropogenic disturbances and disasters were relatively higher in Arunachal Pradesh, Sikkim, Meghalaya, and several isolated patches in southern Assam and Mizoram. The high moisture content was observed in Tripura, Manipur, and a major portion of central Assam representing the low vulnerability of the region, while the low moisture area in north-west of Arunachal Pradesh and west of Assam had high vulnerability. Moreover, high forest cover in the region was attributed to moisture fluctuations (Paul et al., 2018). The fire incidents were high in southwest Mizoram and a few isolated forest patches in south Assam, east Tripura, and south-west Manipur. Areas prone to fire incidences were identified as high-vulnerability areas of the NE region (FSI, 2021). Forest fires in the NEHR accounts for approximately 50% of all forest fire incidences in India (Bahuguna & Upadhyay, 2002), and around 95% of fires are due to *jhum* cultivation (NESAC, 2014).

The pressure governs the condition of the environment of the system in the present assessment. Topographically, around two-thirds of the NEHR is covered with hills and mountains (Choudhury et al., 2018) with a large area under forest cover (Table 6). The present analysis showed that high elevation, varied slopes, curvature, and hill shade are distinctly found in the majority of Arunachal Pradesh, Sikkim Khasi hills of Meghalaya, partial areas of Nagaland, and Manipur states of NEHR. Substantial land-use and land-cover changes are observed in North Sikkim, the majority of West Bengal and Assam, non-forested areas of central Manipur, and hilly regions of Meghalaya due to the increase in agricultural activity, built-up areas, pasture land, and mining areas. The coal mining pits along the Brahmaputra basin is posing a threat to the vegetation and biodiversity of the region. The result showed that non-forested areas of West Bengal, Assam, Manipur, Tripura, and hilly regions of Meghalaya were relatively more vulnerable to ecosystem degradation than the forested areas.

The land-use and land-cover changes (LULC) in India are primarily shaped by anthropogenic activities that affect biophysical and biogeographical attributes of the terrestrial surface (Paul et al., 2018; Lal et al., 2021). Lucan Himalayas results from both the natural and anthropogenic processes leading to increase deforestation, forest fragmentation, and mass wasting events due to unplanned road and habitation (DST, 2019). For example, anthropogenic activities such as the construction of transportation networks in snow-covered regions trap more sunlight and contribute

Table 6 Land-use and land-cover changes (LULCC) for the period from 2002 to 2015 in NEHR

	Area in 2015		Area in 2002		LULCC (w.r.t 2015)	
Land-use and land-cover	$\times10^3$ km^2	%	$\times10^3$ km^2	%	$\times10^3$ km^2	%
Barren	1024.97	3.03	1256.12	3.72	−231.15	−22.55
Croplands	7712.8	22.81	4888.73	14.46	2824.07	36.62
Forest	6493.61	19.21	9513.41	28.14	−3019.8	−46.50
Savannas	14152.19	41.86	15332.13	45.35	−1179.94	−8.34
Shrubland/grassland	3342.96	9.89	2413.63	7.14	929.33	27.80
Snow	104.38	0.31	53.58	0.16	50.80	48.67
Urban	65.32	0.19	0.00	0.00	65.32	100.00
Water bodies	411.31	1.22	79.4	0.23	331.91	80.70
Wetland	503.52	1.49	274.06	0.81	229.46	45.57

significantly to temperature rise, and transformation of snow cover and barren land to forest cover contributes to increase in precipitation due to an increase in leaf area index (LAI). However, increase in agricultural activities, shifting cultivation, and decrease in greenness resulted into a decrease in precipitation, and conversion of forest lands to other land use contributed to reduced latent heat flux and evaporation (Lal et al., 2021). The NEHR is confronting the transition of forested area into cropland and fallow land specially in Mizoram, Meghalaya, Arunachal Pradesh, Assam, and Nagaland. Large-scale LULC such as forests to agriculture (shifting cultivation) and vegetation cover to scrubland is altering soil fertility in the NEHR (Lepcha & Devi, 2020; Choudhury et al., 2021).

Considerable soil erosion and landscape fragmentation were observed across the NEHR over the past two decades and is attributed to forest deforestation, cultivation along the steep slopes in uplands, and extensive open-cast coal mining (Maji et al., 2010; Choudhury et al., 2021). The erosion rates were higher in small isolated patches in the Jaintia hills of Meghalaya, south of Manipur intensely in Churachandpur and Thoubal districts, and sparse patches in Sonitpur, Kokrajhar, and Karbi Anglong districts of Assam state. Jhum cultivation practices along the hill slopes and the abandoned *jhum* lands on steep slopes degrade soil structure and reduce the stability with more prone to runoff resulting into soil erosion and soil loss in this high-intensity rainfall zone (Choudhury et al., 2021). Moreover, erosion permitting crop cultivation under jhum such as maize, ginger, turmeric, and root crops further adds proneness to the soil leading to soil erosion-led topsoil loss (Choudhury et al., 2021; Lepcha & Devi, 2020; Singh et al., 2021). Progressive loss of the top fertile layer can adversely impact nutrient loss and ultimately the productivity of the forest. River flow soil erosion and sedimentation are largely influenced by the Himalayas' steep slopes and depreciated tree cover, thereby adversely impacting the hydrology, soil fertility, and agricultural production (Mishra et al., 2022).

The densely forested area has undergone disproportionate fragmentation making the forest ecosystems more vulnerable from 2000 to 2020. Fragmentation in forests occurs due to shifting cultivation, mining, quarrying, forest fires, overgrazing, and LULC as an increase in agricultural land, expansion of settlements, urbanization, illegal logging, and infrastructure development (Reddy et al., 2013; Lele & Joshi, 2009). Fragmentation is detrimental to the forest biodiversity (Fahrig, 2003), reduces core habitat, and alters microclimatic in the forest edges (Bustamante et al., 2003).

Vegetation structure and health are the key components for EVA, maintaining the resilience of the ecosystem. The assessment observed a decrease in NDVI from 2000 to 2010 across the region due to decreased forest density and low LAI in non-forested areas especially in Assam, north Sikkim, West Bengal, and central Manipur. The NDVI was increased in these areas from 2010 to 2020 probably due to a positive correlation with soil moisture (Thomte et al., 2022). The changes in NDVI in the NEHR is measured by greening and browning of vegetation and noted a predominant greening with browning trend having low intensity and rate than the widespread greening since 1990 in the region due to forest conservation strategies and declining of jhum cultivation (Punitha et al., 2018). Large-scale monoculture plantations (Kumar et al., 2021) and increased CO_2 fertilization and nitrogen deposition due to use of chemical fertilizers for agriculture are also facilitated for the widespread greening (Mishra & Mainali, 2017). Subalpine conifers, broadleaf forest, alpine shrubs, and meadow vegetation (Mishra & Mainali, 2017; Kumar et al., 2021) and farmlands (Nath et al., 2021) are the observed causes of the greenness of the region. However, the browning trend is due to temperature-induced moisture stress (Parida et al., 2020) and deforestation and is majorly due to the shift from savannas to cropland due to agricultural intensification. Moreover, the increasing greening is attributed to an increase in LAI, primarily due to afforestation and use of fertilizers in agriculture in the country (Roy et al., 2007), i.e., from savannas to forests, primarily tea plantations. Moreover, most of the forest area of the NEHR is increasing NDVI with very low forest area attributed to decreasing NDVI (Kumar et al., 2021).

More than 70% of NE population depends on agriculture for their livelihood security, and *jhum* cultivation is predominantly practiced as bio-culture in the states of Arunachal Pradesh, Meghalaya, Mizoram, and Nagaland (NITI Aayog, 2021). The cumulative impact of traditional practices (*jhum*) is degrading the soil for a longer period, thereby increasing the vulnerability. The increasing dependency of humans on the primary sector is exacerbating pressure on forests and the agricultural sector. West Bengal, Assam, Mizoram, and Manipur are highly vulnerable due to rapid land-use changes for agricultural purposes and Arunachal Pradesh, Nagaland, Meghalaya, Mizoram, and Tripura due to excess use of forest resources. Therefore, the development of human settlements must be understood through the lens of ecological dynamics.

5 Conclusion

Ecological vulnerability has increased in North-Eastern Himalayan Region (NEHR) from 2000 to 2020. The north of Brahmaputra basin is showing an increasing trend in very low and low vulnerability areas; however, the southern part of NEHR showed a contradictory result. LULC, forest fire, and temperature rise have significantly affected the vulnerability, indicating that a policy should orient toward checking the land-use change and controlling forest fire. The policy may be oriented as per the hotspot assessment and differential approaches must be applied for various clusters.

Overexploitation of agroecological resources is posing a threat to the natural eco-structure of the NEHR and should be addressed through local community participation for natural resource management. The study also highlights that ecological vulnerability at a larger spatial level must be considered to improve the understanding about the agglomeration of human habitation vis-à-vis ecological system dynamics. The DPSIR framework is suitable for assessing large-scale spatiotemporal vulnerability using remote sensing that will provide assistance for prioritizing regions for conservation interventions. Besides continuous monitoring, change analysis can be performed through integrated remote sensing and GIS.

Shifting cultivation may be a potential reason; therefore, checking the practice may facilitate reduction of vulnerability. Besides, land use should be managed through effective planning by using productive lands for production purposes and waste and unproductive land for habitation purposes. Forest fires should also be paid special attention and must be planned to check for protection and disconnecting the continuous patches through developing pathways between the patch. Moreover, addressing vulnerability of the ecosystem may lead to better productivity of the ecosystem, leading to support for achieving various SDGs such as reducing poverty (SDG 1), alleviating hunger (SDG 2), climate action (SDG 13), and biodiversity conservation and sustainable land management (SDG 15). Moreover, agroforestry systems including strengthening traditional practices can be a potential tool for addressing ecological vulnerability and has the potential to contribute to at least for sustainable energy (SDG 7) and agricultural crop production (SDG 12).

Acknowledgments We would like to acknowledge all the sources from where we have collected data.

Competing Interests The authors declare that they have no competing interests.

Availability of Data and Materials The datasets used during the study were from secondary sources and are available from the corresponding author upon reasonable request.

Author Contribution RP conceived the idea; RP and LT designed the methodology; LT collected the data and analyzed the data. RP developed the table with DV and LT. RP and DV substantially contributed to data interpretation. DV wrote the initial draft with RP and LT. RP modified and critically revised the manuscript with DM. AJN, RK, and AJ provided constructive suggestions for article improvement.

Funding Financial assistance was received from the Department of Science and Technology under the research grant DST/CCP/MRDP/**189/2019.**

References

Adhikari, K., Patten, S. B., Patel, A. B., Premji, S., Tough, S., Letourneau, N., Giesbrecht, G., & Metcalfe, A. (2021). Data harmonization and data pooling from cohort studies: a practical approach for data management. *International Journal of Popular Data Science, 6*(1), 1680.

Adler, C., Wester, P., Bhatt, I., Huggel, C., Insarov, G. E., Morecroft, M. D., Muccione, V., & Prakash, A. (2022). Cross-Chapter Paper 5: Mountains. In H. O. Portner, D. C. Roberts, M. Tignor, E. S. Poloczanska, K. Mintenbeck, A. Alegría, M. Craig, S. Langsdorf, S. Löschke, V. Möller, A. Okem, & B. Rama (Eds.), *Climate change 2022: Impacts, adaptation and vulnerability. Contribution of Working Group II to the sixth assessment report of the Intergovernmental Panel on Climate Change*. Cambridge University Press.

Bahuguna, V. K., & Upadhyay, A. (2002). Forest fires in India: Policy initiatives for community participation. *International Forestry Review, 4*(2), 122–127.

Bhardwaj, A., Wasson, R. J., Chow, W. T., & Ziegler, A. D. (2021). High-intensity monsoon rainfall variability and its attributes: A case study for Upper Ganges Catchment in the Indian Himalaya during 1901–2013. *Natural Hazards, 105*(3), 2907–2936.

Birkmann, J., Cardona, O. D., Carreño, M. L., Barbat, A. H., Pelling, M., Schneiderbauer, S., et al. (2013). Framing vulnerability, risk and societal responses: The MOVE framework. *Natural Hazards, 67*(2), 193–211.

Borah, B. C. (2022). Impact of climate change-induced challenges on fisheries in the North Eastern Region of India and the way ahead. *Aquatic Ecosystem Health & Management, 24*(3), 94–102.

Bradley, P., & Yee, S. (2015). *Using the DPSIR framework to develop a conceptual model: Technical support document* (p. EPA/600/R-15/154). United States Environmental Protection Agency, Office of Research and Development, Atlantic Ecology Division.

Bustamante, R., Serey, I., & Pickett, S. T. (2003). Forest fragmentation plant regeneration and invasion processes in Central Chile. In G. Bradshaw & P. Marquet (Eds.), *How landscapes change: Human disturbance and ecosystem fragmentation in the Americas* (pp. 145–160). Springer.

Chettri, N., & Sharma, E. (2016). Reconciling the mountain biodiversity conservation and human wellbeing: Drivers of biodiversity loss and new approaches in the Hindu-Kush Himalayas. *Proceedings of the Indian National Science Academy, 82*, 53–73.

Chettri, N., Tsering, K., Shrestha, A., & Sharma, E. (2018). Ecological vulnerability to climate change in the mountains: A case study from the Eastern Himalayas. In *Plant Diversity in the Himalaya Hotspot Region* (pp. 707–721). ICIMOD.

Choudhury, B. U., Verma, B. C., Ramesh, T., & Hazarika, S. (2018). Altitude regulates accumulation of organic carbon in soil: Case studies from the hilly ecosystem of Northeastern region of India. In *Advances in crop environment interaction* (pp. 137–149). Springer.

Choudhury, B. U., Ansari, M. A., Chakraborty, M., & Meetei, T. T. (2021). Effect of land-use change along altitudinal gradients on soil micronutrients in the mountain ecosystem of Indian (Eastern) Himalaya. *Scientific Reports, 11*, 14279.

Cutter, S. L., & Finch, C. (2008). Temporal and spatial changes in social vulnerability to natural hazards. *National Academy of Sciences of the United States of America, 105*(7), 2301–2306.

Cutter, S. L., Boruff, B. J., & Shirley, W. L. (2003). Social vulnerability to environmental hazards*. *Social Science Quarterly, 84*(2), 242–261.

Darabi, H., IslamiFarsani, S., & IraniBehbahani, H. (2019). Evaluation of ecological vulnerability in Chelgard mountainous landscape. *Pollution, 5*(3), 597–610.

de Lange, H. J., Sala, S., Vighi, M., & Faber, J. H. (2010). Ecological vulnerability in risk assessment – A review and perspectives. *The Science of the Total Environment, 408*(18), 3871–3879.

De, L. C., & Singh, D. R. (2017). Natural resources in North East region of India. *International Journal of Agricultural Science and Research, 7*(5), 51–66.

Deka, J., & Barua, I. C. (2015). Problem weeds and their management in the North-East Himalayas. *Indian Journal of Weed Science, 47*(3), 296–305.

Depietri, Y., Welle, T., & Renaud, F. G. (2013). Social vulnerability assessment of the Cologne urban area (Germany) to heat waves: Links to ecosystem services. *International Journal of Disaster Risk Reduction, 6*, 98–117.

Ding, Q., Shi, X., Zhuang, D., & Wang, Y. (2018). Temporal and spatial distributions of ecological vulnerability under the influence of natural and anthropogenic factors in an eco-province under construction in China. *Sustainability, 10*(9), 3087.

Dolan, K. A., Hurtt, G. C., Flanagan, S. A., Fisk, J. P., Sahajpal, R., Huang, C., & Masek, J. G. (2017). Disturbance distance: Quantifying forests' vulnerability to disturbance under current and future conditions. *Environmental Research Letters, 12*(11), 1–6.

Dossou, J. F., Li, X. X., Sadek, M., SidiAlmouctar, M. A., & Mostafa, E. (2021). Hybrid model for ecological vulnerability assessment in Benin. *Scientific Reports, 11*(1), 2449.

DST. (2019). *Status of ecosystem health in the Indian Himalayan Region. A report under NMSHE, Climate Change Programme*. DST, MoS&T, GoI.

Eken, S., & Sayar, A. (2021). A map reduce-based distributed and scalable framework for stitching of satellite mosaic images. *Arabian Journal of Geosciences, 14*(18), 1–16.

Eriksson, M., Jianchu, X., Shrestha, A. B., Vaidya, R. A., Nepal, S., & Sandström, K. (2009). *The changing Himalayas: Impact of climate change on water resources and livelihoods in the Greater Himalayas*. ICIMOD.

Fahrig, L. (2003). Effects of habitat fragmentation on biodiversity. *Annual Review of Ecology, Evolution, and Systematics, 34*, 487–515.

FAO. (2015). *Mapping the vulnerability of mountain peoples to food insecurity*. Food and Agriculture Organization of the United Nations.

FSI. (2020). *ATLAS forest types of India*. Forest Survey of India.

FSI. (2021). *India State of Forest Report (ISFR)*. Forest Survey of India.

Gerlitz, Y. J., Brook, N., Macchi, M., Pandey, R., & Jha, S. (2016). The multidimensional livelihood vulnerability index: An instrumental to measure livelihood vulnerability to change in the Hindu Kush Himalayas. *Climate and Development, 9*(2), 124–140.

Guillard-Gonçalves, C., & Zêzere, J. (2018). Combining social vulnerability and physical vulnerability to analyse landslide risk at the municipal scale. *Geosciences, 8*(8), 294.

Guo, B., & Wen, Y. E. (2020). An optimal monitoring model of desertification in Naiman Banner based on feature space utilizing Landsat8 OLI image. *IEEE Access, 8*, 4761–4768.

Gupta, A. K., Negi, M., Nandy, S., Alatalo, J. M., Singh, V., & Pandey, R. (2019). Assessing the vulnerability of socio-environmental systems to climate change along an altitude gradient in the Indian Himalayas. *Ecological Indicators, 106*, 105512.

He, L., Shen, J., & Zhang, Y. (2018). Ecological vulnerability assessment for ecological conservation and environmental management. *Journal of Environmental Management, 206*, 1115–1125.

Hu, X., Ma, C., Huang, P., & Guo, X. (2021). Ecological vulnerability assessment based on AHP-PSR method and analysis of its single parameter sensitivity and spatial autocorrelation for ecological protection—A case of Weifang City, China. *Ecological Indicators, 125*, 107464.

IPCC. (2014). Summary for policymakers. In C. B. Field, V. R. Barros, D. J. Dokken, K. J. Mach, M. D. Mastrandrea, T. E. Bilir, M. Chatterjee, K. L. Ebi, Y. O. Estrada, R. C. Genova, B. Girma, E. S. Kissel, A. N. Levy, S. MacCracken, P. R. Mastrandrea, & L. L. White (Eds.), *Climate change 2014: Impacts, adaptation, and vulnerability. Part A: Global and sectoral aspects. Contribution of Working Group II to the fifth assessment report of the Intergovernmental Panel on Climate Change* (pp. 1–32). Cambridge University Press.

IPCC. (2021). Summary for policymakers. In V. Masson-Delmotte, P. Zhai, A. Pirani, S. L. Connors, C. Péan, S. Berger, N. Caud, Y. Chen, L. Goldfarb, M. I. Gomis, M. Huang, K. Leitzell, E. Lonnoy, J. B. R. Matthews, T. K. Maycock, T. Waterfield, O. Yelekçi, R. Yu, & B. Zhou (Eds.), *Climate change 2021: The physical science basis. Contribution of Working Group I to the sixth assessment report of the Intergovernmental Panel on Climate Change* (pp. 3–32). Cambridge University Press.

IPCC. (2022a). Summary for policymaker. In *Climate change 2022: Mitigation of climate change. Contribution of Working Group III to the sixth assessment report of the Intergovernmental Panel on Climate Change*. Cambridge University Press.

IPCC. (2022b). In P. R. Shukla, J. Skea, R. Slade, A. A. Khourdajie, R. van Dieman, D. McCollum, M. Pathak, S. Some, P. Vyas, R. Fradera, et al. (Eds.), *Climate change 2022: Mitigation of climate change. Contribution of working group III to the Sixth Assessment Report of the Intergovernmental Panel on Climate Change*. Cambridge University Press.

Jamshed, A., Birkmann, J., Feldmeyer, D., & Rana, I. A. (2020). A conceptual framework to understand the dynamics of rural-urban linkages for rural flood vulnerability. *Sustainability, 12*(7), 2894.

Jiang, Y., Shi, B., Li, R., & Guo, L. (2021). A long-term ecological vulnerability analysis of the Tibetan region of natural conditions and ecological protection programs. *Sustainability, 13*, 10598.

Kang, H., Tao, W., Chang, Y., Zhang, Y., Xuxiang, L., & Chen, P. (2018). A feasible method for the division of ecological vulnerability and its driving forces in Southern Shaanxi. *Journal of Cleaner Production, 205*, 619–628.

Kanwar, N., & Kuniyal, J. C. (2022). Vulnerability assessment of forest ecosystems focusing on climate change, hazards and anthropogenic pressures in the cold desert of Kinnaur district, northwestern Indian Himalaya. *Journal of Earth System Science, 131*(1), 1–16.

Kristensen, P. (2004). *The DPSIR framework*. Paper presented at the 27–29 September 2004 workshop on a comprehensive/detailed assessment of the vulnerability of water resources to environmental change in Africa using river basin approach. UNEP Headquarters, Nairobi, Kenya.

Kumar, M., Singh, H., Pandey, R., Singh, M. P., Ravindranath, N. H., & Kalra, N. (2019). Assessing vulnerability of forest ecosystem in the Indian Western Himalayan region using trends of net primary productivity. *Biodiversity and Conservation, 28*(8), 2163–2182.

Kumar, M., Kalra, N., Singh, H., Sharma, S., Rawat, P. S., Singh, R. K., Gupta, A. K., Kumar, P., & Ravindranath, N. H. (2021). Indicator-based vulnerability assessment of forest ecosystem in the Indian Western Himalayan: An analytical hierarchy process integrated approach. *Ecological Indicators, 125*, 107568.

Lal, P., Shekhar, A., & Kumar, A. (2021). Quantifying temperature and precipitation change caused by land cover change: A case study of India using the WRF model. *Frontiers in Environmental Science, 9*, 766328. https://doi.org/10.3389/fenvs.2021.766328

Lele, N., & Joshi, P. K. (2009). Analyzing deforestation rates, spatial forest cover changes and identifying critical areas of forest cover changes in North-East India during 1972–1999. *Environmental Monitoring and Assessment, 156*, 159–170.

Lepcha, N. T., & Devi, N. B. (2020). Effect of land use, season, and soil depth on soil microbial biomass carbon of Eastern Himalayas. *Ecological Processes, 9*(1), 1–14.

Li, W., Rollins, J., & Yan, E. (2018). Web of Science use in published research and review papers 1997–2017: A selective, dynamic, cross-domain, content-based analysis. *Scientometrics, 115*, 1–20.

Li, Q., Shi, X., & Wu, Q. (2020). Exploring suitable topographical factor conditions for vegetation growth in Wanhuigou catchment on the Loess Plateau, China: a new perspective for ecological protection and restoration. *Ecological Engineering, 158*, 106053.

Li, Q., Shi, X., & Wu, Q. (2021). Effects of protection and restoration on reducing ecological vulnerability. *Science of the Total Environment, 761*, 143180.

Li, D., Huan, C., Yang, J., & Gu, H. (2022). Temporal and spatial distribution changes, driving force analysis and simulation prediction of ecological vulnerability in Liaoning Province, China. *Land, 11*, 1025.

Maji, A. K., Reddy, G. P. O., & Sarkar, D. (2010). *Degraded and wastelands of India, status and spatial distribution*. 158 Directorate of Information and Publications of Agriculture. Indian Council of Agricultural Research.

Manzoor, S., & Ahmad Ahanqger, M. (2022). Spatio-temporal trends in precipitation and temperature means/extremes in the Himalayan states of India. *Journal of Water and Climate Change, 13*(7), 2531–2558.

Mir, B. H., Kumar, R., Lone, M. A., & Tantray, F. A. (2021). Climate change and water resources of Himalayan region—Review of impacts and implication. *Arabian Journal of Geosciences, 14*(12), 1088. Available from: https://link.springer.com/10.1007/s12517-021-07438-z

Mishra, N. B., & Mainali, K. P. (2017). Greening and browning of the Himalaya: Spatial patterns and the role of climatic change and human drivers. *Science of Total Environment, 587–588*, 326–339.

Mishra, P. K., Rai, A., Abdelrahman, K., Rai, S. C., & Tiwari, A. (2022). Land degradation, overland flow, soil erosion, and nutrient loss in the Eastern Himalayas, India. *Land, 11*(2), 179.

Nath, A. J., Sileshi, G. W., Laskar, S. Y., Pathak, K., Reang, D., Nath, A., & Das, A. K. (2021). Quantifying carbon stocks and sequestration potential in agroforestry systems under divergent management scenarios relevant to India's Nationally Determined Contribution. *Journal of Cleaner Production, 28*.

NESAC. (2014). Forest fire assessment. In *Northeast India under North Eastern Regional node-disaster risk reduction program* (Document No: NESAC-SR-102-2014, Project Report). NESAC.

NHC (Northwest Hydraulics Consultants). (2006). River flooding and erosion in Northeast India. In *Development and growth in Northeast India: The natural resources, water, and environment nexus*. NHC.

NITI Aayog. (2021). *North Eastern Region District SDG Index: Baseline report 2021–22*. Government of India.

NRSC. (2019). *Land use/land cover database on 1:50,000 scale, natural resources census*. Project, LUCMD, LRUMG, RSAA, National Remote Sensing Centre, ISRO.

Pandey, R., & Jha, S. (2012). Climate vulnerability index-measure of climate change vulnerability to communities: a case of rural Lower Himalaya, India. *Mitigation and Adaptation Strategies for Global Change, 17*, 487–506.

Pandey, R., Meena, D., Aretano, R., Satpathy, S., Semeraro, T., Gupta, A. K., Rawat, S., & Zurlini, G. (2015). Socio-ecological vulnerability of smallholders due to climate change in mountains: Agroforestry as an adaptation measure. *Change and Adaptation in Socio-Ecological Systems, 2*(1).

Pandey, R., Jha, S. K., Alatalo, J. M., Gupta, A. K., & Archie, K. M. (2017). Sustainable livelihood framework-based indicators for assessing climate change vulnerability and adaptation for Himalayan communities. *Ecological Indicators, 79*, 338–346.

Parida, B. R., Pandey, A. C., & Patel, N. R. (2020). Greening and browning trends of vegetation in India and their responses to climatic and non-climatic drivers. *Climate, 8*. https://doi.org/10.3390/CLI8080092

Paul, S., Ghosh, S., Rajendran, K., & Murtugudde, R. (2018). Moisture supply from the Western Ghats forests to water deficit East Coast of India. *Geophysical Research Letters, 45*, 4337–4344. https://doi.org/10.1029/2018GL078198

Punitha, P., Ansari, M. A., Pandey, D. K., Ram, D., Datt, S., Sharma, P. K., Aheibam, M., Jyothi, S. S. P., & Prakash, N. (2018). Shifting cultivation in North East India: Social dimension, cross-cultural reflection and strategies for improvement. *Indian Journal of Agricultural Sciences, 88*, 811–819.

Reddy, C. S., Sreelekshmi, S., Jha, C. S., & Dadhwal, V. K. (2013). National assessment of forest fragmentation in India: Landscape indices as measures of the effects of fragmentation and forest cover change. *Ecological Engineering, 60*, 453–464.

Rehman, S., Hasan, M. S. U., Rai, A. K., Avtar, R., & Sajjad, H. (2021). Assessing flood-induced ecological vulnerability and risk using GIS-based in situ measurements in Bhagirathi sub-basin, India. *Arabian Journal of Geosciences, 14*(15), 1–17.

Roy, S. S., Mahmood, R., Niyogi, D., Lei, M., Foster, S. A., Hubbard, K. G., Douglas, E., & Pielke, R., Sr. (2007). Impacts of the agricultural green revolution-induced land use changes on air temperatures in India. *Journal of Geophysical Research, 112*.

Roy, A., Kolady, D., Paudel, B., Yumnam, A., Mridha, N., Chakraborty, D., & Singh, N. U. (2021). Recent trends and impacts of climate change in North-Eastern region of India – A review. *Journal of Environmental Biology, 42*, 1415–1424.

Saaty, T. L. (1977). A scaling method for priorities in hierarchical structures. *Journal of Mathematical Psychology, 15*(3), 234–281.

Saikia, U. S., & Hazarika, S. (2020). *Climate change – A North East India perspective*. National Innovations in Climate Resilient Agriculture, ICAR Research Complex for NEH Region Umiam.

Sharma, E., Chettri, N., Tsering, K., Shrestha, A. B., Jing, F., Mool, P., & Eriksson, M. (2009). *Climate change impacts and vulnerability in the Eastern Himalayas*. ICIMOD.

Sharma, E., Molden, D., Rahman, A., Khatiwada, Y. R., Zhang, L., Singh, S. P., Yao, T., & Wester, P. (2019). Introduction to the Hindu Kush Himalaya assessment. In *The Hindu Kush Himalaya assessment* (pp. 1–16). Springer.

Shekhar, M. S., Devi, U., Paul, S., Singh, G. P., & Singh, A. (2017). Analysis of trends in extreme precipitation events over Western Himalaya Region: Intensity and duration wise study. *The Journal of Indian Geophysical Union, 21*(3), 225–231.

Singh, B., Jeganathan, C., Rathore, V. S., Behera, M. D., Singh, C. P., Roy, P. S., & Atkinson, P. M. (2021). Resilience of the Central Indian Forest Ecosystem to rainfall variability in the context of a changing climate. *Remote Sensing, 13*(21), 4474.

Singha, S. (2018). Assessing vulnerability and resilience in the face of climate change: A case study of North - East India. *European Journal of Sustainable Development, 7*(3), 53–64.

Thomte, L., Bhagabati, A. K., & Shah, S. K. (2022). Soil moisture-based winter–spring drought variability over West Karbi Anglong region, Assam, Northeast India using tree-rings of Pinuskesiya. *Environmental Challenges, 7*, 100512.

Tsering, K., Sharma, E., Chettri, N., & Shrestha, A. (Eds.). (2010). *Climate change vulnerability of mountain ecosystems in the Eastern Himalayas; Climate change impact and vulnerability in the Eastern Himalayas—Synthesis report*. ICIMOD.

Weibhuhn, P., Müller, F., & Wiggering, H. (2018). Ecosystem vulnerability review: Proposal of an interdisciplinary ecosystem assessment approach. *Environmental Management, 61*, 904–915.

Wester, P., Mishra, A., Mukherji, A., & Shrestha, A. B. (Eds.). (2019). *The Hindu Kush Himalaya assessment-mountains, climate change, sustainability and people*. Springer Nature.

Willroth, P., Revilla Diez, J., & Arunotai, N. (2011). Modelling the economic vulnerability of households in the Phang-Nga Province (Thailand) to natural disasters. *Natural Hazards, 58*(2), 753–769.

Wu, T. (2021). Quantifying coastal flood vulnerability for climate adaptation policy using principal component analysis. *Ecological Indicators, 129*, 108006.

Yang, Y., Chen, G., & Reniers, G. (2020). Vulnerability assessment of atmospheric storage tanks to floods based on logistic regression. *Reliability Engineering & System Safety, 196*, 106721.

Zahan, Y., Mahanta, R., Rajesh, P. V., & Goswami, B. N. (2021). Impact of climate change on North-East India (NEI) summer monsoon rainfall. *Climatic Change, 164*(2).

Zhang, X., Wang, C., Li, E., & Xu, C. (2014). Assessment model of ecoenvironmental vulnerability based on improved entropy weight method. *Scientific World Journal, 2014*, 1.

Part IV
Diversification in Livelihood Opportunities in Mountains: Adapting to Changing Climate

Exploring Alternative Livelihood Opportunities for Mountain People as an Adaptive Strategy to Changing Climate: A Case of Chyulu Hills, Kenya

Albert Gamoe Locham

1 Introduction

Chyulu Hills (otherwise called Chyulus or Chyulu Water Towers) are described as relatively young volcanic ranges consisting of craters and ash cones located in the southern part of Kenya, about 190 km southeast of Nairobi, the capital city of Kenya and 30 km south-west of the Kibwezi town (Kamau & Medley, 2014). The hills rise from an altitude of around 600 m to 2200 m above sea level along the highest summits, thereby creating a natural barrier that redirects the prevailing winds upward resulting in precipitation upon cooling and condensation of water vapor (White et al., 2017). The hills lack obvious permanent surface water. However, rainwater percolating through the porous rocks of the hills feed many permanent freshwater sources in the surrounding plains including the Tsavo and Galana rivers and the Mzima springs that supply water to Mombasa, Kenya's second largest city (KWS, 2008).

The Chyulus forms the boundary between the Kibwezi constituency of Makueni County and the Loitokitok (South Kajiado) constituency of Kajiado County (Kamau & Medley, 2014) and consists of both protected and non-protected areas. The protected areas include the Chyulu National Park administered by the Kenya Wildlife Service (KWS) and Kibwezi Forest Reserve under the management of the Kenya Forest Service (KFS). Ungazetted areas include sections of Chyulus owned by Masai community ranches, namely, Mbirikani and Kuku game ranches. The protected sections of the hills, namely, the national park and the forest reserve, are the least disturbed and provide home to a variety of indigenous floral and faunal species (KWTA, 2017). The floral species include fig trees (*Ficus* spp.), black ironwood (*Olea capensis*), African cherry (*Prunus africana*), and Kenya cedar (*Juniperus*

A. Gamoe Locham (✉)
Kenya Water Towers Agency (KWTA), Nairobi, Kenya

S. Tripathi et al. (eds.), *Warming Mountains*,
https://doi.org/10.1007/978-3-031-62197-0_14

procera), among others, whereas faunal species include African elephant (*Loxodonta africana*), black rhino (*Diceros bicornis*), leopards (*Panthera pardus*), African buffalo (*Syncerus caffer*), and bush pigs (*Potamochoerus larvatus*), among others (KWTA, 2017) (Fig. 1).

Like other forested hills and mountains in Africa, Chyulu Hills provide ecosystem goods and services, namely, water resources, to thousands of people in the lowlands (KWTA, 2017). The two communities that live adjacent to Chyulu Hills and greatly derive their livelihood from the hills are Kamba (crop farmers) and Masai (the pastoralists). Continued dependence on ecosystem goods (food, fuel energy, building materials, fodder and medicine among others) from the water towers for their livelihood has led to environmental degradation due to incremental cutting of trees for firewood and charcoal, opening land for cultivation, the

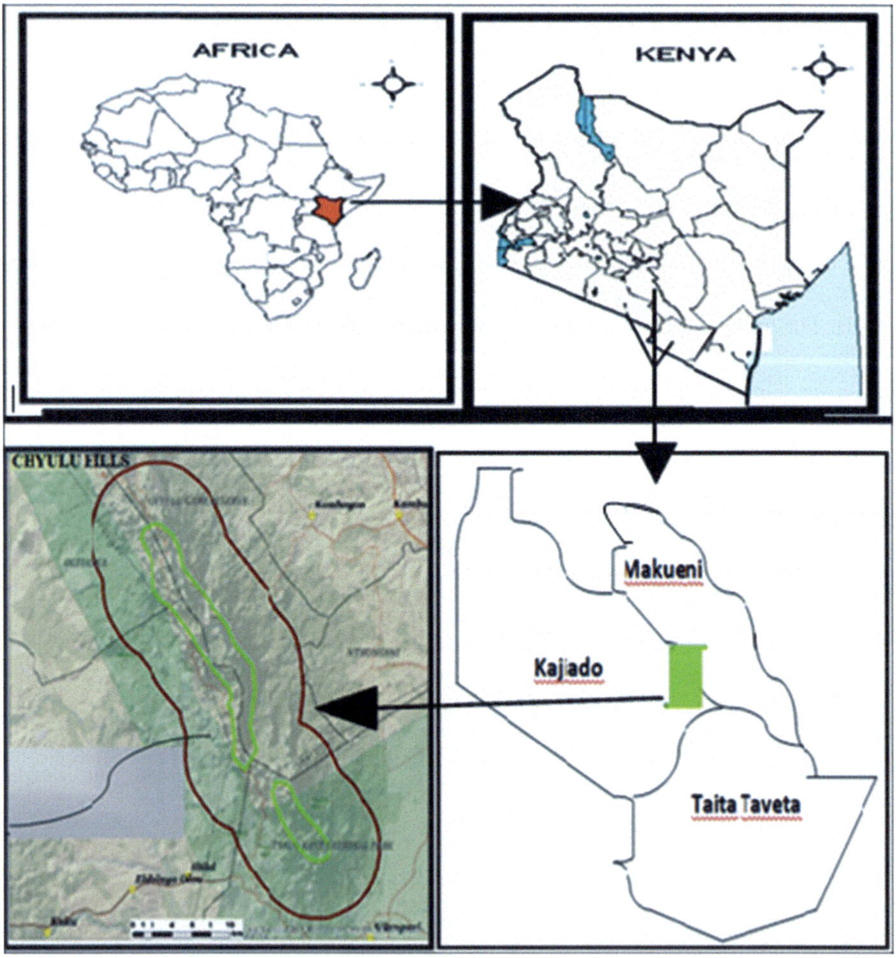

Fig. 1 The map showing the location of Chyulu Hills, Kenya

occurrence of forest or bush fires, and excision of gazetted forest to create human settlement, among others (KWTA, 2017). Climate change has served to compound the above drivers of degradation in Chyulu Hills just like the case elsewhere in Africa with potential impacts on mountain ecosystems and ecosystem services and, therefore, mountain people's well-being and livelihood (Nsengiyumwa, 2019). In Africa, however, information on the impacts of climate change on African mountain landscapes and people is limited (Nsengiyumwa, 2019) except for the insights on linkages between climate change and landslides (Gariano & Guzzetti, 2016). For instance, intense precipitation on African mountain landscape as a result of climate change has been noted to result to flood-mediated diseases (Hay et al., 2002), deadly and destructive landslides on degraded slopes (Gorokhocich et al., 2013), and cryospheric hazards particularly in Lesotho highlands (Grab & Linde, 2014). Changes in rainfall patterns have also been found to affect honey production in Mt. Cameroon where local farmers known for their "white honey" from beehives placed at high altitude reported reduced honey production in drier and wetter years (Cuni-Sanchez et al., 2018). Climate change affects mountain biodiversity in diverse ways including shift in distribution, biomass, and species richness (Hiltner et al., 2016, Ponce-Reyes et al., 2017) and reduced ecosystem services notably water provision to both mountain and lowland communities leading to declined agricultural and livestock productivity (Cuni-Sanchez et al., 2018).

Like other communities in Africa, mountain landscape people such as those in Chyulu Hills are dependent on natural resources and ecosystem services for their livelihoods. So, changes in water resources as a result of altered rainfall patterns will lead to reduced agricultural and livestock productivity, resulting in lower income levels for the community (Alweny et al., 2014).

In light of these climate change-mediated impacts on mountain ecosystems (such as Chyulu) and people, it is therefore fitting for mountain communities to explore alternate livelihood approaches as a strategy to adapt (defined as the process of adjustment to actual or expected climate and its effects, IPCC, 2014) to adverse impacts of climate change. The Government of Kenya through state agencies such as the Kenya Water Towers Agency (KWTA) promotes adoption of alternative sources of livelihood as a strategy to build community-driven resilience against adverse effects of climate change on mountain communities. For instance, livestock keeping populations living within and around water towers such as Chyulu Hills are encouraged to engage on on-farm hay production to feed their livestock as opposed to grazing in the forest in order to reduce grazing pressure in the ecosystem, thus minimizing soil erosion and environment degradation. This chapter discusses some of the alternate livelihood options specifically explored or implemented in Chyulu Hills and their realized impacts on biodiversity, environment, and community livelihoods.

2 Conservation Livelihood

Conservation livelihood is described as an approach that integrates conservation and livelihood, with a perspective of "growth within" through the process of circular economy and by connecting nature, culture, and development with an agenda of empowering the community, skilling and enforcing eco-cultural ethos and ethics (Sarma, 2018). In Chyulu Hills ecosystem, a number of players have mobilized the community to adopt conservation livelihood approach as a pathway to advance conservation while enhancing livelihoods.

Local Masai pastoral communities in Kuku and Rombo ranches are actively encouraged by Masai Wildlife Conservation Trust (MWCT) to take active participation as stewards of the critical ecosystem of Chyulu Hills. MWCT's main conservation programs include wildlife protection and research and the "Wildlife Pays" program (KWTA, 2022). The wildlife protection and research program consists of over 100 rangers drawn from community to combat wildlife crimes including poaching and a further unit of young Masai warriors (otherwise called "Simba" scouts; "Simba" is a Kiswahili name for lion) dedicated to lion monitoring and tracking. This has reduced overreliance on livestock herding as a means of livelihood. On the other hand, "Wildlife Pays" program is concerned with compensating Masai herders on a quarterly basis for losses incurred due to wildlife predation in exchange for their participation in conservation activities. The locals also benefit from the lease payments for conservancy zones, watershed protection, and the provision of ecotourism services (Oduol personal communication).

The establishment of a 90 km electric fence line in the Kibwezi forest of Chyulu Hills by Sheldrick Wildlife Trust (SWT) has served to control forest logging and charcoal burning besides thwarting poachers from the protected area. As a consequence, the secured springs in the forest have continued to supply water to sustain human, livestock, and wildlife (KWTA, 2022). Additionally, fenceline has assisted in preventing marauding wildlife (especially elephants) from raiding farms and subsequent human and wildlife conflict. The fenceline has therefore contributed to food security and harmonious coexistence between wildlife and human beings within and around the fenced area. Members of the community are also employed to maintain and patrol the fenceline, hence a diversified livelihood source.

Promotion of culturally aligned sports to advance threatened species conservation has also been used as an alternative livelihood source since its inception in 2008. In what is generally referred to as Masai Olympics, predominantly young Masai men forego the long-held tradition of hunting lions to demonstrate proof of maturity, reputation, and heroism by engaging in medal hunt in an organized sports event based on traditional Masai warrior skills (BigLife, 2022). The first Masai Olympic was held in 2012 and since then it has been a biennial event where teams from each contesting village are selected through a sequence of tournaments prior to the finals across six categories, namely, javelin and rungu (wooden club) throwing, high jump, and 200 m, 800 m, and 5000 m races (Sambu personal communication). In the lead-up to the sports events, the conservation education initiative takes

Fig. 2 Young Morans warming up during Masai Olympics. (Picture by Sambu, BigLife Foundation)

place around two major focus areas: (1) that killing of lions or any other wildlife species for cultural purposes should be stopped and (2) adopting wildlife conservation for economic benefits and hence a sustainable future for the Masai people. Overall, incidences of culturally induced lions' deaths have reduced, thus increasing local lion population and improving livelihoods (BigLife, 2022) (Fig. 2).

Predator Pay programs by MWCT and Big Life Foundation have been used to successfully protect predator wildlife in Masai community ranches and conservancies adjoining to Chyulu Hills including Elesengei, Rombo, Kuku and Mbirikani group ranches and Kimana Conservancy. Livestock losses due to depredation are compensated alongside incentives for improved livestock protection and husbandry including the upgrade of livestock bomas and improved herding strategies to reduce future depredation incidences. This has contributed to reduced retaliatory killing of lions on account of predation from an average of six per year to one per year (KWTA, 2022).

3 Use of Carbon Finance to Protect Ecosystem and Livelihoods

In 2013, stakeholders of Chyulu ecosystem including Sheldrick Wildlife Trust, Masai Wilderness Conservation Trust (MWCT), Big Life Foundation, Kenya Forest Service (KFS), and Kenya Wildlife Services (KWS) partnered to champion for mitigation and adaptation of climate change impacts and conserve biodiversity as well as to promote climate-smart livelihoods under the United Nations scheme of Reducing Emissions from Deforestation and forest Degradation (REDD+). The Chyulu REDD+ project is validated and verified by Verified Carbon Standard (VCS) and Climate, Community and Biodiversity Standard (CCBS) (Wildlife Works, 2016).

The overarching goal of Chyulu REDD+ project is to prevent release of over 28 million tons of carbon dioxide from Chyulu ecosystem over the 30 years of the project's crediting period through suppression of forest degradation, deforestation, or conversion of forest land to other land uses (KWTA, 2022). Specific activities undertaken to achieve the project's goal include but not limited to strict enforcement of environmental laws, aggressive afforestation and reafforestation programs, control for unplanned agricultural expansion, promotion of sustainable and economically viable livelihoods, and prohibition of unsustainable extractive practices such as charcoal burning and wood collection to make artifacts, among others (KWTA, 2022). Economic benefits arising from carbon financing (of Chyulu REDD+ project) are used to support environmental health and ecologically friendly community livelihoods (Table 1).

4 Community Empowerment Programs

The provision of empowerment programs such as employment, health, and education reduces dependence of local communities on ecosystem services of forested ecosystems such as Chyulu Hills. Conventional health services provision, for instance, reduces the extraction of medicinal plants from Chyulu water tower, whereas provision of education enlightens the community on optional sources of livelihoods that could be adopted to facilitate resilience to climate change impacts in mountain ecosystems such as Chyulu Hills.

MWCT has invested appreciably in the provision of employment, education (both formal and informal), and health services to the communities of Kuku and Rombo ranches through direct and indirect employment of locals alongside its eco lodge partner (Kampi ya Kanzi), hence reducing reliance on livestock keeping that is prone to climate change impacts. Additionally, MWCT promotes the adoption of sustainable agriculture through growth of grass for forage (besides improving water retention) and bee keeping, all of which have little negative impacts to the environment. Also, MWCT supports the tourists' market for beadwork and other local products along the value chain, hence helping in growing the economic base of the local communities in the Chyulus. Similarly, Big Life Foundation through the revenues derived from ecotourism and wildlife conservation supports community education by paying teachers' salaries, developing and implementing conservation prioritized curriculum in schools, and provision of merit and need-based scholarship for local students in Kimana Conservancy, Eselengei and Mbirikani group ranches (Sambu personal communication). By so doing, alternate livelihood sources

Table 1 Milestones realized from Chyulu REDD+ project proceeds by 2022 (MWCT, 2022)

Organization	Milestones realized since 2018 to 2022
Kenya Forest Service (KFS)	Promotion of tree planting in schools and other public institutions Fencing and expansion of Thange tree nursery in Kibwezi Acquisition of 480 beehives to promote community honey production Support for forest operations and patrols Carrying out of outreach meetings to promote bee keeping value chain in Kibwezi East Rehabilitation of degraded wetlands and riparian areas Capacity building in nature-based enterprise development
Kenya Wildlife Service (KWS)	Purchase of 8880 assorted tree seedlings for reforestation Enhanced conservation education awareness through radio broadcast programs Analysis of water quality parameters at wildlife water points Purchase of firefighting equipment including pumps Undertaking of aerial re-seeding Drilling of boreholes and solar pump installation Enhanced vehicle and plant maintenance Increased food rations and water supplies to rangers
Masai Wilderness Conservation Trust (MWCT)	Enhanced ranger recruitment and training Increased supply for ranger rations, water, uniform, and beddings Expanded smartphone and digital radio communication equipment for rangers Enhanced quarterly ranger management meetings
Big Life Foundation	Utility bill payments to support community water access Employment of four forest scouts and five water rangers and issuance with new uniforms Bursaries for school children Enhanced vehicle repairs and maintenance 15 community sensitization meetings conducted Purchase of fire management equipment and response training Installation of the new water point, community water tanks, troughs, and plumbing
Sheldrick Wildlife Trust	Employment of local community members for road clearing Meeting helicopter costs for aerial surveillance Purchase of heavy equipment, fuel, and maintenance of 22 km road/firebreak construction
Eastern Chyulu community	Replacement of ten community water tanks damaged by wildlife Bee keeping training for community Purchase of bee keeping tools and equipment (boots, suits, gloves, smokers, etc.) for the community Selection and award of bursaries to 150 school children Installation of six water tanks and water harvesting infrastructure for schools Community outreach programs undertaken

(continued)

Table 1 (continued)

Organization	Milestones realized since 2018 to 2022
Kuku A group ranch	Employment of new staff: 4 Simba scouts, 15 REDD+ rangers, 15 borehole guards/operators, 1 school cook, 1 early childhood teacher, 2 primary school teachers, 1 secondary school teacher, 1 education deputy coordinator, 1 community health worker coordinator, 4 cleaners for the community dispensary and school, 2 field reporters for Wildlife Pay program Purchase of 2 laptops for group ranch office Outreach meetings for community members and leaders Sustained employment for the initial 14 REDD+ rangers alongside their supplies of uniform and equipment Design, printing, and distribution of 6425 Kuku A group ranch membership cards
Kuku B group ranch	Outreach meetings for group ranch leadership and members Financial support provided to women and youth groups New staff employed: 2 primary school teachers and 3 REDD+ rangers Construction of new staff quarters at Maralal Primary School Installation of new elephant-proof fence at Samai School Sustained employment for the initial 14 REDD+ rangers with enhanced supplies of food rations and water Construction of new latrines for schools and renovation of Moilo Primary School Renovation of borehole and water supply for local community group
Mbirikani group ranch	School fees support for 329 students Employment of 4 forest guards Sustained employment of existing staff: dispensary security guard, nurse assistant, and borehole attendant Support for outreach engagement meetings for the community Purchase of office equipment and supplies for the new group ranch office Construction of 3 new cattle crushes Survey for the ranch boundary
Rombo group ranch	Bursary program for school children Sustainable menstrual supplies for 600 girls Employment of new staff: 1 primary teacher and 3 REDD+ rangers Construction and operationalization of the new Rombo group ranch office Outreach meetings conducted for community members and leadership Supplies of uniforms, ration, and smartphones for rangers 150 beehives provided to support alternative livelihood among women groups Enhanced school supplies and infrastructure improvement Marking of ranch boundary Financial support for vulnerable groups
Emergency allocation	Establishment and equipping of new KWS Ranger Station with food rations, water tanks, and communication radios Acquisition of a vehicle and fuel provision to support operations of the new ranger station Arrangement for firefighting helicopter hire and air tractor Feeding program for 58 schools with students' population in excess of 23,000

Fig. 3 Financial support for school infrastructure development. (Picture by Sambu, Big Life Foundation)

different from traditional livestock keeping are built for local people, hence improving well-being and resilience to climate change shocks (Fig. 3).

The Mabel Foundation and Period Company have also invested remarkably in menstrual health and hygiene among Masai women and girls through the provision of reusable underwear and menstrual educational kit. This has contributed to decreased school dropout among school-going girls during the menstruation period, thus positively contributing to nondisruptive education services in the local community (Nosin personal communication).

5 Conclusion

Different livelihood approaches to promote climate change resilience have different strengths and contexts in which they apply best. While climate-specific interventions such as carbon trading serve to reduce carbon emissions while supporting community livelihoods, other indirect approaches such as health and education provision cultivate attitudes with long-term potential to facilitate both ecosystem health and livelihood improvement. Climate resilience in mountain ecosystems is thus achievable if community livelihoods are diversified and supported alongside efforts to combat carbon emissions into the atmosphere. The enhanced and diversified economic base for mountain communities reduces their reliance on exploitative use of natural resources, whereas the promotion of medal hunting instead of lion hunting assists in sustaining the unique and endangered biodiversity that forms the foundation for nature-based tourism that in turn feeds into the economic pillar of the community.

References

Alweny, S., Nsengiyumva, P., & Gatarabirwa, W. (2014). *African mountains status report: technical report No 1*. Kampala, Uganda: Albertine Rift Conservation Society (ARCOS).

BigLife Foundation. (2022). *The Maasai Olympics, the hunt for medals not lions*. Viewed 3rd February, 2022.

Cuni-Sanchez, A., Omeny, P., Pfeifer, M., Olaka, L., Boru Mamo, B., Marchant, R., & Burgess, D. N. (2018). Climate change and pastoralists: Perceptions and adaptation in montane Kenya. *Climate and Development, 11*(6), 513–524. https://doi.org/10.1080/17565529.2018.1454880

Gariano, S., & Guzzetti, F. (2016). Landslides in a changing climate. *Earth-Science Reviews, 162*, 227–252. https://doi.org/10.1016/j.earscirev.2016.08.011

Gorokhovich, Y., Doocy, S., Walyawula, F., Muwanga, A., & Nardi, F. (2013). Landslides in Bududa, Eastern Uganda: Preliminary assessment and proposed solutions. *Landslide Science and Practice, 4*, 145–149.

Grab, S., & Linde, J. H. (2014). Mapping exposure to snow in a developing African context: Implications for human and livestock vulnerability in Lesotho. *Natural Hazards, 71*(3), 1537–1560. https://doi.org/10.1007/s11069-013-0964-8

Hay, S. I., Cox, J., Rogers, D. J., Randolph, S. E., Stern, D. I., Shanks, G. D., Myers, M. F., & Snow, R. W. (2002). Climate change and resurgence of malaria in East African Highlands. *Nature, 415*, 905–909.

Hiltner, U., Brauning, A., Gebrekirstos, A., Huth, A., & Fischer, R. (2016). Impacts of precipitation variability on the dynamics of a dry tropical montane forest. *Ecological Modelling, 320*, 92–101.

IPCC (Intergovernmental Panel on Climate Change). (2014). Summary for policymakers. In C. B. Field, V. R. Barros, D. J. Dokken, K. J. Mach, M. D. Mastrandrea, T. E. Bilir, M. Chatterjee, K. L. Ebi, Y. O. Estrada, R. C. Genova, B. Girma, E. S. Kissel, A. N. Levy, S. Maccracken, P. R. Mastrandrea, & L. L. White (Eds.), *Climate change 2014: Impacts, adaptation and vulnerability. Part A: Global and sectoral aspects. Contribution of Working Group II to the fifth assessment report of IPCC* (pp. 1–32). Cambridge University Press.

Kamau, P. N., & Medley, K. E. (2014). Anthropogenic fires and local livelihoods at Chyulu Hills, Kenya. *Landscape and Urban Planning, 84*, 76–84.

KWS (Kenya Wildlife Service). (2008). *Tsavo conservation area management plan*. Kenya Wildlife Service.

KWTA (Kenya Water Towers Agency). (2017). *Kenya water towers status report: Chyulu hills* (revised ed.). Kenya Water Towers Agency.

KWTA (Kenya Water Towers Agency). (2022). *Draft ecosystem management plan for Chyulu*. Kenya Water Towers Agency.

MWCT (Masai Wildernes Conservation Trust). (2022). *Tabular summary of milestones realized from Chyulu REDD+ project since 2018*. Masai Wildlife Conservation Trust, viewed on 3rd, December 2022.

Nsengiyumva, P. (2019). *African Mountains in a changing climate. Trends, impacts and adaptation solutions*. International Mountain Society. https://doi.org/10.1659/MRD-JOURNAL-D-19-00062.1

Ponce-Reyes, R., Plumptre, J. A., Segand, D., Samuel Ayebare, S., Fuller, R. A., Hugh, P., Possinghame, P. H., & Watson, E. J. (2017). Forecasting ecosystem responses to climate change across Africa's Albertine Rift. *Biological Conservation, 209*, 464–472.

Sarma, J. K. (2018). Conservation livelihood. A concept of integrating nature conservation and livelihood. *Media for Ecological Sustainability of Northeast India*. Republished from *Biolink, August 2018issue*.

White, C., Aquilina, M., Shepherd, C., & Narvaez, R. (2017). Design and implementation of a payment for ecosystem services scheme in the Chyulu hills: PES scheme principles. In E. Le Cornu, M. Marisa, E. Wakaba, R. Teipelke, S. Smith, & C. Tuite (Eds.), Viewed 3rd July 2023.

Wildlife works. (2016). *Chyulu Hills REDD+ Project Monitoring and Implementation report*.

An Area of Dankness: Food Security of Food Producers When the Kosi Comes Down to the Plains

Amarnath Tripathi, Nilabja Ghosh, and Mayanglambam Rajeshwor

1 Introduction

The effects of the warming of mountains are expected to spill over to the plains that receive rivers descending from the stressed mountains. Himalayas are warming and, arguably, Himalayan glaciers are receding faster than in other parts of the world (Armstrong, 2010) due to warming. Bihar receives a number of rivers coming down the eastern Himalayas that are known to be warming faster than the western parts of the range. The Kosi Basin (KB) of Bihar in India is a zone below the Himalayas that has always struggled with the consequences of devastating floods. This chapter, based on a field visit, discussions and primary surveys, recounts the features of agriculture in Khagaria, a flood-prone district located in KB and the struggles and adaptations of the rural people in terms of livelihood and agro-practices for their food security.

Although the melting of snow might have been a slow and prolonged process from before the Industrial Revolution (Armstrong, 2010; Oerlemans, 2005), evidences of climate change have created concerns over the warming of Himalayas its receding glaciers (Armstrong, 2010). While research is ongoing and the effect of climate change on India's monsoon is yet to be understood (Vaishnavi, 2019), there are signs of accelerated melting of glaciers especially in eastern Himalayas (Bajracharya et al., 2007; Armstrong, 2010). IPCC reports (2014) suggested that

A. Tripathi
Jaipuria Institute of Management, Noida, India

N. Ghosh (✉) · M. Rajeshwor
Institute of Economic Growth, Delhi, India
e-mail: nila@iegindia.org

S. Tripathi et al. (eds.), *Warming Mountains*,
https://doi.org/10.1007/978-3-031-62197-0_15

Hindukush Himalayas are susceptible to temperature increase by 2 °C in a global warming scenario of 1.5°C by the end of the century. and the glaciers of the region may lose 10–30% of their masses by 2030 (Armstrong, 2010; Gupta, 2023; ToI, 2019) supplemented by the albedo effect. The associated replacement of snow by heavy rains in wetter summers especially in eastern regions of Himalayas (Smith & Bookhagen, 2020) may speed up river flows creating catastrophic floods in the plains.

Consisting of about 16 districts, the KB is over-achingly agrarian with little opportunity for jobs in industry. Agriculture, the main occupation and source of sustenance, is rampantly ravaged by heavy local and upstream rainfall. Floods can arise from varied causes, but in Bihar they are largely due to poor drainage stemming from the limitations of river channels to hold the huge volume of water coming from Himalayan rainfall that even the structural bodies cannot always brook (GoI, 2023). Kosi has a transborder mountainous course. Satellite mapping of the entire valley in multilateral initiatives reveals sharp gradients of terrains, precipitation, temperature and ecology (Wang et al., 2012). While helping in zoning, monitoring and preventing floods, mapping has identified Bihar as the second largest in terms of the number of districts (38) flooded in the last 20 years (GoI, 2023). In the recent 3 years, 2020, 2021 and 2022, floods ravaged 39, 35 and 33 districts, respectively, due to the rise in water levels in Kosi, Bagmati and Ganges and other rivers of North Bihar.

Recurrent Kosi floods, caused by the history of its journey, are a constant pressure on livelihood and food security in KB. The Kosi Basin project, jointly managed by the governments of India, Bihar and Nepal, has limited success as a way of coping with excess water. Incessant rainfall causing an upstream breach in 2008 created disputes as the resulting catastrophic flood raised discourse over engineering as a solution and even over the potential benefit of not having an intervention (Somanathan, 2013). As the reality of floods became accepted, nonstructural means of adaptation such as legislated disaster management (IPCC, 2012; NDMA, 2005), weather prediction, inter-country information sharing, climate change awareness, insurance, flood zoning, wetland conservation and timely relief gained emphasis.

The objectives of this paper are to (1) understand the resource problems created by the flood proneness of farms in terms of land, soil cover and water; (2) existing socioeconomic constraints of the vulnerable population and their adaptation to the challenges by migration, occupation and crop choices; (3) their farming practices guided by ecological and economic constraints along with income potentials of crops and the markets they face; (4) sources of food and role of local agriculture as well as public intervention for their survival; and (5) decisions regarding choice of farming and choices of crops to grow in flood-prone areas by means of econometric regression with economic, natural and policy-induced determinants. This introduction is followed by a background to the area under study and its motivation (Sect. 2), the survey methodology.

2 Background of Study Region and Reason for the Choice

River Kosi is a confluence of three rivers arising from Nepal and Tibet in China. While traversing the Nepal Himalayas, it is recharged by seven tributaries before it crosses the Chhatra gorge to enter India near Bheem Nagar and flow down the plains of North Bihar to join River Ganga near Kursela. Coming a long way of 729 km down the unconsolidated terrains of the Himalayas, it encounters tributaries and severely steep slopes to gather increasingly voluminous loads of water and silts (Li et al., 2020; Kumawat et al., 2020). Sharing a common border with the country Nepal, North Bihar's Himalayan highlands gradually become flat alluvial plains or an 'inland delta', drained by the Ganga, Kosi, Gandak, and Son with their tributaries and other seasonal streams.

Kosi River is typically braided in structure with a migratory behaviour due to intermittent sediment deposition. Bihar itself has an annual rainfall of 1192 mm compared to the 1177 mm of India. Drained by other Himalayas rivers such as Gandak, Mahananda and Ganga, Bihar is rich in water. A densely populated state (Table 1), Bihar has the lowest per capita GDP of Rs. 28,127 at 2011–2012 prices compared to India's Rs. 85110 in 2020–2021. Migration is a major characteristic of Bihar's sociology, and KB in Bihar is a major source of unidirectional and short-term migration to other states (Kumar, 2012; Kumar & Singh, 2018). Although not one of the hotspots of the Green Revolution, Bihar is a rural and agrarian state where the average farm size reduced to 0.39 ha (India 1.08 ha) and net sown area per person to 0.05 ha (India at 0.12 ha) which is cultivated intensely (145% vs. India's 141). Natural calamities keep the whole subregion and its agriculture under threat. Maize, grown also in the summer season (Zayad), is gradually gaining a place in human food habits and food processing activities. As a feedstock of ethanol, maize helps curb global warming.

Earlier a part of the district Munger, Khagaria which became a district in 1981, is preferred for study due to its low-lying location, its proximity to the main Kosi channel and its varied cropping pattern. Kosi hydrology affects districts of Supaul, Madhepura, Saharsa, Madhubani, East Champaran, Sitamarhi and Darbhanga, in the upper levels, and the districts of Begusarai, Katihar, Bhagalpur and Khagaria further downstream. Khagaria is highly prone to floods. With seven community development blocks, its geographical area is 1486 sq. km and its total population is 1.7 million with a sex ratio of 886 per 1000 and a population density of 1122 per sq. km.

At the time of the survey, the urban area of Khagaria covered only 12.46 sq. km area and 5% of the population. Only 58% people were literate and a major chunk of workers (58%) were agricultural laborers. Floods did not spare Khagaria between 2000 and 2010 except in 2006. Because of frequent avulsions, Kosi breaks up into many channels flowing into neighboring districts, but this low-lying district is a leading victim of Kosi hydrology. Embankments are constructed on Kosi and along Ganga in KB but in Khagaria they are built more for Ganga than Kosi.

Table 1 Profile of Bihar agriculture (updated data)

	Population	GDP	Average farm size	Sown land	Rural people	Small farms	Cropping intensity	Agriculture worker	Labor
	Per sq. km	Rs./person	Hectare	Person/hectare	%	(%)	(%)	%Worker	% Agri-worker
Year	2011	2020–2021	2015–2016	2011	2011	2015–2016	2015–2016	2011	2011
Bihar	1102	28,127	0.39	0.05	88.71	96.96	145	73.55	71.83
India	382	85,110	1.08	0.12	64.61	86.08	141	54.60	54.87

Source: DES Website, Census Website, Agriculture Census Website

3 Survey and Method

The study design has to bring out the typical elements of the geographical interest of the region with their subtle variations that differentiate spatial units even under similar riverine conditions. The study location for primary data was determined based on the map of Bihar and its rivers; the listing of districts, blocks and villages; and information on cropping patterns. Meant for understanding the lives of the people and their food insecurity, the area was chosen for its vulnerability to Kosi dynamics. At every stage of sampling, the proximity to the rivers is considered subject to overall geographical commonality.

Of the seven blocks, four blocks, which are closest to Kosi and grow maize, comprise the study region. Villages in each block were divided into two groups based on their distances from the river with the maximum distance at 13 km. One village was selected randomly from each group of villages, making up two villages from each sample block and eight villages from the district. From each village, 15 households were sampled giving representation to 5 different classes of farmers. The multi-stage stratified random sampling coincided with the rabi harvest season of 2015, but the reference year being 2014–15, the survey of 120 farm households covered information from both Kharif and Rabi seasons (Table 2).

Farm classes were specified based on farm size holding but keeping the reality of local land distribution in view, so that they were (i) landless (LL) with no land reported (or actually insignificant land) under the operated farm, (ii) marginal (MG) with farm size up to 1 ha, (iii) small (SM) with farm size above 1 ha but up to 2 ha, (iv) medium (MD) with farm size ranging from 2 to 4 ha and (v) large (LG) with farm size above 4 ha. The sample LL class size is 24; for farming activities, only the 96 landed categories of sample are assessed. Categories medium and large differ from official categorization to capture local reality. The lower limit of the large holding category is taken as 4 ha rather than the official 10-ha cutoff which is scarce in the region. The sample villages were located 4–50 km away from the district headquarters and not more than half a kilometer away from a river. Some villages

Table 2 Sampling for primary survey and crop-growing households in Khagaria district, Bihar

			No. of crop-growing households	
Sample size in each category	No. of households	% Share	Rice	Wheat
Sample Khagaria	120	100	44	83
Block [4 blocks]*	30	25	–	–
Village [8 villages]*	15	12.5	–	–
Landless (LL)**	24	20	–	–
Marginal (MG)**	32	27	14	22
Small (SM)**	21	18	8	20
Medium (MD)**	30	25	16	28
Large (LG)**	13	11	6	13

Notes: *[] figures are number of sampling units. **() are abbreviations were proximate to multiple rivers and some were accessible to basic facilities only by boat transport. All villages had at least one primary school, an *anganwadi* for early-age children, and medical clinics often, however without trained professionals but secondary education, were available only in some of them.

Socioeconomic, environmental, market and food-related information were asked of households from structured schedules. Prices at which crops were sold (producer price) were collected along with crop yields, acreages, input use and paid-out costs of cultivation as reported by the households. Farmers reported land holding in the local unit of Bigha which was expressed in hectares for analysis using the appropriate conversion rate. The household castes, scheduled caste (SC), schedule tribe (ST) and other backward classes (OBC), as declared officially, were all added up to make the reserved category for this study. Unreserved castes included *Brahmans*, *Bhumihar*, *Rajput's*, *Kayastha*, etc. A migrant is specified as a member who has moved to places beyond the resident district for at least 1 month with or without spouse in the last 2 years, and a migration-exposed household is one which had at least one migrant member. The working age is taken as 15–60 years. Educated members attained high school levels.

A farm is the land possessed for possible cultivation through ownership, patta, or leasing and can be resown too. This land was seen to be actually sown and harvested at least once in the current crop cycle and fully utilized. Land lost, not included in farm size, is the land that became inaccessible to the farmer before sowing. Lost land may have become part of the river or come under wetland protection as a government-regulated public property. Irrigation ratio is that part of the total cropped area, i.e., both single-cropped and multi-cropped areas, of the farm which is irrigated by tanks, government or private canals, tube wells and other wells, sprinklers, drip, etc. Only rice and wheat were specified as the food crops, while the focus of the study was on rice, wheat and maize.

Observations are presented through tables and graphs based mostly on simple averages of data. Returns in monetary value are measured as production value less the paid-out cost for farmers growing that crop. Access to food is measured by earnings or returns from a hectare under that crop in terms of volumes of food.[1] In the

[1] Food Access

$$PHATF_{ij} = (P_j Y_j A_{j-} Cj / P_i$$

Where PHATF is household access to i-th food crop in tonnes per hectare from market, j is crop for sale namely Maize, Rice, Wheat, P_j = producer price of crop j, C_i is paid cost per hectare in Rs. 000 and Yj is yield of the crop j in the farm (Kg/Hectare) and the product in the numerator is the notional returns from crop j if all crop is sold in market without household retention. Food crops rice and wheat which are purchased either from open market at retail price RP or from PDS shop at price PDSP so that Pi = RP or Pi = PDSP. PHATF is a notional access to rice or wheat assuming all product of any of the three crops grown on farm (Tonnes/Hectare) is sold to buy one of the two food crops.

regression.[2] linear model with quadratic (squared variables) and interactions between variables are specified and estimated for the crops to explain their acreages sown using economic, geographic, infrastructure and demographic explanatory variables. Moreover, since any of the three crops, rice, wheat and maize, are not necessarily grown by all sample farmers, a sample selection bias is corrected with Heckman correction[3] (Heckman, 1979). Alongside the quantitative analysis, perception-based responses were also sought both from the sample households and from village communities, traders and authoritative agents. Households reported their sales outlets and also sources of food purchase, the latter mostly consisting of the retail market and public distribution system (PDS). The study team visited three markets in the Khagaria district located at varying distances from the headquarters and nine maize traders to gain ground-level insight to qualify the data gathered. Retail and PDS prices are obtained from the shops.

---

[2] Regression

$$A_i = a_0 + a_1X_i + a_2X_i^2 + a_3X_iX_j$$

Where $A_i = f(X)$, A is area under crop I in hectares in sample farm and X comprise explanatory variables.

Explain variables

1. Economic: Relative price, Relative revenue, Access to food (Rice and Wheat), Retail price of food relative to PDS rice, producer price of food relative PDS price, Remittance from migrants.
2. Geographic: Distance from river, ratios of land water logged.
3. Demographic: Share of agriculture workers in household, female share of agriculture workers, share of mignats in workers.
4. Agronomic: Net sown area, irrigated share of area.
5. Infrastructure: Distance from Road, Distant from district/headquarters.

[3] Heckman model

The Heckman selection model assumes that there exists an underlying regression relationship,

$$y_j = x_j\beta + u_{1j} \tag{1}$$

The dependent variable, however, is not always observed. Rather, the dependent variable for observation j is observed if

$$z_j\gamma + u_{2j} > 0 \tag{2}$$

where, $u_1 \sim N(0; \sigma)$, $u_2 \sim N(0; 1)$, $corr(u_1; u_2) = \rho$

When $\rho = 0$, standard regression techniques applied to the first equation yield biased results. Hackman

provides consistent, asymptotically efficient estimates for all the parameters in such models.

4 Results

Khagaria is full of water bodies including rivers Kosi, Ganga and Bagmati and their channels. Embankments along the river leave enough space for river to flood into but such vulnerable wet areas were also densely inhabited. Surveyed villages had varied positions between the river and embankment, outside the embankment and between two rivers. While interpreting the findings, the riverine specificity of the sample design is kept in mind because the sample is purposively not directed at being representative of the state or even the district but rather to its dank geography. Also, the sample covers households who were primarily cultivators by their own perceptions both in the landed and landless categories. Floods on account of the movement of Kosi River or heavy rainfall in Nepal damaged crops, property communication and roads. Reportedly, at least 52–80% of the area of a sample village was affected, but the early warning was duly given by the district administration and media and victims received public relief.

The adversity of floods came also from sedimentation. A high-level international workshop on 'River Dynamics and Flood Hazard assessment with special reference to the Kosi River' organized on 23 March 2011 in Patna concluded that sedimentation is a bigger problem than water. A mixture of sand and silt, farmers reported that Kosi sediments made the land uncultivable for nearly a decade depending on its thickness. It was unsuitable as a building material too. The sediments also made canal irrigation the Kosi project offered nearly useless though connecting to canals by field channels was also not affordable. Dependence on wells was stronger but digging wells was depleting groundwater. Historically, droughts were also common, most severe in 2001 and reported in the years 2011, 2012, 2013, or 2014. Two villages faced drought in 4 of the previous 5 years and six villages did not report a drought. Migration was common to all villages.

The survey area was infrastructurally poor with weak connectivity. Electricity and telephone (mobile) were available in the villages but not to all households nor for 24 h. Two villages were not electrified at all. Solar plants, distributed by the state government, were not always effective. Not all villages were accessible by paved all-weather roads and even some paved roads became unusable in the monsoon. Bus was not available to all villages and transportation to the nearest town or district headquarters was by modes such as privately hired vehicles and, if distances were short, by big bicycle or rickshaw. Indeed, cultivation seemed to be the major occupation in all the villages surveyed but not the only one. Wage labour outside agriculture; small businesses as the shop owner, vegetable vendors and hawkers; and semiskilled labour such as masons, drivers, cycle repairing and trading were observed livelihoods of the people, but the economic activity might be conducted outside the village. Dairy was a minor occupation but an organized cooperative 'Sudha dairy' was a buyer. Fodder crops are grown on inferior land.

The average household size was fairly large at 7.6 members due to the culture of the joint family (Table 3). With limited fragmentation of land over generations, the culture was also responsible for the large average farm size (Table 4) compared to

Table 3 Societal characteristics and migration tendency of sample households in Khagaria district

	Size	Educated	Female headed	Reserved caste	Migration exposed	Interstate	Labourer	Interstate	Students
						Migrant worker			
Size class	Members per household		% households			% worker		% labour	% migrant
Landless	7.7	1.7	12.5	87.5	70.8	91.4	54.3	89.5	2.9
Marginal	7.0	2.2	12.5	62.5	31.3	100.0	76.5	100.0	10.0
Small	7.5	2.1	4.8	66.7	66.7	75.0	52.4	100.0	9.5
Medium	7.2	2.2	10.0	76.7	40.0	75.0	37.5	100.0	18.5
Large	9.5	1.8	0.0	76.9	69.2	42.9	28.6	100.0	49.3
Sample	7.6	2.0	9.2	73.3	51.7	86.5	53.1	96.1	15.3

Source: Computed from survey data

Note: Reserved castes include SC, ST and OBC (for details see Sect. 3); educated, up to high school; migration exposed, at least one migrant. Secondary data (Census of India, 2011)) from Bihar: SC, 13%; ST, 1.2%; OBC, 68%. Literacy rate is 63%. Female headed, 7.4%; minority religion, 17%

Table 4 Land endowments of sample households in Khagaria district

	Farm size	Cropped area	Intensity of farming	Leased	Irrigated	Land lost
Size class	Hectare		%	% farm size		
Marginal	0.61	0.93	152	32.1	87	0.00
Small	1.52	2.21	145	33.2	87	9.84
Medium	2.77	3.36	121	15.8	86	0.66
Large	5.86	6.07	104	32.1	98	16.25
Sample	2.19	2.66	121	30.7	88	14.61

Source: Computed from survey data
Note: Land lost is not part of farm size

Bihar (0.39 ha). The difference may be attributable to the purposive sample design, definitional differences or even the lack of information on the legitimate use of leased or common land such as river beds, river banks and waterlogged land. The larger average size of a household at 9.5 members in the LG category among all others that have a uniform size between 7 and 8 members cautions against equating economic strength with farm size which was related rather to family cohesion acting as a security in the disaster-prone region with a close-knit support system.

Over half of the sample households were exposed to migration. Men were more likely to migrate for work than women who bore multiple responsibilities. In relatively longer-term migrations, women in charge became the head of household explaining the 9.2% share of female-headed households and a higher share (12%) among LL and MG farm households (Table 3). Among the castes, other backward castes (OBC) were dominant and together with scheduled castes (14.2%) and tribe (8.3%), 73% of respondent households of the sample were in reserved castes, more so in the LL category. Education required for higher salary jobs was lacking.

4.1 *Migration*

Migration spares no farm size class but not surprisingly, the largest exposure of 71% is observed in the LL category. A sign of regional economic imbalance (much of global migration also occurs also due to distress but which is caused by political unrest), members from severely poor households tend to become periodic migrants. Some of them leave their ties to their roots through women in 'unsettled relationships' resulting in loneliness, hardship and work burden for women (Papademetriou & Martin, 1991). The immensity of the problem of distress-driven migration for the region, migrants, families and society at large is exposed by many incidents (Naregal, 2021).

Households also reported that a high portion (29%) of migrants moved to the National Capital Region (NCR) around Delhi followed by other states in the west such as Punjab, Haryana and Chhattisgarh, while 17%–18% reported intra-state migration within Bihar. Accompaniment of the spouse was confined mostly to intra-state migration or inter-state migration to the NCR, but remittances per household from the NCR were low due to the higher cost of living compared to migration to the south and to the northeast. A high share of LG households also faced exposure to migration, but a sign of debt and poverty was not conveyed by the smaller share of labourers among their migrant workers (Table 3). Of the migrants, 15% were students and their share were largest at nearly half among the LG, although all categories saw student migration including a 3% share of migrants from the LL. Better access to higher education can also, therefore, explain migration among the LG. Field interactions revealed that students went mostly to Patna or other district colleges in Bihar. Overall, 53% of migrant workers of the sample went as labourers and except in the LL category, all labour migration was inter-state.

4.2 Access to Land

Some cultivable land was leased in and much of the land sown was multi-cropped. Farm size as reported in Table 4 is the sum total of land possessed inclusive of land leased in, while total cropped area incorporates sown land and multiple cropped sown land. Also, farmers reported they did not always have access to their owned land which was lost to them. Asked about the extent and reasons of land loss they suffered during the last 10 years, it could be speculated that another 0.47 ha per household could also have been cultivated over and above the 2.2 ha of farmland reported if land loss could be averted. About 16% of the land was lost according to the farmers.

All classes of households were victims of land loss. The LG suffered the most followed by the MD class. Even the LL households reported having degraded lands lying idle under wetland status though this did not happen among the sample households in this period. The largest land loss came from the river devouring land by its movements (Fig. 1). Sedimentation and waterlogging were also culprits. Human encroachment via canals and highways did not farmland. The embankment was the only intervention that took away land even while floods, whose defence they were, brought losses to most classes.

In a historically feudal state, where leasing was nearly prohibited by law, most households reported its incidence in oral form. Unrecorded agreement between the owner and the user of land not being officially recognized explains the divergence from secondary data. LL and MG farmers sometimes report leasing some insignificant quantum of low-quality land for growing fodder crops. Leasing in of land informally by all classes of farmers meant that subsidies and support went to the landowner rather than the cultivator. The unstructured contracts conveyed a general impression that the lease was usually renewable for 1 year at a time. Both cash rents and produce-sharing arrangements on a 50:50 basis were contracted but sharecropping was more popular than fixed rent. Disputes and even violence were ways of redressing violations of terms.

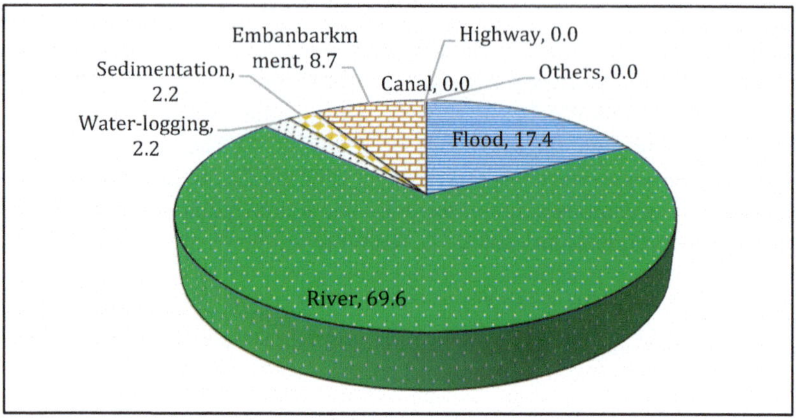

Fig. 1 Factors for land loss in the sample (%)

Table 5 Shares (%) in total cropped area under study crops of all sample households

	Rice	Wheat	Maize	Total
Marginal	21.5	23.7	49.5	94.7
Small	19.9	26.7	44.8	91.4
Medium	23.5	28.9	44.3	96.7
Large	17.8	29.7	50.2	97.7
Sample (Khagaria)	21.1	28.2	47.0	96.3

Source: Computed from survey data

Note: Total cropped area covers the three study crops and other minor crops. Total sample size = 120

Multiple cropping helped expand crop coverage across classes except for the LG class for which farming intensity fell close to one. Interestingly 6% of the landed households also leased out land due to inadequate supervisory power or their interest in off-farm jobs but leasing was not necessarily a localized transaction. Field interactions informed that when poor drainage leading to waterlogging from monsoon rains checked Rabi planting, cropping intensity falling below one was not uncommon. The relation between farm size and intensity of farming clearly seems inverse in this case.

4.3 Farming

Rice, wheat and maize were the major crops making up over 96% of the cropped area in the sample. Others occupied miniscule acreages. The sample households together allocated the largest share (47%) of the total cropped area to maize followed by wheat (28%), while rice the staple food got the least share of 21% (Table 5) and the pattern was similar among the holding classes though not all sample landed household grew all three crops. The greatest preference to maize in the cropping pattern was shown by LG followed by the MG class. The largest number of farmers in the sample also grew maize (89) followed by wheat (83) and rice (44).

Evidently, not all farmers were privileged with land suitable for staple rice. Farmers seemed reluctant to grow kharif crops for fear of floods. Contrarily, among those 44 (only 37% of the sample) households who chose to grow rice, the average area under rice was large at 1.21 ha, while the mean area under rice of all 96 landed sample households (excluding the LL) was only 0.2 ha revealing the underlying preference for growing rice.

The yield rate of the rice for the 44 rice growers at 1.7 tonnes/hectare was less than in Bihar (1.9 tonnes/hectare) and India (2.4 tonnes/hectare) but that of maize among 89 maize growers was (6.8 tonnes/hectare) higher than Bihar and India. Wheat yield was marginally higher than the state average for all classes. Fertilizer consumption of all crops was physically greater than the Bihar level, but among the three, maize was the largest consumer followed by rice in the sample (Fig. 2). A perception of NPK ratio showing an imbalance of nutrients was gained. Use of fertilizer and hiring of labour increased with farm sizes for all three crops but unlike fertilizer, rice required more hiring of labour (Table 6).

Cultivation of maize appeared to be a special feature of this dank, flood-prone district. Across farm classes, it is chosen by the maximum number of farmers, it claims the largest share of cropped area and fertilizer and it is the only crop grown totally for commercial motivation. Farm acreage under maize increases with farm size as the scatter plot of cross-sectional data of maize growers shows (Fig. 3), but a negative quadratic effect is present though statistically insignificant.

4.4 Marketing and Enterprises

Agricultural marketing in India is regulated by the APMC law (MoA&FW, Website), whereby an elected body is supposed to ensure good infrastructure and fair transactions with auctioning to discover prices. The degenerating system was failing to

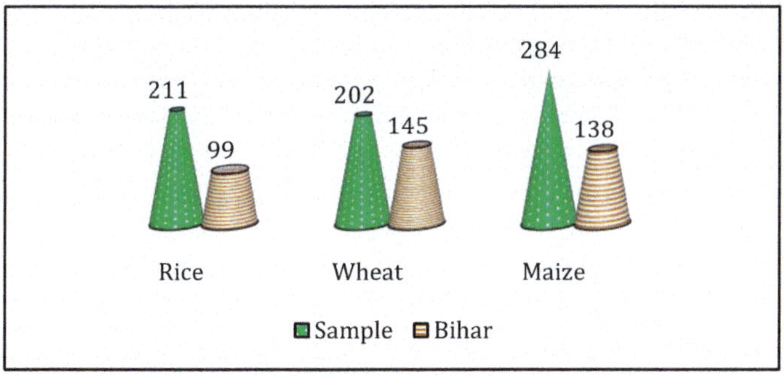

Fig. 2 Fertilizer (NPK) consumption (kg/ha)

Table 6 Average area and yield of study crops grown only by specific crop growing sample farmer households

Crop	Farm class	Marginal	Small	Medium	Large	Total
Rice	Area (hectare)	0.46	1.16	1.47	2.35	1.21
	Yield (tonnes/hectare)	1.65	1.74	1.79	1.34	1.67
	Production (tonnes)	0.76	2.02	2.63	3.15	2.02
Wheat	Area (hectare)	0.32	0.62	1.04	1.80	0.86
	Yield (tonnes/hectare)	2.66	2.45	2.41	2.33	2.47
	Production (tonnes)	0.85	1.52	2.51	4.19	2.12
Maize	Area (hectare)	0.54	1.04	1.54	3.05	1.34
	Yield (tonnes/hectare)	7.24	5.95	6.36	8.24	6.81
	Production (tonnes)	3.91	6.19	9.79	25.13	9.13

Source: Computed from sample data, DES (Website)
Note: Total is for all landed households in the sample (96). From the secondary data of Bihar, the average yield in tonnes per hectare in triennium 2014–2015 of rice, wheat and maize are 1.99, 2.21 and 3.27, respectively. Conversion: 1.5 kg paddy equivalent to 1 kg rice. Parenthesis () figures are number of crops growing farm households

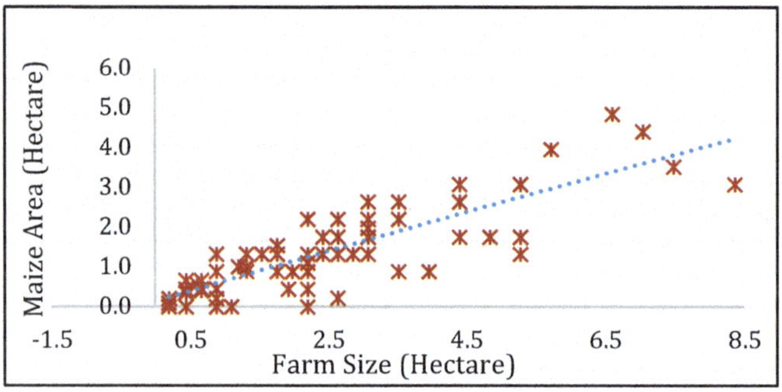

Fig. 3 Scatter plots of maize area against farm size

deliver its mandate in the 1990s as many markets remained under bureaucratic control without elections and became oligopsonistic as licensed traders bargained with disempowered farmers to set producer prices. The government has been trying to reform marketing by allowing easy entry and exit of buyers and allowing farmers to sell to larger more professional and corporate buyers or to organized cooperatives (Ghosh, 2013).

To take advantage of a larger market and improved technology accessible to more organized larger buyers, a law was circulated in 2003, recommending amendment of the state APMC law but success was limited. Bihar did legislate to do away with the old APMC law but failed to pass an amended (model Bill of central government) Act. In 2020, the central government again attempted the reform to allow selling to buyers outside the APMC market premises, but due to pressure from farmers, the Central Farm Act was withdrawn. This study was undertaken prior to this disturbance.

The Bihar government was procuring from farmers using three agencies. First, the Bihar State Food Corporation (BSFC) was procuring paddy and wheat at MSP under the decentralized procurement scheme of India followed by milling and distribution through the PDS. Cooperative movement at the grass root level was being revived with the Primary Agricultural Credit Society (PACS), the basic cooperative credit institution, forming a second procuring agency. Even in remote villages, they were equipped with storage infrastructure and computer linkage enabling early assessment of the crop situation. The Food Corporation of India (FCI) was the third agency that respondents complained was apathetic towards the purchase and storage of grains.

PACS was seen as a popular marketing outlet but there were complaints of 'high-handedness, collusion with local traders and also with banks' to delay procurement forcing farmers to sell at distressed prices. Table 7 shows that while maize was grown totally for commercial reasons, rice and wheat were also marketed by farmers, village traders being the leading buyer. Local powers bought maize from all classes and was the dominant buyer of rice and wheat from the LG class. The LG class sold almost 40% of its marketed maize in the market which was otherwise not a common outlet. Strangely, sale to private and also public organized buyers or processor was not reported by any respondent, but there were reports of a recent confrontation between farmers and PACS over payment default, delayed procurement and fraud. For each crop, the retail price at which the residents purchased from shops was the highest followed by the producer price at which the sample households sold. Subsidized PD price (PDSP) was the lowest (Fig. 4)

All three markets visited governed by ineffective committees before the APMC law was repealed had no official supervision at the time of the survey. Two of them found management from local agencies. A local food business managed the nearest market situated on the main road in the headquarters to transact only in rice and maize daily. Managed by a local charity organization the second market was further away from the headquarters and had a smaller size and catchment than the first one. Apart from daily rice and maize seasonal sale of wheat was reported. The third market having unutilized premises and no administrative offices and trader stalls was the most remote. Unsupervised it transacted in rice wheat and maize in a small paved yard with the smallest catchment. All three markets functioned on a daily basis but without refrigerators freezers grading equipment canteens and cow sheds. Traders complained of staff shortage and arranged their loading platforms and open storage. Field insight exposed the difficulties of the trader as also a sense of freedom brought by the lack of regulations

Maize is the major commercial crop in the area; nine maize traders in particular were interviewed. Between 25 and 65 years in age, the traders were male and educated but not beyond school level, except for the two largest traders who were both graduates in Commerce. Working for farmers' groups, they were sole proprietors of their businesses, operating in competition with other agencies without complaint. Both large traders had the business model of a company to which farmers and farmer groups delivered. Normally traders purchased maize with cash mostly from farm-gate from a catchment of 3–5 km, but large traders, some of whom were

Table 7 Marketed and sales outlets (%) of crop growing households of the sample

Type	Rice	Wheat	Maize hybrid
Marginal			
Marketed (% produced)	21.3	20.4	98.1
Outlet sold to (% marketed)			
Local chiefs	0.0	0.0	8.3
Village trader	50.0	83.3	66.7
Market	0.0	0.0	20.8
Private buyer	0.0	0.0	0.0
Public buyer	0.0	0.0	0.0
Others	50.0	16.7	4.2
Small			
Marketed (% produced)	51.6	22.8	66.3
Outlet sold to (% marketed)			
Local chiefs	0.0	25.0	20.0
Village trader	100	75.0	60.0
Market	0.0	0.0	15.0
Private buyer	0.0	0.0	5.0
Public buyer	0.0	0.0	0.0
Others	0.0	0.0	0.0
Medium			
Marketed (% produced)	37.9	46.7	95.1
Outlet sold to (% marketed)			
Local chiefs	25.0	22.2	22.2
Village trader	62.5	61.1	44.4
Market	12.5	0.0	33.3
Private buyer	0.0	0.0	0.0
Public buyer	0.0	0.0	0.0
Others	0.0	16.7	0.0
Large			
Marketed (% produced)	43.4	58.0	76.4
Outlet sold to (% marketed)			
Local chiefs	66.7	40.0	30.8
Village trader	0.0	30.0	23.1
Market	33.3	20.0	38.5
Private buyer	0.0	0.0	0.0
Public buyer	0.0	0.0	0.0
Others	0.0	10.0	7.7
All			
Marketed (% produced)	39.3	43.1	86.9
Outlet sold to (% marketed)			
Local chiefs	22.2	23.8	19.1
Village trader	61.1	59.5	51.2
Market	11.1	4.8	26.2

(continued)

Table 7 (continued)

Type	Rice	Wheat	Maize hybrid
Private buyer	0.0	0.0	1.2
Public buyer	0.0	0.0	0.0
Others	5.6	11.9	2.4

Source: Authors computation from survey data
Note: For production of crop growing farms, see Table 6. Outlets are Local chiefs, landlord, Sahukar and Mahajan; village trader, village trader; market, Mandi; private buyer, producer company; public buyer, state trader and PACS; others, others

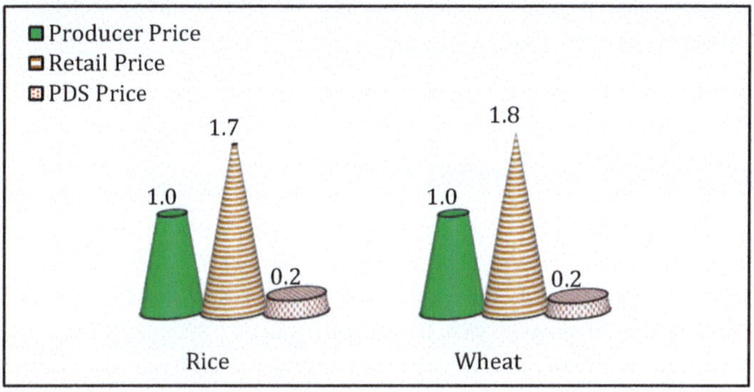

Fig. 4 Indices of producer price in crop wise in the sample

themselves landlords, reported buying large volumes at their stores from farmers even over 100 km away and also from other smaller traders. They paid both cash and cheque to farmers and only cheques to traders.

In farm-gate purchases, the transport cost, affected by road quality, was borne by the traders. Perceived as low, losses in transportation were attributed to the moisture content in the grain, faulty packaging, fracture in the journey and careless practices of farmers. A weighing machine (Dharma-kata) was available. Resourceful traders also had open or semi-permanent structured storage facilities. Loading and offloading, stationery, storage and license cost were reported expenses but not fumigation.

Using diesel threshing machines and local labour, enterprising households engaged in threshing maize for farmers for a fee. In villages, rice and wheat mills operated using electricity from public supply and private generators. Micro-enterprises were found only in the town where rice products (*Chura*, *Muhri*), wheat products (biscuits, bread), oils (mustard oil), poultry and dairy feed (maize and mustard) were manufactured in the sheds of family enterprises sometimes led by women entrepreneurs using family labour. Loans were taken from private sources and from bank via self- help groups.

4.5 Profitability

The total cost of cultivation relative to maize also shows the preference to maize (Fig. 5), the ratio exceeding one only for wheat among the MG class, but the relative cost came down and went up with farm sizes for wheat and rice, respectively. The LG class spent only 75% and 66% respectively on wheat and rice of what they spent on maize. Maize emerged as the only profitable crop across the classes, although for rice too the notional revenue from production exceeded the cost in the first three classes but was highly loss-making in commercial terms for the LG rice. Wheat suggested negative income potential except for the MG class but at the best, the returns were 20% of maize returns from a hectare.

4.6 Consumption and Access to Food

Rice was the main staple but the risk of flood limited its cultivation. Whatever rice produced served to protect the food security of the vulnerable households, supplemented by wheat which was grown in the rabi (safer) season if the soil remained suitable for planting. Farm production amounting to subsistence agriculture and wages as food were the dominant source of consumption. Yet 39% and 43% of rice and wheat, respectively, produced per household were also marketed (Table 7). Table 8, however, shows that the marketed share hardly represents the marketable surplus as households also purchased food grains for consumption and in fact the rice surplus was negative as the average household consumption of the total sample fell short of the farm production. Only some part of the farm production was retained for home consumption and a larger share was bought again from the market.

The LL and MG classes emerge as food deficient on the overall average even if cash needs for other purposes are ignored. Moreover, even the surplus at the average (Table 8) hides the plight of farmers who do not produce rice or wheat. The government steps in to counter the low purchasing power over open market supplies through subsidized PDS. In a scheme operational from 2007, coupons specifying

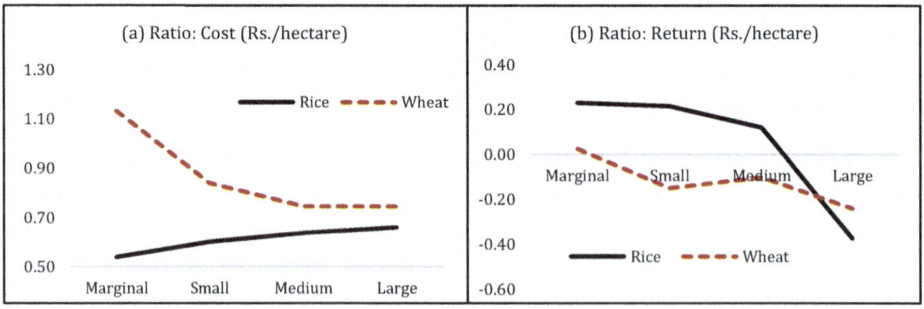

Fig. 5 Ratios relative to Maize

Table 8 Sources of food consumption of sample households in Khagaria

Average annual quantity (kg)/ household		Landless	Marginal	Small	Medium	Large	Sample
Rice	Production	–	336.7	778.0	1466.7	1312.7	734.7
	Consumption	813.6	628.8	704.4	759.6	890.4	740.4
	Farm produced	–	213.6	357.6	360.0	332.4	245.5
	Wage for food	115.2	–	–	–	–	23.0
	Public distribution	194.4	138.0	110.4	108.0	45.6	127.2
	Market	505.2	277.2	237.6	289.2	512.4	344.4
	Surplus	–	−86.8	9.5	48.2	32.2	−0.8
	Consumption/production (%)	–	186.8	90.5	51.8	67.8	100.8
Wheat	Production	–	594.0	1340.0	2381.0	3912.0	1412.0
	Consumption	843.6	698.4	844.8	872.4	1075.2	837.6
	Farm produced	–	369.6	685.2	758.4	1062.0	522.9
	Wage for food	124.8	–	–	–	–	25.0
	Public distribution	141.6	88.8	72.0	88.8	14.4	88.8
	Market	577.2	240.0	86.4	26.4	–	201.6
	Surplus	–	−17.6	37.0	63.4	72.5	40.7
	Consumption/production (%)	–	117.6	63.0	36.6	27.5	59.3

Source: Calculated from survey data

Table 9 Households (%) in farm classes holding ration card for food in PDS

		Types of card (%)			
Farm class	Ration card	APL	BPL	Antyodaya	Others
Landless	88	24	71	5	0
Marginal	94	50	43	7	0
Small	76	63	37	0	0
Medium	77	78	22	0	0
Large	85	100	0	0	0
Sample	84	58	39	3	0

Source: Authors computation from survey data

annual entitlements to wheat, rice and kerosene and their prices were distributed through village camps, organized by the local gram panchayat and block development office for beneficiaries to redeem a coupon each month. Mid Day Meal Scheme (MDM) and PDS were functional though marked by the high level of discrimination (Thorat & Lee, 2005). Dependence on PDS was more for rice than wheat which had a marketable surplus and was marketed as well as consumed at home to a greater extent (Table 8). In reality, farmers were found to resort to all three sources, namely, farm production, open market and PDS to meet food needs. Maize too served as a way of acquiring access to food through the market (Table 9).

PDS is seen as a critical support for food security through ties with the open market which was kept strong. Ration card was possessed by 84% of sample households, but 58% of them had APL cards. Majority of the ration card possessors was the MG class, 7% of whom were very poor with Antyodaya cards and 43% of their cards were for BPL for whom prices were highly subsidized. The LG class was third among the ration card holders but all of them were in the APL class paying the high price (cost price of distribution), while cards possessed by SM and MD classes were distributed between APL and BPL charging varied prices.

The economics of the food security of the climatically compromised community having access also to government support is calculated by the notional access to food assuming the entire volume of a crop produced in a hectare is converted to cash to buy food and comparing with the potential production of the food crops as an alternative. The relative producer prices and differentials across the three prices were already seen (Table 10). For MG rice grown and sold could buy rice for consumption of 3.4 tonnes per hectare, higher than its yield at 1.65 tonnes accessible if rice was grown on the same land. This is not so in all cases and the access was negative in the LG class. At the retail price, the access was much lower but maize gave greater access in all cases. However, farming was no means the only source of livelihood. Off-farm jobs in the state and migration to other states added to household incomes and some households also supplemented earnings by participating in NREGA, but it appeared that all the supplementation was compelled by the shortage of income from farming to meet daily needs.

Table 10 Access to food grain from market (tonnes/hectare)

			Marginal	Small	Medium	Large
Access to rice	PDS	Rice	3.42	1.35	0.86	−3.93
		Wheat	0.39	−0.93	−0.72	−2.52
		Maize	14.88	6.28	7.17	10.57
	Retail	Rice	0.37	0.14	0.09	−0.42
		Wheat	0.04	−0.10	−0.08	−0.27
		Maize	1.59	0.67	0.77	1.13
	Yield	Rice	1.65	1.74	1.79	1.34
Access to wheat	PDS	Rice	5.13	2.03	1.30	−5.90
		Wheat	0.59	−1.40	−1.08	−3.79
		Maize	22.33	9.42	10.75	15.86
	Retail	Rice	0.44	0.17	0.11	−0.50
		Wheat	0.05	−0.12	−0.09	−0.32
		Maize	1.90	0.80	0.91	1.35
	Yield	Wheat	2.66	2.45	2.41	2.33

Source: Authors' computation from survey data

5 Motivation to Produce Crop

Selection equations suggest (Table 11) the tendency of taking up any of the three

Table 11 Regression results of area equations of sample (with Heckman correction): sample period—2014–2015 (cross-sectional data)

	Rice		Wheat		Maize	
Independent variables	Coef.	Z-stat	Coef.	Z-stat	Coef.	Z-stat
Economic						
Incentive	4.42	2.03**	2.39	2.42**	2.68	1.73*
Food (rice) security					−0.002	−2.04**
Distance from river*incentive	−0.51	−1.73*	−0.67	−2.07**	−1.15	−2.13**
Distance from road*incentive			−0.77	−2.77***		
Remittance	−0.000	−1.00	0.000	1.80*		
Geography						
Land waterlogged dummy			−0.83	−3.14***	0.59	2.05**
Distance from river	0.67	2.03**	0.79	1.97**	0.90	2.05**
Distance from road	0.69	4.87***	1.06	2.94***		
Agronomic						
Farm size			0.31	9.46***	0.16	2.62**
Irrigation^^	0.32	10.28***			0.32	6.08***
Demographic						
Female workers^					0.70	1.03
Migration ratio^	4.92	3.20***	−1.55	−2.40**		
Constant	−5.47	−2.21**	−2.36	−1.89*	−1.89	−1.37
Select						
Farm size	0.40	4.08***	0.87	5.40***	1.14	5.20***
Squared farm size	−0.020	−3.22***	−0.04	−3.59***	−0.05	−3.85***
Female workers^	−1.81	−2.31**	−1.19	−1.23		
Land waterlogged – dummy	−0.65	−1.86*				
Migration^					−3.52	−2.47**
Distance from river			−0.04	−1.10		
Distance from road			−0.23	−1.43		
Constant	−0.46	−1.12	−0.30	−0.56	−0.95	−2.99***
Lambda	−0.46	−0.90	−0.43	−1.04	0.20	0.41
Rho	−0.46		−0.48		0.20	
Sigma	1.000		0.89		0.97	
Number of observations	116		115		109	
Censored observations	40		78		78	
Uncensored observations	76		37		31	

Source: Authors' computation from survey data

Note: Incentives—producer price of crop relative to identified substitute crop. Substitute crops—rice, kharif maize; wheat, rabi maize; maize, wheat. Food security: producer price/PDS price of food crop. ***, ** and * denote the level of significance at 1%, 5% and 10%. ^ Ratio to workers, ^^ ratio to total cropped area crops for cultivation is positively related with farm size (representing access to cultivable land) though the response was relatively less for rice. The quadratic effect of the sown area was noted in all cases suggesting that the relation with land available was not monotonic. Female dominance among working members was a discouragement for growing both rice and wheat, while migration intensity of workers discouraged maize cultivation for earning reflecting labour and food security implications. Possession of waterlogged land was a deterrent to rice cultivation perhaps because of the seasonal attribute. Other chosen variables including economic incentives were not relevant for crop selection.

Although not every farm household had the privilege of growing a crop, once a crop is chosen for cultivation, varied factors come into play in deciding the acreage of a chosen crop. Producer price ratios served as incentive variables for explaining acreage in all cases only if an interaction term with the distance for the river is allowed. The substitution effect is shown by the incentive having a positive significant effect with expectedly maize and wheat substituting each other in the rabi season and rice and maize in the kharif season but for maize the rabi season effect dominated. In all cases, the incentive was stronger for farms located closer to the river, so the incentive did not really emerge as the prime driver of acreage, while geography was a determinant even for the incentive effect. In fact, food security in terms of the availability of rice seemed important for the commercial crop maize which was made attractive or even imperative by the falling producer price of rice relative to its PDS price. The use of the price variable for rice was enabled by the variation of PDS prices across households as seen from the card holding pattern (Table 9), but the limited variation could be a reason for not finding the effects of the wheat price variable. Remittance is an economic factor affecting wheat acreage.

Distance from the river expectedly had a positive effect on the acreage of each crop. Distance from the road has a positive effect on acreage though only for rice and wheat, probably for food sufficiency. Farm size was important for only wheat and irrigation intensity had a strong positive effect on rice and maize acreages. A higher migration ratio that can undermine labour availability reduced wheat acreage but contrary to expectation enhanced rice acreage. Unlike in the selection equation, the prevalence of waterlogging has no effect on rice area but it reduced wheat acreage but helped maize acreage.

6 Conclusion

The vulnerability of households of the Kosi Basin contributes to the poverty of the state of Bihar. This study conducted in a district in the Kosi-Ganga Basin of Bihar observed that producers regularly bear the damages of land lost and soil sedimentation due to flooding and adapt to the hardships by mass migration to distant places by growing commercial crop maize while also securing their food security with the help of subsidized public support.

The data suggests that larger farm size is not an indicator of economic well-being in this dank environment where much of farming is not remunerative. All farm size classes suffer floods, waterlogging and loss of land. In fact, the large average farm size is a way of coping with adversity through family cohesion and mutual support in large joint families with undivided land. Growing food crops is risky, especially rice in the monsoon season, but with inadequate drainage, the effect of excess water might even creep into the rabi season too. Maize, a commercial crop that can be grown in drier seasons, is increasingly grown as an income buffer for generating higher yields, larger cash returns and greater access to food from the market.

Traders and local power agents are dominant sales outlets of the market that producers encounter. Open market and farm production both serve as sources of consumption, but PDS has been a more powerful support to their food security especially as small holding classes do not generate enough food or income to meet their needs and not all farmers are privileged with land suitable for growing rice and wheat. In fact, nearness to the river is a strong deterrent to growing of the food crop rice. Rice as a staple food appears as a risky crop to cultivate in the region. An econometric model with Heckman correction shows that despite the constraints, economic incentives are important for crop choice with maize being a major alternative to food crops rice and wheat, but the location of a farm with respect to the river and the road along with other agronomic and demographic factors are also determinants of crop choice.

The misery brought by natural conditions may aggravate if the eastern Himalayas further warm up as apprehended by scholars and glaciers melt releasing sequestered water down the unconsolidated steep slopes. This is a problem that requires public planning and government intervention. Weather forecasts and market intelligence can be a minimum need, but the government may continue and intensify support for food security and devote research to identify alternative farm or non-farm livelihood which will be viable in the condition. Social support for health and education will improve human resources for a safer future. Collaboration among river-sharing countries can help to improve weather forecast, trade and food security, drainage of land and data recording regarding land, water, power and climate.

Acknowledgements This paper draws from a study conducted under the Koshi Basin Programme at the International Centre for Integrated Mountain Development (ICIMOD), which is supported by the Australian government through the Sustainable Development Investment Portfolio for South Asia, as well as core funds of ICIMOD contributed by the governments of Afghanistan, Australia, Austria, Bangladesh, Bhutan, China, India, Myanmar, Nepal, Norway, Pakistan, Switzerland and the United Kingdom. The authors especially acknowledge Dr. Golam Rasul, Dr. Shahriar Wahid and Dr. Nilhari Neupane for their support. Ms. Rashmi, Mr. Ruchin Verma, Mr. Shankar Kumar and Ms. Amritanshi Preeti are acknowledged for their contribution in data collection/management and drafting.

References

Agriculture Census (Website). https://agcensus.nic.in/. Ministry of Agriculture and Farmers Welfare, GoI.

Armstrong, L. R. (2010). *The glaciers of the Hindu Kush-Himalayan Region. A summary of the science regarding glacier melt/retreat in the Himalayan, Hindu Kush, Karakoram, Pamir, and Tien Shan mountain ranges* (Technical Paper). ICIMOD, USAID.

Bajracharya, S. R., Mool, P. K., & Shrestha, B. R. (2007). *Impact of climate change on Himalayan glaciers and glacial lakes: Case studies on GLOF and associated hazards in Nepal and Bhutan*. ICIMOD.

Census India (Website). Office of the Registrar General and Census Commission, Ministry of Home Affairs, Government of India. http://censusindia.gov.in/

Directorate of Economic & Statistics (DES) (Website). https://eands.dacnet.nic.in/. Ministry of Agriculture and Farmers Welfare, GoI.

Ghosh, N. (2013). *India's agricultural marketing: Market reforms and emergence of new channels – India studies in business and economics*. Springer India.

Government of India (GoI). (2023). *Flood affected area atlas of India – Satellite based study*. NRSC, ISRO, DoS, NDMA, MoHA.

Gupta, J. (2023). *IPCC: 'Severe consequences' of further warming in disaster-hit Himalayas*. The Third Pole.

Heckman, J. (1979). Sample selection bias as a specification error. *Econometrica, 47*, 153–161.

IPCC. (2012). *Managing the risks of extreme events and disasters to advance climate change adaptation*. A Special Report of Working Groups I and II of the Intergovernmental Panel on Climate Change [C. B. Field, V. Barros, T. F. Stocker, D. Qin, D. J. Dokken, K. L. Ebi, M. D. Mastrandrea, K. J. Mach, G.-K. Plattner, S. K. Allen, M. Tignor, and P. M. Midgley (eds.)] (582 pp). Cambridge University Press.

Intergovernmental Panel on Climate Change (IPCC). (2014). *Climate change 2014–Impacts, adaptation and vulnerability: regional aspects*. Cambridge University Press.

Kumar, M. (2012). *Governing flood, migration, and conflict in North Bihar*. Mahanirban Calcutta Research group. Top of Form. chrome-extension://efaidnbmnnnibpcajpcglclefindmkaj/http://www.mcrg.ac.in/pp45.pdf

Kumar, A., & Singh, A. K. (2018). Flood-induced migration and its socioeconomic impact on women – A case study of Kosi Flood in North-Bihar. *International Journal of Research in Social Sciences, 8*(9) September, ISSN: 2249-2496 Impact Factor: 7.081.

Kumawat, A., et al. (2020). Soil and water conservation measures for agricultural sustainability. In R. S. Meena & R. Datta (Eds.), *Soil moisture importance*. IntechOpen. https://doi.org/10.5772/intechopen.92895

Li, X., Gao, J., Guo, Z., Yin, Y., Zhang, X., Sun, P., & Gao, Z. (2020). A study of rainfall-runoff movement process on high and steep slopes affected by double turbulence sources. *Scientific Reports, 10*(1), 9001. https://doi.org/10.1038/s41598-020-66060-3. PMID: 32488023; PMCID: PMC7265536.

Naregal, V. (2021). Labour 'Invisibility' during COVID-19 Times. *Economic & Political Weekly, 56*(34).

National Disaster Management Act (NDMA). (2005). Government of India. chrome-extension://efaidnbmnnnibpcajpcglclefindmkaj/https://ndma.gov.in/sites/default/files/PDF/DM_act2005.pdf

Oerlemans, J. (2005). Extracting a climate signal from 169 glacier records. *Science, 308*, 675–677.

Papademetriou, D. G., & Martin, P. L. (1991). *The unsettled relationship: Labor migration and economic development*. Greenwood Publishing Group.

Smith, T., & Bookhagen, B. (2020). Remotely sensed rain and snowfall in the Himalaya. In A. Dimri, B. Bookhagen, M. Stoffel, & T. Yasunari (Eds.), *Himalayan weather and climate and their impact on the environment*. Springer. https://doi.org/10.1007/978-3-030-29684-1_8

Somanathan, E. (2013). Are embankments a good flood control strategy? A case study on Kosi River. *Water Policy, 15*, 75–88.

Thorat, S., & Lee, J. (2005). Caste discrimination and food security programmes. *Economic and Political Weekly, 40*(39), 24–30.

Times of India (ToI). (2019). *64% glaciers in Hindu Kush Himalayan region maybe lost by 2100: IPCC report*. September 25.

Vishnavi, C. (2019). *As the monsoon and climate shift, India Faces Worsening Floods*. Yale School of the Environment. https://e360.yale.edu/features/as-the-monsoon-and-climate-shift-india-faces-worsening-floods

Wang, J., Zhang, Z., Su, B., & Zhang, L. (2012). A case research on economic spatial distribution and differential of agriculture in China. *Agricultural Science, 3*(8), 996–1006.

Enhancing Carbon Stocks in Agricultural and Forest Soils: A Crucial Step Toward Climate Change Mitigation and Ecological Security

Rishita Pakhira and Hukum Singh

1 Introduction

The global challenge of climate change has necessitated exploring sustainable and effective solutions to mitigate its impacts (Singh et al., 2022a, b). Enhancing soil carbon stocks in agricultural and forest soils has emerged as a crucial and promising approach among various strategies. This practice involves increasing the amount of carbon stored in the soil, which helps combat climate change and contributes to ecological security (Joshi & Singh, 2020).

Soil carbon sequestration has gained significant attention due to its potential to offset greenhouse gas emissions, particularly carbon dioxide, which is a major contributor to global warming (IPCC, 2014). Increasing the carbon content in soils is a practical approach to removing carbon dioxide from the atmosphere and storing it in a stable form, hence mitigating the greenhouse effect and reducing the adverse impacts of climate change.

Agricultural soils play a vital role in this process. Implementing sustainable farming practices, such as cover cropping, crop rotation, and conservation tillage, can enhance soil organic matter content and stimulate the accumulation of carbon in the soil. These practices improve soil fertility and productivity and offer additional benefits such as increased water retention, reduced erosion, and improved nutrient cycling (Kumar & Singh, 2024; Malik et al., 2024; Kanta et al., 2024).

Similarly, forest soils, particularly those in sustainably managed forests, present a significant opportunity for carbon sequestration. Sustainable forestry practices, including reduced logging intensity, afforestation, and reforestation, contribute to

R. Pakhira · H. Singh (✉)
Plant Physiology Discipline, ICFRE-Forest Research Institute, PO New Forest, Dehradun, Uttarakhand, India

S. Tripathi et al. (eds.), *Warming Mountains*,
https://doi.org/10.1007/978-3-031-62197-0_16

carbon accumulation in the soil. Forests act as valuable carbon sinks, absorbing atmospheric carbon dioxide and storing it in their biomass and soil (Kumar et al., 2020a, b, c).

The importance of enhancing soil carbon stocks extends beyond climate change mitigation. It also plays a crucial role in ensuring ecological security (Devi et al., 2023). Carbon-rich soils support diverse and resilient ecosystems by promoting biodiversity, enhancing soil fertility, and improving the overall health of plants and organisms. Such soils are better equipped to withstand extreme weather events, maintain water quality, and provide essential ecosystem services.

Encouraging adoption of sustainable land management practices through incentives and support can lead to widespread implementation and contribute to long-term climate change mitigation and ecological security.

Enhancing soil carbon stocks in agricultural and forest soils in the Himalayan region represents a crucial step toward addressing climate change and ensuring ecological security (Singh & Kumar, 2022; Singh et al., 2024b). Increasing carbon sequestration in soils could be a better measure to simultaneously mitigate greenhouse gas emissions, enhance ecosystem resilience, promote sustainable land management practices, and work toward a more sustainable and secure future (Joshi et al., 2020).

Besides, addressing soil abuse or degradation is essential for enhancing climate change adaptability (Egli et al., 2021). One crucial measure to reduce atmospheric CO_2 is increasing carbon storage in soils. Therefore, it becomes imperative to conserve soils and enhance their capacity to store organic carbon. By preventing and controlling soil abuse and promoting sustainable land uses, it is possible to improve soil fertility, increase carbon storage in soil and biomass, and enhance agricultural productivity and food security. Preserving soil nutrients is also crucial for mitigating the impacts of climate change (Elbasiouny & Elbehiry, 2020).

2 Soil Organic Carbon Distribution: Global Scenario

The distribution of soil organic carbon (SOC) stocks in the context of climate change scenarios varies significantly among global biosphere models. The underlying controlling factors can be categorized into various classes, each assigned a probability and linked to an output variable. This framework establishes a network that incorporates a set of equations governing the input and output of carbon to and from the soil carbon pool. An analysis based on probability-weighted results revealed that global climatic influences on net primary productivity (NPP) had the most pronounced impact on SOC stocks (Köchy et al., 2015).

In contrast, the colder temperature in temperate regions exhibited a less pronounced reduction in degradation rates. Notably, variations in degradation rates between current and future climates were most significant in drier subtropical areas due to projected substantial increases in precipitation (Davidson & Janssens, 2006).

Tropical highland forests emerged as hotspots with imminent, yet relatively secure, carbon losses across several NPP scenarios. If NPP rises through carbon dioxide fertilization, there is a 75% chance of a 1% global increase in soil carbon content. However, if the availability of fertilizers limits the CO_2-induced increase in NPP, there is a 75% probability that the soil carbon mass will remain unchanged (Köchy et al., 2015).

3 Soil Carbon Storage

Soil serves as the world's largest and most significant carbon reservoir, accounting for more than 75% of carbon stocks in terrestrial ecosystems. It possesses a substantial storage capacity, estimated at approximately 1500–1600 petagrams (Pg) of carbon, surpassing the combined carbon content of the atmosphere (780 Pg C) and vegetation (620 Pg C) (Elbasiouny et al., 2014; Hsu et al., 2021). Soil organic carbon (SOC) plays a pivotal role in soil biogeochemistry, supporting essential ecosystem services such as nutrient cycling, soil health, microbiome regulation, and erosion control (Liang et al., 2018).

The soil organic carbon (SOC) reservoir represents the largest terrestrial carbon pool, with estimates ranging from 1220 to 2000 Pg C in the top 1 m of soil and 2376–2456 Pg C in the top 2 m (Liang et al., 2019). Within the soil profile, approximately 46–61% of the SOC is retained within the top 30 cm layer. Understanding SOC content and soil carbon sequestration (SCseq) at the national or regional scale, particularly in agricultural and forest soils, is paramount for research and practical applications. These studies provide valuable insights into the potential for carbon storage in soils and inform strategies to enhance soil carbon sequestration in agricultural systems.

Agricultural management plays a critical role in improving soil organic carbon (SOC) levels by either enhancing soil organic matter (SOM) input, reducing SOM degradation and oxidation, or employing a combination of both approaches (Yadav et al., 2019). Increasing SOM stocks in agricultural soils is beneficial for soil fertility as the mineralization of organic matter releases essential nutrients for plant growth. However, achieving this requires implementing agricultural practices tailored to local conditions that enhance soil carbon inputs while maintaining or reducing carbon output, thereby promoting soil carbon sequestration (Dignac et al., 2017).

Agricultural ecosystems are significant contributors to the global carbon stock and can potentially reduce CO_2 emissions by 5.5–6.0 gigatons per year by 2030, with approximately 89% of this reduction attributed to SOC storage. However, various agricultural practices such as cultivation, irrigation, and fertilization can profoundly influence the net carbon exchange in agricultural ecosystems. It is important to note that overfertilization can lead to soil degradation and water pollution, counteracting the benefits of carbon sequestration (Niu et al., 2016).

Soil organic matter (SOM) encompasses all organic elements in the soil, including soil humus, plant debris, dead roots, dead tissue, and living biomass such as plant, animal, and microbial tissues. The soil organisms interact with plants and other small animals, maintaining soil fertility (Laban et al., 2018).

3.1 Soil Carbon Accumulation or Depletion

Intensive agricultural practices have resulted in increased carbon losses due to reduced plant inputs and higher levels of erosion and degradation (Knops & Tilman, 2000). The equilibrium between carbon inputs from photosynthesis and deposition and carbon losses from respiration, leaching, and erosion plays a crucial role in determining soil organic carbon (SOC) levels (Liang et al., 2019). Factors such as lower organic inputs, higher degradation, changes in soil temperature and moisture regime, leaching, and soil erosion contribute to SOC depletion in agroecosystems (De Oca, 2009). Tillage has been shown to increase soil carbon losses, ranging from 28% to 77%, depending on climate and soil type (Abdelhakim & Tahar, 2010). The breakdown of macroaggregates into nutrient-poor microaggregates through tillage can lead to reduced organic matter levels and declining nutrient supply efficiencies in cultivated soils (Cambardella & Elliott, 1992).

Agricultural soils have the potential to store between 0.4 and 0.9 Pg C globally annually (Su et al., 2009). Long-term impacts on emissions and sequestration from agricultural soils are evident (Wu et al., 2021). The delicate balance between carbon uptake and loss can influence the biological terrestrial carbon sink, although detecting even minute changes in the geographically diverse supply of SOC is challenging (Don et al., 2007). Notably, leaching, volatilization, absorption, erosion, and denitrification are significant nitrogen loss processes (Zapata, 2008). Given the agricultural and ecological importance of soils, maintaining increased carbon stocks in the soil is of growing importance, as soil disturbances have accelerated carbon loss from agricultural soils (Adhikari et al., 2017).

The distribution of SOC into labile and recalcitrant stocks with varying residence times underlies the SOC levels. A healthy balance between labile and recalcitrant organic carbon stocks is essential for long-term soil sustainability and improved crop performance (Shahid et al., 2017; Choudhury et al., 2018). Effective SOC management is necessary to maintain soil productivity and prevent degradation (Dotaniya et al., 2016). Balanced application of inorganic fertilizer has been found to promote soil productivity and improve SOC storage, either with or without organic manure applications, as it influences soil dynamics and enhances soil microbial biomass (Shahid et al., 2017).

Plant inputs through photosynthesis and the subsequent decomposition of plant tissues contribute to carbon sequestration in the soil. The potential to enhance soil carbon storage and serve as a significant atmospheric CO_2 sink varies among management strategies. Soil degradation has resulted in a loss of 16–20% of worldwide soil carbon stock over the past 1000 years, and the conversion of forests to agricultural land and mechanization of agriculture have further decreased soil carbon by 22% and 78.12 Tg, respectively (Srinivasarao et al., 2015).

Conservation agriculture practices, such as reduced tillage, organic and inorganic mulching, cover cropping, nutrient balancing, compost application, and crop diversification, can increase SOC storage in agricultural soils (Mahmud et al., 2021; Upadhyay & Edrisi, 2021). Additionally, incorporating agricultural residues and the balanced use of organic and inorganic fertilizers can contribute to soil carbon storage (Choudhury et al., 2018).

3.2 The Influence of Climate Change on Soil Organic Carbon (SOC) Stocks

The impact of climate change on SOC stocks has been the subject of limited research. Wales revealed that SOC was depleted from soils at a mean annual rate of 0.6% between 1978 and 2003, which was attributed to climate change (Bellamy et al., 2005). However, no significant changes in SOC stocks under climate change were reported (Hopkins et al., 2009).

Additionally, changes in land management practices and carbon inputs, such as the application of manure, during the period between 1978 and 2003 have also contributed to changes in SOC stocks (Smith et al., 2007). The estimates suggested that the terrestrial biosphere in Europe absorbs 7–12% of anthropogenic CO_2 emissions, and it is believed that soils across Europe may currently act as net sinks of CO_2 (Janssens et al., 2003).

3.3 The Impact of Climate Change on Soil Organic Matter and Soil Health: Understanding the Linkages

Soil organic matter, a complex component of soils, encompasses a diverse array of living and nonliving elements that exhibit variations in properties, functions, and turnover rates (Weil & Magdoff, 2004; Kumar et al., 2021a, b). Its function as a carbon and nitrogen source and sink, capacity to interact with polyvalent ions and organic compounds, role in providing habitat and sustenance for microbial and animal life, and impact on aggregate stability determine soil property (Haynes, 2008; Singh et al., 2022a, b).

Soil organic carbon accounts for approximately 50% of SOM, is essential for plant productivity, and is closely linked to organic nitrogen. Easily mineralizable carbon and nitrogen are key indicators used to assess the status of SOM (Haynes, 2008). Decreased SOM levels can compromise soil water-holding capacity, elevate erosion risks, increase bulk density, and lead to soil compaction (Weil & Magdoff, 2004).

Land use and management practices that promote SOM development contribute to the sequestration of atmospheric carbon dioxide (CO_2), thereby mitigating global warming. It plays a significant role in reducing the impacts of flooding by enhancing water storage during heavy rainfall events. During periods of drought, SOM aids in storing water, enhancing soil resilience. However, the response of SOM to increased temperatures remains a subject of debate with no consensus (Ågren &

Wetterstedt, 2007). SOM indicators have been employed in the long term to assess the effects of climate change on the soil system. Nonetheless, elevated temperatures, increased precipitation, CO_2 fertilization, and atmospheric nitrogen deposition can enhance crop production and soil organic matter input, subsequently leading to increased SOM levels (Singh et al., 2018). Generally, rising temperatures have been shown to accelerate SOM degradation. However, the extent of SOM losses depends on its accessibility and availability to microorganisms rather than being solely governed by climate-driven factors (Kuzyakov & Gavrichkova, 2010).

3.4 Impacts of Climate Change on Soil Nutrient Availability and Soil Erosion

The availability of soil nutrients exhibits significant variation among plant species and is susceptible to the effects of climate change (Gong & Gao, 2019). Increased atmospheric CO_2 levels induce changes in soil water, temperature, and nutrient competition, directly and indirectly affecting soil properties and processes (Dotaniya et al., 2016). Hatfield and Prueger (2011) highlight the linkages between climate change and plant nutrition, particularly in range and grazing systems. Nitrogen often acts as a limiting factor for plant growth in certain habitats. The response of rangelands to global climate change differs for organic and inorganic nitrogen compounds, influenced by the rate of nitrogen cycling. Organic matter breakdown by soil fauna and microflora produces minerals and accessible nitrogen forms, benefiting plant growth (Kumar et al., 2022). Legume species exhibit varying responses to increased CO_2 emissions, with uncertainties in the effects of dietary restrictions on simulated growth (Seneweera & Norton, 2011). Additionally, certain circumstances indicate that phosphorus and molybdenum can limit the response to increased CO_2, which may worsen with increased carbon and nitrogen supply.

Elevated atmospheric CO_2 concentrations, accompanied by temperature and precipitation changes, have profound effects on key soil processes, leading to various forms of degradation such as physicochemical degradation, salinization, erosion, reduced soil nutrient storage, decreased soil biodiversity, and reduced availability of soil water (Sarfraz et al., 2019). Wang et al. (2016) found that combined CO_2 enrichment and increased heat improved the availability of several soil micronutrients, including copper, iron, and zinc. Rising temperatures also impact nutrient cycles and availability in natural ecosystems (Dotaniya et al., 2016). Warmer temperatures are expected to lengthen the growing season and enhance net mineralization, increasing nutrient availability and uptake (Singh et al., 2018; Arndal et al., 2013; Kumar et al., 2017). At higher altitudes, temperature increases associated with climate change occur more rapidly, promoting soil microbial activity and altering microbial physiology (Kumar et al., 2019).

Predicted temperature increases are likely to impact nutrient availability in terrestrial ecosystems throughout the twenty-first century, as biological mechanisms such as soil organic matter degradation, nitrification, and mineralization tend to increase with temperature (Rennenberg et al., 2009).

Climate change accelerates soil erosion and nutrient leaching rates, leading to nutrient degradation. Higher temperatures stimulate microbial activity, resulting in increased breakdown of soil organic matter and when exceeding plant uptake, leaching of nutrients. This scenario is more likely to occur when low plant requirements and nitrogen mineralization rates increase due to elevated soil temperatures (Lotze-Campen, 2011). Phosphorus availability in soil is influenced by climate change, as higher temperatures generally enhance its mineralization in the litter. Mineralization rates increased by an average of 48%, with temperatures ranging from 0.3 to 6.0 °C. Increased nitrogen amounts stimulate phosphatase excretion and enhance phosphorus uptake by plants, potentially leading to elevated soil phosphorus levels.

Soil water availability, closely related to nutrient uptake, may be influenced by climate change and altered precipitation patterns (Khan et al., 2021). It is projected that erosion will increase by 1.5–2% for every 1% rise in precipitation (Khan et al., 2021).

Additionally, intense rainfall and the transition from snow to rain contribute to accelerated erosion. Vegetative cover reduces soil erosion rates by mitigating the intensity of rainfall. Therefore, changes in plant biomass can exacerbate nutrient leaching and soil erosion effects. Conversely, in arid and semiarid regions, dry soils are vulnerable to erosion caused by wind and rain (Lotze-Campen, 2011). Low water potential remains a common constraint on soil microbial activity, with dry soils experiencing a complete suppression of microbial activity. Drought conditions elicit various physiological responses from soil microorganisms, including decreased cell osmotic potential, dehydration, and mortality (Rosenblatt & Schmitz, 2016).

3.5 *The Influence of Climate Change on Necromass Dynamics*

Necromass refers to the dead organic matter in ecosystems, including dead plants, animals, and microorganisms. It plays a crucial role in nutrient cycling and carbon sequestration in soils. One of the primary impacts of climate change on necromass is the alteration of decomposition rates. Warmer temperatures and increased precipitation can accelerate the decomposition process, leading to a faster breakdown of necromass. Conversely, in some cases, increased moisture may promote anaerobic conditions that slow down decomposition rates and can lead to necromass accumulation in waterlogged areas.

Changes in precipitation patterns, such as more frequent and intense droughts or heavy rainfall events, can also influence necromass dynamics. Drought conditions can inhibit decomposition by limiting microbial activity and nutrient availability, potentially leading to ecosystem necromass accumulation. On the other hand, extreme rainfall events can cause erosion and displacement of necromass, affecting its distribution and availability in different parts of the ecosystem.

Elevated atmospheric CO_2 levels can alter plant physiology and chemistry, leading to changes in plant necromass's quality and nutritional value. For example, increased carbon dioxide concentrations can lead to higher lignin content in plants, which is more decomposition-resistant.

Shifts in temperature and moisture regimes can influence microbial community composition and activity, affecting decomposition processes and nutrient cycling. Changes in the availability of necromass can also impact the diversity and abundance of decomposer organisms, potentially altering soil microbial dynamics.

Climate change can disrupt the balance of necromass decomposition and accumulation in ecosystems, with implications for carbon and nutrient cycling. Understanding these dynamics is crucial for predicting the feedback between climate change and ecosystem processes and developing strategies for mitigating its impacts on ecosystem functioning.

4 Carbon Storage and Soil Carbon Pools

Soil and plant systems contain the highest carbon. Organic carbon derived from dead plant matter and microbes makes up most of the soil carbon. Bacteria and fungi break down dead plant matter, releasing carbon through microbial metabolism (Jangra et al., 2024; Singh 2024; Singh et al., 2024a; Kumar et al., 2024). Soil can store a substantial amount of carbon, estimated at around 2500 gigatons (Gt), more than three times the atmospheric carbon.

The pools of SOM differ in their chemical structure, degree of decomposition, and contributions to soil function and health. Understanding how carbon pools change in response to management actions provides valuable insights into soil functioning and health. Each carbon pool decomposes or changes at different rates and durations, involving multiple soil processes. The active and passive carbon pools in soil are the two main components of the total carbon pool. The active pool, called labile carbon, consists primarily of living organisms, agricultural residues, and manure, exhibiting seasonal and annual turnover rates. This pool is essential for maintaining soil structural integrity and serves as a food source for soil microbes. Tillage and cultivation practices can rapidly impact the levels of active soil carbon.

On the other hand, the passive pool has a longer turnover time, ranging from hundreds to thousands of years. Clay binding in this pool affects cation exchange capacity and water holding capacity, making it stable and physically protected from microbial activity. Loss of passive carbon mainly occurs through soil erosion by wind and water. Understanding the dynamics and importance of carbon pools in the soil is crucial for managing soil health and comprehending their role in global carbon cycling and mitigating climate change.

4.1 Soil Carbon Sequestration and Climate Change

Soil carbon sequestration is vital in mitigating climate change by reducing the concentration of greenhouse gases, particularly carbon dioxide (CO_2), in the atmosphere (Raturi et al., 2022). The carbon stored in soils through sequestration helps offset the carbon emissions from human activities and natural processes, effectively acting as a carbon sink (Lal et al., 2007).

When plants undergo photosynthesis, they absorb CO_2 from the atmosphere and convert it into organic carbon compounds through carbon fixation. A significant portion of this organic carbon is then transferred to the soil through plant roots, decaying plant material, and the activity of soil microorganisms. These carbon inputs to the soil contribute to the buildup of organic matter, which is rich in carbon (Joshi et al., 2021). The sequestration of carbon in soil has several positive impacts on climate change mitigation:

Carbon Removal: Soil carbon sequestration actively removes CO_2 from the atmosphere, reducing its concentration and helping to mitigate the greenhouse effect.

Carbon Storage: Carbon stored in soils can persist for long periods, from decades to centuries. This provides a long-term storage solution for atmospheric carbon, preventing it from contributing to climate change.

Carbon Offsetting: Soil carbon sequestration can offset carbon emissions from other sources, such as fossil fuel combustion and deforestation. It helps balance the carbon budget by absorbing CO_2 that would otherwise remain in the atmosphere.

Co-benefits for Agriculture: Increasing soil organic carbon through sequestration improves soil fertility, water-holding capacity, and nutrient availability. This enhances agricultural productivity, making farming systems more resilient to climate variability and reducing the need for synthetic inputs.

Biodiversity and Ecosystem Health: Soil carbon sequestration promotes soil health and biodiversity by creating favorable conditions for soil microorganisms and other organisms that play crucial roles in ecosystem functioning.

Various management practices can be adopted to maximize the potential of soil carbon sequestration. Conservation agriculture techniques such as reduced tillage, cover cropping, and crop rotation help maintain or increase soil organic carbon levels. Agroforestry systems combine trees and crops and contribute to soil carbon sequestration. Efforts are underway to incentivize soil carbon sequestration through policy mechanisms, such as carbon markets and payment schemes for ecosystem services. These initiatives provide economic benefits to farmers and landowners who adopt practices that enhance soil carbon storage. Henceforth, soil carbon sequestration is a valuable strategy for mitigating climate change. By storing carbon in the soil, it would be possible to reduce greenhouse gas emissions, enhance agricultural sustainability, and promote ecosystem health (Kumar et al., 2020a, b, c, 2021a, b). It is a win-win solution for addressing climate change and supporting resilient and productive agricultural systems.

5 Management Practices and Their Impact on Soil Carbon Storage: Enhancing Climate Change Mitigation Through Sustainable Approaches

Forest soils are vital in storing approximately 50% of the organic carbon in terrestrial ecosystems. The way forests are managed directly influences the amount of carbon stored in the soil, as it affects the rates of carbon addition and removal. The following forest management practices can significantly impact soil carbon storage.

5.1 Afforestation for Climate Change Mitigation: Harnessing the Power of Forests for a Sustainable Future

Afforestation has the potential to mitigate climate change via sequestering atmospheric carbon in plants and soil systems.

1. Afforestation is a crucial strategy for climate change mitigation by establishing forests in areas lacking significant tree cover.
2. Afforestation helps sequester carbon dioxide from the atmosphere and store it in biomass and soil.
3. Forests act as carbon sinks, absorbing and storing carbon, thus reducing greenhouse gas emissions and combating climate change.
4. Afforestation has multiple environmental benefits, including soil protection, biodiversity conservation, and water management.
5. Trees provide shade, cool the environment through evapotranspiration, and regulate local climate conditions.
6. Afforestation initiatives contribute to ecosystem resilience, making landscapes more adaptable to climate change impacts.
7. Forests created through afforestation can provide habitats for numerous plant and animal species, promoting biodiversity.
8. Afforestation programs offer socioeconomic advantages, such as employment opportunities and sustainable land use practices.
9. Trees in afforested areas provide renewable resources such as timber and non-timber forest products.
10. Afforestation is crucial in fostering a sustainable future by addressing climate change, protecting ecosystems, and supporting local communities.

5.2 Tree Species Selection for Enhanced Carbon Sequestration and Ecosystem Resilience

Tree species selection is crucial in maximizing carbon sequestration and enhancing ecosystem resilience. Tree ability could be optimized by choosing tree species adaptive to specific environmental conditions capturing and storing carbon in the soil and mitigating the impacts of climate change. Here are some key points regarding the importance of tree species selection:

1. Carbon sequestration potential: Different tree species have varying capacities to sequester carbon due to their growth rates, biomass production, and longevity. Some tree species have higher carbon storage potential than others, making them more effective in mitigating climate change.
2. Adaptability to climate and soil conditions: Selecting tree species well-adapted to local climate and soil conditions is crucial for their successful establishment and growth. Native species are often better suited to local environments and have better resilience against pests, diseases, and climate variability.
3. Ecosystem services: Tree species selection should consider the broader ecosystem services the forest provides, such as water filtration, biodiversity support, and soil conservation. Some species may have specific ecological roles, such as nitrogen fixation or soil stabilization, that contribute to the overall health and functionality of the ecosystem.
4. Timber and non-timber products: Tree species chosen for their commercial value, such as high-quality timber or non-timber forest products, can provide economic incentives for sustainable forest management. Balancing economic benefits with ecological considerations is crucial in selecting tree species.
5. Species diversity and resilience: Planting a mix of tree species with varied canopy layers and rooting patterns can enhance forest resilience by reducing the risk of catastrophic losses from pests, diseases, or extreme events. Diverse forests are often more adaptable to changing environmental conditions and can better withstand disturbances.
6. Local ecological knowledge: Incorporating local ecological knowledge and engaging local communities in tree species selection processes can lead to more informed and context-specific decisions. Indigenous and traditional knowledge can offer valuable insights into the suitability and management of specific tree species.
7. Long-term planning: Considering tree species' long-term growth and development is essential. Some species may exhibit rapid early growth but decline in productivity over time, while others may have slower initial growth but become more productive and resilient in the long run.

Forest managers can optimize the ecological and climate benefits of afforestation and reforestation efforts by carefully selecting tree species based on their carbon sequestration potential, adaptation to local conditions, ecosystem services, and long-term sustainability.

5.3 Forest Fire Management for Climate Change Mitigation: Strategies for Preventing and Controlling Wildfires

Forest fire management plays a crucial role in mitigating the impacts of climate change. Effective strategies for preventing and controlling wildfires are essential for reducing carbon emissions and preserving the health and resilience of forest ecosystems. The following are the strategies for forest and wildfire management:

1. Implementing proactive measures such as fuel management, prescribed burning, and early detection systems can help prevent the occurrence and spread of wildfires. Reducing fuel loads and creating firebreaks can minimize the risk of large-scale, destructive fires, thereby reducing carbon emissions from forest fires.
2. Additionally, strategic firefighting efforts and rapid response systems are crucial for controlling wildfires when they do occur. By suppressing fires promptly and efficiently, the extent of carbon emissions can be minimized, protecting both the atmosphere and the carbon stored within the forest ecosystem.
3. Integrated approaches that combine forest management practices, community engagement, and advanced technologies can enhance the effectiveness of forest fire management. Promoting sustainable forest practices, raising awareness, and utilizing tools such as remote sensing and predictive modeling can mitigate the impact of wildfires on climate change.
4. Furthermore, restoring fire-adapted ecosystems through prescribed burns and ecological restoration projects can promote healthier forest landscapes. Forest resilience can be enhanced by reintroducing natural fire regimes, enabling ecosystems to withstand future climate challenges better.
5. Collaborative efforts among government agencies, local communities, and forest stakeholders are essential for successful forest fire management. It is suggested to work toward a more sustainable and resilient future by fostering partnerships, sharing knowledge and resources, and implementing comprehensive fire management plans.
6. Forest fire management not only mitigates climate change but also provides additional co-benefits. Reduction in the occurrence of large-scale fires is a suggested measure to protect biodiversity, safeguard water resources, and preserve valuable ecosystem services vital for human well-being.
7. Education and public awareness campaigns are critical in promoting responsible behavior and preventing human-caused wildfires. Fostering a fire safety culture and encouraging responsible land use practices could significantly reduce the occurrence of preventable fires and minimize their environmental impact.
8. Climate change is expected to increase the frequency and severity of wildfires in many regions. Therefore, investing in research and innovation to develop advanced fire management techniques, early warning systems, and fire-resistant tree species can further enhance ability to mitigate the impacts of climate change through effective forest fire management.

9. Comprehensive forest fire management strategies that prioritize prevention, rapid response, and ecosystem restoration are vital for mitigating climate change. Protecting forests from destructive fires can preserve their carbon-storing capacity, promote biodiversity, and contribute to a sustainable future for ecosystems and human communities.

5.4 Agricultural Management Strategies for Soil Carbon Sequestration: Enhancing Carbon Storage in Agricultural Land

The following are the suggested strategies for enhancing soil carbon sequestration in the agricultural system:

1. Conservation tillage: Implementing conservation tillage practices, such as no-till or reduced tillage, can help sequester soil carbon. By minimizing soil disturbance, these practices preserve organic matter and promote carbon accumulation in the soil.
2. Cover cropping: Planting cover crops during fallow periods or between cash crop rotations can enhance soil carbon sequestration. Cover crops protect the soil from erosion, enhance organic matter inputs, and promote microbial activity, leading to increased carbon storage.
3. Crop rotation and diversification: Implementing diverse crop rotations and intercropping systems can enhance soil carbon sequestration. Different crops have varying root structures and residue quality, contributing to increased organic matter inputs and carbon storage in the soil.
4. Nutrient management: Balanced nutrient management practices, including optimized fertilizer application, can positively influence soil carbon sequestration. Proper nutrient management promotes plant growth, increasing biomass production and organic matter inputs to the soil.
5. Agroforestry systems: Introducing trees into agricultural systems through agroforestry practices can significantly enhance soil carbon sequestration. Trees contribute to increased biomass production, leaf litter inputs, and root exudates, which enhance organic carbon accumulation in the soil.
6. Organic farming: Adopting organic farming practices, such as using organic fertilizers, compost, and crop residue incorporation, can promote soil carbon sequestration. Organic farming methods enhance soil health, microbial activity, and organic matter inputs, increasing carbon storage.
7. Manure management: Proper livestock manure management can contribute to soil carbon sequestration. Organic carbon can be effectively stored by applying manure to agricultural fields at appropriate rates and incorporating it into the soil.
8. Water management: Implementing water management strategies, such as irrigation optimization and water conservation practices, can indirectly influence

soil carbon sequestration. Efficient water management promotes optimal crop growth and organic matter accumulation, leading to increased carbon storage in the soil.

9. Precision agriculture: Precision agriculture technologies, such as remote sensing and variable rate applications, can optimize input usage and enhance soil carbon sequestration. Precise nutrient and irrigation management improve crop productivity and organic matter inputs, thereby increasing carbon storage potential.
10. Integrated crop-livestock systems: Integrating crops and livestock in agricultural systems can enhance soil carbon sequestration through improved nutrient cycling and organic matter inputs. Livestock grazing and manure deposition increase carbon inputs and nutrient availability, benefiting soil health and carbon storage.

Farmers and land managers can actively contribute to soil carbon sequestration, mitigating climate change and promoting sustainable agricultural practices by adopting these agricultural strategies.

6 Conclusion

It is concluded that enhancing soil carbon stocks in agricultural and forest soils is a crucial step toward climate change mitigation and ecological security. Both agricultural and forest management practices play a significant role in sequestering carbon and promoting sustainable land use practices. Implementing strategies such as conservation tillage, cover cropping, crop rotation, and agroforestry are better options to increase organic matter inputs and promote carbon storage in the soil. These practices contribute to climate change mitigation by reducing greenhouse gas emissions and enhancing soil health, water retention capacity, and nutrient cycling, leading to improved ecological security. Furthermore, adopting organic farming methods, proper manure management, water management strategies, precision agriculture techniques, and integrated crop-livestock systems can further optimize soil carbon sequestration potential in agricultural landscapes. Sustainable forest management practices in forest ecosystems, including tree species selection, proper harvesting techniques, and fire management, can help maintain and enhance soil carbon stocks. Forests act as significant carbon sinks, and their preservation and effective management are crucial for climate change mitigation and biodiversity conservation. Overall, recognizing the importance of soil carbon sequestration and implementing these management strategies is essential for achieving a sustainable future. By harnessing the power of agriculture and forests to enhance soil carbon stocks, it would be possible to mitigate climate change, ensure ecological security, and foster resilient ecosystems for future generations.

References

Abdelhakim, B., & Tahar, G. (2010). Land use effect on soil and particulate organic carbon, and aggregate stability in some soils in Tunisia. *African Journal of Agricultural Research, 5*(8), 764–774.

Adhikari, K. R., Dahal, K. R., Chen, Z. S., Tan, Y. C., & Lai, J. S. (2017). Rice–wheat cropping system: tillage, mulch, and nitrogen effects on soil carbon sequestration and crop productivity. *Paddy and Water Environment, 15*(4), 699–710.

Ågren, G. I., & Wetterstedt, J. M. (2007). What determines the temperature response of soil organic matter decomposition? *Soil Biology and Biochemistry, 39*(7), 1794–1798.

Arndal, M. F., Merrild, M. P., Michelsen, A., Schmidt, I. K., Mikkelsen, T. N., & Beier, C. (2013). Net root growth and nutrient acquisition in response to predicted climate change in two contrasting heathland species. *Plant and Soil, 369*(1), 615–629.

Bellamy, P. H., Loveland, P. J., Bradley, R. I., Lark, R. M., & Kirk, G. J. (2005). Carbon losses from all soils across England and Wales 1978–2003. *Nature, 437*(7056), 245–248.

Cambardella, C. A., & Elliott, E. T. (1992). Particulate soil organic-matter changes across a grassland cultivation sequence. *Soil Science Society of America Journal, 56*(3), 777–783.

Choudhury, S. G., Yaduvanshi, N. P. S., Chaudhari, S. K., Sharma, D. R., Sharma, D. K., Nayak, D. C., & Singh, S. K. (2018). Effect of nutrient management on soil organic carbon sequestration, fertility, and productivity under rice-wheat cropping system in semi-reclaimed sodic soils of North India. *Environmental Monitoring and Assessment, 190*(3), 1–15.

Davidson, E. A., & Janssens, I. A. (2006). Temperature sensitivity of soil carbon decomposition and feedbacks to climate change. *Nature, 440*(7081), 165–173.

De Oca, P. C. M. (2009). Effect of land use, climate and soil structure on soil organic carbon in Costa Rican Ecoregions (Doctoral dissertation, The Ohio State University).

Devi, N., Thakur, A., & Singh, H. (2023). Allometric equations for evaluating above-ground biomass and carbon storage capability of Indian bamboos: Review approach. *Ecology, Environment and Conservation, 29*(2), 451–457.

Dignac, M. F., Derrien, D., Barre, P., Barot, S., Cécillon, L., Chenu, C., et al. (2017). Increasing soil carbon storage: mechanisms, effects of agricultural practices and proxies. A review. *Agronomy for Sustainable Development, 37*(2), 1–27.

Don, A., Schumacher, J., Scherer-Lorenzen, M., Scholten, T., & Schulze, E. D. (2007). Spatial and vertical variation of soil carbon at two grassland sites—implications for measuring soil carbon stocks. *Geoderma, 141*(3–4), 272–282.

Dotaniya, M. L., Rajendiran, S., Meena, B. P., Meena, A. L., Dotaniya, C. K., Meena, B. L., Dotaniya, C. K., Jat, R. L., & Saha, J. K. (2016). Elevated carbon dioxide (CO_2) and temperature vis-a-vis carbon sequestration potential of global terrestrial ecosystem. In *Conservation agriculture* (pp. 225–256). Springer.

Egli, M., Bösiger, M., Lamorski, K., Sławiński, C., Plötze, M., Wiesenberg, G. L., Tikhomirov, D., Musso, A., Hsu, S.-Y., & Raimondi, S. (2021). Pedogenesis and carbon sequestration in transformed agricultural soils of Sicily. *Geoderma, 402*, 115355.

Elbasiouny, H., & Elbehiry, F. (2020). Soil carbon sequestration for climate change mitigation: Some implications to Egypt. In *Climate change impacts on agriculture and food security in Egypt* (pp. 151–181). Springer.

Elbasiouny, H., Abowaly, M., Abu_Alkheir, A, & Gad, A. (2014). Spatial variation of soil carbon and nitrogen pools by using ordinary Kriging method in an area of north Nile Delta, Egypt. *Catena, 113*, 70–78.

Gong, H., & Gao, J. (2019). Soil and climatic drivers of plant SLA (specific leaf area). *Global Ecology and Conservation, 20*, e00696.

Hatfield, J. L., & Prueger, J. H. (2011). Agroecology: Implications for plant response to climate change. *Crop Adaptation to Climate Change*, 27–43.

Haynes, R. J. (2008). Soil organic matter quality and the size and activity of the microbial biomass: their significance to the quality of agricultural soils. In *Soil mineral microbe-organic interactions* (pp. 201–231). Springer.

Hopkins, D. W., Waite, I. S., McNicol, J. W., Poulton, P. R., Macdonald, A. J., & O'donnell, A. G. (2009). Soil organic carbon contents in long-term experimental grassland plots in the UK (Palace Leas and Park Grass) have not changed consistently in recent decades. *Global Change Biology, 15*(7), 1739–1754.

Hsu, C. C., Tsai, H., Huang, W. S., & Huang, S. T. (2021). Carbon storage along with soil profile: An example of soil chronosequence from the fluvial terraces on the Pakua Tableland, Taiwan. *Land, 10*(5), 447.

IPCC. (2014). *Climate change 2014: Synthesis report.* Contribution of Working Group I, II and III to the Fifth Assessment Report of the Intergovernmental Panel on Climate Change [Core Writing Team, R.K.Pachauri and L.A.Meyer (eds)] (p. 151). IPCC.

Jangra, A., Kumar, K., Maikhuri, S., Bhandari, M. S., Pandey, S., Singh, H., & Barthwal, S. (2024). Unveiling stress-adapted endophytic bacteria: Characterizing plant growth-promoting traits and assessing cross-inoculation effects on *Populus deltoides under abiotic stress. Plant Physiology and Biochemistry, 210, 108610.* https://doi.org/10.1016/j.plaphy.2024.108610

Janssens, I. A., Freibauer, A., Ciais, P., Smith, P., Nabuurs, G. J., Folberth, G., et al. (2003). Europe's terrestrial biosphere absorbs 7 to 12% of European anthropogenic CO2 emissions. *Science, 300*(5625), 1538–1542.

Joshi, R., & Singh, H. (2020). Carbon sequestration potential of disturbed and non-disturbed forest ecosystem: A tool for mitigating climate change. *African Journal of Environmental Science and Technology, 14*(11), 385–393.

Joshi, R., Singh, H., Chhetri, R., & Yadav, K. (2020). Assessment of carbon sequestration potential in degraded and non-degraded community forests in Terai Region of Nepal. *Journal of Forest and Environmental Science, 36*(2), 113–121.

Joshi, R., Singh, H., Chhetri, R., Poudel, S. R., & Rijal, S. (2021). Carbon sequestration potential of community forests: A comparative analysis of soil organic carbon stock in community managed forests of Far-Western Nepal. *Eurasian Journal of Soil Science, 10*(2), 96–104. https://doi.org/10.18393/ejss.825066

Kanta, C., Kumar, A., Chauhan, A., Singh, H., & Sharma, I. P. (2024). The interplay between plant functional traits and climate change. In N. Kumar & H. Singh (Eds.), *Plant functional traits for improving productivity. Springer.* https://doi.org/10.1007/978-981-97-1510-7_3

Khan, S., Anwar, S., Sun, M., Shafiq, F., Khalilzadeh, R., & Gao, Z. Q. (2021). Characterizing differences in soil water content and wheat yield in response to tillage and precipitation in the dry, normal, and wet years at the Loess Plateau. *International Journal of Plant Production, 15*(4), 655–668.

Knops, J. M., & Tilman, D. (2000). Dynamics of soil nitrogen and carbon accumulation for 61 years after agricultural abandonment. *Ecology, 81*(1), 88–98.

Köchy, M., Don, A., van der Molen, M. K., & Freibauer, A. (2015). Global distribution of soil organic carbon–Part 2: Certainty of changes related to land use and climate. *The Soil, 1*(1), 367–380.

Kumar, N., & Singh, H. (2024). *Plant functional traits for improving productivity. Springer Nature.* https://doi.org/10.1007/978-981-97-1510-7

Kumar, N., Suyal, D. C., Sharma, I. P., Verma, A., & Singh, H. (2017). Elucidating stress proteins in rice (Oryza sativa L.) genotype under elevated temperature: a proteomic approach to understand heat stress response. *3 Biotech, 7*, 205. https://doi.org/10.1007/s13205-017-0856-9

Kumar, N., Jeena, N., & Singh, H. (2019). Elevated temperature modulates rice pollen structure: a study from the foothill of the Himalayan agro-ecosystem in India. *3 Biotech, 9*(5), 175. https://doi.org/10.1007/s13205-019-1700-1

Kumar, A., Dwivedi, G. K., Tewari, S., Paul, J., Anand, R., Kumar, N., Kumar, P., Singh, H., & Kaushal, R. (2020a). Carbon mineralization and inorganic nitrogen pools under *Terminalia chebula* Retz.-based agroforestry system in Himalayan foothills, India. *Forest Science, 66*(5), 634–643. https://doi.org/10.1093/forsci/fxaa012

Kumar, A., Kumar, M., Pandey, R., ZhiGuo, Y., & Cabral-Pinto, M. (2020b). Forest soil nutrient stocks along altitudinal range of Uttarakhand Himalayas: An aid to nature based climate solutions. *Catena, 207, 105667.*

Kumar, P., Singh, R., Singh, H., Chand, T., & Bala, N. (2020c). Assessment of soil carbon dioxide efflux and its controlling factors in moist temperate forest of West Himalayas. *Current Science, 19*(4), 661–669. https://doi.org/10.18520/cs/v119/i4/661-669

Kumar, M., Phukon, S. N., Paygude, A. C., Tyagi, K., & Singh, H. (2021a). Mapping Phenological Functional Types (PhFT) in the Indian Eastern Himalayas using machine learning algorithm in Google Earth Engine. *Computers and Geosciences, 158*, 104982. https://doi.org/10.1016/j.cageo.2021.104982

Kumar, A., Tewari, S., Singh, H., Kumar, P., Kumar, N., Bisth, S., Kushwah, S., & Nidhi, K. R. (2021b). Biomass accumulation and carbon stocks in different agro forestry system prevalent in Himalayan foothills, India. *Current Science, 120*(6), 1083–1188. https://doi.org/10.18520/cs/v120/i6/1083-1088

Kumar, A., Singh, H., Kumari, G., Bisht, S., Malik, A., Kumar, A., Singh, M., Raturi, A., Barthwal, S., Thakur, A., & Kaushal, R. (2022). Adaptive resilience of roadside trees to vehicular emissions via leaf enzymatic, physiological, and anatomical trait modulations. *Environmental Pollution, 313.* https://doi.org/10.1016/j.envpol.2022.120191

Kumar, N., Singh, H., Giri, K., Kumar, A., Joshi, A., Yadav, A., Singh, R., Bisht, S., Kumari, R., Jeena, N., Khairakpam, R., & Mishra, G. (2024). Physiological and molecular insights into the allelopathic effects on agroecosystems under changing environmental conditions. *Physiology and Molecular Biology of Plants, 30, 417–433.* https://doi.org/10.1007/s12298-024-01440-x

Kuzyakov, Y., & Gavrichkova, O. (2010). Time lag between photosynthesis and carbon dioxide efflux from soil: A review of mechanisms and controls. *Global Change Biology, 16*(12), 3386–3406.

Laban, P., Metternicht, G., & Davies, J. (2018). *Soil biodiversity and soil organic carbon: Keeping drylands alive* (p. 10). IUCN.

Lal, R., Follett, R. F., Stewart, B. A., & Kimble, J. M. (2007). Soil carbon sequestration to mitigate climate change and advance food security. *Soil Science, 172*(12), 943–956.

Liang, F., Li, J., Zhang, S., Gao, H., Wang, B., Shi, X., Huang, S., & Xu, M. (2018). Two-decade long fertilization induced changes in subsurface soil organic carbon stock vary with indigenous site characteristics. *Geoderma, 337*, 853–862.

Liang, C., Amelung, W., Lehmann, J., & Kästner, M. (2019). Quantitative assessment of microbial necromass contribution to soil organic matter. *Global Change Biology, 25*(11), 3578–3590.

Lotze-Campen, H. (2011). Climate change, population growth, and crop production: An overview. *Crop Adaptation to Climate Change*, 1–11.

Mahmud, A. A., Upadhyay, S. K., Srivastava, A. K., & Bhojiya, A. A. (2021). Biofertilizers: A Nexus between soil fertility and crop productivity under abiotic stress. *Current Research in Environmental Sustainability, 3*, 100063.

Malik, A., Kumar, K., Kumari, G., Kumar, N., & Singh, H. (2024). Application of functional traits in modelling productivity and resilience under climate change. In N. Kumar & H. Singh (Eds.), *Plant functional traits for improving productivity. Springer.* https://doi.org/10.1007/978-981-97-1510-7_5

Niu, L., Manxia, C., Xiumei, G., Xiaohua, L., Hongbo, S., Zhao, L., & Zed, R. (2016). Carbon sequestration and Jerusalem artichoke biomass under nitrogen applications in coastal saline zone in the northern region of Jiangsu, China. *Science of the Total Environment, 568*, 885–890.

Raturi, A., Singh, H., Kumar, P., et al. (2022). Characterizing the post-monsoon CO_2, CH_4, N_2O, and H_2O vapor fluxes from a tropical wetland in the Himalayan foothill. *Environmental Monitoring and Assessment, 194, 50.* https://doi.org/10.1007/s10661-021-09721-8

Rennenberg, H., Dannenmann, M., Gessler, A., Kreuzwieser, J., Simon, J., & Papen, H. (2009). Nitrogen balance in forest soils: Nutritional limitation of plants under climate change stresses. *Plant Biology, 11*, 4–23.

Rosenblatt, A. E., & Schmitz, O. J. (2016). Climate change, nutrition, and bottom-up and top-down food web processes. *Trends in Ecology & Evolution, 31*(12), 965–975.

Sarfraz, R., Hussain, A., Sabir, A., Ben Fekih, I., Ditta, A., & Xing, S. (2019). Role of biochar and plant growth promoting rhizobacteria to enhance soil carbon sequestration—A review. *Environmental Monitoring and Assessment, 191*(4), 1–13.

Seneweera, S., & Norton, R. M. (2011). Plant responses to increased carbon dioxide. *Crop Adaptation to Climate Change*, 198–217.

Shahid, M., Nayak, A. K., Puree, C., Tripathi, R., Lal, B., Gautam, P., et al. (2017). Carbon and nitrogen fractions and stocks under 41 years of chemical and organic fertilization in a sub-humid tropical rice soil. *Soil and Tillage Research, 170*, 136–146.

Singh, H. (2024). Exploring adaptive modulation in plant functional traits and its impact on the productivity of Acacia auriculiformis under CO_2-enriched environment. *Industrial Crops & Products, 210, 118186.* https://doi.org/10.1016/j.indcrop.2024.118186

Singh, H., & Kumar, M. (2022). Climate change and its impact on Indian Himalayan forests: Current status and research needs. In S. Rani & R. Kumar (Eds.), *Climate Change (pp. 223–242). Springer Climate.* https://doi.org/10.1007/978-3-030-92782-0_11

Singh, H., Sharma, R., Savita, S. M. P., Kumar, M., Verma, A., Ansari, M. W., & Sharma, S. K. (2018). Adaptive physiological response of *Parthenium hysterophorus* to elevated atmospheric CO_2 concentration. *Indian Forester, 144*, 1–14.

Singh, H., Bandyopadhyaya, S., Kumar, A., Singh, M., Malik, A., Kumari, G., Raturi, A., Yadav, S. K., Kumar, M., Barthwal, S., Thakur, A., & Rawat, P. S. (2022a). Understanding the physiological and biophysical response of urban roadside plantations for assessing adaptation and mitigation mechanisms toward vehicular emissions. *Urban Climate, 44*, 101183. https://doi.org/10.1016/j.uclim.2022.101183

Singh, C., Iyer, S., New, M. G., Few, R., Kuchimanchi, B., Segnon, A. C., & Morchain, D. (2022b). Interrogating 'effectiveness' in climate change adaptation: 11 guiding principles for adaptation research and practice. *Climate and Development, 14*(7), 650–664.

Singh, M., Singh, H., Kumar, A., Kumar, M., Barthwal, S., & Thakur, A. (2024a). Soil nitrogen availability determines the CO_2 fertilization effect on tree species (Neolamarckia cadamba): growth and physiological evidence. *Environmental Sustainability, 7, 53–60.* https://doi.org/10.1007/s42398-023-00300-w

Singh, H., Kumar, N., & Kumar, A. (2024b). Enhancing resource use efficiency in crops through plant functional traits. In N. Kumar & H. Singh (Eds.), *Plant functional traits for improving productivity. Springer.* https://doi.org/10.1007/978-981-97-1510-7_6

Smith, P., Chapman, S. J., Scott, W. A., Black, H. I., Wattenbach, M., Milne, R., et al. (2007). Climate change cannot be entirely responsible for soil carbon loss observed in England and Wales, 1978–2003. *Global Change Biology, 13*(12), 2605–2609.

Srinivasarao, C., Lal, R., Kundu, S., & Thakur, P. B. (2015). Conservation agriculture and soil carbon sequestration. *Conservation Agriculture*, 479–524.

Su, Y. Z., Liu, W. J., Yang, R., & Chang, X. X. (2009). Changes in soil aggregate, carbon, and nitrogen storages following the conversion of cropland to alfalfa forage land in the marginal oasis of northwest China. *Environmental Management, 43(6), 1061–1070.*

Upadhyay, S. K., & Edrisi, S. A. (2021). Developing sustainable measures to restore fly-ash contaminated lands: Current challenges and future prospects. *Land Degradation & Development, 32*(17), 4817–4831.

Wang, J., Zhang, X., Li, L., Cheng, K., Zheng, J., Zheng, J., et al. (2016). Changes in micronutrient availability and plant uptake under simulated climate change in winter wheat field. *Journal of Soils and Sediments, 16*(12), 2666–2675.

Weil, R. R., & Magdoff, F. (2004). Significance of soil organic matter to soil quality and health. *Soil Organic Matter in Sustainable Agriculture, 1–43.*

Wu, H., Tang, J., Huang, H., Chen, W., & Meng, Y. (2021). Net carbon sequestration performance of cropland use in China's principal grain-producing area: An evaluation and spatiotemporal divergence. *Land, 10*(7), 714.

Yadav, G. S., Lal, R., Meena, R. S., Babu, S., Das, A., Bhowmik, S. N., Datta, M., Layak, J., & Saha, P. (2019). Conservation tillage and nutrient management effects on productivity and soil carbon sequestration under double cropping of rice in northeastern region of India. *Ecological Indicators, 105*, 303–315.

Zapata, F. (2008). Introduction to nitrogen management in agricultural systems. In *Guidelines on nitrogen management in agricultural systems* (p. 1).

Warming Mountains: Gendered Implications for Livelihood and Sustainability

Kosar Bano and Rozina Begum

1 Introduction

Climate change is a global phenomenon which is causing extreme weather conditions. These conditions are influencing the lives and livelihoods of people in numerous ways (Goodrich et al., 2022; ICIMOD, 2019; IPCC, 2021). Mountains and mountain people are frequently more vulnerable to these changes and events. Warming of the mountains is one of the phenomena of climate change. Consequently, glaciers are melting rapidly, causing floods in the river, soil erosion, droughts, and sedimentation, thereby influencing negatively the lives of mountain people. The repercussions of these events, combined with other socioeconomic and political realities, severely impact people. It affects not only human lives but also the livelihood attached to them (Goodrich et al., 2017a, b, 2022). It results in food insecurity, resource damage, halting communication, damaging infrastructure, and biodiversity loss (Goodrich et al., 2022; ICIMOD, 2019; Hussain et al., 2016, 2021; Rasul et al., 2014; Schild, 2008). It is evident from the research that the mountainous regions, particularly South Asia and the Hindukush Himalaya (HKH) region, are more vulnerable to climate change-related risks(ICIMOD, 2019; IPCC, 2019, 2021). Due to climate change and increase in temperature, glacial lakes in HKHs areas have increased over a short period, and according to the ICIMOD study in 2018 currently, there are 25,614 glacial lakes covering an area of 1444 km. Nepal alone has 1466 glacial lakes, out of which 21 have been identified as potentially dangerous. So far, 24 events have been recorded in Nepal (ICIMOD, 2011), posing a flood threat and

K. Bano (✉)
ICIMOD Headquarters, Lalitpur, Nepal

R. Begum
National Council of Arts, Lahore, Pakistan

S. Tripathi et al. (eds.), *Warming Mountains*,
https://doi.org/10.1007/978-3-031-62197-0_17

damage to downstream communities. Likewise, Pakistan, China, and India are also affected by glacial outbreak floods (GLOFs), damaging the lives and livelihood of communities. According to the Government of Pakistan, 3044 glacial lakes have formed in Gilgit-Baltistan (GB) and Khyber Pakhtunkhwa (KP). Over 7.1 million people in GB and KP are vulnerable, and these areas continuously witness floods affecting human lives and livelihoods (Government of Pakistan, 2022; ICIMOD, 2019, 2022; Samaa, 2022). Apart from GLOFS, there is some evidence from Nepal and Pakistan that due to rising temperatures, decline in soil moisture levels, droughts, fires, and possible pest outbreaks that are affecting crop yields (Regmi & Adikhari, 2007) and evidence of the extinction or migration to higher altitudes of some plant species occur (ICIMOD 2011; Singh 2011; Regmi, & Adhikari 2007). However, in the longer term, such temperature rise could lead to an increase in the snowline and the disappearance of many glaciers, causing severe impacts on the populations relying on the seven main rivers in Asia fed by meltwater from the Himalayas; therefore, throughout Asia, one billion people could face water shortage leading to drought and land degradation by the 2050s (Goodrich et al., 2017a, b; IPCC, 2019; ICIMOD, 2019; Goodrich et al., 2022). Last year, heat wave in India and Pakistan resulted in 62% less rain than regular wildfires raging across forested regions of Punjab, Kashmir, Khyber Pakhtunkhwa, and Balochistan for days which destroyed biodiversity and livelihood options of the residents (Saeed, 2022). For instance, it damaged pine trees which bring millions to the people of the area (Saeed, 2022).

Warming mountain-related events impact different segments of society differently due to preexisting differences based on gender, caste, class, ethnicity, (dis) ability, and geographical location. When these factors intersect with gender due to preexisting roles, responsibilities, conditions, and positions within the family, community, and the wider society, they make people privileged or vulnerable during and after disasters and emergencies. For instance, the world's majority of resources and poor are women (70%) (Rashid & Cyprian, 2011), and disasters pose more consequences for women who are predominantly at risk with no or less capacity to respond (Goodrich et al., 2017a, b, 2021, 2022; Bano & Chimi, 2022). Disasters often intensify inequalities and reinforce disparities among women and marginalized groups regarding exposure and capacities for preparedness and response. Moreover, women are more vulnerable before, during, and after disasters because they are disadvantaged on political, legal, economic, and social fronts, and lack of access and ownership over productive resources, information, and technology aggravates the sufferings (IPCC, 2014; Goodrich et al., 2017a, b; Goodrich et al., 2022; Patel et al., 2019; UNFCCC, 2019).

Rigid and ingrained norms and traditions restrict women's mobility, a voice in decision-making and women's due status, particularly in remote and mountainous areas. In such a scenario, continuous climatic events have intensified the miseries leaving women and vulnerable groups more at risk. Research and anecdotal data acknowledge that it is critical to consider gender dynamics in disaster for effective relief and rehabilitation processes. Hence, it is essential to understand and recognize realities and sensitivities for better planning and response-related strategies (Bano & Chimi, 2022; Amir, 2022; Unwoman, 2022).

2 Objective of the Chapter

The main objective of this chapter is to highlight the differential impact of warming mountains on different segments of society based on gender, (dis)ability, and people with different orientations. Through this chapter, writers have brought some cases from the most vulnerable areas of South Asia and the Hindu Kush Himalayas to consider problems and propose actions to overcome issues.

3 Methodology

3.1 *Conceptual Framework*

In this chapter, we have used the social relation approach coined by Naila Kabeer (1994) and feminist political ecology (Rocheleau et al., 1996), using the gender and intersectionality framework (Crenshaw, 1989, 1991, 1992) to assess the differential vulnerabilities to and impacts of climate change based on gender and social differences. The social relation approach helps us identify the nature of inequality as it determines responsibilities, distribution of resources, and power dynamics in society. It also allows us to understand how these dynamics influence institutional mechanisms (formal and informal) at various levels, from household to state and beyond (Kabeer, 1997). We used the feminist political ecology and gender and intersectionality framework in South Asia and HKH; due to social, religious, political, and cultural settings, people experience numerous forms of inclusion and exclusion, privilege, and discrimination. Therefore, we argue that when gender intersects with other social fault lines such as sex, class, caste, ethnicity, race, religion, age, (dis) ability, education, and profession, as well as geographical location, people become vulnerable to the situation of climate change events and their comping abilities vary due to their condition (ICIMOD, 2022). Moreover, feminist political ecology emphasizes using a bottom-up approach to analyze people's embodied experiences and connections of environmental degradation, mobility, displacements, and another impact on power relations and decision-making processes. Both approaches complement each other in various ways and help us understand and analyze the situation (Goodrich et al., 2019a, b, 2021, 2022; ICIMOD, 2019).

3.2 *Area of Coverage and Gender Disparities in South Asian Countries*

South Asia and HKH region are the southern subregion of Asia, defined in geographical and ethnic-cultural terms. The region consists of the countries of Afghanistan, Bangladesh, Bhutan, India, Maldives, Nepal, Pakistan, Sri Lanka, and China.

South Asia and the HKH region are immensely diverse and densely populated areas. One-quarter of the population lives in these regions (i.e., almost 1.5 billion people). At the same time, more than millions of people are living below the poverty line (Islam et al., 2021; Kumar & Gurung, 2015; Treadway, 2019). South Asian region has almost 51.18% male population and 48.82% female population. Gender disparities exist in various forms across South Asia and HKH countries due to strong patriarchal norms and traditions. Gendered roles, responsibilities, positioning, and conditioning are shaped by the patriarchal systems that have placed women and girls at the margins (Arora-Jonsson, 2011; Goodrich et al., 2019a, b, 2022; Joshi, 2014; Nelleman et al., 2011; Sultana, 2010). Intersections shape these realities between multiple identities of gender, class, caste, ethnicity, culture, and religion by making women and men and girls and boys vulnerable or privileged in different settings. These realities are not stagnant but change from place to place and culture to culture. In most the societies of HKH, women are restricted from acquiring control over resources, freedom in choices and mobility, participation in decision-making, and having fundamental rights and power with an increased burden of unpaid care work (Goodrich et al., 2017a, b). In the last couple of years, the COVID-19 pandemic and continuous climatic events have execrated preexisting gender inequalities making choice lessors and pushing them into poverty and violence (Goodrich & Bano, 2020; UN Women, 2020a). The recent Gender-Equality Index (GEI) based on women in leadership and political participation, labor force participation, educational attainments, and access to health has placed many South Asian countries such as Pakistan at the bottom with Afghanistan (Goodrich et al., 2022; World Economic Forum, 2022).

3.3 Policy Landscape and Institutional Mechanisms

At the international level, various institutions have put efforts into including Gender Equality and Social Inclusion (GESI) considerations in policies and programs. The UN's Sustainable Development Goals have considered gender equality as one of the commitments. Various international forums and institutions such as the United Nations (UNDRR, 2015; UNWomen & UNICEF, 2016), World Bank (World Bank 2019), and Asian Development Bank (ADB, 2020) have commitments to involve women and marginalized groups in processes. Such institutions have made governments accountable for considering GESI as the primary outcome. Sendai framework also has a clear vision and mission of GESI and supports institutions to achieve its goals (UNDRR, 2015). Most HKH countries have included GESI concerns in policies and strategies related to climate change. National development commitments also include GESI; however, implementation is a big concern. The institutional mechanisms in almost all countries show a clear bias toward women and marginalized groups in structures and leading roles. Although governments have fixed quotas for women for various sociocultural and religious reasons, women do not join formal sectors if they want. They do not get jobs due to gender bias and the glass ceiling. According to the World Bank, 2018, in South Asia, women's participation in positions is not very encouraging; the least is in India, where 20% of women are doing jobs, and the highest is in Nepal, with 70% of jobs being assigned to women in various sectors. Disaster risk reduction is an area where we do not see many women at a different level. Moreover, inclusive financial allocation is also a concern to achieve goals and commitments. At the regional level, the South Asian Association for Regional Cooperation and the Bay of Bengal Initiative for Multi-Sectoral Technical and Economic Cooperation (BEMSTIC) also operate; however, their role in engaging with governments is limited due to political and financial issues (Goodrich et al., 2021, 2022).

At the regional level, ICIMOD has succeeded so far in giving South Asia and HKH countries a platform to come together and take measures to tackle climate issues with the support of various local and international donors. Under multiple initiatives such as the Koshi River Basin initiative, Indus River Basin initiative, and SWARMA, ICIMOD is undertaking research and co-creating knowledge shared with national, regional, and international stakeholders to influence policies related to climate change. ICIMOD is part of various international forums such as IPCC 2019; COP 2021 and 2022, and COP (biodiversity), in which ICIMOD explicitly highlights mountain issues and advocates for climate action related to unheard mountain areas at such forums.

At the national level, countries have devised policies and national action plans. Most of the policies include gender perspectives. However, in sector-specific policies, there is a strong disconnect between ground scenarios and policy formulation. Primarily, women are seen as recipients instead of active partners, stakeholders, and leaders (Goodrich et al., 2022). The role of women in water diplomacy at the transboundary level and water cooperation at the country level, at the sub-basin level, and the basin level and regional and international level is also limited (Sehring et al., 2022). Nepal's National Water Policy has only one clause regarding women's role in managing water services (Ahlers & Zwarteveen, 2009).

Likewise, countries have devised a gender-inclusive disaster risk reduction policy that acknowledges the gendered vulnerabilities of women and children and how the active role of women during and after disasters is overlooked (Goodrich et al., 2021). Moreover, the rehabilitation of affected communities and implementation of this policy in its true spirit are debatable due to gendered institutional arrangements, lack of transparency, low or no financial allocations, and existing sociocultural norms toward women and vulnerable groups (Goodrich et al., 2021, 2022).

3.4 Gendered Impacts of Warming Mountains

Climate change has different impacts on men and women due to the gender roles and responsibilities that are prevalent in many societies. Women and girls are often more vulnerable to the effects of climate change, such as natural disasters, disease outbreaks, and food shortages, due to their preexisting inequalities that restrict their rights and access to resources, assets, and power, consequently excluding them from decision-making processes (Goodrich et al., 2022; IPCC, 2014; Goodrich et al.,

2017a, b; Patel et al., 2019; UNFCCC, 2019). Likewise, women collect water and fuel in many agricultural communities and grow and process food. Climate change can exacerbate water scarcity and food insecurity, increasing women's workload and reducing their time for other activities (Gurung & Bisht, 2014; VOA, 2022). Women and girls are more susceptible to health impacts related to climate change, such as malnutrition, increased risk of disease transmission, and exposure to toxic chemicals. Climate change has disrupted traditional livelihoods, such as agriculture, fishing, and forestry, affecting the economic security of both men and women. However, women are often more vulnerable to poverty and unemployment and may have limited access to alternative sources of income (Bano & Chimi, 2022; Goodrich et al., 2022). Also, climate change can lead to migration as people seek to escape the impacts of natural disasters, such as floods and droughts. Women and girls are often at increased risk of exploitation and abuse during migration and may face additional barriers to accessing resources and services in their new communities. Therefore, it is essential to consider the gendered impacts of climate change in policy and decision-making and to ensure that women and girls have equal access to resources and support to adapt to the changing climate. This can include investments in gender-sensitive agricultural programs, disaster risk reduction initiatives, and health and livelihood support. On the other hand, such a situation can increase women's resilience, adaptative capacities, and leadership (IUCN, 2015; ICIMOD, 2019).

Women and youth are resilient and are essential in preparedness, response, recovery, and adaptation. They participate in preparedness activities, arrange training as practitioners, and give awareness about hazards. They are the first responders to face events since they spend most of the time in their houses and villages; they know the family needs and vulnerabilities of older people, children, and people with disabilities for being primarily responsible for it (ICIMOD, 2019).

During the events of disasters, they take care of the families, neighbors, and injured and disabled persons by taking up leadership roles. They help identify losses and damages if consulted, provide relief during disasters, and participate in solidarity, providing food to volunteers and fund-raising events. It is seen that when women are empowered and given responsibilities, then they manage floods better than their male counterparts (Goodrich & Bano, 2020; Bano & Chimi, 2022). Women in administrative jobs are less in Pakistan; however, a deputy commissioner (DC) in Nowshera KPK has done a magnificent job of visiting door to door and evacuating people from their houses resulting in no human loss (Samaa 2022). Likewise, women doctors and nurses provide equipment and health facilities. Women social activists and workers arrange food and non-food items, particularly hygiene kits and natal care-related supplies, in flood-affected areas (Bano & Waqar, 2020). It is seen that women can play a better role in rescue operations if they are given training and equipment through extended community engagement activities.

After the disasters, women play an essential role in rehabilitation. They support building houses and rebuilding livelihood options. If given a chance, they participate in the decision-making process, coordinating with recovery process institutions and distributing the funds. However, at a high level, women's participation in decision-making related to the environment remained limited, restricting the integration of women's issues or gender perspectives into any policymaking about the climate (ICIMOD, 2019; Bano & Chimi, 2022).

3.5 *Human Losses and Physical Damages*

South Asia has been continuously facing floods, heavy rains, droughts, and avalanches, due to which thousands of people have lost their lives. Recent floods in Pakistan estimated 1700 deaths, and 1.1 million homes were damaged or destroyed, causing $9.7 billion in damages in 46 of the country's 135 districts (Aljazeera, 2022; Climate Analytics, 2022). According to the Ministry of Home Affairs in Nepal, between June and October 27, 2021, 673 people lost their lives, 69 people were missing, and 181 were injured due to the water-induced disaster in Nepal (Kathmaandu Post, 2022). Research has proved that the number of death counts for women, children, and the elderly are more than for men during and after disaster events due to physical and sociocultural norms (Bano & Chimi, 2022). Women, the elderly, children, and disabled persons cannot run due to their physical conditions during flash floods in the mountains. In some parts of KPK and Gilgit Baltistan, more women and children died under the debris of flash floods.

It is evident from research that during these events, women are more likely than men to die from an outbreak or a disaster because of less access to information and participation in disaster learning skills, mobility issues, and Purdah (women's mandatory isolation from men except siblings) (ADB, 2014). In contrast, men who provide ends meet at home usually migrate for livelihoods. This norm has put women, children, and the elderly at high risk of floods, especially in mountainous regions (Goodrich & Bano, 2020). It is reported that more women and children died in recent flash floods in Gilgit Baltistan and Chitral since they were at home when floods entered their houses (Dawn, 2016). During the tsunami in Japan and floods in India and Nepal, more deaths among women and children were reported. Men usually lost lives while traveling to their offices and business, visiting riverbanks, and collecting timber from the flooded water. Various events also noted that getting gender-disaggregated data was almost impossible in multiple countries.

When we look at physical losses caused by climatic conditions, people from mountains and plains suffer loss of connectivity from the rest of the world for months and days. The response-related activities were very strenuous in such situations. It damages highways, roads, airports, and communication systems. Moreover, thousands of schools, banks, hospitals, and shops were underwater for months in Sindh and Balochistan (ADB Government of Pakistan, 2022). Apart from human losses, floods, avalanches, and landslides have damaged various indigenous settings, spiritual places, and historical places. Recent floods in Pakistan have damaged many such areas, including ancient and historical sites such as Mohenjo-Daro, the oldest civilizations, mosques, Shirin's, and graveyards (Government of Pakistan, 2022; Maharjan et al., 2021). In such a scenario, women and vulnerable groups suffer the most and remain dependent on local services due to traditional roles. Men easily migrate and settle elsewhere in the cities and towns or abroad (DevoKota, 2013; Pearson et al., 2021; Government of Pakistan, 2022).

3.6 Economic Losses

Climate change causes significant economic losses, both in the short term and in the long term. For instance, climate change leads to changes in precipitation patterns and temperatures, affecting crop yields and livestock productivity. This resulted in food shortages and increased prices for basic staples, leading to economic losses for farmers and consumers. Extreme weather events, such as hurricanes, droughts, and floods, cause significant damage to buildings, roads, bridges, and other infrastructure, leading to substantial economic losses. Changes in weather patterns and the availability of natural resources, such as water and wildlife, can impact the tourism industry, leading to financial losses for hotels, airlines, and other businesses that rely on tourism. Climate change disrupts supply chains, causing companies to incur additional costs and leading to losses in productivity. Preliminary estimates suggest that, as a direct consequence of the floods in Pakistan, the national poverty rate will increase by 3.7–4.0 percentage points, pushing between 8.4 and 9.1 million people into poverty. According to the Asian Development Bank, Nepal faced 10% of its GDP from glacial melt and flooding.

The economic impact of climate change is more complicated and harder for women. They are already resourceless, mainly associated with the informal sector as producers or daily wage workers. Such events halt economic activities. For instance, during floods in Pakistan, women laborers in agricultural sectors, private teachers, and health workers did not get salaries for months as the agricultural land, schools, and their homes were underwater for months (Buzdar & Ali, 2011; ILO, 2022). In Nepal, most women carry household, livelihood, and community activities without male family members to other countries. Apart from care-related activities, women are employed or running tourism-related businesses for additional income. Such events increase their vulnerability by impacting the livelihoods of women and vulnerable groups (IFCR, 2021).

3.7 Agriculture and Food Security

Agriculture and food security are the most significant sectors affected by climate change. Climate change leads to various issues that can impact agricultural productivity, food security, and global food systems (IPCC, 2014; Rasul, 2014; Bano et al., 2023). Some of the major problems include the following:

Agriculture is the largest sector of the HKH economy that accounts for many South Asian and HKH countries. This sector is the most vulnerable and prone to the impacts of climate change, with pressure on agricultural lands, scarcity of water, and changing weather patterns putting the risk of food insecurity. According to the UNDP, this sector is also the major contributor to carbon emissions (Khan & Imran, 2021; Rasul, 2014; Hussain et al., 2018; Goodrich et al. 2022).

Agriculture is central to livelihood, especially in rural areas, and women constitute a significant labor force in this sector. The role and responsibility of women in agriculture vary from region to region and community to community. In the mountain areas where women have no land rights and limited access to capital, double responsibility makes them more vulnerable. Women's role is central in crop production, horticulture, livestock rearing, crop harvesting, collecting and drying fruits and vegetables, collecting and storing fodder for winters, collecting wood for fuel, fetching drinking water in harsh winters, and protecting the new trees from livestock by painting it with cow dunk in winters. Women are not direct beneficiaries of the land, but they have work for their survival and livelihood (Hussain et al., 2018; Goodrich et al., 2022). Due to a lack of employment opportunities, men migrate to the urban part of the country or abroad. Still, women stay back and take all the responsibility from domestic chores to agricultural activities, which is a double burden on women and has remained unpaid and unrecognized.

Extreme weather conditions, floods, melting of glaciers, and landslides are some of the impacts of climate change which adversely affect the agricultural sector (Goodrich et al., 2022; IPCC, 2014; Rasul 2016, 2021). Summers are the time for collecting fruits and fodder for winter; due to the change in the weather pattern and rainfall, it is becoming difficult for women to dry their fruits, vegetables, and grass for winter for the livestock. Flood damages the water sources halting water supplies for agricultural activities. Flood also washes out the infrastructure as roads that connect the far-flung areas with the rest of the country, making it challenging to transport crops and fruits from the site to the market, which is a loss. An example of a climate-induced disaster is the land sliding of Attabad in 2010, Hunza, Pakistan. The incident happened in the Sarat village of Attabad, where a massive landslide followed by boulders and rocks slid the town of Attabad into the Hunza River. Twenty people died, 1600 were displaced, and about 4 km of the Karakorum were blocked, forming a lake. Twenty-five thousand people were cut off from the rest of the country. Although the area is remote, this climate change-induced disaster made it more vulnerable, where boats were used to travel to connect with the other part of the country. This disaster made the life of the residents very difficult, but women and children were affected the most. The agricultural activities were slowed down or stopped because of the blockage, making transporting goods to the market complex. The difficulties and challenges faced by women during this disaster are neglected in the studies. This disaster shifted the dependence from agriculture to other sources of income till the area was again connected to the lower parts of the country (Hussain et al. 2018; Goodrich, 2021, 2022).

Addressing these issues requires a gender-sensitive approach that considers the unique needs and perspectives of women and girls and provides them with the resources and support they need to adapt to the changing climate. Additionally, agricultural research and development investments and sustainable and resilient food systems are critical for ensuring food security in a changing climate.

Pakistan has launched a food security policy in 2018 which aims to make the agricultural sector more productive, profitable, and resilient to climate change. Sixty-five percent of women in Pakistan are involved in agricultural activities as landless, unpaid laborers, and only 4% of women own land (Hussain et al., 2018). Keeping this figure in mind, women are the most vulnerable segment to the impacts of climate change.

3.8 Water Availability and Water-Induced Hazards

Water availability is a growing concern due to the impacts of climate change. For instance, some key issues include increased temperatures and changes in precipitation patterns that can lead to drought in some regions, reducing water availability for human and agricultural use. Moreover, glaciers and ice caps provide water for millions of people, but as the mountains warm, they are melting at an unprecedented rate. This can increase water scarcity in areas that depend on glacial meltwater. Freshwater sources, leading to increased salinity and reduced water quality and higher temperatures, can increase the rate of evaporation, reducing the amount of water in lakes and rivers and exacerbating water scarcity in regions that are already arid or semiarid (ICIMOD 2019; Goodrich et al., 2019a, b, 2022; Molden et al., 2022).

These changes in water availability due to climate change can have far-reaching impacts on food security, human health, and economic development. We must take action to mitigate and adapt to these changes to ensure a sustainable future (ICIMOD, 2019).

Hence, too little and too much water is a reality for upstream and downstream communities. Too much water causes damage, human losses, and disruption in livelihood and too little water is also the reason why communities feel stress, especially during winter. Six months of winter become very tough for mountain communities of HKH, which is a cultivation season (IPCC, 2014; Rasul, 2021 ICIMOD, 2019; Goodrich et al. 2021, 2022).

In South Asia's rural and mountainous areas, women are the primary caretaker of household chores; it is their responsibility to fill the immediate needs of water, food, care, and the family. In Indian streets, women spend almost 4 hours every day fetching water. Young girls spend more time getting clean drinking water than on productive work such as education and skill enhancement. Women, being a vulnerable group, even get more susceptible when they become pregnant or have a disability (ICIMOD, 2019; Bano & Chimi, 2022; Goodrich et al., 2021, 2022; Joshi, 2014).

ICIMOD commissioned a study in the Gilgit-Baltistan and Leh-Ladakh subregions of the Upper Indus Basin (UIB) in 2022. It revealed that water resources, agriculture, and livelihoods are highly vulnerable to climate change-induced hazards and risks. There are direct and indirect impacts on water availability, agriculture, livelihoods, and water variability in traditional irrigation systems. As a result, primarily women, as they are the ones who do all activities in the absence of men, face difficulties in meeting the water needs. Regarding water, men are responsible for irrigation water and women for household use. During an FGD in Hasan Abad Hunza, a water channel was damaged, and the drinking water supply was disconnected. Since most of the time, men's needs are met while rehabilitation and irrigation channel were rehabilitated, the drinking water supply was on halt even after years of flooding (Goodrich et al. 2021, 2022; Bano & Chimi, 2022).

Villages and towns dependent on springs and groundwater face water shortages, and women spend 3–4 h for water collection. In the case of Nepal, changing climate has impacted water availability, and it has increased the burden on women directly as they spend their time collecting water to fulfill household water needs. Also, in India and Nepal, due to drying springs, villages were displaced (ICIMOD, 2019; Goodrich et al., 2021, 2022; Bano & Chimi, 2022).

3.9 Displacement, Migration, and Social Issues

Climate change is causing increasing numbers of people to migrate and be displaced from their homes, villages, and settlements (Kohler & Maselli, 2009). These migrations and displacements can have significant social, economic, and political consequences, including increased poverty, conflict, and instability. There is an urgent need to address the root causes of climate change, as well as provide support to communities and countries affected by migration and displacement due to climate change. Millions of people, including women, children, the elderly, and the disabled, were forced to migrate and leave their houses due to less or more water in India, Bangladesh, China, Pakistan, and Nepal (IFRC, 2021). Flooding in southern Nepal in August (2022) affected over a million people, displacing tens of thousands. The destruction was widespread in Rautahat district. In the Kosi Basin, 1500 families were displaced (NDTV, 2022). Recently, in Pakistan, around 33 million people have been affected by the floods, including nearly 8 million displaced. Rain-induced floods, accelerated glacial melt, and resulting landslides devastated millions of homes and critical infrastructure, submerging entire villages and destroying livelihoods. In such a scenario, women, elderly, children, and disabled people suffer the most as they become more vulnerable without a roof. They live in temporary shelters or with relatives or in flooded houses (ICIMOD, 2019; Government of Pakistan, 2022; Bano & Chimi, 2022; NDTV, 2022).

Studies have also indicated that in Nepal increase in the collection time for fodder and firewood and a decline in agricultural productivity increase the probability of local (within district) out-migration of women, closely linking climate change-induced migration for women with deterioration of natural resources as they are both the primary collectors and users (Goodrich et al., 2017a, b; Massey et al., 2010).

Women, men, and different segments of the society face different experiences and challenges during natural disaster-led migration and displacement. These differences are shaped by gender roles, social norms, and unequal power relationships between men and women, resulting in unequal access to resources and opportunities. During recent floods in the countries of HKH region, women and girls have faced exploitation and abuse, including sexual exploitation and violence, during migration and displacement (The Guardian 2022). Women and girls had limited access to healthcare, education, and other essential services in migration and displacement contexts, putting them at increased risk of poor health and reduced opportunities. Moreover, these camps do not provide good space for children and women with proper health and hygiene provisions, especially during menstrual cycle (Government of Pakistan, 2022; UN Women, 2022; UNFPA, 2022). It is observed that women mostly avoid drinking liquids/water in camps to minimize toilet use, as they must share the toilet facility with men. Floods in Pakistan have triggered many issues for girls and women. Early marriages were increased for economic terms and other benefits. Temporary shelters were provided to the flood-affected people; however, the safety and security of mainly women and children were not considered in most cases, making them vulnerable to violence and abuse (Bano & Chimi, 2022; Friday Times, 2022; The Guardian, 2022; UNFPA, 2022).

Sociocultural norms often compel women to eat less or at the last after providing for the family (Pick & Jenna, 2010). Most of the women and children in Pakistan, Nepal, and India are already malnourished, which aggravates during floods since they do not get enough or proper nutrients. This weakens their immune system, and hence they recover slowly from flood-borne diseases. It was found in the gender need assessment done by UN women during and after the 2010 floods in Pakistan that it is the men who have access to food and distribute food during and after disasters. The situation worsens for pregnant and new mothers and their babies as many health facilities were damaged or collapsed. According to United Nations Fund for Population Activities (UNFPA), "almost 650,000 pregnant women in the flood-affected areas require maternal health services to ensure a safe pregnancy and childbirth. Up to 73,000 women expected to deliver next month will need skilled birth attendants, newborn care, and support" (UNFPA, 2022; Relief Web, 2022).

Most of the times, institutions and researchers ignore some of the pertinent issues people, particularly women, suffer during distress, primarily mental health-related issues. Before, during, and after climatic events, women and vulnerable groups face more grief and mental problems due to their social conditions as they lose belongings, shelter, and sometimes their loved ones. Research also validates those stresses and increases the chances of mental issues. Most countries lack the appropriate mechanisms to overcome such situations (Vujanovic & Gallagher, 2017; Goodrich et al., 2022; Beno & Chimi, 2022).

It is essential to address gender-related migration and displacement issues to ensure that needs and experiences of both men and women are recognized and discussed in policies and programs inclined toward migration and displacement.

3.10 *Increased Burden of Reproductive Activities*

Women's workload and burden increase during and after disasters and floods since they are primarily responsible for reproductive activities, children, and elderly care (Bano & Chimi, 2022). For instance, in mountain areas, women spend all their time doing household activities, agricultural activities, children care, and elderly care. In the absence of clean energy options, they collect firewood from distances and collect water for household use and cattle. They are also responsible for managing fodder for winter. In summer, they go to pastures for fodder and wood. The floods, droughts, and rains further magnify the burden when floods damage power supply systems and roads, and women must go the extra mile to manage resources for cooking and lighting (Gurung & Bisht, 2014; Bano & Chimi, 2022).

Likewise, floods and disasters leave men with limited employment and business opportunities, crops, and livestock, making it difficult to earn enough for their families. Furthermore, women are solely responsible for fixing minor damages such as cleaning and drying wet places and items and providing food to family members in any circumstances. During current floods, young men were seen at the forefront of saving lives by risking their own lives without proper training (Goodrich & Bano, 2020).

4 Summary and the Way Forward

Climate change is exacerbating the existing discrimination and inequalities based on gender, cast, class, and many other factors in South Asia. Warming mountains is one of the scenarios of climate change. Due to warming mountains, glaciers are melting, causing floods in the upstream and downstream rivers. Forest fires are another issue caused by warming mountains. These hazards impact the lives and livelihood of people, particularly those living in the mountains, in poverty, and with disabilities. In most mountain societies and elsewhere, women and marginalized community segments have no control over resources and access to decision-making places. Such disparities make them more vulnerable to the disaster and hazards caused by warming mountains. Such events increase women's workload, restrict mobility, and limit access to social, cultural, and economic services. Women are crucial in preparedness, mitigation, and rehabilitation during and after disasters. However, their contributions are not acknowledged, and they do not get into decision-making places at various levels.

Moreover, their specific needs are often neglected before, during, and after disasters. Although there are policies in place in all countries of HKH, the implementation is a crucial issue, all the mechanisms are gendered, and the percentage of women in elated institutions is deficient. Sociocultural norms play a vital role in perpetuating these inequalities, making them more vulnerable; therefore, it is essential to consider below mentioned suggestion to overcome warming mountain-related hazards and risk mitigation and adaptation more inclusive and responsive:

- There is a need for a holistic approach to unpacking these issues, strategies, and careful consideration to overcome these issues. Moreover, women should be given equal space in decision-making regarding the environment at all levels. They can play a very productive role in preventing, coping, mitigating, and ensuring/developing adaptive strategies during and in the aftermath of disasters. This reality has often been neglected, and women are only seen as victims of these events. Climate change must be taught in schools, and training must be provided to all to cope with the disasters.
- Gender integration in integrated water resource management (IWRM) approach should be ensured for equitable access to water and benefit different social groups, i.e., women and men and rich and poor.
- Regional countries of HKH need to ensure that policies, institutions, and projects address the needs of the most affected, vulnerable women and men, and introduce social protection programs that must cover the reproductive activities associated with women.
- Robust mechanisms should be devised at the local, provincial, and national levels to ensure the safety and security of women and children during and after disasters, and organizations must ensure the resources.
- Women should be given land and asset entitlements during rehabilitation processes.
- More needs to be done to increase the status and position of women by providing equal access to information and life skills, improving socioeconomic opportunities, and safeguarding and educating men to acknowledge women's equality in rehabilitation processes.

References

Ahlers, R., & Zwarteveen, M. (2009). The water question in feminism: Water control and gender inequities in a neo-liberal era. *Gender, Place and Culture, 16*(4), 409–426.

Aljazeera. (2022). *Death toll in devasting Pakistan floods crosses 1300*. https://www.aljazeera.com/news/2022/9/5/death-toll-in-devastatingpakistan-floods-crosses-1300

Amir, F. (2022). *Pakistan floods; women should be at the forefront of relief effort and future climate policies*. Stimon https://www.stimson.org/2022/pakistan-floods-women-should-be-at-the-forefront-of-relief-efforts-and-future-climate-policies/

Arora-Jonsson, S., (2011). Virtue and vulnerability: Discourses on women, gender and climate change. *Global Environmental Change, 21*(2), pp.744–751.

Asian Development Bank (ADB). (2014). Gender Inclusive risk management. https://www.adb.org/sites/default/files/institutionaldocument/34130/files/gender-inclusive-disaster-risk-management-0.pdf

Asian Development Bank (ADB). (2020). *Gender equality and social inclusion diagnostic of selected sectors in Nepal*. Asian Development Bank. https://doi.org/10.22617/TCS200291-2

Bano, K., & Chimi, S.. (2022). *Gendered dynamics and disaster events in Hasan Abad Hunza*. https://www.icimod.org/article/gender-dynamics and-disaster-events-in-hasanabad-Pakistan/

Bano, K., & Waqar, K. (2020). *The long road-to-gender-equality in Pakistan's labour force*. ICIMOD. https://www.icimod.org/article/the-long-road-to-gender-equality-in-pakistans-labour-force/

Bano, K., Waqar, K., & Ali, A. (2023). Contributions of women's collective farming to women's agency in the Upper Indus Basin in the face of climate change. *Journal of Cleaner Production, 432*(November), 139734. https://doi.org/10.1016/j.jclepro.2023.139734

Buzdar, A., & Ali, A. (2011). *Socio-economic effects of floods on female teachers in Jampur*. The Islamia University of Bahawalpur.

Climate Analytic. (2022). *Pakistan's floods must be a wakeup call on climate action ahead of Sharm El-Sheikh*.

Crenshaw K. (1989). *Demarginalizing the intersection of race and sex: A black feminist critique of antidiscrimination doctrine, feminist theory, and antiracist politics*, 1989 University of Chicago Legal Forum, 139.

Crenshaw, K. (1991). Mapping the margins: Intersectionality, identity politics, and violence against women of color. *Stanford Law Review, 43*(6), 1241–1299.

Crenshaw, K. (1992). Whose story is it anyway? Feminist and antiracist appropriations of Anita Hill. In T. Morrison (Ed.), *Raceing justice, engendering power*. Pantheon.

Datey, A., Sharma, M., Sinha, R. & Kumar, P., (2021). Climate change and its impact on health in South Asia. *Environmental Health Perspectives, 129*(4), p.047001.

DAWN. (2016). *Rain, flash floods claim 60 lives in KPK and GB*. https://www.dawn.com/news/1249891

Devkota, F. (2013). Climate change and its sociocultural impact in the Himalayan region of Nepal—A visual documentation. https://doi.org/10.4000/anthrovision.589

Goodrich, C. G., & Bano, K. (2020, April 20). *Impacts of COVID-19 could heighten gender inequalities in the HKH: Unless responses focus on women*. https://www.icimod.org/article/impacts-of-covid-19-could-heighten-genderinequalities-in-the-hkh-unless-responses-focus-on-women/

Goodrich, C. G., Udas, P. B., Gurung, M. B., Shrestha, N. T., & Shrestha, S. (2017a). *Gender and social inclusion in local water planning: Lessons from water use master plan practices in Nepal*. International Centre for Integrated Mountain Development (ICIMOD). https://doi.org/10.53055/ICIMOD.668

Goodrich, C. G., Mehta, M., & Bisht, S. (2017b). *Status of gender, vulnerabilities, and adaptation to climate change in the Hindu Kush Himalaya: Impacts and implications for livelihoods, and sustainable mountain development* (ICIMOD Working Paper 2017/3). International Centre for Integrated Mountain Development (ICIMOD). https://doi.org/10.53055/ICIMOD.692

Goodrich, C. G., Prakash, A., & Udas, P. B. (2019a). Gendered vulnerability and adaptation in Hindu-Kush Himalayas: Research insights. *Environmental Development, 31*, 1–8. https://doi.org/10.1016/j.envdev.2019.01.00

Goodrich, C. G., Pranita, B. U., & Harriet, L. S. (2019b). Conceptualizing gendered vulnerability to climate change: Contextual conditions and drivers of change. In C. G. Goodrich, P. B. Udas, & A. Prakash (Eds.), *Special issue of environmental development* (pp. 9–18). Elservier. https://doi.org/10.1016/j.envdev.2018.11.003

Goodrich, G. C., Gurung, D. D., & Bastola, A. (2021). *State of gender equality and climate change in Nepal*. ICMOD, UNEP and UN Women. https://doi.org/10.53055/ICIMOD.790

Goodrich, G. C., Gurung, D. D., Hussain, A., Pasakhala, P., Bano, K., Bhuchar, S., Chitale, V. S., Bastola, A., Bisht, S., & Silpakar, S. (2022). *State of gender equality and climate change in Nepal*. ICMOD, UNEP and UN Women.

Government of Pakistan. (2022). *Pakistan floods 2022, PDNA main report*. https://thedocs.worldbank.org/en/doc/4a0114eb7d1cecbbbf2f65c5ce0789db-0310012022/original/Pakistan-Floods-2022-PDNA-Main-Report.pdf

Gurung, D. D., & Bisht, S. (2014). *Women's empowerment at the frontline of adaptation: Emerging issues, adaptive practices, and priorities in Nepal* (ICIMOD working paper 2014/3). ICIMOD.

Hussain, A., Rasul, G., Mahapatra, B., & Tuladhar, S. (2016). Household food security in the face of climate change in the Hindu-Kush Himalayan region. *Food Security, 8*, 921–937. https://doi.org/10.1007/s12571-016-0607-5

Hussain, A., Rasul, G., Mahapatra, B., Wahid, S., & Tuladhar, S. (2018). Climate change – induced hazards and local adaptations in agriculture: A study from Koshi River basin, Nepal. *Natural Hazards, 91*(3), 1365–1383. https://doi.org/10.1007/s11069-018-3187-1

Hussain, A., Qamar, F. M., Adhikari, L., Hunzai, A. I., Rehman, A. U., & Bano, K. (2021). Climate change, mountain food systems, and emerging opportunities: A study from the Hindu Kush Karakoram Pamir landscape, Pakistan. *Sustainability, 13*(6), 3057. https://doi.org/10.3390/su13063057

Hussain, S., Khan, A., & Hussain, A. (2023). The Intersection of Patriarchal Chains and The Natural Disasters: Exploring the Impact of Flood on Women in The Rural Areas of Dera Ghazi Khan, Pakistan. *Contemporary Issues in Social Sciences and Management Practices, 2*(3), 276–291.

ICIMOD. (2011). Status of Gender, Vulnerabilities and Adaptation to Climate Change in the. www.icimod.org/hicap

ICIMOD. (2019). *The Hindukush Himalaya assessment: Mountains, climate change, sustainability and people*. Springer.

ICIMOD. (2020). *The HKH call to action to sustain mountain environments and improve livelihoods in the Hindu Kush Himalaya*. International Centre for Integrated Mountain Development. https://doi.org/10.53055/ICIMOD.1

ICIMOD. (2022). *GLOF hazard and exposure risk assessment methodology* (Unpublished methodological framework). ICIMOD.

IFCR. 2021. *Climate change impacts on health and livelihoods: Nepal assessment*. https://www.climatecentre.org/wp-content/uploads/RCRC_IFRC-Country-assessments-NEPAL-3.pdf

Intergovernmental Panel on Climate Change (IPCC). (2014). Climate change 2014: Impacts, adaptation, and vulnerability. Part A: Global and sectoral aspects. In C. B. Field et al. (Eds.), *Contribution of Working Group II to the fifth assessment report of the Intergovernmental Panel on Climate Change*. Cambridge University Press.

Intergovernmental Panel on Climate Change (IPCC). (2019). Summary for policymakers. In P. R. Shukla et al. (Eds.), *Climate change and land: An IPCC special report on climate change, desertification, land degradation, sustainable land management, food security, and greenhouse gas fluxes in terrestrial ecosystems*.

Intergovernmental Panel on Climate Change (IPCC). (2021). Summary for Policymakers. In *Climate change 2021: The physical science basis. Contribution of Working Group I to the sixth assessment report*. Intergovernmental Panel on Climate Change.

International Labour Organization. (2022). *Pakistan floods 2022 post-disaster needs assessment*. https://www.ilo.org/wcmsp5/groups/public/ed_emp/documents/publication/wcms_863195.pdf

Islam, T. M. T., Newhouse, D., & Yanez-Pagans, M. (2021). International comparisons of poverty in South Asia. *Asian Development Review, 38*(1), 142–175. https://doi.org/10.1162/adev_a_00161

IUCN. 2015. Gender and climate change: Strengthening climate action by promoting gender equality. IUCN Issue Brief.

Joshi, D., (2014). Feminist solidarity? Women's engagement in politics and the implications for water management in the Darjeeling Himalaya. *Mountain Research and Development, 34*(3), pp.243–254.

Kabeer, N. (1994). *Reversed realities: Gender hierarchies in development thought*. Verso.

Kabeer, N. (1997). *Reversed realities: Gender hierarchies in development thought*. Verso.

Khan, S. & Imran, M., (2021). Assessing the impact of climate change on agricultural productivity in Pakistan: A gender perspective. *Journal of Environmental Management, 285*, 112–174.

Khan, Z. A., Koondhar, M. A., Khan, I., Ali, U., & Tianjun, L. (2021). Dynamic linkage between industrialization, energy consumption, carbon emission, and agricultural products export of Pakistan: an ARDL approach. *Environmental Science and Pollution Research, 28*, 43698–43710.

Kohler, T., & Maselli, D. (2009). *Mountains and climate change. From understanding to action*. Geographica Bernensia.

Kumar, A., & Gurung, T. R. (2015). SAARC outlook on water-energy-food nexus in SAARC region.

Maharjan, S. B., Steiner, J. F., Shrestha, A. B., Maharjan, A., Nepal, S., Shrestha, M., Bajracharya, B., Rasul, G., Shrestha, M., Jackson, M., & Gupta, N. (2021). *The Melamchi flood disaster: Cascading hazard and the need for multihazard risk management – ICIMOD*. ICIMOD.

Massey, D., Axinn, W., & Ghimire, D. (2010). Environmental change and out-migration: Evidence from Nepal. *Population and Environment, 32*, 109–136.

Molden, D. J., et al. (2022). The great glacier and snow-dependent Rivers of Asia and climate change: Heading for troubled waters. In A. K. Biswas & C. Tortajada (Eds.), *Water security under climate change* (pp. 223–250). Springer Singapore. https://doi.org/10.1007/978-981-16-5493-0_12

NDTV. (2022). *Women forced to stay in flooded village reason a matter of honour*. https://www.ndtv.com/world-news/women-forced-to-stay-in-flooded-pakistan-village-reason-a-matter-of-honour-3342838

Nellemann, C., Verma, R. & Hislop, L. (eds.), (2011). *Women at the frontline of climate change: Gender risks and hopes. A Rapid Response Assessment. Arendal*: United Nations Environment Programme (UNEP), GRID-Arendal.

Patel, S.K., Sharma, V. & Tiwari, M., (2019). Impact of climate change on agriculture and its mitigation strategies: A review. *Journal of Cleaner Production, 232*, 849–857.

Pearson, J., Jackson, G., & McNamara, K. E. (2021). Climate-driven losses to indigenous and local knowledge and cultural heritage. *The Anthropocene Review, 0*(0). https://doi.org/10.1177/20530196211005482

Pick, S., & Jenna, S. (2010). Socio-cultural norms as impediments to individual and social change: The case of Mexico. *Breaking the poverty cycle: The human basis for sustainable development* (2010; online edn, Oxford academic, 1 Sept. 2010). https://doi.org/10.1093/acprof:oso/9780195383164.003.0002. Accessed 15 June 2023.

Rashid, S., & Cyprian, J. (2011). Impact of poverty on women (A worldwide overview). *Nice Research Journal, 4*, 85.

Rasul, G. (2014). Food, water, and energy security in South Asia: A nexus perspective from the Hindu Kush Himalayan region. *Environmental Science and Policy, 39*, 35–48. https://doi.org/10.1016/j.envsci.2014.01.010

Rasul, G. (2016). Managing the food, water, and energy nexus for achieving the sustainable development goals in South Asia. *Environmental Development, 18*, 14–25.

Rasul, G. (2021). Twin challenges of COVID-19 pandemic and climate change for agriculture and food security in South Asia. *Environmental Challenges, 2*, 100027.

Rasul, G., Hussain, A., Khan, M. A., Ahmad, F., & Jasra, A. W. (2014). *Towards a framework for achieving food security in the mountains of Pakistan* (Working Paper 2014/5). International Centre for Integrated Mountain Development (ICIMOD). https://doi.org/10.53055/ICIMOD.596

Regmi, B. R., & Adhikari, A. (2007). Human Development Human Development Report Office Country Case Study-Nepal. www.libird.org

Relief Web. (2022). *Women and girls need urgent health and protection services amid epic flood disaster*. https://reliefweb.int/report/pakistan/women-and-girls-pakistan-need-urgent-health-and-protection-services-amid-epic-flood-disaster

Rocheleau, D., Thomas-Slayter, B., & Wangari, E. (1996). *A feminist political ecology perspective. Feminist political ecology: Global issues and local experiences* (pp. 3–26).

Saeed, R. (2022). *Long-term impacts of "'terrible' fire season in ' 'Pakistan's mountains*. Third pole.

Samaa. (2022). When Nowshera's elders failed, they turned a woman for help. https://www.samaaenglish.tv/news/40015756

Schild, A. (2008). ICIMOD's position on climate change and mountain systems. *Mountain Research and Development, 28*(3), 328–331.

Sehring, J., ter Horst, R., & Zwarteveen, M. (2022). Gender dynamics in transboundary water governance. In *Feminist perspectives on water conflict and cooperation* (p. 226). Taylor & Francis.

Singh, S. P. (2011). *Climate change in the Hindu Kush-Himalayas: the state of current knowledge.* International Centre for Integrated Mountain Development.

Sultana, A. M. (2010). Socio-Cultural Dimensions of Women's Discriminations in Rural Communities. *Ozean Journal of Social Sciences 3*(1).

The Friday Times. (2022). *Flood crisis women at risk of sexual abuse violence and harassment.* https://www.thefridaytimes.com/2022/09/19/flood-crisis-women-at-risk-of-sexual-abuse-violence-and-harassment/

The Guardian. (2022). *The-hospital-has-nothing-pakistans-floods-put-pregnant-women-in-danger.* https://www.theguardian.com/world/2022/sep/14/the-hospital-has-nothing-pakistans-floods-put-pregnant-women-in-danger

The United Nations Population Fund (UNFPA), (2022) UNFPA Pakistan Situation Report #3 - Monsoon Flood. https://pakistan.unfpa.org/en/publications/unfpa-pakistan-situation-report-3-monsoon-floods

Treadway, A. (2019). Summary of the HKH Assessment Report Summary of the Hindu Kush Himalaya Assessment Report. www.icimod.org/himaldoc

Unwoman. (2022). *Pakistan floods 2010 rapid gender needs assessment.* https://www.unwomen.org/sites/default/files/Headquarters/Media/Publications/UNIFEM/PakistanFloods2010RapidGenderNeedsAssessmenten.pdf

Un Women UNICEF. (2016). *Gender and age inequality of disaster risks, policy brief.*

UN Women, (2020a). *Gender equality: Women's rights in review 25 years after Beijing.* New York: UN Women.

UNDRR. (2015). *Sendai framework 2015–2030.* https://www.undrr.org/quick/11409

UNFCCC. (2019). *Climate change is a global challenge that requires urgent and ambitious action.* Bonn: United Nations Framework Convention on Climate Change.

Voice of America. (2022). *Women and children affected most by Pakistan floods.* https://learningenglish.voanews.com/a/women-and-children-affected-most-by-pakistan-s-floods/6753486.html

Vujanovic, A. A., & Gallagher, M. W. (2017). *Psychological impact of natural disasters in adults.* University of Houston.

World Bank. (2018). *South Asia' 's prosperity will require more women to work for pay.*

World Bank. (2019). *Nepal gender brief.*

World Economic Forum. (2022). *Global gender gap report 2022.* https://www.weforum.org/reports/global-gender-gap-report-2022/

Analysis of Normalized Difference Vegetation Index (NDVI) of the Chittagong Hill Tracts from 1991 to 2021 Using Remote Sensing Tools

Tasnim Rahman, G. N. Tanjina Hasnat, Md Iftaker Chowdhury, Liton Chandra Voumik, Mohd Nazish Khan, and Muhammad Ashar Ayub

1 Introduction

Vegetation cover maintains natural ecosystems and various ecosystem services for human well-being (Haines-Young & Potschin, 2010; Liu et al., 2021; Wu, 2013; Xu et al., 2022). Plant cover also serves as the habitat of diversified fauna. Green covers both in urban and rural areas ameliorate air; regulate water and nutrient cycle; sequester carbon; absorb greenhouse gases; reduce the soil erosion, land degradation, and surface temperature; and balance microclimate (Ghorbanian et al., 2022; Xu et al., 2022). The reduction of healthy vegetation in an area negatively impacts the local climate as well as taking part in global climate change and global warming (Haines-Young & Potschin, 2010; Thomson et al., 2012; Zhao et al., 2016). Vegetation cover is changing due to the change of land use and land cover by human beings. Developing countries are facing the changes mostly as less land area against the huge population (Berihun et al., 2019; Hassan & Nazem, 2016; Talukdar et al., 2020). A systematic and planned change is crucial for the development of a nation,

T. Rahman · G. N. T. Hasnat (✉) · M. I. Chowdhury
Institute of Forestry and Environmental Sciences, University of Chittagong, Chattogram, Bangladesh
e-mail: gnthasnat@cu.ac.bd

L. C. Voumik
Department of Economics, Noakhali Science and Technology University, Noakhali, Bangladesh

M. N. Khan
Transformation Department, Samarkand State University, Samarkand City, Samarkand Region, Uzbekistan

M. A. Ayub
Institute of Agro-Industry and Environment, The Islamia University of Bahawalpur, Punjab, Pakistan

S. Tripathi et al. (eds.), *Warming Mountains*,
https://doi.org/10.1007/978-3-031-62197-0_18

but a drastic change in green cover causes an imbalance in the environment for living. Small-scale vegetation cover change can be analyzed by field survey, but large-scale vegetation cover changes are difficult, most of which are in remote and inaccessible areas. Information on large-scale vegetation cover changes is essential for planning for climate change and global warming. As a solution, different remote sensing tools are used nowadays for large-scale vegetation cover change monitoring and analysis, like Enhanced Vegetation Index (EVI), Normalized Difference Vegetation Index (NDVI), and Soil-Adjusted Vegetation Index (SAVI). Among these, NDVI proved more dependable (Ghorbanian et al., 2022; Zhen et al., 2021).

Normalized Difference Vegetation Index (NDVI) is a widely used remote sensing tool for large-scale vegetation cover analysis due to its high accuracy level. The tool is very much usable and reliable for remote areas like the mountainous regions. This index is widely used for monitoring and trend analysis of vegetation density, vegetation health, and land use land cover change over time (Ghorbanian et al., 2022; Zhen et al., 2021).

The Chittagong Hill Tracts (CHT), a distinctive area of Bangladesh, is recognized for its abundant biodiversity and beautiful scenery. The area covers only 10% of Bangladesh's total land mass but comprises almost 40% forest cover (Ahammad et al., 2022) and 80% of the total biodiversity of the country (Datta, 2015; Jashimuddin & Inoue, 2012). The area is renowned for its rich cultural heritage and variety of natural spaces (Hossain, 2013; Uddin, 2010). The hilly region is also the most well-known tourist destination of Bangladesh due to its natural environment (Haque et al., 2021; Hasan et al., 2020). There are 12 different ethnic communities in 3 districts, Rangamati, Khagrachari, and Bandarban, which make the region more diversified (Hasan et al., 2020; Rasul, 2015). Nonetheless, the area has experienced a massive land use land cover changes during the last few decades (Hasan et al., 2020). Among the major reasons, land crises and land conflicts between the stakeholders were important. Due to land disputes, many natural covers have been destroyed (Ahammad et al., 2022; Muhammed et al., 2012). The most recent reason is an upsurge in tourism, which has had a tremendous impact on the ecology and land use (Ahmed, 2017; Chakma et al., 2019). The other recent reason is horticulture or monoculture by deforesting natural covers (Hasnat & Bhadouria, 2023). Most of the hills which were highly forested once are found already barren in recent days. In the present study, the changes in green cover and other land cover of CHT are analyzed using NDVI.

The Normalized Difference Vegetation Index (NDVI) values for Rangamati, Khagrachari, and Bandarban districts were computed during the previous three decades using remote sensing data. The changes in healthy vegetation were detected to determine the impacts of land cover changes. The places where vegetation has increased or reduced were discovered and investigated with the potential causes of these changes by comparing the NDVI values of various time periods. Previously, some other studies were done to detect the land use and land cover changes in CHT due to tourism, but that was only to track the changes in the land cover and to analyze the dynamics and change in the built-up area of CHT (Anjum et al., 2021). Several studies have also been done by different researchers on NDVI analysis

(Ishtiaque et al., 2017; Islam & Mamun, 2015; Nath & Acharjee, 2013; Rafa et al., 2021; Rashid et al., 2023; Shimu et al., 2019) over many other areas of Bangladesh except overall CHT. In this analysis, it was attempted to find out how the NDVI of the CHT region has changed between 1991 and 2021, as well as how those changes are brought on by different land use land cover changes. The present study will provide a comprehensive understanding of the land use and land cover changes in the Chittagong Hill Tracts region over the past three decades, which can be useful for future land management, land conservation, and policy-making purposes.

2 Methodology

2.1 Study Site

The study was carried out in Rangamati, Khagrachari, and Bandarban, hilly areas of Bangladesh, which are situated at the country's southeast part (Fig. 1). The Chittagong Hill Tracts covers 13344.3 sq. km. in total, with Rangamati covering 6116.1 sq. km., Khagrachari 2749.2 sq. km., and Bandarban 4479 sq. km., (BBS, 2022). India surrounds them to the north and east, Myanmar to the southeast, and

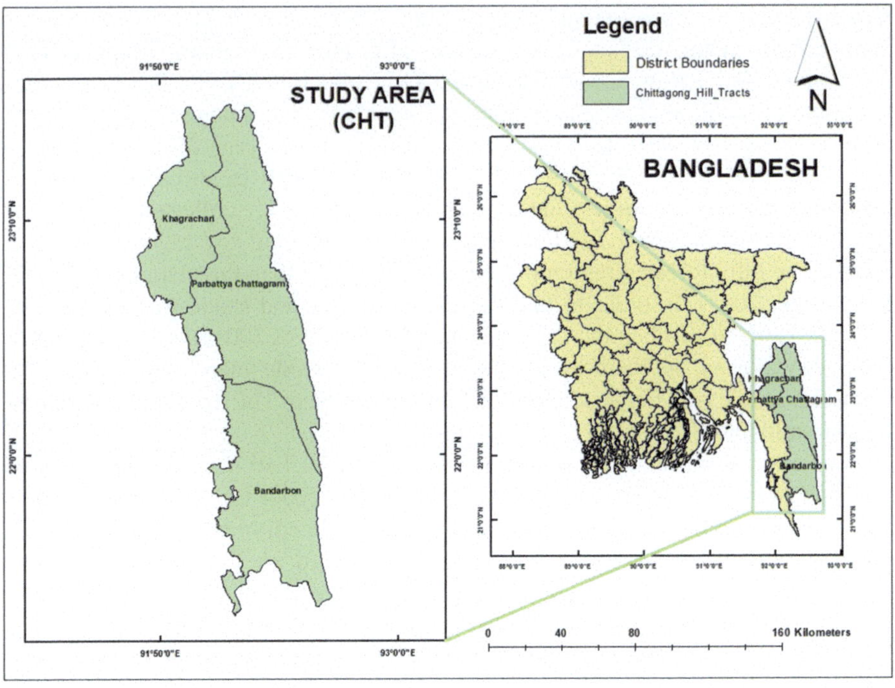

Fig. 1 Location of study area

Chattogram and Cox's Bazar districts of Bangladesh to the west and southwest, respectively. The area lies between 21°25′ N to 23°45′ N latitude and 91°54′ E to 92°50′ E longitude (Khisha et al., 2012).

The weather of this region is characterized by a tropical monsoon climate with a mean annual rainfall of nearly 2540 mm in the north and east and 2540–3810 mm in the south and west. The dry and cool season is November–March; the pre-monsoon season is April–May which is very hot and sunny; and the monsoon season is June–October which is warm, cloudy, and wet (Banglapedia, 2021). A mean monthly maximum temperature between 25 and 34 °C characterizes the tropical to subtropical nature of the area (Ahammad, 2019). The hill soils are mainly yellowish-brown to reddish-brown loams which grade into broken shale or sandstone as well as mottled sand at a variable depth. The soils are very strongly acidic. The hills are unsuitable for cultivation, but natural vegetation remains widely. Jhum cultivation is being practiced on the hill slopes. Cotton, rice, tea, and oilseeds are raised in the valleys between the hills (Banglapedia, 2021).

2.2 *Satellite Image Collection and Processing*

Images were collected from the United States Geological Survey (USGS) Earth Explorer (USGS, 2022). For the year 2021, images of Landsat 8 Collection 2 Tier 1 Raw Scenes were collected and filtered from 1 November 2021 to 31 December 2021. For the years 2001 and 1991, images of Landsat 5 Thematic Mapper (TM) Collection 2 Tier 1 Raw Scenes (Landsat 5 TM Collection 2 Tier 1 DN values represent scaled, calibrated at-sensor radiance) were collected. Imageries were filtered from 1 November 2001 to 31 December 2001 for the year 2001 and from 1 November 1991 to 31 December 1991 for the year 1991. In all cases, unwanted shade-free imageries with less than 1% cloud coverage were collected to get accurate data for classification. Finding imageries with no cloud coverage was difficult because the hilly forested region produces more clouds than the plain land (Basnet & Vodacek, 2015). In order to find out the cloud-free and shade-free images, the winter season (November and December) was chosen. For 2001, the use of Landsat 7 images was tried, but there were some stains on the whole image, due to which the Landsat 7 image was replaced with the Landsat 5 image. The details of the images are shown in Table 1.

The vector layer shape files of Rangamati, Khagrachari, and Bandarban were collected from DIVA-GIS country data. The required codes for the desired Landsat images were entered in the Google Earth Engine code editor tab. Then the dataset was filtered by date and cloud cover. It was determined whether or not the images covered the Area of Interest (AOI). When the image covered properly the study area, the shape files were imported into the asset section, and after that, the AOI was clipped. The image collection and processing were conducted in the Google Earth Engine (GEE).

Table 1 Details of the collected Landsat imageries

Features	1991	2001	2021
Spacecraft ID	USGS Landsat 5	USGS Landsat 5	USGS Landsat 8
Sensor type	TM	TM	OLI/TIRS
Collection	Collection 2 Tier 1 Raw Scenes	Collection 2 Tier 1 Raw Scenes	Collection 2 Tier 1 Raw Scenes
Filter date	1 November 1991 to 31 December 1991	1 November 2001 to 31 December 2001	1 November 2021 to 31 December 2021
Resolution	30 m	30 m	30 m
Cloud cover	Less than 1	Less than 1	Less than 1
Band	Band 3, Band 4	Band 4, Band 5	Band 4, Band 5

2.3 NDVI Calculation

NDVI was calculated in GEE by subtracting the near-infrared (NIR) band from the red band and dividing the results by the sum of these two bands. Therefore, the red and NIR bands from the satellite dataset needed to be selected. The equation was also needed to be given in the code. Finally, the NDVI data was exported as a *.tif* file with a spatial resolution of 30 m, and the ArcGIS software 10.4 was used for further analysis and processing, like area calculation, interpretation of the NDVI data, and final map preparation (Fig. 2).

2.4 NDVI-Based Land Use Land Cover (LULC) Classification and Area Calculation

The total landscape was divided into six different major classes in terms of use or cover: (i) water (−0.28–0.015); (ii) built-up area (0.015–0.14); (iii) barren land (0.14–0.18); (iv) agriculture and grassland (0.18–0.27); (v) sparse vegetation (0.27–0.36); and (vi) dense vegetation (0.36–0.7) (Akbar et al., 2019). After defining the NDVI values, the classifications were done based on the defined values. The areas for the classes were calculated in square kilometers for the years 1991, 2001, and 2021 using NDVI maps, and land cover percentages were also analyzed for each class. Land cover changes between the years and the districts were compared with each other. The land cover classification and area calculation were done by using ArcGIS Pro.

2.5 Accuracy Assessment and Ground Truthing

The results of land cover classification were compared with geospatial data in this process. To test the errors, the resulting land cover classes were compared with Google Earth maps for the study years 1991, 2001, and 2021. A total of 30 random

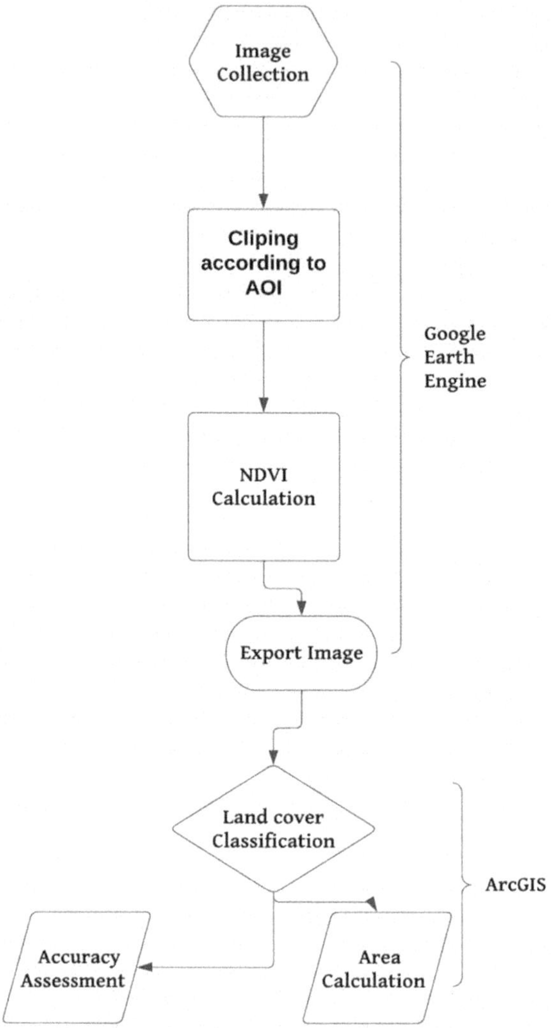

Fig. 2 Procedure of NDVI calculation

points for every year map were taken to conduct the assessment. Every classified map was compared with the same location points on the geospatial map of the corresponding year. Thus, accuracy assessment provides both overall map accuracy and preciseness for each class in the map.

Besides the accuracy assessment, ground truthing was also conducted for the classified map of Rangamati of 2021 for proper verification. Whether or not the classified land cover of the map existed in reality, the corresponding map was cross-checked with recent ground truth data. For this process, 16 points in Rangamati District were chosen randomly (Fig. 3), and coordinates were taken on those points with Global Positioning System (GPS). The data on the location of various ground

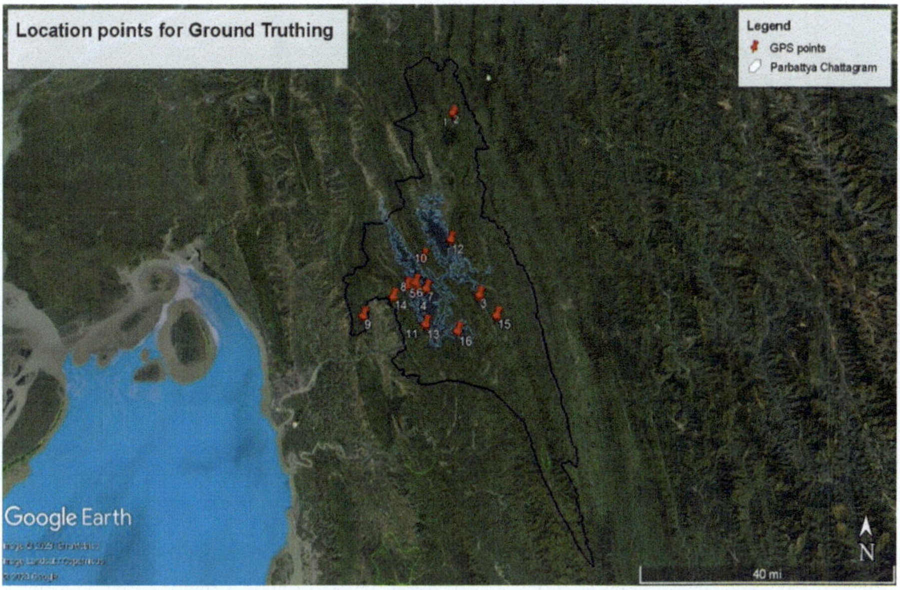

Fig. 3 Ground truthing locations

features were collected and marked on the reference map, and then comparison between the image elements and ground features was conducted.

3 Results

3.1 NDVI-Based LULC Classification

NDVI of all three districts was calculated to find out the vegetation density and changes in plant health over time. Furthermore, based on the NDVI values, a supervised classification for six land cover classes was derived for a distinct and defined demonstration of the classes. After classification, it was found that the dominant land use category in all three districts was dense vegetation during the study period but in a reducing rate, and it is surprising that most of the forest reduction occurred in the first decade from 1991 to 2001 than the next two decades from 2001 to 2021. Figure 4 is clear evidence of diminishing dense forest cover land from 1991 to 2021.

In Rangamati, the dense forest cover decreased from 84.10% in 1991 to 54.33% in 2021. In Khagrachari, its extent has decreased from 91.71% in 1991 to 49.69% in 2021, and in Bandarban, it reduced from 96% in 1991 to 63.40% in 2021. The results also revealed that the most healthy and dense vegetation was observed in Bandarban among the three districts. On the contrary, the sparse vegetation increased significantly in all three districts and was found as the second most dominant land

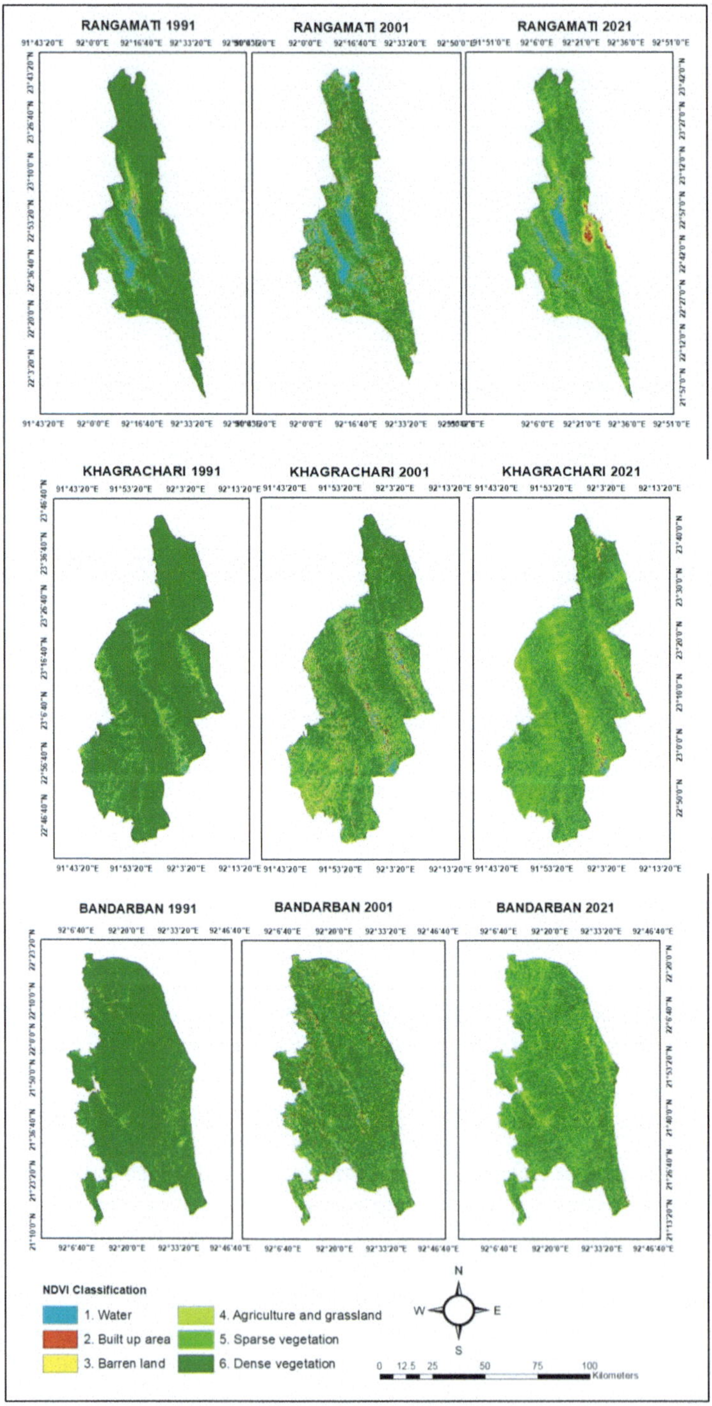

Fig. 4 Normalized Difference Vegetation Index (NDVI)-based land use land cover (LULC) of Rangamati, Khagrachari, and Bandarban districts in 1991, 2001, and 2021

use category. In Rangamati, the percentage of sparse vegetation increased from 5.43% to 26.23%; in Khagrachari, from 5.63% to 40.49%; and in Bandarban, from 2.40% to 29.20% between 1991 and 2021 (Fig. 5).

In terms of changes over time in Rangamati, it was observed that the built-up area has increased from 1.42% in 1991 to 5.50% in 2021, indicating a degree of development. Barren land has also increased over time, from 0.59% in 1991 to 1.96% in 2021. This may be the result of deforestation or other land degradation processes. Agriculture and grassland have also increased from 2.39% in 1991 to 7.70% in 2021, indicating that there has been an expansion of agricultural activities in the area (Fig. 5). Overall, the findings suggest that Rangamati has undergone a variety of changes in land use land cover over time, with an increase in built-up areas, agricultural land, barren land, and sparse vegetation. However, dense vegetation continues to cover a significant portion of the land area, although its extent has decreased over time (Figs. 4 and 5).

Considering the land use land cover changes in Khagrachari during the three decades, the built-up area increased from 0.10% in 1991 to 3.76% in 2021, indicating an urban development. Barren land has also increased over time, from 0.29% in 1991 to 0.62% in 2021. This is the result of deforestation or other land degradation processes. Agriculture and grassland have also shown a significant increase from 1.83% in 1991 to 9.93% in 2021, indicating that there has been an expansion of agricultural activities in the area (Fig. 5). We can visualize the obvious changes in land classes from the different time lines, in this case in 1991, 2001, and 2021. Overall, the results suggest that Khagrachari has undergone a significant change in land use land cover during the time period, with an increase in built-up areas, agricultural land, and barren land. However, dense vegetation continues to cover the major portion of the land area, although its extent has decreased over time. Sparse vegetation has also increased significantly (Fig. 5).

In terms of changes in Bandarban, we can see that the built-up area has increased from 0.9% in 1991 to 1.7% in 2021, indicating some degree of urban development. Barren land has also increased, from 0.2% in 1991 to 0.7% in 2021. This is the result of deforestation or other land degradation processes. Agriculture and grassland have seen a significant increase from 1% in 1991 to 6.2% in 2021, indicating that there has been some expansion of agricultural activities in the area. If we compare them side by side using a bar diagram (Fig. 5), we can visualize the obvious changes in land classes from different time lines, in this case in 1991, 2001, and 2021. Overall, the data suggest that Bandarban has undergone some changes in land use over time, with an increase in built-up areas and agricultural land. However, dense vegetation continues to cover the largest area in the region.

Overall, area changes in the three districts over the period of 1991–2021 are shown graphically in Fig. 6. In Rangamati, there has been an increase in the area of built-up land (101.49 sq. km.). There has also been an increase in the area of barren land (86.14 sq. km.), and agriculture and grassland (334.94 sq. km.). The area covered by sparse vegetation has increased (1311.28 sq. km.), while the area of dense vegetation has decreased (−1876.97 sq. km.). The area covered by water has increased slightly (43.13 sq. km.).

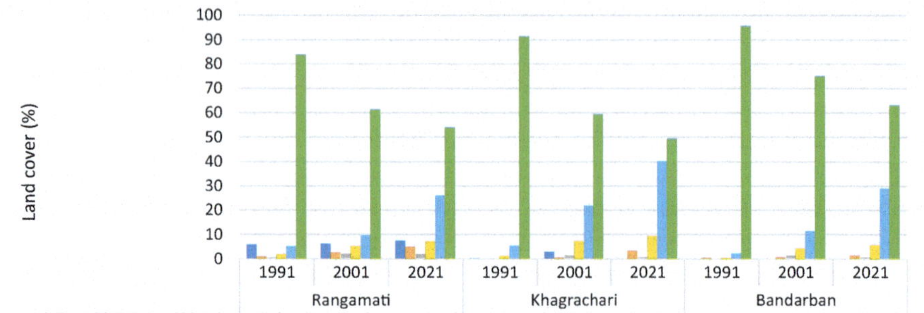

	Rangamati 1991	Rangamati 2001	Rangamati 2021	Khagrachari 1991	Khagrachari 2001	Khagrachari 2021	Bandarban 1991	Bandarban 2001	Bandarban 2021
Water	6.08	6.53	7.76	0.44	3.21	0.26	0.09	0.08	0.05
Built-up	1.42	3.03	5.50	0.10	1.00	3.76	0.90	1.20	1.70
Barren land	0.59	2.16	1.96	0.29	1.39	0.62	0.20	1.37	0.70
Agriculture and grassland	2.39	5.81	7.70	1.83	7.93	9.93	1.00	4.88	6.20
Sparse vegetation	5.43	9.97	26.23	5.63	21.97	40.49	2.40	11.64	29.20
Dense vegetation	84.10	61.54	54.33	91.72	59.72	49.69	96.00	75.32	63.40

Fig. 5 Land use land cover (LULC) of Rangamati, Khagrachari, and Bandarban districts from 1991 to 2021

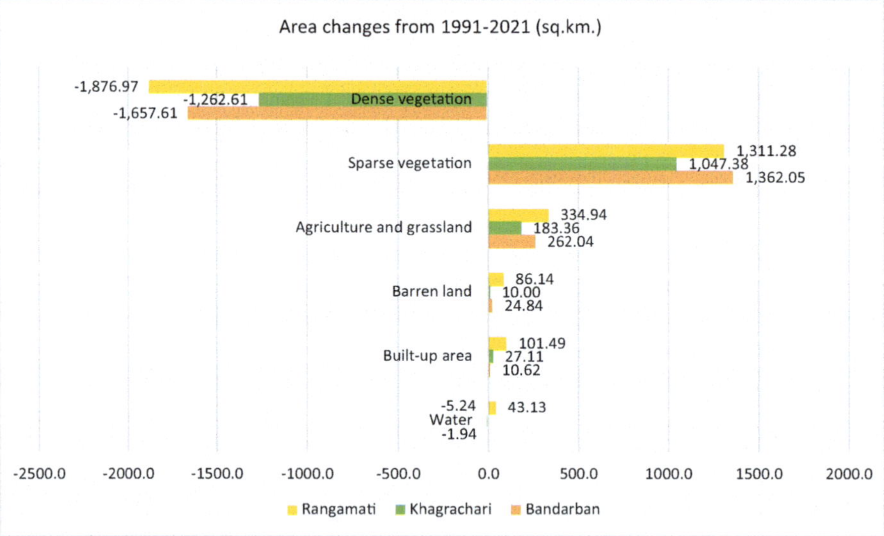

Fig. 6 Relative land use changes in Rangamati, Khagrachari, and Bandarban from 1991 to 2021

In Khagrachari, the area covered by built-up land has increased significantly (27.11 sq. km.). There has also been an increase in the area of barren land (10 sq. km.), and agriculture and grassland (183.36 sq. km.). The area covered by sparse vegetation has increased (1047.38 sq. km.), while the area of dense vegetation has decreased significantly (−1262.61 sq. km.). The area covered by water has decreased

(−5.24 sq. km.) also in Khagrachari (Fig. 6). In Bandarban, there has been a significant increase in the area of sparse vegetation (1362.05 sq. km.). There has also been an increase in the area of agriculture and grassland (262.04 sq. km.). The area covered by built-up land has increased slightly (10.62 sq. km.), and there has also been an increase in the area of barren land (24.84 sq. km.). However, a significant decrease in the dense vegetation (−1657.61 sq. km.) is observed in Bandarban and also a slight decrease in the water-covered area (−1.94 sq. km.).

3.2 Accuracy Assessment

The user accuracy and producer accuracy for the three districts in three different time frames (1991, 2001, and 2021) are presented in Table 2. The highest user accuracy in the table is for the water class in all three districts, which was 100% in 1991 and remained high in 2021 with 100% accuracy in all districts except for a significant decrease to 66.7% in Bandarban. This indicates that the classification method was able to accurately identify water bodies in these districts over the years. The lowest user accuracy in the table is for the sparse vegetation class found in Khagrachari, 50% in 1991, 20% in 2001, and 75% in 2021. This suggests that the classification method had more difficulty identifying sparse vegetation in the area compared to other land cover classes and in other areas. As for the producer accuracy, the highest accuracy in the table is for the built-up class, which was classified with almost 100% accuracy in all three districts in the study period. This indicates that the classification method was able to accurately identify built-up areas in these districts at that time. The lowest producer accuracy in the table is for the sparse vegetation class in 1991, with only 50% accuracy in all three districts and 33.3% accuracy in Khagrachari in 2001. This suggests that the classification method had more difficulty identifying sparse vegetation in these districts at that time, and the accuracy improved in later years.

The summary of overall classification accuracy and Kappa coefficient for the three different time frames (1991, 2001, and 2021) for the three different regions (Rangamati, Khagrachari, and Bandarban) is shown in Table 3. The highest accuracy is found for the year 2021 in the Bandarban area (92%) and the lowest for the year 1991 in the Khagrachari area (72%). The Kappa coefficient values range from 66.09% to 90.06% for different regions and years. In the year 1991, the Kappa coefficient was the lowest for Khagrachari (66.09%) and the highest for Bandarban (85.44%). However, in the year 2001, the Kappa coefficient was the lowest for Rangamati (70.06%) and the highest for Bandarban (84.97%). Further, in the year 2021, the Kappa coefficient was the lowest for Khagrachari (74.44%) and the highest for Bandarban (90.06%). Moreover, the Kappa coefficient values for all three regions have generally improved from 1991 to 2021, indicating an improvement in the classification algorithm's performance.

Table 2 User accuracy for Rangamati, Khagrachari, and Bandarban

Classes	1991 (%)			2001 (%)			2021 (%)		
	Rangamati	Khagrachari	Bandarban	Rangamati	Khagrachari	Bandarban	Rangamati	Khagrachari	Bandarban
	User accuracy								
1. Water	100	100	100	83.3	75	66.7	100	100	100
2. Built-up area	66.7	50	100	75	100	66.7	50	50	66.7
3. Barren land	66.7	80	100	50	75	100	100	100	100
4. Agriculture and grassland	100	75	40	66.7	100	100	100	83.3	100
5. Sparse vegetation	100	50	100	100	20	100	100	75	66.7
6. Dense vegetation	77.8	66.7	100	71.3	100	100	83.3	75	100
	Producer accuracy								
1. Water	66.7	100	100	100	100	100	60	100	100
2. Built-up area	100	50	100	100	100	100	100	100	100
3. Barren land	100	80	85.7	100	80	100	100	75	100
4. Agriculture and grassland	100	50	100	50	57.1	100	100	71.4	80
5. Sparse vegetation	50	50	50	75	33.3	80	87.5	75	100
6. Dense vegetation	100	100	100	62.5	100	77.8	100	100	87.5

Table 3 Overall accuracy and Kappa coefficient of the three hilly districts: Rangamati, Khagrachari, and Bandarban

	Overall accuracy			Kappa coefficient		
Year	Rangamati	Khagrachari	Bandarban	Rangamati	Khagrachari	Bandarban
1991	84%	72%	88%	80.08%	66.09%	85.44%
2001	76%	76%	88%	70.06%	71.15%	84.97%
2021	88%	80%	92%	83.96%	74.44%	90.06%

Table 4 Ground truthing points accuracy assessment for Rangamati

Classes	Water	Built-up area	Barren land	Agriculture and grassland	Sparse vegetation	Dense vegetation	Total (%)
Matched proportion (%)	100	50	66.7	100	100	100	81.25

3.3 Ground Truthing

The data in Table 4 shows the matched proportion of the ground truthing points at Rangamati District measured by the percentage of different land use land cover pixels on the classified maps that correspond to the actual land use land cover types on the ground. The table lists six land cover types: water, built-up area, barren land, agriculture and grassland, sparse vegetation, and dense vegetation. The total number of ground points taken in the study was 16, out of which 13 points were matched on the classified map corresponding to 81.25%. The land cover type with the highest matched proportion found water, agriculture and grassland, sparse vegetation, and dense vegetation with 100% accuracy. The land cover type with the lowest matched proportion were built-up area and barren land, with 50% and 66.7% accuracy, respectively.

4 Discussion

The results of the study indicate significant changes in land use patterns in CHT between 1991 and 2021. NDVI analysis revealed a decrease in dense vegetation cover and increase in sparse vegetation, which is consistent with the increasing tourism and population growth in the region. This trend is further supported by the results of tourist and local surveys, which indicated a perceived increase in infrastructure development and a decrease in forest cover. A study by Kafy et al. (2017) reported a significant loss of hill forest coverage (18%) from 76% to 58% that occurred from 1995 to 2015. At the same time, the crop and shrub land and the built-up area have increased by almost double for each (Kafy et al., 2017). These findings support the present study where the dense vegetation cover decreased and built-up area increased at the same time in all three districts. The increased tourism, urban development, and land conversion are found as the main reasons of the changes. Another study (Rahman, 2019) on land use change at Bandarban showed that the water class has shrunk from 149 square kilometers in 2000 to 45 square kilometers in 2019, whereas the bare land class has grown from 168 square kilometers in 2000 to 184 square kilometers in 2019, which also support the findings of the present study.

A study (Rahman et al., 2021) of the land use land cover changes in Rangamati stated that the highest value of NDVI was discovered in 1977 (0.88), which means healthy vegetation. Thereafter, it followed a trend of decline valued at 0.79 in 1989, 0.74 in 2000, 0.71 in 2011, and 0.53 in 2019, which demonstrates a quick shift in the studied area's plant cover. Another research (Hasnat & Bhadouria, 2023) on the prediction of land cover changes in Khagrachari stated that vegetation coverage increased by nearly 24.24% between 1990 and 2020. This report also supports the findings of the present study where the sparse vegetation coverage increased by 34.86% during 1991–2021.

5 Conclusion

The Chittagong Hill Tracts (CHT) in Bangladesh has experienced considerable changes in vegetation cover and quality over the past three decades, as revealed by an analysis of the Normalized Difference Vegetation Index (NDVI). In places with considerable tourism activities and extension of built-up area, the healthy vegetation coverage in the CHT has decreased during the time period. The effects of land conversion on vegetation differ across the regions, according to a comparative investigation of NDVI variations in the districts of Rangamati, Khagrachari, and Bandarban. The unique natural and cultural heritage of the CHT must be preserved, and this requires the development of sustainable urban development and tourism. To fully comprehend how land conversion and tourism affect the vegetation and ecosystems of the CHT, research gaps must yet be filled. More investigations are needed in a number of areas, including long-term monitoring, regional variability, establishing causal linkages, community viewpoints, and policy implications. Hence, a policy framework that encourages sustainable urban development while preserving the environment and safeguarding the welfare of local communities is needed to ensure the long-term sustainability of the region. Besides CHT, *Sylhet* hilly region of Bangladesh and other mountainous areas of the world are under huge pressure from population density, tourism industries, land grabbing, and land conversion. The NDVI and LULC analysis for these areas are also essential for proper land management.

References

Ahammad, R. (2019). *Recent trends in forest and livelihood relationships of rural communities in the Chittagong Hill Tracts region, Bangladesh* (Doctoral dissertation, Charles Darwin University (Australia)). ProQuest Dissertations Publishing. 28911666. https://doi.org/10.25913/5e4f60e93cb55.

Ahammad, R., Stacey, N., Sunderland, T., & Sangha, K. K. (2022). Land use preference for ecosystem services and well-being in Chittagong Hill Tracts of Bangladesh. *Forests, 13*(12), 2086. https://doi.org/10.3390/f13122086

Ahmed, H. S. (2017). *Tourism and state violence in the Chittagong Hill Tracts of Bangladesh* (Doctoral dissertation, The University of Western Ontario (Canada)). Electronic Thesis and Dissertation Repository. 4840. https://ir.lib.uwo.ca/etd/4840

Akbar, T. A., Hassan, Q. K., Ishaq, S., Batool, M., Butt, H. J., & Jabbar, H. (2019). Investigative spatial distribution and modelling of existing and future urban land changes and its impact on urbanization and economy. *Remote Sensing, 11*(2), 105. https://doi.org/10.3390/rs11020105

Anjum, F., Zilany, H. M., Rakin, S. S. A., Hoque, M. A., Ania, A. N., Zaman, M. A., Moni, J. N., & Rahman, R. M. (2021). Land use/land cover change analysis due to tourism in the Chittagong Hill Tracts of Bangladesh. In *Advances and trends in artificial intelligence. From theory to practice: 34th international conference on industrial, engineering and other applications of applied intelligent systems, IEA/AIE 2021, Kuala Lumpur, Malaysia, July 26–29, 2021, proceedings, part II 34* (pp. 173–184). Springer International Publishing. https://doi.org/10.1007/978-3-030-79463-7_15

Banglapedia. (2021). *Chittagong Hill Tracts.* https://en.banglapedia.org/index.php/Chittagong_Hill_Tracts. Viewed on: 28.03.2023.

Basnet, B., & Vodacek, A. (2015). Tracking land use/land cover dynamics in cloud prone areas using moderate resolution satellite data: A case study in Central Africa. *Remote Sensing, 7*(6), 6683–6709. https://doi.org/10.3390/rs70606683

BBS. (2022). *Statistical yearbook Bangladesh 2021* (41th ed.). Bangladesh Bureau of Statistics (BBS) Statistics & Informatics Division (SID), Ministry of Planning Government of the People's Republic of Bangladesh Dhaka, Bangladesh. https://bbs.portal.gov.bd/sites/default/files/files/bbs.portal.gov.bd/page/b2db8758_8497_412c_a9ec_6bb299f8b3ab/2022-06-15-10-49-3cf641425dd693f9e954de5ae9470775.pdf. Viewed on: 28.03.2023.

Berihun, M. L., Tsunekawa, A., Haregeweyn, N., Meshesha, D. T., Adgo, E., Tsubo, M., Masunaga, T., Fenta, A. A., Sultan, D., & Yibeltal, M. (2019). Exploring land use/land cover changes, drivers and their implications in contrasting agro-ecological environments of Ethiopia. *Land Use Policy, 87*, 104052. https://doi.org/10.1016/j.landusepol.2019.104052

Chakma, S., Chakma, G., & Chowdhury Nani, N. (2019). Growth of tourism and hospitality industry: Impact on environment and demographic trend of Chittagong Hill Tracts. *Asian Journal of Economics, Business and Accounting, 12*(4), 1–8. https://doi.org/10.9734/ajeba/2019/v12i430159

Datta, R. (2015). *Land-water management and sustainability: An indigenous perspective in Laitu Khyang community, Chittagong Hill Tracts (CHT), Bangladesh.* (Doctor of Philosophy (Ph. D.)). University of Saskatchewan, Saskatoon, SK, Canada.

Ghorbanian, A., Mohammadzadeh, A., & Jamali, S. (2022). Linear and non-linear vegetation trend analysis throughout Iran using two decades of MODIS NDVI imagery. *Remote Sensing, 14*(15), 3683. https://doi.org/10.3390/rs14153683

Haines-Young, R., & Potschin, M. (2010). The links between biodiversity, ecosystem services and human well-being. In D. G. Raffaelli & C. L. Frid (Eds.), *Ecosystem ecology: A new synthesis* (Vol. 1, pp. 110–139). Cambridge University Press.

Haque, M. N., Baroi, A., Gomes, J., Toppo, A., Dasand, R. S., & Hossain, M. K. (2021). Food security in Chittagong Hill Tracts (CHT), Bangladesh and way forward for achieving through sustainable agricultural practices. *International Journal of Information Research and Review, 8*(9), 7330–7340.

Hasan, S. S., Sarmin, N. S., & Miah, M. G. (2020). Assessment of scenario-based land use changes in the Chittagong Hill Tracts of Bangladesh. *Environmental Development, 34*, 100463. https://doi.org/10.1016/j.envdev.2019.100463

Hasnat, G. N. T., & Bhadouria, R. (2023). Prediction of land cover changes of Khagrachhari Hilly Upazila using artificial-neural-network-based cellular automata model. In *Understanding soils of mountainous landscapes* (pp. 325–343). Elsevier. https://doi.org/10.1016/B978-0-323-95925-4.00003-0

Hassan, M. M., & Nazem, M. N. I. (2016). Examination of land use/land cover changes, urban growth dynamics, and environmental sustainability in Chittagong city, Bangladesh. *Environment, Development and Sustainability, 18*, 697–716. https://doi.org/10.1007/s10668-015-9672-8

Hossain, D. M. (2013). Socio-economic condition of the indigenous people in the Chittagong Hill Tracts (CHT) of Bangladesh. *Middle East Journal of Business, 8*(2), 22–30. https://doi.org/10.5742/MEJN.2013.82231

Ishtiaque, T. A., Tasin, Z. T., & Akter, K. S. (2017). Urban heat Island intensity assessment through comparative study on land surface temperature and normalized difference vegetation index: A case study of Chittagong, Bangladesh. *International Journal of Urban and Civil Engineering, 11*(1), 37–42. https://doi.org/10.5281/zenodo.1128987

Islam, M. M., & Mamun, M. M. I. (2015). Variations of NDVI and its association with rainfall and evapotranspiration over Bangladesh. *Rajshahi University Journal of Science and Engineering, 43*, 21–28. https://doi.org/10.3329/rujse.v43i0.26160

Jashimuddin, M., & Inoue, M. (2012). Management of village common forests in the Chittagong Hill Tracts of Bangladesh: Historical background and current issues in terms of sustainability. *Open Journal of Forestry, 2*(3), 121–137. https://doi.org/10.4236/ojf.2012.23016

Kafy, A. A., Rahman, M. S., & Ferdous, L. (2017). Exploring the association of land cover change and landslides in the Chittagong hill tracts (CHT): A remote sensing perspective. In *Proceedings of the international conference on disaster risk mitigation, Dhaka, Bangladesh* (Vol. 23).

Khisha, T., Karim, R., Chowdhury, S. R., & Banoo, R. (2012). Ethnomedical studies of Chakma communities of Chittagong hill tracts, Bangladesh. *Bangladesh Pharmaceutical Journal, 15*(1), 59–67.

Liu, R., Dong, X., Wang, X. C., Zhang, P., Liu, M., & Zhang, Y. (2021). Study on the relationship among the urbanization process, ecosystem services and human well-being in an arid region in the context of carbon flow: Taking the Manas river basin as an example. *Ecological Indicators, 132*, 108248. https://doi.org/10.1016/j.ecolind.2021.108248

Muhammed, N., Hossain, M., Chakma, S., Masum, F. H., von Detten, R., & Oesten, G. (2012). Conflict and corollaries on forest and indigenous people: Experience from Bangladesh. *Sustainable Forest Management—Case Studies.*

Nath, B., & Acharjee, S. (2013). Forest cover change detection using normalized difference vegetation index (NDVI): A study of Reingkhyongkine lake's adjoining areas, Rangamati, Bangladesh. *Indian Cartographer, 33*(2), 348–403. https://www.researchgate.net/publication/271908557

Rafa, N., Nuzhat, S., Uddin, S. M. N., Gupta, M., & Rakshit, R. (2021). Ecotourism as a forest conservation tool: An NDVI analysis of the Sitakunda Botanical Garden and Ecopark in Chattogram, Bangladesh. *Sustainability, 13*(21), 12190. https://doi.org/10.3390/su132112190

Rahman, I. (2019). *A report on the comprehensive geological field work at Bandarban Anticline, Bandarban, Bangladesh* (Report, University of Dhaka). Dhaka, Bangladesh. p. 94.

Rahman, M. M., Hossain, K. M., Islam, M. T., & Zahid, D. (2021). Land use/land cover changes monitored by NDVI index in Rangamati, Bangladesh for the last four decades. *Bangladesh Journal of Agriculture, 2019–2021, 44–46*, 127–140. https://doi.org/10.3329/bjagri.v46i1-6.59980

Rashid, M. B., Sheik, M. R., Haque, A. E., Siddique, M. A. B., Habib, M. A., & Patwary, M. A. A. (2023). Salinity-induced change in green vegetation and land use patterns using remote sensing, NDVI, and GIS techniques: A case study on the southwestern coast of Bangladesh. *Case Studies in Chemical and Environmental Engineering, 7*, 100314. https://doi.org/10.1016/j.cscee.2023.100314

Rasul, G. (2015). *A strategic framework for sustainable development in the Chittagong Hill Tracts of Bangladesh*. ICIMOD Working Paper, (2015/3). Ministry of Chittagong Hill Tracts Affairs Government of the People's Republic of Bangladesh. Kathmandu: ICIMOD. https://doi.org/10.13140/RG.2.1.1196.2963.

Shimu, S. A., Aktar, M., Afjal, M. I., Nitu, A. M., Uddin, M. P., & Al Mamun, M. (2019). NDVI based change detection in Sundarban Mangrove Forest using remote sensing data. In *2019 4th international conference on electrical information and communication technology (EICT)* (pp. 1–5). IEEE. https://doi.org/10.1109/EICT48899.2019.9068819

Talukdar, S., Singha, P., Mahato, S., Praveen, B., & Rahman, A. (2020). Dynamics of ecosystem services (ESs) in response to land use land cover (LU/LC) changes in the lower Gangetic plain of India. *Ecological Indicators, 112*, 106121. https://doi.org/10.1016/j.ecolind.2020.106121

Thomson, J. R., Bond, N. R., Cunningham, S. C., Metzeling, L., Reich, P., Thompson, R. M., & Mac Nally, R. (2012). The influences of climatic variation and vegetation on stream biota: Lessons from the Big Dry in southeastern Australia. *Global Change Biology, 18*(5), 1582–1596. https://doi.org/10.1111/j.1365-2486.2011.02609.x

Uddin, N. (2010). Politics of cultural difference: Identity and marginality in the Chittagong hill tracts of Bangladesh. *South Asian Survey, 17*(2), 283–294. https://doi.org/10.1177/097152311201700206

USGS. (2022). *United States Geological Survey (USGS)*. Land use land cover images. https://earthexplorer.usgs.gov/. Access on 23.12. 2022.

Wu, J. (2013). Landscape sustainability science: Ecosystem services and human well-being in changing landscapes. *Landscape Ecology, 28*, 999–1023. https://doi.org/10.1007/s10980-013-9894-9

Xu, Y., Yang, Y., Chen, X., & Liu, Y. (2022). Bibliometric analysis of global NDVI research trends from 1985 to 2021. *Remote Sensing, 14*(16), 3967. https://doi.org/10.3390/rs14163967

Zhao, S., Liu, S., & Zhou, D. (2016). Prevalent vegetation growth enhancement in urban environment. *Proceedings of the National Academy of Sciences, 113*(22), 6313–6318. https://doi.org/10.1073/pnas.1602312113

Zhen, Z., Chen, S., Yin, T., Chavanon, E., Lauret, N., Guilleux, J., Henke, M., Qin, W., Cao, L., Li, J., & Gastellu-Etchegorry, J. P. (2021). Using the negative soil adjustment factor of soil adjusted vegetation index (SAVI) to resist saturation effects and estimate leaf area index (LAI) in dense vegetation areas. *Sensors, 21*(6), 2115. https://doi.org/10.3390/s21062115

Communal Arrangements for Sustainably Managing Soil Quality and Adapting to Climate Change: A Case Study in the Peruvian Central Andes

Federica Ravera and David Tarrasón

1 Introduction

Mountain systems show specific characteristics, including not only a high degree of inaccessibility, fragility, marginality and diversity (or heterogeneity) but also unique production opportunities with comparative advantages over other regions (Jodha, 1992). The fragility of mountain systems is due to their biophysical, topographical and edaphic characteristics, which are responsible for the great sensitivity of their renewable resources and possibly irreversible degradation (Jodha, 1998). The latest report from the Intergovernmental Panel on Climate Change (IPCC, 2022) on impacts, adaptation and vulnerability confirms that climate change increases challenges to the good sand services that mountains offer. Specifically, mountain soils are highly susceptible to land use pressures at different intensities, and this may lead to their degradation and erosion, as demonstrated in several studies around the globe (see, for instance, Correa et al., 2016; Mishra et al., 2022; Bhadouria et al., 2023). In turn, this may affect their multiple functions and the provision of key services for the livelihoods of mountain people (Food and Agriculture Organization [FAO], 2015; Martín-López et al., 2019). Furthermore, extreme climate events may also enhance soil erosion and degradation (Egli & Poulenard, 2016).

Although mountain systems have a limited capacity to respond to climate changes, mainly due to high natural risks and the structural conditions of social sensitivity, some authors recognize that local communities have differential capacities for adaptation (Marston, 2008; McDowell et al., 2013). Examples of the strong resilience and adaptation of mountain communities in the face of multiple

F. Ravera (✉)
Geography Department, University of Girona, Girona, Catalunya, Spain
e-mail: Federica.ravera@udg.edu

D. Tarrasón
Indipendent Researcher, Barcelona, Catalunya, Spain

S. Tripathi et al. (eds.), *Warming Mountains*,
https://doi.org/10.1007/978-3-031-62197-0_19

pressures, including extreme climatic hazards, are mainly based on adjustments and learning processes, the use of indigenous and local knowledge and self-organization and multiscale institutional arrangements at several scales. Specifically, mountain communities have shown the capacity to adapt and develop strategies and practices to maintain mountain soil functionality and, thus, contribute to the resilience of mountain systems in the face of climate change (FAO, 2015; Mishra et al., 2021). Further research is needed to understand how local mountain communities in specific contexts have addressed climate variability and other pressures over time and particularly how their land use and management arrangements have evolved and how this may affect the final conditions of the conservation of mountain soil key functions. Through a specific case study, this chapter addresses this gap of context-specific knowledge in the Central Peruvian Andes.

Soil degradation, one of the most frequent natural threats to Peruvian Andean Mountain systems, is related to abrupt slopes, scarce vegetation cover and irregular but highly intense rainfall, which, together with thin soil depths and lower fertility than flatter areas, increases the risk for soil erosion (Correa et al., 2016). Moreover, according to Torres-Guevara and Velásquez (2007), climatic instability in Peruvian high mountain ecosystems makes weather forecasting more difficult. Despite these limitations, historically, local Peruvian Andean Mountain communities have exhibited the ability to overcome climate variability and natural risks (Sietz & Feola, 2016), maintaining ecological stability and productivity, and ensure many functions and benefits, i.e. the multifunctionality of the agroecosystem, through ingenious soil management and conservation practices (Magrin et al., 2014). Examples of adaptive capacity strategies are the implementation of a series of soil and water conservation practices, such as the system of terraces and self-organized irrigation systems (Koohafkan & Altieri, 2010); the use of run-off control and soil management (Kessler, 2006); the adoption of conservative tillage technologies like the Andean foot plough, also known as the *chaquitaclla* (Gade & Rios, 1972); and the functioning of traditional communal land management organization such as the *Laymi* system. Orlove and Godoy (1986) defined the *Laymi system* as a complex pattern of sectoral rotations and associations of crops and pasture use, where some sectors are allocated by the community for cultivation and other sectors are left unseeded for a period (fallow soils) and used for community grazing (Kendall & Drew, 2019). This form of communal resource management is a successful method for restoring soil fertility (Styger & Fernandes, 2006) and requires specific behavioural norms, management principles, time orientation and social relations (Brush & Guillet, 1985). In addition, the Peruvian Andean populations have been characterized by a remarkable capacity to overcome the climatic risks of frost and drought thanks to articulated vertical control of ecological zoning, the adaptation of the agricultural calendar and the management of soil humidity (Morlon, 1996; Dollfus, 1991; Murra, 1985, 2002).

Moreover, abrupt changes to political and socio-economic systems in the Peruvian Andes over time have influenced land tenure, food-agrarian systems and local institutions, leading to land use and management changes and progressive erosion of traditional ecological knowledge and practices (Halloy et al., 2005). These changes have disrupted communal rules, which has affected the multifunctionality of agroecosystems. This has led to land degradation and limited the sustainable maintenance of ecosystem services (Madrigal-Martínez & Miralles i García, 2019). Our work links to the literature on community-based arrangements and communal systems of management and the analysis of conditions that successfully (re)produced and sustained in the face of disturbances, such as climate or political changes. Various studies on pastoral management, fishery, forestry and irrigated farming have shown that communal system arrangements are not a panacea and successes are specific to the local contexts and histories of each case (Ostrom & Cox, 2010). Despite the heterogeneity of Peruvian cases, we focused on the process of "commoning" or the act of "doing the common things" on a local scale (Bollier & Helfrich, 2019). This is defined as a process of common people undertaken in specific circumstances when they bond with each other to create as much wealth and wellbeing as possible for everyone. It converges with the concept of "communalization" proposed by Díez Hurtado (1998) in the case of Peruvian peasant communities. Therefore, this research specifically acknowledges that in the Peruvian Andes, the historical community as an institution and the communal management, i.e. the *Laymi*, have changed in response to socio-economic and environmental drivers, surprises and crises. We worked with the hypothesis that communities, in some cases, have maintained their communal organization to guarantee social and ecological benefits, for example, those linked to the capacity to maintain ecological key functions related to the conservation of soil quality over time.

To understand the interactions between socio-institutional drivers of change and ecological impacts in a complex adaptive system, we developed an interdisciplinary and participatory framework to identify how communal use and management of agropastoral lands has evolved in the face of different drivers of change and compare the effects of such evolution on soil quality in two communities in Huancavelica, Central Peruvian Andes. Soil quality was considered a proxy for the description of the capacity of soil to provide the multiple key functions of the agroecosystem, and it was assessed by a set of sensitive soil attributes (Bünemann et al., 2018). These attributes may include biochemical indicators, and those indicators are based on indigenous knowledge and stakeholders' assessment (Barrera-Bassols & Zinck, 2003).

This study aimed to (1) explore land use and communal management changes over time through document analysis and interviews, (2) bridge local and scientific knowledge to assess soil quality and (3) identify adaptive strategies to respond to new challenges for mountain areas.

2 Methods

2.1 Study Area

The study area (Fig. 1) is located in the highlands of the Huancavelica Department in Yauli District (12°45′57.01″S, 74°47′30.75″W). The climate is characterized by two differentiated seasons: *Usyai ukuu* (April to October) is characterized by sunny days, very cold nights (with frequent frosts) and the absence of rain, and *Pukuy uku* (November to March) is characterized by rainfall and warmer temperatures (Valladolid Rivera, 1994). A feature that characterizes the Andean climate is its great interannual variability, with untimely periods of drought, frost and excess rainfall. Two communities, Paltamachay (hereafter A) and Cceroncancha (hereafter B), were selected after field visits to the area and considering information from an international not-governmental organization (NGO), i.e. Agronomes et Vétérinaires Sans Frontières (AVSF), who had worked in the area for a long time. These communities, located in a micro-basin of the Ichu River, present similar biophysical conditions but different institutional systems and management practices.

In the Peruvian context, a "Community" has been defined by Díez Hurtado (1998: 9) as a "group of commoners who share the same territory and maintain social relationships of kinship and reciprocity, and they are organized by communal decision-making institutions and arrangements". According to the last census (INEI, 2018), in Peru, there were 7267 communities (6138 officially recognized), and they occupied 42% of the surface of the country. Specifically, in the Huancavelica region, 64% of inhabitants recognized themselves as commoners. Despite their social and demographic importance, they have been largely invisible in public and political debate. The two selected communities have around 30 families each.

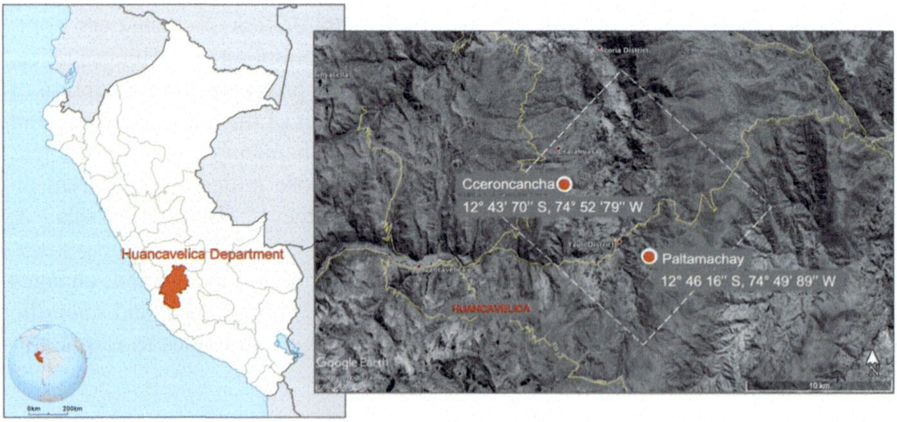

Fig. 1 Study area. The location of the two sampled communities in the Huancavelica Department (Central Andes). (Own elaboration)

Three agro-ecological regions are recognized in the Andes based on topography (Murra, 2002). The study area is located in the regions of Suni (between 3500 and 4000 m asl) and Puna (>4000 m asl). The main crops of the Suni region are various species and many varieties of potatoes (e.g., *Solanum tuberosum*, *S. andigenum*, *S. curtilobum*), Andean tubers (mashua or *Tropaeolum tuberosum*, oca or *Oxalis tuberosa* and olluco or *Ullucus tuberosus*), cereals (e.g., barley or *Hordeum vulgare L.*, wheat or *Triticum sp.*, quinoa or *Chenopodium quinoa*) and legumes (e.g., broad bean or *Vicia faba L.* and pea or *Pisum sativum*). In this zone, agriculture is accompanied by raising cattle, *Chusco* sheep and small livestock, e.g. poultry, pigs and guinea pigs, for family consumption used occasionally for sale. In the Puna region, the main crops are tubers (e.g., yellow potatoes or *Solanum goniocalyx* and bitter potatoes or *Solanum juzepczukii*) and barley. Natural pastures also prevail, with semi-extensive sheep and alpaca herds.

Most of the soils in the study area are poorly developed and derived from limestone and shallow rocks. They are strongly washed inceptisols and entisols (<1 mm lithics). These soils were developed under a xeric regime, with a base saturation of <50%, acid soil (pH < 6 in all studied soils) and other characteristics, including percentage of sand, percentage of organic matter (OM), horizon diagnosis, clay accumulation, decarbonization and colour. To protect this layer of fragile fertile soil and minimize soil erosion and landslides, the Andean foot plough, also known as a *chaquitaclla* (*chaqui* = foot and *taclla* = plough), is widely used in the Andean region.

2.2 Methodological Process and Tools

The participatory and integrated assessment process used in this research combined methods and tools from different disciplines that bridged local and scientific knowledge in three main steps, as illustrated in Fig. 2.

Step I: Historical and institutional analysis and agroecosystem characterization

This first step allowed us to validate the hypothesis suggested by the literature review and design the experimental sampling for the soil quality assessment in Step II.

On the basis of information provided by AVSF, we selected two communities in the same area with different historical socio-institutional changes. We applied Ostrom and Hess's (2008) "bundle of rights" classification to identify the organization of resource access and withdrawal, management, alienation and exclusion within communities. According to Ribot and Peluso (2003), access is extended to not only ownership but also how people gain benefits from resources and shape their control and power through authorities, social identities and social relationships (as in the case of reciprocity norms); rights of management refers to how people decide on the use and transformation of resources; alienation refers to the right of selling/buying resources; and exclusion refers to the right of transferability and the inheritance of access rights.

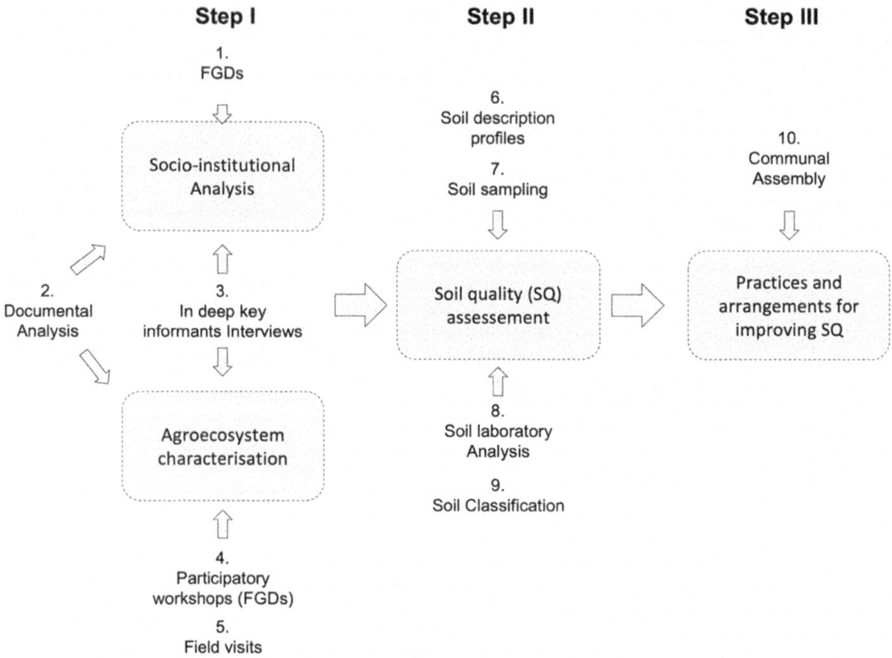

Fig. 2 Methodological framework adopted in the study. (Own elaboration)

A documentation analysis was conducted to systematize historical information and identify linkages between historical events and social-institutional changes in the two communities.

We first organized two focus group discussions (hereafter FGDs), one per community ($N = 17 \pm 4$), to introduce the research; identify potential key informants, i.e. older people are recognized for their knowledge and are interested in being interviewed (see Fig. 2, box 1); and validate initial information about the local history.

To explore oral histories and identify the main ecological and socio-institutional changes experimented with by the two communities in recent decades, we conducted a series of in-depth interviews with key informants ($N = 5$). After an initial coding of key themes, we undertook a content analysis. The interviews also included questions about agroecosystem characterization related to the management of fields; types of soils; crop calendars; types of tillage, i.e. full ploughing, ridges and timing; irrigation practices and soil conservation practices; and the evolution of agroecosystem management. The interviews were also very useful for compiling terminology in the Quechua language and traditional soil classifications. We worked with local interpreters to translate from Quechua to Spanish (see Fig. 2, boxes 2 and 3). This information was triangulated in two FGDs or participatory workshops, one per community, where we adopted participatory tools to facilitate the discussion. The participants ($N = 27 \pm 4$) collectively (a) identified local soils; (b) characterized field management, e.g. fallow land, type of ploughing and crop rotations; (c) mapped soils and crops within communities; and (d) assessed indicators for soil quality evaluation from the local knowledge. After the workshop, we carried out a field visit to select the areas and plots to be sampled (see Fig. 2, boxes 4 and 5).

We adopted Mayer's (2002) scheme to explain the segmentation process of the *Laymi* system over time.

Step II: Soil quality assessment

After sampling the types of soils in the two communities for their classification according to *Keys to Soil Taxonomy* (SSS, 2010) and to local knowledge, we performed a soil quality assessment (see Fig. 2, boxes 6 and 7) in the Suni agroecological area. A number of plots were selected as representatives of different management practices: although different practices are carried out in the area, only fallow time was considered in this study (see Fig. 3). For each treatment, three 50 × 50 m^2 plots about 50 m apart from each other were selected, and three replications were performed.

Soil profile characterization was carried out, including location, orientation, slope, surrounding vegetation, type of management, land use history and a description of the soil profile. The characterization was made by integrating scientific knowledge, i.e. soil characterization and sampling by horizons, with local knowledge, using Quechua terminology according to Tapia (1996). For each plot, ten cylindrical 5-cm-diameter × 20-cm-depth soil cores for laboratory analysis were randomly collected and mixed to obtain a composite sample to minimize spatial variability (see Fig. 2, boxes 8 and 9). The samples were air-dried, and the gravel (>2 mm) was removed and weighed, and any visible roots were removed. The following biochemical parameters of the soil were measured as scientific indicators of soil quality, chosen among the most frequent ones, according to the literature (Bünemann et al., 2018): soil organic carbon (SOC), measured by the wet oxidation method (Nelson & Sommers, 1982); total nitrogen (N), measured by the Kjeldahl method (Bremmer & Mulvaney, 1982); assimilable phosphorus (P), measured by the Olsen method (Watanabe & Olsen, 1965); and pH (1:2.5 suspension of soil in water) and texture, measured by the Bouyoucos method (1962). During the fieldwork visits, we also identified and assessed visual parameters (e.g. colour, porosity, structure).

Data analysis was performed to observe for differences in the tested parameters according to the selected management practices. A comparison of the different management practice means was carried out by a one-factor ANOVA analysis, i.e. management practices, with a 95% level of significance ($p < 0.05$). Additionally, the fallow effect was analysed using the independent-samples *t*-test ($p < 0.05$) to compare the means of fields with and without fallows for community A. All analyses were performed using SPSS software, Version 12.0.

Step III: Discussion of adaptive management practices

To conclude, we conducted community assemblies with both communities (with old/young people and men/women fairly represented) to collectively discuss the results of Step I and the scientific soil quality assessment. Specifically, using

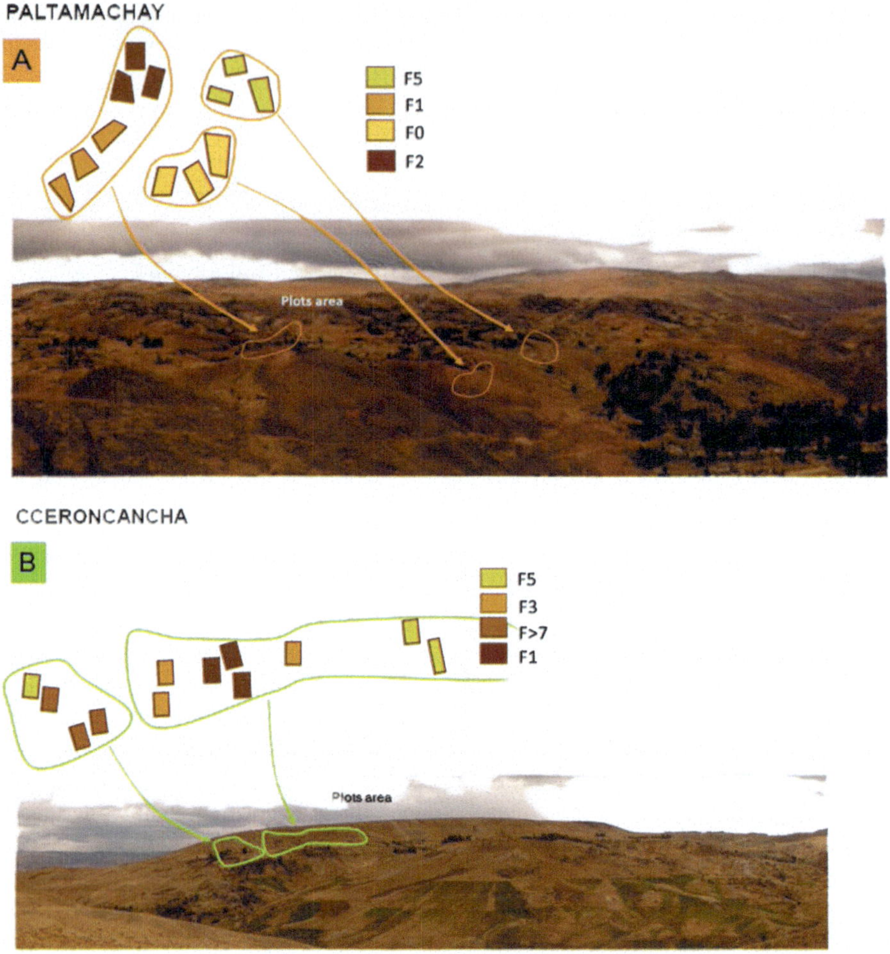

Fig. 3 Distribution of the experimental plots in the study area in both communities Paltamachay (**a**) and Cceroncancha (**b**). Fx being the number of fallow years. (Own elaboration)

participatory tools, we asked the participants to identify ethnoecological indicators of soil quality (e.g. soil depth, colour, penetration resistance, soil agricultural aptitude), to list the main criteria that defined the critical problems of maintenance of soil functions and their causes and to propose possible adaptive management practices (see Fig. 2, box 10). We used the term adaptive management practices to refer to the development of specific practices, technologies and knowledge for the management of the agroecosystem elements that responded to the uncertainty of environmental changes, e.g. changes in land use, crop diversification and irrigation (Berkes et al., 2003). We also included in the term the specific adjustments or changes to institutional arrangements, e.g. the functioning of assembly and communal decisions on crop management (Olsson et al., 2006).

3 Results

3.1 History of Agroecological and Institutional Systems During the Twentieth Century

We recognized different institutional arrangements for the two communities (Table 1). In community A, the rights exercised by the families corresponded, although not always completely, to three of the five criteria proposed by Ostrom and Hess (2008: 16). These were (1) the access to the land, (2) the withdrawal or possibility of benefiting from the resources of these parcels and (3) the management of the land. In community B, the use and management were collectively decided. The rights of alienation and exclusion still partially depend on community control.

Our findings defined the most significant socio-institutional changes and their links with the maintenance of differentiated bundles of rights and power between the community and the families (see Fig. 4). Before colonization, the local population was organized into ayllu, i.e. a kinship group with social linkages in a defined territory. They lacked individual ownership of the land; the collective periodically redistributed land according to the needs of each family and shaped the use and management (Rostworowski, 1962).

At the beginning of the twentieth century, in community A the land tenure of the hacienda disappeared. Through a new constitution, the *Constitution of the Republic of Peru* (1933; art. 209), indigenous communities were juridically recognized, and their lands, in addition to being unattachable, also become inalienable and imprescriptible. Legal status and a register of communities were established. In the case of community A, they decided to split from the ancient "mother community" it belonged from, and the lands were shared among ancient commoners. By the 1960s, in the Quechua and Suni regions, the pressure on the land due to population growth increased the process of land parcellation, i.e. families had more power to make access and control decisions for land use and management. This led to a decrease in the use of the fallow land system, and by the 1980s, the *Laymi* was completely dismantled in the Quechua and Suni regions. Under a fair inheritance between sons regime, the fragmentation of the land increased, with the average area of arable land per family decreasing to one hectare, resulting in crop intensification and the use of mineral fertilizers and continuous cultivation. Each family, having major control over management decisions, could implement new fertilization practices fomented by the Green Revolution. Pastoral lands were organized by sectors, and herds and grazing lands decreased, displaced by an agricultural frontier.

In community B, the traditional subsistence, community agroecosystem management system of the hacienda regime was maintained until the 1964 Agrarian Reform. The land was owned privately under a regime of employment and submission for commoners and *huachilleros*, a new shepherd figure without land employed for wool production. With population growth controlled in the 1960s, the transformation of the hacienda regime into a cooperative land regime, promoted by the *Agrarian Reform Law* (1964), created conditions for the recognition of control of

Table 1 Traditional institutional arrangements in the area based on local interviews and literature (Díez Hurtado, 2003; Duval, 2005; Castillo et al., 2007; CEPES, 2018)

	Institutional arrangements
Reciprocity norms	
Ayni	Reciprocal exchange of work among households. In the area the ayni refers to agricultural activities. Being part of the community shapes the access to work and solidarity networks The majority of families of community B participates in the ayni A number of families of community A participates in the ayni
Minka	Not reciprocal exchanges of services among households, paying with another service or products. For instance, building houses is realized with the assistance of other families. It's a form of work exchange The majority of families of both communities participate in the minka
Other reciprocity forms	*Trueque* = exchange of products in the markets Loans among households Exchanges during ceremonies/festivity They persist in both communities
Institutional arrangements for decision-making over land and resources access	
Main unit of decision-making: a polycentric system	The system of decisions is polycentric. A complex mix of individual and collective decision-making arrangements co-exists and is specific for each community Historically, the main entities for decision-making are the assemblies of the community. In the assemblies commoners decide together about who has access to lands and resources and what each field should produce and how to organize the work and activities. Inclusion and exclusion of access is a customary right-based decision, and it also depends on social identities (gender, age etc.). However, in several cases the assembly has been losing its function, and families take decisions over lands autonomously
Customary authorities	The consuetudinary juridic system is recognized by the Law (Ley General de comunidades Campesinas N° 24656, 1987) (CEPES, 2018). The ancestral (customary) authority is the President of the community. He responds to the assembly and supervises the decisions taken. The assemblies are the body that mediates and solves social conflicts (internally and externally). In ancient times, the *varayoc* was the leader in charge of conflict resolution. Today in some communities, a Council of elders or the community authorities may solve conflicts. The community is recognized as the unit that dialog with externals and other levels of administration
Bundle of rights over land ownership, access and use/management[a]	

(continued)

Table 1 (continued)

	Institutional arrangements
Right of ownership and access	From a legal point of view, the constitution of 1993 recognizes the community and its right of land ownership, even if not all the communities have requested and obtained it (CEPES, 2018). In reality, two parallel systems may co-exist: the household ownership and the collective ownership *Agricultural and pastoral land*: According to the customary land access, each family may receive a quantity of land that it can work (access and use defined by belonging from a group). In many cases, who shapes the control and maintains the rule of access and management of productive cycles (see below) may be both individual (HH) and communitarian, depending on the case. In community A mainly HH have the control over the access and the use of lands without intervention of the community in decisions. In community B, the access and use is individual, but the community still has a role of supervision to avoid conflicts. Several decisions are taken over resources as common pools *Water bodies*: The State recognizes the right of access and use of communities and indigenous people to water bodies which flow in their territories. In some cases they are organized in comités for water management, who shape the control Ley No. 29785 (2011) establishes that for activities or decisions that may affect the natural resources, land and water in community territory, the communities have to be informed and consulted (CEPES, 2018)

[a]We use here the definition of Ribot and Peluso (2003) that differentiates "access" to "property"

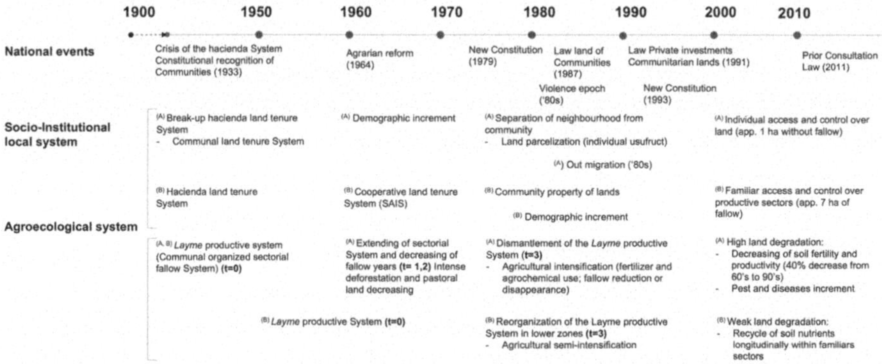

Fig. 4 Timeline of the most significant socio-institutional changes affecting the two studied communities (t = 0; t = 1, t = 2; t = 3 correspond to the temporal evolution shown in Fig. 5). (Own elaboration)

land and its management by commoners organized as members of a cooperative. However, the cooperative system entered a crisis during the 1970s, and with the implementation of the new *Constitution of the Republic of Peru* (1979), the community redistributed the land among its families under a regime of unattachable, imprescriptible and inalienable common land. In 1987, the community received official recognition, and land titles were issued under a new law (*Law No 24657*, 1987). At the same time, a process of self-definition and the reinforcement of internal security in the face of political conflict between the state and Sendero Luminoso had begun. Pressed by renewed demographic growth in the 1980s, the *Laymi* system was reorganized by sectors while maintaining common ownership, which allowed access to and control of the productive sectors at the family level but under communal decision-making. This arrangement ensured the maintenance of fallow periods for agricultural land, the use of animal dung as fertilizer and crop rotations.

During the 1990s, the land protection regime was modified under a new policy of land and economic liberalization (Burneo, 2013). If a communities' territory was still imprescriptible, under new private investment and economic development laws for communitarian lands (*Law 26505*, 1991) and the 1993 constitution, the community lost its unattachable and inalienable nature. Since then, despite different measures in favour of the recognition of indigenous and rural communities, for example, *Law 29785* (2011) for prior consultation, the situation has not decisively changed, particularly in the Sierra region (Hall, 2017), where a context of weakening the communal institution has been created. Additionally, in the last two decades, a progressive out-migration of young commoners to urban centres, particularly from community A, due to the impossibility of accessing land and increased environmental and productive crises, has contributed to the undermining of the social cohesion of communities.

How the *Laymi* system has been dismantled over time is illustrated in Fig. 5. In the case of community A, collective production zones have been transformed into very small areas, and the grazing of fallow land had been progressively reduced (no more than 3 years fallow) or almost disappeared. In comparison, for Community B, the grazing of fallow sectors persists (from 4 to 7 fallow years), and the community is in charge of most of the decision-making around access and management.

3.2 Soil Quality Assessment

During the field trip, the participants recognized different soil types according to two characteristics: texture and colour. According to their contributions, we classified the soils, as shown in Table 2. Texture is related to soil productivity, where loamy soils are recognized as being better for cultivation, e.g. *allin hallp'a*, *uspa hallp'a* and *oqeq hallp'a*. In addition, the participants linked the friability of the soil, its water retention and ease of working to the texture and recognized that sandy soils (*aqoq hallp'a*) and stony soils (*rangra rangra*) are poorer and less suitable for cultivation.

Figure 6 only shows the biochemical indicators of soil quality from the different plots that comparatively allowed us to determine any differences attributable to the type of management. In community B, with a scheme organized by sectors and communal land management rules and practices, i.e. a fallow system, the soils in all plots had higher SOC content than community A, where individuals manage the land. Likewise, in plots with 1 year fallow ($F = 1$) in community B, we observed

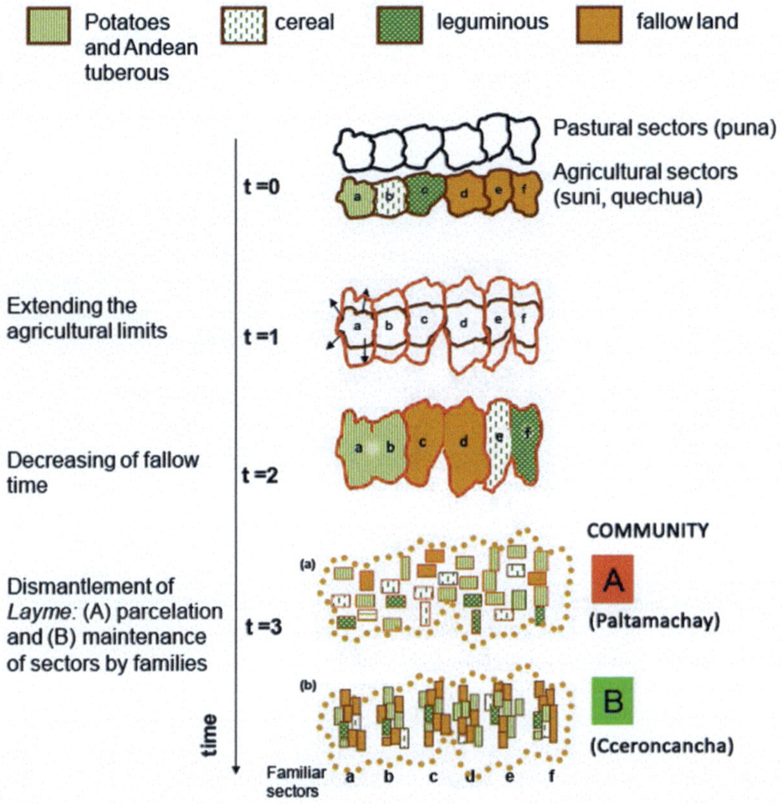

Fig. 5 Description of changes of the fallow sectoral system (Laymi) in both communities: A, Paltamachay and B, Cceroncancha. The textures in boxes represent the type of crop. The letters (a–f) identify the different sectors of Laymi. Changes that occurred in Laymi system: ($t = 0$) initial situation before starting to dismantle the Laymi system; ($t = 1$) the sectors persist, but there is an extension of the agricultural limits embodying the pastoral sectors; ($t = 2$) the sectors persist, but there is a decrease on fallow time; ($t = 3$) there is a different evolution between both communities—in community A, common fallow sectors disappeared, and the management became individual with a short time of fallow, while in community B, common fallow sectors persist with communal management. (Own elaboration; adapted by Mayer and de la Cadena (1989) through local interviews)

how the SOC content was slightly lower (about 1% less) than soils fallow for more years ($F > 7$). This reflected a natural return of fertility.

Total N followed the same pattern as SOC, indicating a strong relationship between these parameters. This was demonstrated by a simple regression analysis, where the total N content depends on the SOC ($R^2 = 0.99$; $p < 0.001$). In soils without fallow management, this relationship did not exist.

We observed differences in the P content of the soils, with higher assimilable P in soils not left fallow compared to fallow soils. These differences were particularly significant in community A's soils ($F = 17{,}293$; $p = 0.002$). This could indicate that

Table 2 Classification of soils in the studied communities

Community	Agroecological zone	Soil aptitude[a]	Texture[b]	Soil colour[a]	Name in Quechua[a]	Soil classification[c]
A	Suni	+++	Clay sandy	Red	*llink'i hallp'a puka*	*Lithic Mollic Haploxeralf*
A	Suni	++	Sandy loam	Black	*Aqoq hallp'a*	*Mollic Haploxeralf*
A	Suni	+++	Loam	White	*Allin hallp'a*	*Typic Xeropsamment*
A	Suni	++++	Silty clay loam	Red	*uspa hallp'a*	*Ultic Haploxeralf*
A	Puna	+	Stony soils	–	*rangra rangra*	*Lithic Entisol*
B	Suni	++	Sandy clay loam	Yellow	*Aqoq hallp'a*	*Mollic Haploxeralf*
B	Suni	++++	Loam	Grey	*oqeq hallp'a*	*Typic Dystroxerept*
B	Puna	+++	Silty clay	Lead	*ccalla hallp'a*	*Lithic Rhodoxeralf*

Source: Interviews and field work; own elaboration
[a]Soil aptitude according to: ++++ good; +++ regular; ++ bad; + very bad (no crops); According to Quechua's community members during the field walk
[b]Soil texture field test of horizon A
[c]Soil classification based on *Keys to Soil Taxonomy* (SSS, 2010)

the source of the P was mainly inorganic and came from the mineral fertilization of intensified cropping. There were no differences in the P content of community B's soils, where no mineral fertilizer has ever been applied.

When analysing the effects of fallow management on community A soils, using the independent-samples *t*-test and grouping the plots as fallow (F5 and F2) and not fallow (F0), we observed significant differences in the SOC and total N ($F = 7.547$, $p = 0.021$ and $F = 7.453$, $p = 0.021$, respectively), with higher levels in the fallow plots.

3.3 Adaptive Management Practices

During the collective discussion with the community, the participants recognized that over the past few decades, local knowledge and institutions, i.e. a community decision-making unit and communal arrangements, have been slowly eroded, with grave repercussions for both ecosystem health and human adaptive resilience. This process is especially critical for community A, where the participants emphasized the progressive dismantling of communal controls and the increasing independence of each family during agricultural decision-making.

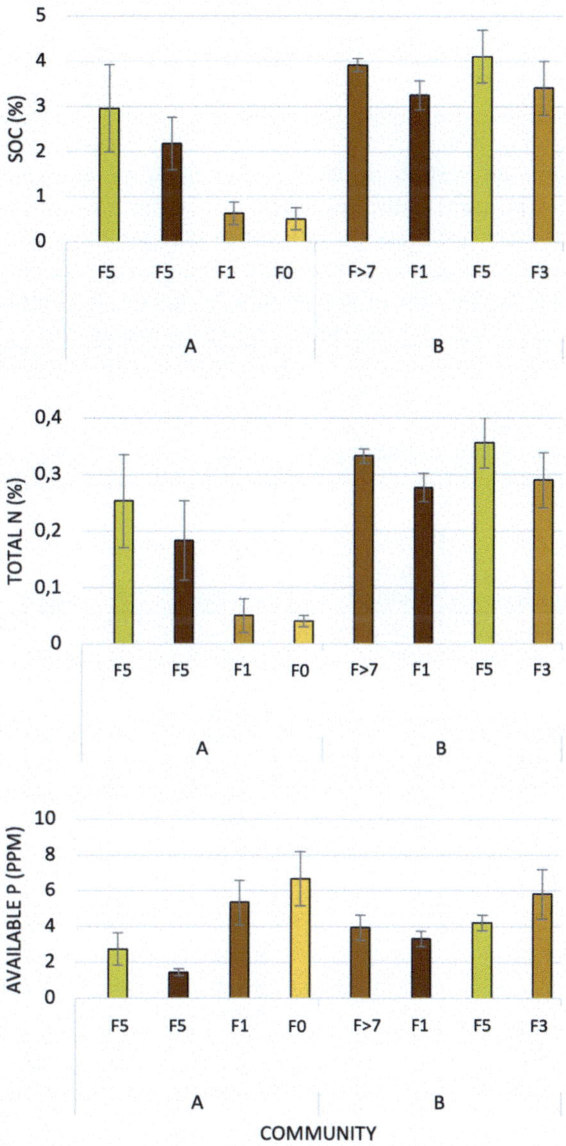

Fig. 6 Soil organic matter (SOM) and total *N* on soils of both communities: Paltamachay (**a**) and Cceroncancha (**b**). Tas/Fx = tillage after seeding with fallow; CR/Fx = crop rotation with fallow; CR/F0 = crop rotation without fallow; Fx = field in current fallow; "x" means number of years of fallow land. (Own elaboration)

In both communities, the participants perceived climate change as mainly related to the higher intensity of extreme frost and hailstorm events, shifts in seasons and the unpredictability of precipitation and droughts. They identified criteria and indicators that they recognized as warning signals of the three main problems identified with and linked to an exacerbation of climate risks. They also discussed adaptive management practices, such as new institutional arrangements, to improve soil fertility and conserve crop-available nutrients and water, reduce soil erosion and avoid land parcellation and intensification with the progressive loss of communal control over the production cycle. The identified technologies and practices are shown in Table 3.

The participants stressed the need to introduce changes in agricultural management and add value to the available resources. They concluded that a transition to farming that does not depend on external agrochemical inputs is necessary. They proposed optimizing the use of animal manure (guano) through its stabilization (composting) or the development of other manure treatment practices and organizing times and areas for fallow and crop rotations. However, a combination of these two actions requires institutional arrangements because of the communities' crucial roles. In the case of community A, due to the complexity of recovering the *Laymi* system, they proposed reorganizing the management of cultivated fields (or *chacras*) between families to allow crop rotation with intermediate fallow years. They also

Table 3 Criteria identified by participants for each problem (crop productivity, soil erosion, dismantling of Laymi system), indicators that act as warning signals and adaptive management practices adopted

Problem	Criteria identified	Indicators (acting as warning signals)	Adaptive strategies	Expected impacts
Crop productivity	Decreasing crop yield	Soil fertility: SOM content, biological activity (i.e. earthworm activity)	Use of animal manure (guano): composting, green manure, etc. Grazing during fallow period	Restoration soil fertility Improve crop yields
Soil erosion	Shallower soil Shallower organic layer Soil compaction	Soil penetration resistance (SPR) to foot plough (*chaquitaclla*)	Soil conservation systems: Slow-forming terraces Live fences	Soil protection (against heavy rainfall) Increased water availability Easier tillage with *chaquitaclla*
Parcelization	Decreasing grazing fallow and intensification	Fallow duration Number of crops' rotations Use of agrochemicals in some fields	Temporal user right (*aynoka*): communal decisions on rotations to allow grazing in fallow period and keep the animals outside the land under cultivation Recovering crops' rotation according to the typology of soil Communal decision on management of inputs Knowledge transfer through farmer-to-farmer extension	Restoration of soil fertility during the fallow period and *guano* (animal manure) under communal norms

Source: Final communal meeting; own elaboration

proposed improving manure management by mixing animal manure with vegetable crop residues and forage crops and optimizing the supply time and quantity of manure according to the needs of the crops.

4 Discussion

4.1 Institutional Changes and Community Resistance

This study sheds light on the contemporary dynamics of local communities (Urrutia Cerruti, 2014) that have received little research attention. The integrated and participatory approach we adopted enabled us to understand how historical socio-institutional changes have led to different adaptation pathways for communal organization and specifically affected the protection or disappearance of customary rules governing land use and traditional systems of management, such as the *Laymi* system. The main consequence of the disappearance of the *Laymi* system is the replacement of traditional sectored communal agriculture by the progressive segmentation of production zones and more intense and permanent cultivations, i.e. the use of mineral fertilizers and continuous cultivation. This also implies a progressive fine-tuning of increased labour inputs. This tendency has been observed in other Peruvian Andean regions (Díez Hurtado, 2012) and other Andean regions, such as the Paramo in Ecuador (López-Sandoval & Maldonado, 2019). Our comparison of two Peruvian highland Quechua peasant communities illustrated this process of intensification very well and reflected the importance of historical changes when seeking to understand differences between communities. Facing the challenge of recovering their identity as a community after the colonial epoch and the collapse as cooperatives, community B opted for territorial and social restructuration, focused on strengthening social cohesion and rules, e.g. informal exchange of goods and services and mutual aid and reciprocity, and resisting the dismantling of communal land access and management based on their traditional knowledge (Fauvre, 1976; Martí & Pimbert, 2007). Similar results were found by Hall (2017), where some communities in southern Peru officially recognized after the colonial and cooperative epochs have reinforced their internal social rules, the protection of alienation and exclusion rights and control over land access and management as part of a commoning process, i.e. the way of doing common things, and to resist environmental changes, globalization and liberalization. In contrast, in community A the hacienda system disappeared early and the demographic pressure and commercialization of agriculture during the 1980s (Orlove & Godoy, 1986) and the liberalization programmes and outmigration of the last two decades have accelerated the progressive dismantling of communal control. This change has important social and ecological effects, in particular, on soil fertility, which relies on this mix of polycentric and individual institutional arrangements for land management (Zulkafli et al., 2017).

In summary, this case study has demonstrated how the everyday management of commons land is the basis of the resistance of some Peruvian Andean communities to enclosure. It confirms Kervyn's (1992) hypothesis, which recognized that the maintenance of communal systems has proved greater efficiency, guaranteed the security of commoners' access to land, defended their exclusivity in the face of external pressure, supported conflict resolution and enhanced positive ecological effects related to the coordination of land management.

4.2 *Impacts of Dismantling Communal Management on Soil Quality*

Several authors have demonstrated that agroecosystems in the Andean highlands are experiencing rapid transformation related to land use changes and intensification (De Valença et al., 2017). Similarly to Jiménez Álvarez et al. (2021), we collected local knowledge on physical and biological indicators of soil quality (colour, texture, resistance to penetration etc.), and we integrated it with scientific knowledge to have more robust information on the impacts of intensification.

Specifically, our results indicated that differences in soil carbon, nitrogenous and phosphorous were attributable to the type of land management system, i.e. presence of fallow time. The length and timing of fallow periods are crucial variables for the regeneration of soil fertility and improving soil properties (Pestalozzi, 2000; Aguilera et al., 2013). We observed that soil fertility was positively influenced by fallow periods, suggesting that there is a natural return of fertility. Such increases in soil fertility have been attributed to the decline of the above-ground and root biomasses of fallow vegetation, the presence of native leguminous species and the regrowth of evergreen vegetation that controls soil erosion (Aguilera et al., 2013). Additionally, during fallow years, the deep roots of perennial grasses reach deeper levels of soil than annual crops and take up nutrients from these inferior strata, which, during the preparation of fallow fields for planting, are reincorporated into the soil, thereby increasing the concentration of nutrients in the upper soil layers (Orlove & Godoy, 1986). Of note, other studies carried out on these agricultural systems aimed at explaining the loss of fertility during cultivation and its progressive recovery during fallow periods have not detected soil changes, which is probably due to the spatial heterogeneity of mountain landscapes (Lindell et al., 2010). Nonetheless, in our study, we detected clear differences when analysing the effect of community A's fallow periods. In all cases, we observed higher levels of SOC and total N in fields managed with grazing fallow periods. Similarly, Sarmiento and Bottner (2002) confirmed initial significant differences in biological soil properties between adjacent plots that only differed by their fallow intervals. Sarmiento and Bottner (2002) also concluded that the removal of fallow periods could be linked to a progressive reduction of organic soil matter, which can compromise the sustainability of agricultural systems.

Future research should explore how communal arrangements or their disaggregation in specific contexts impact key agroecosystem functions, such as genetic and species agrobiodiversity, pollination and dispersion and water conservation.

4.3 Managing the Future of Andean Mountain Communities in the Face of Climate Change

Climate change models are mainly applied at global and regional scales, but impacts and adaptation strategies are context-specific and need to be explored at local scales (Smit & Wandel, 2006). Historically, Andean mountain communities have adapted to climate variability by adopting production practices and technologies, e.g. maintaining high diversity and wild crops, terracing and rotational systems (contour planting, terraces, etc.), conservation tillage and improved crop residues management, increased organic fertilization, water harvesting etc. (see, for instance, Zimmerer, 1996: Perez et al., 2010). Additionally, scholars studied context-specific strategies in response to the impacts of climate change and other simultaneous drivers (Regalsky & Hosse, 2009; FAO, 2015).

Several authors suggested that communal arrangements of land management and formal and informal social relationships and networks for work-sharing and cooperation may play a fundamental role in empowering mountain communities and their resilience (Hock et al., 2019; Vallejo-Rojas et al., 2022). For instance, in the central Andes, Crespeigne et al. (2010) studied the *allapakuy*, an informal social system where commoners who have experienced a crop loss may offer their labour to others. In other studies, the livelihood strategies based on sharing resources and information along a vertical gradient have been demonstrated as robust adaptation strategies, e.g. the *Challayplasa*, or barter markets, in Peru that are an example of cooperative exchange of goods and services (Martí & Pimbert, 2007). In our study, the local communities suggested adaptation practices and future policy agendas for agroecosystem management, where the objectives of sustainability focus on soil quality management and the restoration and restructuring of governance arrangements. Specifically, our discussions concerning the dismantling of the *Laymi* system and its effects on soil degradation and erosion suggested the progressive vulnerability of individual farmers, especially the poorest ones with limited access to land, and the importance of improving communal arrangements.

Different factors constrain farmers to implement the adaptive strategies proposed in the face of the critical environmental changes they are experimenting, such as socio-economic vulnerability, a great reliance on external actors and outside interventions and insufficient access to information, among others. To deal with them, first of all, NGOs, universities and research centres need a change in their approach to research and action with greater articulation between them, a direct engagement of farmers and a clear valorization of local knowledge and practices in decision-making processes. Second, a suggestion for an adequate implementation of agroecological strategies is to work with local experimenter farmers who can become promoters of farmer to farmer exchanges with visits to successful experiences. Finally, as suggested by similar studies, further research and attention on policy programs is needed in those social and institutional factors that may foster adaptation and resilience to global environmental change. For instance, while out-migration is currently one of the main strategies to cope with hazards, future policies and research should jointly flesh out solutions for people to stay, support community initiatives and strengthen communal institutions for the sustainable management of resources in the face of climate change.

5 Conclusion

In the Andean Peruvian mountains, local indigenous knowledge and practices for dealing with uncertainty and risks related to climate variability are threatened by several factors and aggravated by the speed and intensity of changes. Nevertheless, local people's perspectives, experiences and voices have been silenced or ignored in climate change policy design, which usually follows top-down and scientific-based approaches.

This chapter adopts a participatory and inclusive research and explores how the dismantling versus the conservation of community-led knowledge and practices of land use management have differently influenced the maintenance over time of key indicators of ecological resilience of local agroecosystems, such as soil quality. After discussing with local communities the indicators to follow early warning signals of change, the study compiles possible strategies and argues that the collective reorganization of local families on the vertical and longitudinal local management plays a crucial role for more efficient and less destitute adaptation pathways to climate change.

In-depth empirical research on effective models of governance in mountain areas is needed to propose multiscale management interventions that foster conservation while sustaining local livelihoods and reducing poverty in local communities. The study recommends that successful experiences of adaptation should be more accessible and shared with others through communities of practices. Moreover, it suggests that the adaptation plans at national and regional level should be informed by context-specific local traditional knowledge systems for a more robust identification of adaptation measures and their monitoring/evaluation over time.

Acknowledgements We thank the local communities of Cceroncancha and Paltamachay (Huancavelica region), who participated in the research and welcomed us. We also acknowledge the NGOs CEPES and AVSF who gave us the means for carrying out the research. During the 2 years of writing, the first author was funded by AXA Research Fund (2016) and by the Ramón y Cajal fellowship from the Minister of Science and Innovation (RYC2018-025958-I).

References

Aguilera, J., Motavalli, P., Valdivia, C., & Gonzales, M. A. (2013). Impacts of cultivation and fallow length on soil carbon and nitrogen availability in the Bolivian Andean 583 Highland Region. *Mountain Research and Development, 33*, 391–403. https://doi.org/10.1659/MRD-JOURNAL-D-12-00077.1

Barrera-Bassols, N., & Zinck, J. A. (2003). Ethnopedology: A worldwide view on the soil knowledge of local people. *Geoderma, 111*(3–4), 171–195. https://doi.org/10.1016/s0016-7061(02)00263-x

Berkes, F., Folke, C., & Colding, J. (2003). *Navigating social-ecological systems: Building resilience for complexity and change* (p. 388). Cambridge University Press.

Bhadouria, R., Singh, S., Tripathi, S., & Singh, P. (2023). *Understanding soils of mountainous landscapes: Sustainable use of soil ecosystem services and management*. Elsevier.

Bollier, D., & Helfrich, S. (2019). *Free, fair and alive: The insurgent power of the commons* (p. 448). New Society Publishers.

Bouyoucos, G. J. (1962). Hydrometer method improved for making particle size analyses of soils. *Agronomy Journal, 54*, 464–465.

Bremmer, J. M., & Mulvaney, C. S. (1982). Nitrogen – Total N. In R. H. Miller & D. R. Keeney (Eds.), *Methods of soil analysis* (pp. 595–624). American Society of Agronomy, Soil Science Society of America.

Brush, S. B., & Guillet, D. W. (1985). Small-scale agro-pastoral production in the Central Andes. *Mountain Research and Development, 5*(1), 19–30. https://doi.org/10.2307/3673220

Bünemann, E. K., Bongiorno, G., Bai, Z., Creamer, R. E., De Deyn, G., de Goede, R., Fleskens, L., Geissen, V., Kuyper, T. W., Mäder, P., Pulleman, M., Sukkel, W., van Groenigen, J. W., & Brussaard, L. (2018). Soil quality – A critical review. *Soil Biology and Biochemistry, 120*, 105–125. https://doi.org/10.1016/j.soilbio.2018.01.030

Burneo, Z. (2013). Propiedad y tenencia de la tierra en comunidades campesinas. Revisión de literatura reciente en el Perú. In P. Castillo, A. Diez, B. Zulema, J. Urrutia, & P. del Valle (Eds.), *Qué sabemos de las comunidades campesinas?* (pp. 153–258). ALLPA, Comunidades y Desarrollo.

Castillo, P., Díez Hurtado, A., Burneo, Z., Urrutia, J., & del Valle, P. (2007). *Qué sabemos de las comunidades campesinas?* (p. 364). CEPES, ALLPA, Comunidades y Desarrollo.

CEPES. (2018). *Derechos de las comunidades campesinas y nativas: Principales leyes y reglamentos*. https://larevistaagrariaperu.files.wordpress.com/2018/02/derechos_comunidades.pdf. Accessed 10/05/2022.

Correa, S. W., Mello, C. R., Chou, S. C., Curi, N., & Norton, L. L. D. (2016). Soil erosion risk associated with climate change at Mantaro River basin, Peruvian Andes. *CATENA, 147*, 110–124.

Crespeigne, E., Olivera, E., Ccanto, R., & Scurrah, M. (2010). Exploración de las estrategias y prácticas de una comunidad campesina de los Andes Centrales frente a los riesgos extremos asociados al cambio climático. In P. Ames & V. Caballero (Eds.), *Perú: El Problema Agrario en Debate* (SEPIA XIII) (pp. 260–290). Seminario Permanente de Investigación Agraria (SEPIA).

De Valença, A. W., Vanek, S. J., Meza, K., Ccanto, R., Olivera, E., Scurrah, M., Lantinga, E. A., & Fonte, S. J. (2017). Land use as a driver of soil fertility and biodiversity across an agricultural landscape in the Central Peruvian Andes. *Ecological Applications, 27*(4), 1138–1154. https://doi.org/10.1002/eap.1508

Díez Hurtado, A. (1998). *Comunes y haciendas: Procesos de comunalización en la Sierra de Piura (siglos XVIII al XX)* (p. 262). CICPA.

Díez Hurtado, A. (2003). *Titulación de las comunidades campesinas* (Debate Agrario 36) (pp. 89–103). CEPES.

Díez Hurtado, A. (2012). *Gobierno comunal: Entre la propiedad y el control territorial. El caso de la comunidad de Catacaos* (Perú: El problema agrario en debate. SEPIA, 14) (pp. 115–148). SEPIA.

Dollfus, O. (1991). *Territorios andinos, reto y memorias* (p. 221). IFEA y IEP.

Duval, J. (2005). *Agriculture et élevage dans les Andes Centrales du Pérou: Diagnostic agraire des systèmes de production du district de Huando en prévision de l'irrigation* [MSc dissertation]. CICDA-CNEARC Montpellier, France.
Egli, M., & Poulenard, J. (2016). Soils of mountainous landscapes. In D. Richardson, N. Castree, M. F. Goodchild, A. Kobayashi, W. Liu, & R. A. Marston (Eds.), *International encyclopedia of geography: People, the earth, environment and technology* (pp. 1–10). Wiley.
Fauvre, H. (1976). La evolución y la situación de las haciendas en la región de Huancavelica. In J. M. Mar (Ed.), *Hacienda, Comunidad y campesinado en el Perú* (pp. 105–138). IEP Ediciones.
Food and Agricultural Organization of the United Nations [FAO]. (2015). *Understanding mountain soils: A contribution from mountain areas to the international year of soils 2015*. FAO.
Gade, D. W., & Rios, R. (1972). Chaquitaclla: The native footplough and its persistence in central Andean agriculture. *Tools & Tillage, 2*(1), 3–15. https://doi.org/10.11588/diglit.48999.4
Hall, I. (2017). De la colectividad a la comunidad. Reflexiones acerca del derecho de propiedad en Llanchu, Perú. *Revista de Antropología Social, 26*(2), 379–398. [fecha de Consulta 28 de Julio de 2022]. ISSN: 1131-558X. https://www.redalyc.org/articulo.oa?id=83853471009
Halloy, S. R. P., Seimon, A., Yager, K., & Seymon, A. (2005). Traditional Andean cultivation. Systems and implications for sustainable land use. *Acta Horticulturae, 670*, 31–55. https://doi.org/10.17660/ActaHortic.2005.670.4
Hock, R., Rasul, G., Adler, C., et al. (2019). High mountain areas. In H. O. Pörtner, D. C. Roberts, V. Masson-Delmotte, et al. (Eds.), *IPCC special report on the ocean and cryosphere in a changing climate* (pp. 131–202). Cambridge University Press. https://www.ipcc.ch/srocc/chapter/chapter-2/
INEI. (2018). *Censos Nacionales XII de Población y VII de Vivienda*. Resultados Definitivos. Available at: https://www.inei.gob.pe/estadisticas/indice-tematico/poblacion-y-vivienda/. Accessed May 20, 2023.
IPCC. (2022). *Climate change 2022: Impacts, adaptation, and vulnerability*. Contribution of Working Group II to the Sixth Assessment Report of the Intergovernmental Panel on Climate Change (H.-O. Pörtner, D. C. Roberts, M. Tignor, E. S. Poloczanska, K. Mintenbeck, A. Alegría, M. Craig, S. Langsdorf, S. Löschke, V. Möller, A. Okem, & B. Rama, Eds.; 3056 pp). Cambridge University Press.
Jiménez Álvarez, L. S., Andrade, E., Capa-Mora, D., Fierro, N., Quichimbo, P., Jiménez, W., & Carrión-Paladines, V. (2021). Traditional knowledge on soil management and conservation in the inter-Andean region, northern Ecuador. *Spanish Journal of Soil Science, 11*(1), 55–71. https://doi.org/10.3232/SJSS.2021.V11.N1.05
Jodha, N. S. (1992). Mountain perspective and sustainability: A framework for development strategies. In N. S. Jodha, M. Banskota, & T. Partap (Eds.), *Sustainable agriculture* (pp. 42–82). Oxford & IBH Publishing.
Jodha, N. S. (1998). Poverty and environmental resource degradation: An alternative explanation and possible solutions. *Economic and Political Weekly, 33*, 2384–2390.
Kendall, A., & Drew, D. (2019). The rehabilitation of pre-Hispanic agricultural infrastructure to support rural development in the Peruvian Andes. In C. Isendahl & D. Stump (Eds.), *The Oxford handbook of historical ecology and applied archaeology* (pp. 421–440). Oxford University Press.
Kervyn, B. (1992). Mercado de tierras: Argumentos para un debate. *Lima, Sepia, Cuadernos de Trabajo, 1*, 39.
Kessler, A. (2006). *Moving people – Towards collective action in soil and water conservation. Experiences from the Bolivian mountain valleys* [PhD thesis]. https://edepot.wur.nl/121852. Accessed March 1, 2023.
Koohafkan, P., & Altieri, M. A. (2010). *Globally important agricultural heritage systems: A legacy for the future*. UN-FAO.

Lindell, L., Åström, M., & Öberg, T. (2010). Land-use versus natural controls on soil fertility in the Subandean Amazon, Peru. *Science of the Total Environment, 408*, 965–975. https://doi.org/10.1016/j.scitotenv.2009.10.039

López-Sandoval, M. F., & Maldonado, P. (2019). Change, collective action, and cultural resilience in páramo management in Ecuador. *Mountain Research and Development, 39*(4), R1–R9. https://doi.org/10.1659/MRD-JOURNAL-D-19-00007.1

Madrigal-Martínez, S., & Miralles i García, J. L. (2019). Land-change dynamics and ecosystem service trends across the central high-Andean Puna. *Scientific Reports, 9*, 9688.

Magrin, G. O., Marengo, J. A., Boulanger, J. P., Buckeridge, M. S., Castellanos, E., Poveda, G., Scarano, F. R., & Vicuña, S. (2014). Central and South America. In V. R. Barros, C. B. Field, D. J. Dokken, M. D. Mastrandrea, K. J. Mach, T. E. Bilir, M. Chatterjee, K. L. Ebi, Y. O. Estrada, R. C. Genova, B. Girma, E. S. Kissel, A. N. Levy, S. MacCracken, P. R. Mastrandrea, & L. L. White (Eds.), *Climate change 2014: Impacts, adaptation, and vulnerability. Part B: Regional aspects* (Contribution of Working Group II to the Fifth Assessment Report of the Intergovernmental Panel on Climate Change). Cambridge University Press.

Marston, R. A. (2008). Land, life, and environmental change in mountains. *Annals of the American Association of Geographers, 98*(3), 507–520.

Martí, N., & Pimbert, M. (2007). Barter markets for the conservation of agro-ecosystem multifunctionality: The case of the chalayplasa in the Peruvian Andes. *International Journal of Agricultural Sustainability, 5*(1), 51–69. https://doi.org/10.1080/14735903.2007.9684813

Martín-López, B., Leister, I., Cruz, P. L., Palomo, I., Grêt-Regamey, A., Harrison, P. A., Lavorel, S., Locatelli, B., Luque, S., & Walz, A. (2019). Nature's contributions to people in mountains: A review. *PLoS One, 14*, e0217847.

Mayer, E. (2002). *The articulated peasant. Household economies in the Andes*. Westview Press.

Mayer, E., & de la Cadena, M. (1989). *Cooperación y conflicto en la comunidad Andina. Zonas de producción y organización social*. Instituto de Estudios Peruanos.

McDowell, G., Ford, J. D., Lehner, B., Berrang-Ford, L., & Sherpa, A. (2013). Climate-related hydrological change and human vulnerability in remote mountain regions: A case study from Khumbu, Nepal. *Regional Environmental Change, 13*(2), 299–310. https://doi.org/10.1007/s10113-012-0333-2

Mishra, P. K., Rai, A., Abdelrahman, K., Rai, S. C., & Tiwari, A. (2021). Analysing challenges and strategies in land productivity in Sikkim Himalaya, India. *Sustainability, 13*, 11112.

Mishra, P. K., Rai, A., Abdelrahman, K., Rai, S. C., & Tiwari, A. (2022). Land degradation, overland flow, soil erosion, and nutrient loss in the Eastern Himalayas, India. *Land, 11*, 179.

Morlon, P. (1996). *Comprender la agricultura campesina en los Andes Centrales, Perú-Bolivia* (p. 495). IFEA, CBC.

Murra, J. V. (1985). El archipielago vertical revisited. In S. Masuda, I. Shimada, & C. Morris (Eds.), *Andean ecology and civilization* (pp. 3–14). University of Tokyo Press.

Murra, J. V. (2002). *El mundo andino: Población, medio ambiente y economía* (p. 511). IEP/Pontificia Universidad Católica de Perú. Historia Andina.

Nelson, D. W., & Sommers, L. E. (1982). Total carbon, organic carbon, and organic matter. In A. L. Page, R. H. Miller, & D. R. Keeney (Eds.), *Methods of soil analysis, Part 2* (Agronomy Monograph) (Vol. 9, 2nd ed., pp. 539–579). ASA and SSSA.

Olsson, P., Gunderson, L. H., Carpenter, S., Ryan, P., Lebel, L., Folke, C., & Holling, C. S. (2006). Shooting the rapids: Navigating transitions to adaptive governance of socio-ecological systems. *Ecology and Society, 11*(1), 18. http://www.ecologyandsociety.org/vol11/iss1/art18/

Orlove, B. S., & Godoy, R. (1986). Sectoral fallowing systems in the Central Andes. *Journal of Ethnobiology, 6*(1), 169–204.

Ostrom, E., & Cox, M. (2010). Moving beyond panaceas: A multitiered diagnostic approach for social-ecological analysis. *Environmental Conservation, 37*(4), 451–463. https://doi.org/10.1017/S0376892910000834

Ostrom, E., & Hess, C. (2008). Private and common property rights. In *Encyclopedia of law & economics*. Edward Elgar. http://papers.ssrn.com/sol3/papers.cfm?abstract_id=1304699

Perez, C., Nicklin, C., Dangles, O., Vanek, S., Sherwood, S., Halloy, K. S., Garrett, A., & Forbes, G. (2010). Climate change in the high Andes: Implications and adaptation strategies for small-scale farmers. *International Journal of Environmental, Cultural, Economic and Social Sustainability, 6*(5), 71–88. https://doi.org/10.18848/1832-2077/CGP/v06i05/54835

Pestalozzi, H. (2000). Sectoral fallow systems and the management of soil fertility: The rationality of indigenous knowledge in the high Andes of Bolivia. *Mountain Research and Development, 20*(1), 64–71. https://doi.org/10.1659/0276-4741(2000)020[0064:SFSATM]2.0.CO;2

Regalsky, P., & Hosse, T. (2009). *Estrategias campesinas andinas de reducción de riesgos climáticos*. CENDA-CAFOD.

Ribot, J., & Peluso, N. L. (2003). A theory of access. *Rural Sociology, 68*(2), 153–181. https://doi.org/10.1111/j.1549-0831.2003.tb00133.x

Rostworowski, M. (1962). Nuevos datos sobre tenencia de tierras reales en el incario. *Revista del Museo Nacional, XXXIXXXI*, 130–159.

Sarmiento, L., & Bottner, P. (2002). Carbon and nitrogen dynamics in two soils with different fallow times in the high tropical Andes: Indications for fertility restoration. *Applied Soil Ecology, 19*, 79–89. https://doi.org/10.1016/S0929-1393(01)00178-0

Sietz, D., & Feola, G. (2016). Resilience in the rural Andes: Critical dynamics, constraints and emerging opportunities. *Regional Environmental Change, 16*, 2163–2169. https://doi.org/10.1007/s10113-016-1053-9

Smit, B., & Wandel, J. (2006). Adaptation, adaptive capacity and vulnerability. *Global Environmental Change, 16*, 282–292. https://doi.org/10.1016/j.gloenvcha.2006.03.008

Soil Survey Staff. (2010). *Keys to soil taxonomy* (11th ed.). USDA-NRCS.

Styger, E., & Fernandes, E. C. M. (2006). Contribution of managed fallows to soil fertility recovery. In N. Uphoff, A. S. Ball, E. Fernandes, H. Herren, O. Husson, M. Liang, C. Palm, J. Pretty, P. Sanchez, N. Sanginga, & J. Thies (Eds.), *Biological approaches to sustainable soil systems* (pp. 425–437). Taylor & Francis.

Tapia, M. E. (1996). *Ecodesarrollo en los Andes altos* (p. 196). Friedrich Ebert Stiftung.

Torres-Guevara, J., & Velásquez, D. (2007). Successful experiences of sustainable land use in hyperarid, arid and semiarid zones from Peru. In M. V. K. Sivakumar & N. Ndiang'ui (Eds.), *Climate and land degradation* (pp. 501–521). Springer.

Urrutia Cerruti, J. (2014). Las comunidades campesinas nuevamente en la agenda pública. In *Comunidades indígenas y campesinas del Perú: Una visión histórica* (pp. 9–20). Comisión de derechos humanos – COMISEDH.

Valladolid Rivera, J. (1994). Visión Andina del Clima. In F. Grillo et al. (Eds.), *Crianza Andina de la chacra* (pp. 183–232). PRATEC.

Vallejo-Rojas, V., Rivera-Ferre, M. G., & Ravera, F. (2022). The agri-food system (re)configuration: The case study of an agroecological network in the Ecuadorian Andes. *Agriculture and Human Values, 39*, 1301–1327. https://doi.org/10.1007/s10460-022-10318-1

Watanabe, F. S., & Olsen, S. R. (1965). Test of an ascorbic acid method for determining phosphorus in water and NaHCO3 extracts from the soil. *Soil Science Society of America Journal, 29*, 677–678.

Zimmerer, K. S. (1996). *Changing fortunes: Biodiversity and peasant livelihood in the Peruvian Andes* (p. 309). Univ of California Press.

Zulkafli, Z., Perez, K., Vitolo, C., Buytaert, W., Karpouzoglou, T., Dewulf, A., De Bièvre, B., Clark, J., Hannah, D. M., & Shaheed, S. (2017). User-driven design of decision support systems for polycentric environmental resources management. *Environmental Modelling & Software, 88*, 58–73. https://doi.org/10.1016/j.envsoft.2016.10.012

Consequences of Climate Change on Medicinal and Aromatic Plants in the Indian Himalayan Region: An Overview

Harshita Pandey

1 Introduction

Climate change, which can be described by substantial shifts in average values of environmental components over a prolonged period (including rainfall, temperature, exposure to sunlight, humidity, wind, precipitation, and so on), is one of the greatest challenges facing the earth today (WMO, 1992; Gavrić et al., 2023). The average temperature across the world has risen by 0.15–0.20 °C every decade since 1975, and this trend is anticipated to continue in the coming decade (Malhi et al., 2021; Gavrić et al., 2023). Apart from rising temperatures, climate change also results in severe droughts, variation in rainfall patterns, and a rise in the frequency of hailstorms (Dejanovic et al., 2019). Many experts are worried that the negative effects of climate change will worsen and become more frequent in the future, especially if environmental circumstances introduced due to climate change have an extreme impact on public health (Harish et al., 2012). The incidence of more often adverse weather conditions (e.g. increasing temperature, a lack of rainfall, an increase in disasters, altered agricultural areas) results in the loss of vegetation, alterations to habitat setup, distribution of species, and diversity of plants (Rakic & Johnson, 2008; Vucijak et al., 2014). Medicinal and aromatic plants (MAPs) that are widely distributed in the Indian Himalayan Region (IHR) may also be impacted from these extreme climate events.

Based on the previous studies, the influence of climate change additionally impacts the essential phyto-molecules of MAPs (Gairola et al., 2010; Mahajan et al., 2020; Abbass et al., 2022; Yeshi et al., 2022). Globally these phyto-molecules serve as key constituents in the personal healthcare sector (Gairola et al., 2010; Mahajan et al., 2020). Plant-based secondary metabolites contribute significantly to

H. Pandey (✉)
Department of Horticulture, Sikkim University, Gangtok, India

S. Tripathi et al. (eds.), *Warming Mountains*,
https://doi.org/10.1007/978-3-031-62197-0_20

the process of developing new drugs, but due to their lower yield in the source plant and the uncertainty in supply as a result of various intervening factors, especially global warming and climate change, biotechnological intervention is required to develop suitable and economical alternative production methods (Gairola et al., 2010; Negi et al., 2022; Sun et al., 2023).

There is no doubt that the rate of global warming is accelerating unprecedentedly. As the climate is changing more speedily and the biosphere warms faster than plant species can adapt in IHR, there is an extreme risk of biodiversity extinction (Tripathi et al., 2022). To understand the pattern of climate change, one of the most significant global environmental threats, various consequences must be documented and evaluated, vulnerabilities need to be mitigated, and approaches to adaptation must be explored (Das et al., 2016; Malhi et al., 2021). To reassure governments for the protection of plant and animal species, the UN announced 2010 as the International Year of Biodiversity and acknowledged the Convention on Biological Diversity (CBD). However, indigenous populations who have used MAPs for eras believe that certain species are disappearing from the planet at an alarming rate (Cragg & Newman, 2013). Furthermore, they claimed that it is getting harder to find these species, which could be due to climate change.

According to the previous literature, higher-elevation areas are more vulnerable to the adverse consequences of variations in the climate (Mehta et al., 2021; Abbass et al., 2022). One of the world's biodiversity hotspots, the IHR is also a rich repository of valuable MAPs and possesses a wide range of traditional and modern therapeutic knowledge, and it offers a wealth of ecological services to people. The IHR spans about 5.3 lakh km^2 area in the northern boundary of India which represents 16.2% of the population's total geographical area (Acharya et al., 2022). MAPs' diversity in IHR is under severe stress and may eventually go detrimental because of climate change (Ratha et al., 2012; Mehta et al., 2021).

Variations in the mountainous ecosystem, habitat fragmentation, shifting distribution ranges, phenology patterns, secondary metabolites, and the emergence of new species have been associated to climate change and impose an adverse effect on the existing species of MAPs (Mahajan et al., 2020; Walia et al., 2022). Several ecologically significant MAPs species, including *Rhododendron species*, *Myrica esculenta*, *Pinus wallichiana*, *P. roxburghii*, etc., have undergone phenological alterations as a result of changing rainfall and temperature patterns in the Himalayan range. Some species are moving to higher elevations as a result in order to adapt to the changing weather (Tripathi et al., 2022).

Recently, the potential sources of biological resources have already been exposed to climate change and global warming, which are already in existence throughout the planet, putting pressure on people's resource practices and living styles (Raza et al., 2019). According to a recent publication (Applequist et al., 2020), scientists gave "Warning to Humanity" to highlight the accelerated rates of anthropogenic damage to the environment in 1992, including the climate, biodiversity, freshwater, soil, forests, ocean resources, etc. Nowadays it is widely recognized that MAPs will be going to face major changes due to climate variability, making people's situation even worse than formerly estimated. Numerous researchers endorsed the involvement of MAPs in modern and traditional medicines (Applequist et al., 2020). Climate variation may have an impact on the phytochemical composition and yield of MAPs, especially in IHR, possibly also altering their pharmaceutical properties (Gairola et al., 2010; Ratha et al., 2012; Das et al., 2016).

The United Nations Framework Convention on Climate Change (UNFCCC) came up with an important agreement to address climate change and improve the actions and investments required to achieve a sustainable low-carbon future at the Conference of the Parties (COP 21) in Paris on December 12, 2015 (Sharma et al., 2020a; Chien et al., 2021). It is currently believed that climate alteration is examined due to anthropogenic activities (human-caused disturbance) (Murshed, 2022). Excessive agricultural operations, which also involve an elevated level of fuel-based systems, the combustion of agricultural waste, the use of fossil fuels, deforestation, etc., are some examples of anthropogenic activities (Abbass et al., 2022). Climate change illustrates the ecosystems that are already under extreme stress from growing urbanization, industrialization, and economic growth. As a result, these human-caused climate changes harm infrastructures, human health, and the overall yield of plant species (Ishikawa-Ishiwata & Furuya, 2022). Ever since, the majority of energy produced in emerging countries comes from fossil fuels, and energy consumption has increased greenhouse gas (GHG) levels in response to global warming (Balsalobre-Lorente et al., 2020).

Overexploitation and habitat destruction have diminished MAPs' population, and some plant species in the IHR are facing extinction. In order to support the MAP sector's long-term sustainable growth, it needs urgent intervention and requests for suitable guidelines as well as indicators (Negi et al., 2018; Shrestha et al., 2022). According to Willis (2017), there are currently 28,000 plant species worldwide that have medicinal uses (World Bank, 2018). Approximately, more than 25% of recently released medications come from inherent sources such as medicinal plants, which are also widely used to develop more than 70% of anticancer drugs (Newman et al., 2003; Cragg & Newman, 2013; Pandey et al., 2015, 2020; Singh et al., 2018).

Around 10,000 genera of medicinal plants are vulnerable, according to the IUCN Species Survival Commission's Medicinal Plants Specialist Group (Mehta et al., 2021). Global warming is arising more rapidly, and the distribution of some IHR vegetation is already being impacted by modifications in snow patterns, ice cover, and temperatures. Some specialists postulate that the existence of some MAPs and their chemical composition may differ due to climate change in the IHR. According to published research, secondary metabolites and other molecules derived from MAPs may be impacted by climatic change (Wang et al., 2019; Applequist et al., 2020; Acharya et al., 2022).

Besides providing an in-depth analysis of how climate change affects global agriculture, biodiversity, MAPs in the IHR, and secondary metabolites of plants, this review also aims to suggest some precautionary actions and mitigation approaches that might be applied as trustworthy alternatives to deal with the consequences of climate change (Malhi et al., 2021; Abbass et al., 2022; Yeshi et al., 2022). To reduce climate change, the following steps have been taken by several advisory teams: conserving valuable plants, providing harvesters with conservation training and authorizing commercial material, and maintaining traditional knowledge (Raza et al., 2019; Malhi et al., 2021). This study aims to emphasize the implications of climate change from a socio-scientific perspective by examining the published literature that have an impact on the environment, especially on the MAPs distributed in the Himalayan region.

2 Climate Change Impacts on the Biodiversity of Medicinal and Aromatic Plants

The Indian Himalayan Region is one of the major hotspots in the world and is well known for its habitation of therapeutically important medicinal plants, according to modern studies (Negi et al., 2018; Mehta et al., 2021; Tripathi et al., 2022). The IHR also contains a dense flora of MAPs, with 1748 species so far being recorded to be of medicinal value (Ratha et al., 2012; Tripathi et al., 2022). The IHR is not an exception to the experts' warning that plant species that are native to geographical locations which are remarkably vulnerable to climate change might soon be in serious jeopardy. The Indian forests are already under a lot of pressure from overexploitation, habitat loss, forest fires, livestock grazing, tourism, fuel wood harvesting, and other factors (Negi et al., 2018). An additional source of stress will be the fluctuating environment that might have a significant impact on plant variety, while other pressures (such as the prevalence of insects and other pests, diseases, etc.) could cause these species to become extinct in their natural habitats (Yoon, 1994; Mahajan et al., 2020; Negi et al., 2022). The world has experienced an increase in heat waves, droughts, floods, fluctuating precipitation patterns, and other extreme occurrences as a result of unprecedented temperature rise (Zobayed et al., 2005). Climate change-induced rapid desertification and nutrient-deficient soils are the results of a peculiarly high rate of land degradation, which is getting worse gradually (Abbass et al., 2022; Negi et al., 2022).

The concept of "climate change" refers to alterations in the global temperature with time. By 2100, temperatures are predicted to rise by another 1.40–5.80 °C, which will undoubtedly have acute implications on biodiversity. Biodiversity and ecosystems are already being forced by climate change to adapt to changing habitats, altered life cycles, and the emergence of novel physical features (Raza et al., 2019). Since the middle of the nineteenth century, it has been noted that the Himalayas and its surroundings have warmed by about 0.68 °C (Ratha et al., 2012). Furthermore, climate change (atmospheric greenhouse gases, high temperature, extreme rainfall, flood, land degradation, drought, storm, etc.) events are increasing intensely, and these numerous climatic factors have influence on MAPs in the Indian Himalayan Region as shown in Table 1. In addition, the abovementioned climatic change variables have an impact on plants' morphological/phenological traits, molecular/biochemical functions, habitat destruction, alteration in secondary

Table 1 Some instances of the effect of climatic change on diversity, secondary metabolites, and morpho-physiological patterns of Himalayan MAPs

S. no.	Medicinal and aromatic plants (Family)	Impacts of climate change	References
1.	*Pinus taeda* (Pinaceae)	***Ozone stress*** Increases tannins without any increase in total phenolic content	Jordan et al. (1991); Gupta et al. (2019); Mahajan et al. (2020)
2.	*Cymbopogon nardus* (Gramineae)	***Drought stress*** Reduces plant height, leaf length, leaf weight, leaf area, and constituents of essential oils (citral and geraniol)	Singh-Sangwan et al. (1994); Mahajan et al. (2020)
3.	*Cymbopogon pendulus* (Gramineae)	***Drought condition*** Decreases plant height, leaf length, leaf weight, and leaf area, while essential oil content (citral and geraniol) is enhanced	Singh-Sangwan et al. (1994); Mahajan et al. (2020)
4.	*Hypericum brasiliense* (Guttiferae)	***Drought condition*** The content of phenolic compounds and betulinic acid is enhanced	de Abreu and Mazzafera (2005)
5.	*Hypericum perforatum* (Hypericaceae)	***Elevated CO_2*** Hypericin concentration is reduced significantly by 22.0% ***Elevated temperature*** Hypericin concentration was 13.76% higher Dry biomass including root was higher under both conditions	Sharma et al. (2020b)
6.	*Aconitum heterophyllum* (Ranunculaceae)	***Habitat loss, overexploitation, and environmental stress*** Significant loss of diversity	Malcolm et al. (2006); Das et al. (2016)
7.	*Decalepis hamiltonii* (Periploaceace)		
8.	*Picrorhiza kurroa* (Plantaginaceae)		
9.	*Taxus wallichiana* (Taxaceae)		
10.	*Panax quinquefolius* (Araliaceae)	***Elevated temperature*** Early onset of leaf senescence, less root biomass, and reduction in photosynthesis while enhancing ginsenoside content	Jochum et al. (2007); Mahajan et al. (2020)

(continued)

Table 1 (continued)

S. no.	Medicinal and aromatic plants (Family)	Impacts of climate change	References
11.	*Ginkgo biloba* (Ginkgoaceae)	***Ozone (O_3) elevation*** Increases the concentration of bilobalide and ginkgolide but decreases the content of tannins and isorhamnetin	Xingyuan et al. (2009)
		Elevated CO_2 Decreases the content of kaempferol aglycon by 10% and isorhamnetin aglycon by 15%; elevated O_3 decreases the content of isorhamnetin aglycon by 7% but increases the concentration of quercetin aglycon by 6%	Huang et al. (2010)
12.	*Salvia officinalis* (Lamiaceae)	***Drought stress*** Increases the content of monoterpenes	Nowak et al. (2010)
		Ozone stress Reduces photosynthetic efficiency and chlorophyll loss while increasing phenolic content, remarkably gallic acid, catechinic acid, caffeic acid, and rosmarinic acid	Pellegrini et al. (2015); Gupta et al. (2019)
13.	*Salvia sclarea* (Lamiaceae)	***Cold stress*** Reduces individual leaf area but increases the length and number of spikes with higher percentage of linalool (52.9%) and sclareol (39.4%) in essential oil	Kaur et al. (2015)
		Elevated CO_2 Total dry biomass production was higher, while the number of leaves per plant and leaf area per plant decreased Increases photosynthetic pigments and glutathione content (18.3%) ***Elevated temperature*** Increases dry biomass production, number of leaves per plant, and leaf area per plant Violaxanthin and neoxanthin were found to be decreased by 23% and 18%, respectively	Kumar et al. (2017)
		Drought stress Reduction in plant biomass and secondary metabolite content	Mahajan et al. (2020)
14.	*Salvia miltiorrhiza* (Lamiaceae)	***Drought stress*** Rosmarinic acid decreases and salvianolic acid increases	Yeshi et al. (2022)
15,	*Artemisia annua* (Asteraceae)	***Increasing CO_2 level*** Leads to an increased concentration of artemisinin	Zhu et al. (2015); Applequist et al. (2020)

(continued)

Table 1 (continued)

S. no.	Medicinal and aromatic plants (Family)	Impacts of climate change	References
16.	*Valeriana jatamansi* (Caprifoliaceae)	***Elevated CO_2*** Enhances the essential oil content (patchouli alcohol, globulol, β-patchoulene, germacrene D, etc.) by 17.7% ***Elevated temperature*** Decreases essential oil content by 4.3%	Kaundal et al. (2018)
		Drought stress Increases valerenic acid content (1.0–6.9-fold) and reduces the stomatal conductance and transpiration rate ***Elevated temperature*** Content of valerenic acid is higher (1.0–1.2-fold) in belowground part, while it is decreased (1.1–1.3-fold) in aerial part, and the stomatal conductance and transpiration rate were found to be increased	Pandey et al. (2021)
17.	*Digitalis lanata* (Plantaginaceae)	***High levels of CO_2*** Increases digoxin (3.5-fold), whereas the concentrations of digitoxin and digitoxigenin were significantly decreased	Gupta et al. (2019)
18.	*Hedychium spicatum* (Zingiberaceae)	***Drought stress*** Linalool content slightly increased (1.0–1.5-fold) ***Elevated temperature*** Linalool content significantly increased (6.2–6.5-fold)	Pandey et al. (2021)
19.	*Rhododendron arboretum* (Ericaceae)	Early blooming in cold zone	Acharya et al. (2022)
20.	*Salix sikkimensis* (Salicaceae)	In cold stress early leaf emergence	Acharya et al. (2022)

metabolites and physiological differences, etc. (Fig. 1). The devastating impacts of climate change are shown in plant growth and yield. Due to the significance of MAPs for traditional medicine and extensive trade collection, high-altitude MAPs are a major concern throughout the IHR. Important MAPs (such as *Aconitum*, *Rhododendron*, *Juniperus*, *Taxus*, *Bergenia*, *Ephedra*, *Dactylorhiza*, *Paeonia*, *Fritillaria*, *Valeriana*, *Asparagus*, *Cinnamomum*, *Polygonatum*, *Podophyllum*, *Swertia*, *Picrorhiza*, *Nardostachys*, *Saussurea*, *Artemisia* species, etc.) and many additional high-altitude Himalayan MAPs have immense market potential but are currently confined to either inaccessible locations or protected habitats as a result of overexploitation and several other human and environmental circumstances (Ratha et al., 2012; Joshi et al., 2016; Mehta et al., 2021; Murshed, 2022). The range and

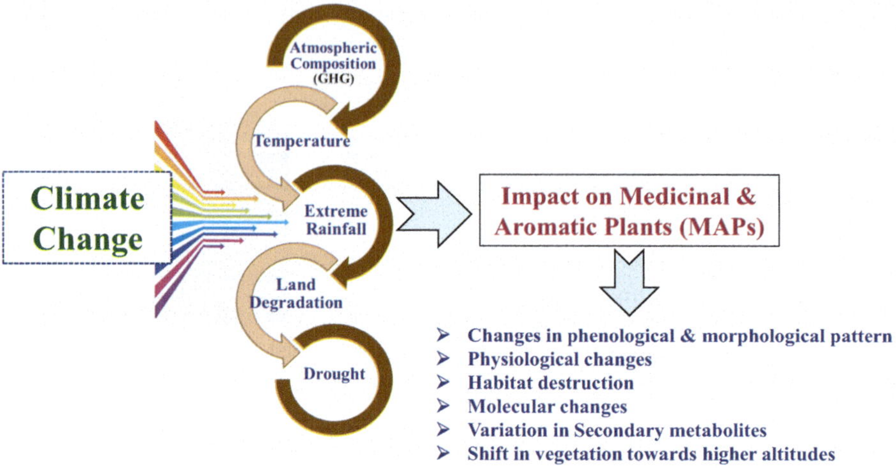

Fig. 1 Schematic representation of prospective impacts of climate change on MAPs

phenology of some plant species are already being impacted by modifications in snow patterns and temperatures in the IHR, especially in alpine regions (Nautiyal et al., 2004).

2.1 Impact of Climate Change on the Secondary Metabolites of MAPs

The world is alarmed about the changes in climate since they are affecting MAPs and its secondary metabolites. Global warming is a result of industrialization and toxic gas emissions, which eventually damages the ecosystem. Changes in climate have a negative impact on the distribution, quality, quantity, and diversity of MAPs. In addition to the aforesaid effects, climate change might have an impact on the chemical constituents of MAPs distributed at high-altitude regions (Gairola et al., 2010; Mahajan et al., 2020). The secondary metabolites and other compounds produced from MAPs can be impacted by temperature stress (Zobayed et al., 2005; Acharya et al., 2022). In stressed conditions, usually the production of secondary metabolites may alter frequently (Acharya et al., 2022). Earlier studies revealed that increased temperature may influence the secondary metabolites of MAPs in the IHR (Loreto et al., 2006; Liao et al., 2017; Wang et al., 2019; Applequist et al., 2020) (Table 1). Some researchers claim that secondary metabolites increased when exposed to elevated temperature (Litvak et al., 2002; Harsh et al., 2016; Negi et al., 2022), while other researchers contradict these results (Snow et al., 2003; Acharya et al., 2022). Overall, the most prominent adverse variations on Himalayan MAPs caused by the climatic change are examined, and some instances of the effect of climatic change on diversity, secondary metabolites, and morpho-physiological patterns of Himalayan MAPs are given in Table 1. Due to climate change, plants move towards a new range, towards higher elevation, for proper growth and development. This circumstance could result in the extinction of numerous native species worldwide due to habitat loss and migratory difficulties associated with climate change (Gupta et al., 2019; Pandey et al., 2022). In the previous studies, elevated CO_2 levels precisely impact the biomass and morpho-physiological patterns of MAPs as well as the content of their bioactive molecules (Huang et al., 2010; Kumar et al., 2017; Sharma et al., 2020b). Scientifically, it is proven that negligence of human activities on earth have significantly raised the level of the greenhouse gas O_3 in the atmosphere. It causes oxidative stress in plants, slows down photosynthetic efficiency and respiration rate, and changes the secondary metabolites of MAPs (Jordan et al., 1991; Xingyuan et al., 2009; Pellegrini et al., 2015; Gupta et al., 2019; Mahajan et al., 2020). According to a theory, elevated CO_2 levels can trigger MAPs' growth cycles, and that physiological changes in MAPs may lead to alterations in their active components (Sharma et al., 2020b). Plant secondary metabolites are important in the interactions between plants and their environment and have a significant impact on human health (Gairola et al., 2010). Although a thorough knowledge of such factors is currently missing, it is anticipated that alterations in the overall climate will have a remarkable impact on plant secondary metabolism (Sun et al., 2023).

2.2 *Adaptation Approaches for Climate Change*

According to published observations (Ratha et al., 2012; Negi et al., 2022), some important MAPs showed the following remarks in the perspective of climate change:

- The elevation of the vegetation shifts from lower to higher.
- Early blooming and changes in phenological patterns were noticed in most contexts, e.g. *Lantana camara* and *Rhododendron arboretum*.
- High-altitude plants have been shown to have long seed dormancy periods, e.g. *Saussurea obvallata*.
- The high-altitude medicinal plant's seeds have a low germination rate, e.g. *Picrorhiza kurroa* and *Polygonatum species*.
- In some situations, the erect behaviour of plants which grows vertically upwards might alter into prostate behaviour (horizontally grow e.g. creepers).
- Plant species from the Alpine Region are migrating to cold desert conditions.

It is important to investigate how medicinal and aromatic plants behave both in cultivated areas and in their natural environments. There is also an escalating opportunity to start research on how climate change affects phenology, species range alterations, habitat change, and secondary metabolite production of MAPs (Fig. 1). A network of long-term ecological research (LTER) stations needs to be set up immediately in India's various eco-regions. Knowing about the prospective impacts of increased CO_2, warmer temperatures, and increased UV radiation on plant's secondary metabolites is the key assignment for the future (Negi et al., 2018; Sharma et al., 2020b). Monitoring the long-term consequences of climate change variables and their interactions with secondary metabolites should be part of future research on MAPs. The consequences of climate change can be seen in ecosystems all around the world, including MAPs. All stakeholders' input is required for the development of an effective action, conservation, and management plan when acquiring a policy on the sector of medicinal and aromatic plants (Fig. 2). Due to the quickly changing atmosphere, MAPs in the IHR face challenges, and some researchers are concerned about the potential genetic diversity and the decline in native plant populations in the IHR (Mehta et al., 2021). It has been noted that plant phenologies and distributions are changing all over the world. These variables may eventually endanger wild MAP species, exposing certain early-blooming MAP species and posing additional concerns including migratory issues (Negi et al., 2022). Extreme weather is already having an impact on the production and supply of MAPs, and rising weather is predicted to have an adverse effect on MAP yields in the future.

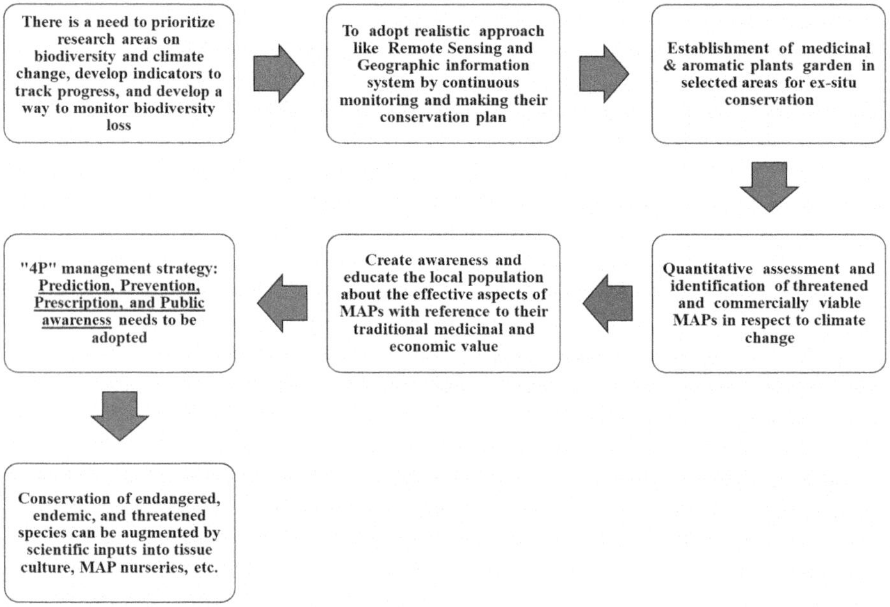

Fig. 2 The appropriate mitigation and conservation strategies to overcome climate variability on MAPs

2.3 *The Appropriate Mitigation and Conservation Strategies to Overcome Climate Variability on MAPs*

The diversity of MAPs in the Himalayan region is facing numerous risks. According to literature (Ratha et al., 2012; Negi et al., 2018; Yeshi et al., 2022) and the governments' promotional schemes for conservation and cultivation of MAPs, here are some recommendations (Fig. 2) to overcome the impact of climate change in IHR:

- There is a requisite to arrange research areas on biodiversity and climate change, develop indicators to track progress, and develop a way to monitor biodiversity loss.
- To adopt representative approach like remote sensing (RS) and geographic information system (GIS) by constant examining and making strategies for their conservation plan.
- Creation of MAP gardens, nurseries for supply of quality planting material, cultivation of threatened high-value plant species in specific regions.
- Particular attention should be given to plant species that are essential to human subsistence and necessitate quantitative analysis and identification of medicinal plants that are both threatened and commercially viable in the context of climate change.
- Improved community awareness regarding biodiversity conservation, advantages of MAPs, and climate change and emphasis on their practical and financial benefits.
- "4P" management strategy—Prediction, Prevention, Prescription, and Public awareness—should be implemented.
- Conservation of endangered, endemic, and threatened species can be augmented by scientific inputs into tissue culture, MAP nurseries, etc. The creation of herbaria indirectly helps in the identification, evaluation, and investigation of valuable MAP species which must be classified according to the invasion stage because this will affect management.

3 Future Strategies for Research

Climate change refers to a shifting target for plant growth adaptation and includes variations in rainfall, temperature, and composition of the environment (greenhouse gases), within other variables. The greenhouse gases (GHGs) are hydrofluorocarbons, carbon dioxide, methane, nitrous oxide, and chlorofluorocarbons. Daily emissions of CO_2 into the atmosphere are over 70 million tonnes (Harish et al., 2012; IPCC, 2014; Sharma et al., 2020b). The burning of fossil fuels and deforestation are two examples of human actions that have increased the concentration of GHGs in the biome, causing climatic warming and disruptions to hydrologic cycles. By the end of the century, atmospheric CO_2 is likely to reach 730–1000 parts per million (ppm), which will contribute to the expected increases in average global surface temperature of 1.0–3.7 °C throughout this time (IPCC, 2014; Gray & Brady, 2016; Sharma et al., 2020a).

According to the literature, the following steps should be taken to mitigate climate change:

- Reduce the consumption of chlorofluorocarbons and fossil fuels.
- Conserve energy.
- Expand tree plantations in order to consume more CO_2 from the surrounding environment.
- Follow sustainable cultivation of MAPs.
- All the countries must truly adhere to the Kyoto Protocol to reduce CO_2 emissions by 5.2%.

Research on the majority of high-altitude Himalayan medicinal plants has been concentrated on assessing the status and wild distribution patterns, looking into the financial consequences of the rich identification (in the context of biomass and secondary metabolite production), in vitro propagation protocols, mass propagation using conventional methods, and developing agro-techniques (Harish et al., 2012; Raza et al., 2019; Malhi et al., 2021; Sun et al., 2023). To examine how phenology, species range shifts, habitat change, and the synthesis of secondary metabolites in high-altitude MAPs are impacted by climate change, there is an increasing potential to start investigations. Figure 2 depicts the appropriate mitigation and conservation strategies to overcome climatic change on MAPs. Scientists have advised that because of the further challenges provided by climate change, certain significant species of MAPs, which are already threatened by overharvesting, may go extinct.

The scientists from the Intergovernmental Panel on Climate Change (IPCC) claim that limiting atmospheric CO_2 concentration to 400 ppm will significantly minimize the probability of global warming above 2 °C. Conclusively, this perspective understands the knowledge of plant developmental response to climate change (Gray & Brady, 2016; Sharma et al., 2020a; Negi et al., 2022).

Research conducted in other regions of the world revealed that high-altitude plants' life cycles and distribution are being significantly impacted by climate change. To establish conservation strategies (e.g. propagation, plant herbarium, germplasm, gene bank, seed bank, etc.) and agro-technologies for cultivation (Srivastava et al., 2009; Banerjee et al., 2012), more research is required on the habitat range and production efficiency of secondary metabolites of vulnerable Himalayan MAPs in the context of global change. These investigations are crucial because variations in secondary metabolites can have significant impacts on plants' ecology and evolution as noted by previous research (Mahajan et al., 2020). People should support the maintenance of conventional knowledge and the utilization of Himalayan MAPs and anticipate further initiatives to defend and preserve the Himalayan ecology. People strongly urge non-governmental organizations, governments, and communities involved in ethnobotany and public health to take action to support all populations in maintaining the availability of superior conventional remedies, especially those that depend on MAPs for their health concerns or income.

It is also important to protect and appreciate traditional knowledge about plants and their sustainable uses. Finally, it is important to train harvesters in sustainable practices that may support MAPs diversity in mountainous regions.

4 Conclusion

Due to its wide variety in elevation, deep glaciers, water valleys, places of abundant rainfall, and high desert regions, the Himalayas is a rich area of flora and fauna with significant diversity. There is an emerging consensus that the populations, infrastructure, ecology of cities, and plant biodiversity are at risk from the impacts of climate change. Climate change will alter plant growth and secondary metabolites, which will have significant impacts on the function of MAPs in natural ecosystems. Higher temperatures and more often ultimate weather incidents, such as heat waves and droughts, will be the outcome of future climatic variations and changes to the atmospheric chemical composition. Diverse factors include variations in annual rainfall, average temperature, heat waves, global changes in atmospheric gases, and alterations in sea level, and the MAPs of the Indian Himalayan range are being impacted by climate change. Although MAPs are more at risk from climate change, the impacts of climate change on MAPs have not been thoroughly explored. Influences of climate change on medicinal plants may become a more vital issue as this situation develops for the herbal community, users, producers, and species of MAPs. It is necessary to conduct extensive and ongoing field measurements at representative sites to have more awareness of how climate change affects medicinal plants and how it is responsible for such alterations. Important MAPs may go extinct from the Himalayan region soon if the rate of exploitation rate is not checked and conservation efforts are not taken seriously. Further research must be focused on the affected areas of the Himalayan region for developing conservation strategies, and threatened MAPs can be preserved by the establishment of a gene bank.

Acknowledgement I wish to express my sincere thanks to Prof. Niladari Bag, for his mentorship to carry out the research work at the Sikkim University. I am thankful to the Science and Engineering Research Board (SERB), Govt. of India for financial support in the form of SERB-National Post Doctoral Fellowship (PDF/2023/000111). I also wish to express my sincere thanks to the editors for providing the opportunity to contribute to this edited volume.

References

Abbass, K., Qasim, M. Z., Song, H., Murshed, M., Mahmood, H., & Younis, I. (2022). A review of the global climate change impacts, adaptation, and sustainable mitigation measures. *Environmental Science and Pollution Research, 29*, 42539–42559.

Acharya, S., Sigdel, S., Kandel, G., & Adhikari, P. (2022). Climate change: Threats on the medicinal plants in Nepal. *Asian Journal of Pharmacognosy, 1*, 41–48.

Applequist, W. L., Brinckmann, J. A., Cunningham, A. B., Hart, R. E., Heinrich, M., Katerere, D. R., & Van Andel, T. (2020). Scientists' warning on climate change and medicinal plants. *Planta Medica, 86*, 10–18.

Balsalobre-Lorente, D., Driha, O. M., Bekun, F. V., Sinha, A., & Adedoyin, F. F. (2020). Consequences of COVID-19 on the social isolation of the Chinese economy: Accounting for the role of reduction in carbon emissions. *Air Quality, Atmosphere and Health, 13*, 1439–1451.

Banerjee, S., Singh, S., Pandey, H., Pandey, P., & Rahman, L. U. (2012). Conservation and storage of *Curcuma amada* Roxb synseeds on Luffa sponge matrix and RAPD analysis of the converted plantlets. *Industrial Crops and Products, 36*, 383–388.

Chien, F., Anwar, A., Hsu, C. C., Sharif, A., Razzaq, A., & Sinha, A. (2021). The role of information and communication technology in encountering environmental degradation: Proposing an SDG framework for the BRICS countries. *Technology in Society, 65*, 101587.

Cragg, G. M., & Newman, D. J. (2013). Natural products: A continuing source of novel drug leads. *Biochimica et Biophysica Acta (BBA) – General Subjects, 1830*, 3670–3695.

Das, M., Jain, V., & Malhotra, S. (2016). Impact of climate change on medicinal and aromatic plants. *Indian Journal of Agricultural Sciences, 86*, 1375–1382.

de Abreu, I. N., & Mazzafera, P. (2005). Effect of water and temperature stress on the content of active constituents of *Hypericum brasiliense* Choisy. *Plant Physiology and Biochemistry, 43*, 241–248.

Dejanovic, T., Trbić, G., & Popov, T. (2019). Hail as a natural disaster in Bosnia and Herzegovina. In W. Leal Filho, G. Trbic, & D. Filipovic (Eds.), *Climate change adaptation in eastern Europe* (Climate Change Management) (pp. 245–266). Springer. https://doi.org/10.1007/978-3-030-03383-5_17

Gairola, S., Shariff, N. M., & Bhatt, A. (2010). Influence of climate change on production of secondary chemicals in high altitude medicinal plants: Issues needs immediate attention. *Journal of Medicinal Plants Research, 4*, 1825–1829.

Gavrić, T., Marković, S., & Čengić, L. (2023). Factibility of growing basil as an alternative crop for adaptation to climate change in Bosnia and Herzegovina. *Chilean Journal of Agricultural Research, 83*, 43–51.

Gray, S. B., & Brady, S. M. (2016). Plant developmental responses to climate change. *Developmental Biology, 419*, 64–77.

Gupta, A., Singh, P. P., Singh, P., Singh, K., Singh, A. V., Singh, S. K., & Kumar, A. (2019). Medicinal plants under climate change: Impacts on pharmaceutical properties of plants. In *Climate change and agricultural ecosystems* (pp. 181–209). Elsevier/Woodhead Publishing.

Harish, B. S., Dandin, S. B., Umesha, K., & Sasanur, A. (2012). Impact of climate change on medicinal plants–A review. *Ancient Science of Life, 32*, S23.

Harsh, A., Sharma, Y. K., Joshi, U., Rampuria, S., Singh, G., Kumar, S., & Sharma, R. (2016). Effect of short-term heat stress on total sugars, proline and some antioxidant enzymes in moth bean (*Vigna aconitifolia*). *Annals of Agricultural Science, 61*, 57–64.

Huang, W., He, X. Y., Liu, C. B., & Li, D. W. (2010). Effects of elevated carbon dioxide and ozone on foliar flavonoids of *Ginkgo biloba*. *Advanced Materials Research, 113*, 165–169.

IPCC. (2014). *Climate change 2014: Synthesis report*. Contribution of Working Groups I, II and III to the Fifth Assessment Report of the Intergovernmental Panel on Climate Change (Core Writing Team, R. K. Pachauri, & L. A. Meyer, Eds.; p. 151). IPCC. https://doi.org/10013/epic.45156.d001

Ishikawa-Ishiwata, Y., & Furuya, J. (2022). Economic evaluation and climate change adaptation measures for rice production in Vietnam using a supply and demand model: Special emphasis on the Mekong River Delta region in Vietnam. In *Interlocal adaptations to climate change in east and southeast Asia: Sharing lessons of agriculture, disaster risk reduction, and resource management* (pp. 45–53). Springer International Publishing.

Jochum, G. M., Mudge, K. W., & Thomas, R. B. (2007). Elevated temperatures increase leaf senescence and root secondary metabolite concentrations in the understory herb *Panax quinquefolius* (Araliaceae). *American Journal of Botany, 94*, 819–826.

Jordan, D. N., Green, T. H., Chappelka, A. H., Lockaby, B. G., Meldahl, R. S., & Gjerstad, D. H. (1991). Response of total tannins and phenolics in loblolly pine foliage exposed to ozone and acid rain. *Journal of Chemical Ecology, 17*, 505–513.

Joshi, R. K., Satyal, P., & Setzer, W. N. (2016). Himalayan aromatic medicinal plants: A review of their ethnopharmacology, volatile phytochemistry, and biological activities. *Medicine, 3*, 6.

Kaundal, M., Bhatt, V., & Kumar, R. (2018). Elevated CO2 and temperature effect on essential oil content and composition of *Valeriana jatamansi* Jones. with organic manure application in a western Himalayan region. *Journal of Essential Oil Bearing Plants, 21*, 1041–1050.

Kaur, T., Bhat, H. A., Bhat, R., Kumar, A., Bindu, K., Koul, S., & Vyas, D. (2015). Physio-chemical and antioxidant profiling of *Salvia sclarea* L. at different climates in north-western Himalayas. *Acta Physiologiae Plantarum, 37*, 1–10.

Kumar, R., Kaundal, M., Sharma, S., Thakur, M., Kumar, N., Kaur, T., Vyas, D., & Kumar, S. (2017). Effect of elevated [CO_2] and temperature on growth, physiology and essential oil composition of Salvia sclarea L. in the western Himalayas. *Journal of Applied Research on Medicinal and Aromatic Plants, 6*, 22–30.

Liao, P., Liu, D., Xu, T. R., Yang, Y., & Cui, X. (2017). Soil water stress attenuate the growth and development but enhance the saponin synthesis of *Panax notogesing* during flowering stage. *Industrial Crops and Products, 108*, 95–105.

Litvak, M. E., Constable, J. V., & Monson, R. K. (2002). Supply and demand processes as controls over needle monoterpene synthesis and concentration in Douglas fir [*Pseudotsuga menziesii* (Mirb.) Franco]. *Oecologia, 132*, 382–391.

Loreto, F., Barta, C., Brilli, F., & Nogues, I. (2006). On the induction of volatile organic compound emissions by plants as consequence of wounding or fluctuations of light and temperature. *Plant, Cell & Environment, 29*, 1820–1828.

Mahajan, M., Kuiry, R., & Pal, P. K. (2020). Understanding the consequence of environmental stress for accumulation of secondary metabolites in medicinal and aromatic plants. *Journal of Applied Research on Medicinal and Aromatic Plants, 18*, 100255.

Malcolm, J. R., Liu, C., Neilson, R. P., Hansen, L., & Hannah, L. (2006). Global warming and extinctions of endemic species from biodiversity hotspots. *Conservation Biology, 20*, 538–548.

Malhi, G. S., Kaur, M., & Kaushik, P. (2021). Impact of climate change on agriculture and its mitigation strategies: A review. *Sustainability, 13*, 1318.

Mehta, P., Bisht, K., & Sekar, K. C. (2021). Diversity of threatened medicinal plants of Indian Himalayan region. *Plant Biosystems – An International Journal Dealing with all Aspects of Plant Biology, 155*, 1121–1132.

Murshed, M. (2022). Pathways to clean cooking fuel transition in low and middle income Sub-Saharan African countries: The relevance of improving energy use efficiency. *Sustainable Production and Consumption, 30*, 396–412.

Nautiyal, M. C., Nautiyal, B. P., & Prakash, V. (2004). Effect of grazing and climatic changes on alpine vegetation of Tungnath, Garhwal Himalaya, India. *Environmentalist, 24*, 125–134.

Negi, V. S., Kewlani, P., Pathak, R., Bhatt, D., Bhatt, I. D., Rawal, R. S., Sundriyal, R. C., & Nandi, S. K. (2018). Criteria and indicators for promoting cultivation and conservation of medicinal and aromatic plants in Western Himalaya, India. *Ecological Indicators, 93*, 434–446.

Negi, V. S., Tiwari, D. C., Singh, L., Thakur, S., & Bhatt, I. D. (2022). Review and synthesis of climate change studies in the Himalayan region. *Environment, Development and Sustainability, 24*, 10471–10502.

Newman, D. J., Cragg, G. M., & Snader, K. M. (2003). Natural products as sources of new drugs over the period 1981–2002. *Journal of Natural Products, 66*, 1022–1037.

Nowak, M., Kleinwächter, M., Manderscheid, R., Weigel, H. J., & Selmar, D. (2010). Drought stress increases the accumulation of monoterpenes in sage (*Salvia officinalis*), an effect that is compensated by elevated carbon dioxide concentration. *Journal of Applied Botany and Food Quality, 83*, 133–136.

Pandey, H., Pandey, P., Singh, S., Gupta, R., & Banerjee, S. (2015). Production of anti-cancer triterpene (betulinic acid) from callus cultures of different *Ocimum* species and its elicitation. *Protoplasma, 252*, 647–655.

Pandey, H., Pandey, P., Singh, S., Negi, A. S., & Banerjee, S. (2020). Unveiling the future reservoir of anti-cancer molecule—Combretastatin A4 from callus and cell aggregate suspension culture of flame creeper (*Combretum microphyllum*): Growth, exudation and elicitation studies. *Plant Cell, Tissue and Organ Culture, 143*, 681–691.

Pandey, V., Tiwari, D. C., Dhyani, V., Bhatt, I. D., Rawal, R. S., & Nandi, S. K. (2021). Physiological and metabolic changes in two Himalayan medicinal herbs under drought, heat and combined stresses. *Physiology and Molecular Biology of Plants, 27*, 1523–1538.

Pandey, J., Singh, L., Saxena, G., & Devkota, H. P. (2022). Environmental challenges for himalayan medicinal plants. In *Environmental challenges and medicinal plants: Sustainable production solutions under adverse conditions* (pp. 29–47). Springer International Publishing.

Pellegrini, E., Francini, A., Lorenzini, G., & Nali, C. (2015). Ecophysiological and antioxidant traits of *Salvia officinalis* under ozone stress. *Environmental Science and Pollution Research, 22*, 13083–13093.

Rakic, Z., & Johnson, C. B. (2008). Influence of environmental factors (including UV-B radiation) on the composition of the essential oil of *Ocimum basilicum*-sweet basil. *Journal of Herbs Spices & Medicinal Plants, 9*, 157–162.

Ratha, K. K., Sruajani, M. S., Arya, J. C., & Joshi, G. C. (2012). Impact of climate change on diversity of himalayan medicinal plant: A threat to ayurvedic system of medicine. *International Journal of Research in Ayurveda and Pharmacy, 3*, 1–5.

Raza, A., Razzaq, A., Mehmood, S. S., Zou, X., Zhang, X., Lv, Y., & Xu, J. (2019). Impact of climate change on crops adaptation and strategies to tackle its outcome: A review. *Plants, 8*, 34.

Sharma, R., Sinha, A., & Kautish, P. (2020a). Examining the impacts of economic and demographic aspects on the ecological footprint in South and Southeast Asian countries. *Environmental Science and Pollution Research, 27*, 36970–36982.

Sharma, S., Walia, S., Rathore, S., Kumar, P., & Kumar, R. (2020b). Combined effect of elevated CO2 and temperature on growth, biomass and secondary metabolite of *Hypericum perforatum* L. in a western Himalayan region. *Journal of Applied Research on Medicinal and Aromatic Plants, 16*, 100239.

Shrestha, U. B., Lamsal, P., Ghimire, S. K., Shrestha, B. B., Dhakal, S., Shrestha, S., & Atreya, K. (2022). Climate change-induced distributional change of medicinal and aromatic plants in the Nepal Himalaya. *Ecology and Evolution, 12*, e9204.

Singh, S., Pandey, P., Ghosh, S., & Banerjee, S. (2018). Anti-cancer labdane diterpenoids from adventitious roots of *Andrographis paniculata*: Augmentation of production prospect endowed with pathway gene expression. *Protoplasma, 255*, 1387–1400. https://doi.org/10.1007/s00709-018-1211-7

Singh-Sangwan, N., Abad Farooqi, A. H., & Singh Sangwan, R. (1994). Effect of drought stress on growth and essential oil metabolism in lemongrasses. *New Phytologist, 128*, 173–179.

Snow, M. D., Bard, R. R., Olszyk, D. M., Minster, L. M., Hager, A. N., & Tingey, D. T. (2003). Monoterpene levels in needles of Douglas fir exposed to elevated CO2 and temperature. *Physiologia Plantarum, 117*, 352–358.

Srivastava, V., Khan, S. A., & Banerjee, S. (2009). An evaluation of genetic fidelity of encapsulated microshoots of the medicinal plant: *Cineraria maritima* following six months of storage. *Plant Cell Tissue and Organ Culture, 62*, 235–238.

Sun, Y., Alseekh, S., & Fernie, A. R. (2023). Plant secondary metabolic responses to global climate change: A meta-analysis in medicinal and aromatic plants. *Global Change Biology, 29*, 477–504.

Tripathi, A. K., Pandey, P. C., Sharma, J. K., Triantakonstantis, D., & Srivastava, P. K. (2022). Climate change and its impact on forest of Indian Himalayan region: A review. In *Climate change: Impacts, responses and sustainability in the Indian Himalaya* (pp. 207–222). Springer Nature.

Vucijak, B., Kupusović, T., Midžić-Kurtagić, S., Silajdžić, I., & Ćerić, A. (2014). Evaluation of the climate change effects to the precipitation patterns in the selected Bosnia and Herzegovina cities. *Thermal Science, 18*(3), 787–798. https://doi.org/10.2298/TSCI1403787V

Walia, S., Rathore, S., & Kumar, R. (2022). Elucidating the mechanisms, responses and future prospects of medicinal and aromatic plants to elevated CO2 and elevated temperature. *Journal of Applied Research on Medicinal and Aromatic Plants, 26*, 100365.

Wang, S., Liang, W., Yao, L., Wang, J., & Gao, W. (2019). Effect of temperature on morphology, ginsenosides biosynthesis, functional genes, and transcriptional factors expression in Panax ginseng adventitious roots. *Journal of Food Biochemistry, 43*, e12794.

Willis, K. J. (2017). *State of the world's plants report-2017*. Royal Botanic Gardens.

WMO. (1992). International meteorological vocabulary. In *World Meteorological Organization (WMO)* (2nd ed.).

World Bank. (2018). *Medicinal and aromatic plants* (Strategic Segmentation Analysis: Nepal). http://documents.worldbank.org/curated/en/496421556737648658/pdf/Medicinal-and-AromaticPlants.pdf

Xingyuan, H. E., Huang, W., Wei, C. H., Tian, D. O., Changbing, L. I., Zhenju, C. H., Sheng, X. U., & Yanan, R. U. (2009). Changes of main secondary metabolites in leaves of *Ginkgo biloba* in response to ozone fumigation. *Journal of Environmental Sciences, 21*, 199–203.

Yeshi, K., Crayn, D., Ritmejerytė, E., & Wangchuk, P. (2022). Plant secondary metabolites produced in response to abiotic stresses has potential application in pharmaceutical product development. *Molecules, 27*, 313.

Yoon, C. K. (1994, June 21). Warming moves plants up peaks, threatening extinction. *The New York Times*, p. C4.

Zhu, C., Zeng, Q., McMichael, A., Ebi, K. L., Ni, K., Khan, A. S., Zhu, J., Liu, G., Zhang, X., Cheng, L., & Ziska, L. H. (2015). Historical and experimental evidence for enhanced concentration of artemisinin, a global anti-malarial treatment, with recent and projected increases in atmospheric carbon dioxide. *Climatic Change, 132*, 295–306.

Zobayed, S. M., Afreen, F., & Kozai, T. (2005). Temperature stress can alter the photosynthetic efficiency and secondary metabolite concentrations in St. John's wort. *Plant Physiology and Biochemistry, 43*, 977–984.

Part V
Meeting Sustainable Development Goals (SDGs) in Mountains

Landslide's Recovery and Restoration: A Way to Sustainable Development Goal (SDG) 15

Deepesh Goyal and Varun Joshi

1 Introduction

Landslides, often triggered by heavy rainfall, are a type of land deterioration that can have devastating impacts on human life and economic activities in mountains. These gravity-induced downslope movements of earth materials, including debris, soil, and rock, are a significant environmental threat in mountainous regions worldwide, leading to loss of life and economic assets (Hung et al., 2016). Landslides are a complex natural hazard, occurring under various geomorphological and hydrogeological conditions (Pal et al., 2019), and can be triggered by prolonged precipitation, seismic activities, and human interventions. "In India, about 4.2 lakh km^2, excluding snow covered area, is vulnerable to landslide hazard," out of which 1.8 lakh km^2 lies in North-Eastern Himalayan states; 1.4 lakh km^2 in North-Western Himalayas; 0.9 lakh km^2 in Western Ghats and Konkan Hills; and 0.1 lakh km^2 in Eastern Ghats in Andhra Pradesh. The Himalayan Region, which is prone to landslides, is also located in the highest earthquake risk zones (Zones IV and V; BIS, 2002), where earthquake tremors of intensity VIII–IX can trigger. As a result, this region is also vulnerable to landslides triggered by earthquakes (Rawat et al., 2015).

The most significant forms of land deterioration in the Himalayan Region are landslips, mass movement, rock falls, rill and sheet erosion, and broadening of riverbanks (Mishra et al., 2022). According to Hung et al. (2016), landslides have the potential to harm biodiversity, erode the upper soil (fertile), disrupt the environment and ecosystems, and impede traffic flow in hilly regions. They can have enduring effects on the ecosystem and environment since they alter the landscape by

D. Goyal (✉) · V. Joshi
University School of Environment Management, Guru Gobind Singh Indraprastha University, New Delhi, India

S. Tripathi et al. (eds.), *Warming Mountains*,
https://doi.org/10.1007/978-3-031-62197-0_21

damaging wildlife, forests, and natural resources. Geertsema et al. (2009) stated that large forested areas can be destroyed, animal habitats can be eliminated, and productive soils can be eroded from hillsides due to landslides. Furthermore, the International Centre for Integrated Mountain Development (ICIMOD, 1994) noted that the degradation of land in the Himalayan Region is primarily caused by a decline in forest and pastureland, as well as landslides, mudslides, terrace failures, and steep slope erosion. Apart from this, the intensification of mass movements (landslips) in the Himalayan region can be imputed to human activities such as the cutting of trees, infrastructural developments, mining, and the development of hydropower projects (Bhattacharyya et al., 2015; Bhadouria et al., 2023). Currently, there is a lack of empirical evidence demonstrating that changing climatic patterns are responsible for landslides, although there is strong theoretical evidence for the rise in landslide frequency caused by intensifying rainfall (Crozier, 2010; Huggel et al., 2012; Gariano & Guzzetti, 2016). However, the total amount of rainfall is expected to rise up to 80% due to the combined effects of amplified precipitation intensity and larger weather systems (Prein et al., 2017). This increase in precipitation poses a considerable risk for land deterioration in terms of landslips, flash floods, and severe erosional activities, among others.

The restoration of degraded land is a crucial strategy for enhancing the well-being of communities and safeguarding their environment. To address this issue efficiently, rapidly, and cost-effectively, new approaches are essential (Cohen-Shacham et al., 2016; Escobedo et al., 2019). Nature-based solutions (NbSs) can offer a long-term, cost-efficient method for restoring ecosystems and support decision-makers in addressing geological hazards and land degradation within their jurisdiction. The International Union for Conservation of Nature (IUCN) has defined NbS as: "Actions to protect, sustainably manage and restore natural or modified ecosystems that address societal challenges effectively and adaptively, simultaneously providing human well-being and biodiversity benefits" (Cohen-Shacham et al., 2016; Dhyani et al., 2018). NbS is considered as a comprehensive concept for ecosystem-related strategies (Fig. 1). In recent years, there has been an increasing emphasis on incorporating the concept of ecosystem services in decision-making procedures and international agreements (Alexander et al., 2011). Various intergovernmental organizations (e.g., Intergovernmental Panel on Biodiversity and Ecosystem Services [IPBES], 2013), conventions (e.g., Convention on Biological Diversity, 2012), and policy initiatives (e.g., IUCN, 2014) have acknowledged the significance of ecological restoration in addressing global challenges such as combating desertification and land deterioration and reducing the effects of human-induced changing climate (Aronson & Alexander, 2013). Assessment studies on land degradation, overland flow, and loss of soil and nutrients are crucial for ensuring food security and livelihoods of inhabitants in mountainous regions (Mishra et al., 2022).

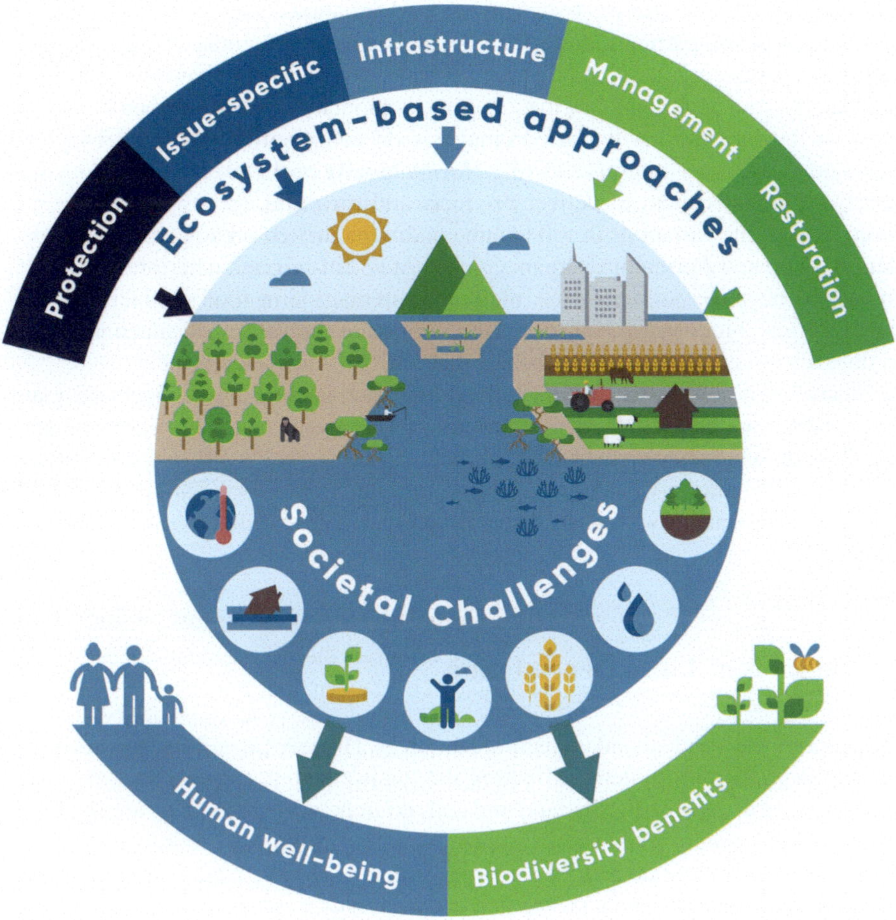

Fig. 1 Nature-based solutions (NbSs) as an umbrella concept for ecosystem-related approaches. (Modified from Cohen-Shacham et al. (2016))

1.1 Defining Land Degradation

"Land degradation" is a significant worldwide issue characterized by the decline or loss of land productivity, both biologically and economically, and a reduction in carbon sink in vegetation and soil. Deterioration of agricultural lands, forest area, and rangelands constitutes a significant obstacle to sustainable development, conservation of biodiversity, and mitigation to changing climate (IUCN, 2014). Furthermore, it contributes significantly to the loss of habitats and species, ultimately leading to a decline in ecosystem services. According to the Global Assessment of Human-induced Soil Degradation (Oldeman et al., 1991), approximately 15% of the earth's area suffers from land degradation.

Land degradation encompasses various land situations, such as surface erosion, compaction of soil particles, desertification and increase in salinity of land, decrease in soil nutrient content, and biodiversity (Gupta, 2019). This universal condition is affected by both natural and socioeconomic components and is described in various ways in the literature, with varied emphasis on conservation of biodiversity and ecological processes and services (e.g., Montanarella et al., 2018). The diminished ability of an ecosystem to deliver products and resources for its beneficiaries is referred to as the variations in soil health conditions, which is known as land degradation (Food and Agriculture Organization [FAO], 2014). Land deterioration results in various negative consequences, including a reduction in food productivity, less water storage, biodiversity loss, depletion of organic carbon, and the decline in ecosystem resources (IUCN, 2015; Gilbey et al., 2019).

According to Olsson et al. (2019), land degradation is the declining state of surface quality and productivity due to various anthropogenic factors, including changing climate, which have a detrimental impact on biological health and productivity and ecological integrity, leading to long-term reduction or loss. Soil quality degradation is a subcategory of land deterioration that directly affects soil quality and productivity.

2 Impacts of Land Degradation

India is endowed with abundant natural resources. However, it suffers from a variety of land degradation problems, affecting the quality and quantity of available land (Sreenivas et al., 2021). Fertile land and soil, nonrenewable resources, are the foundation for all primary production systems. As urbanization continues to expand and farm sizes shrink, the demand for food is rising. However, land degradation may result in a decrease in land's productive capacity, which can be either temporary or permanent depending on the extent of the damage. Environmental degradation is causing a noticeable decline in the productivity of natural resources such as fertile land, water, and grasslands, leading to increased health costs. This widespread impact is particularly evident in land-hunger, agriculture-dependent, and developing economies such as India (Reddy, 2003).

Approximately 20% of cultivated areas, 30% of forest land, and 10% of grasslands around the world are presently undergoing some form of degradation, with detrimental effects on the lives of roughly 1.5 billion individuals (Bai et al., 2008). Half of the planet's topsoil has been depleted in the last 150 years, according to the World Wildlife Fund (WWF). In India, nearly 146.8 million hectares of land is degraded, according to the NBSS&LUP (2004). Unsustainable and nonconventional cultivation practices, overgrazing, deforestation, soil erosion, and climate fluctuations are among the various factors contributing to land degradation (Bai et al., 2008). Land degradation can cause a reduction in productivity and ecosystem services and land impoverishment due to excessive use, a discrepancy between land quality and usage, and physical, biological, and chemical processes induced by human activity and natural hazards (Gupta, 2019) (Fig. 2).

As mentioned earlier, landslides are one of the most severe forms of land deterioration in mountainous areas due to the increased frequency and intensity in the area. They are triggered due to instability of slopes resulting from natural activities (seismic activities, intense precipitation, etc.) and/or anthropogenic activities (unplanned land use, deforestation, etc.) (Guariguata, 1990; Walker et al., 2009; Bhardwaj et al., 2019). Landslides are recognized as major processes of degradation to land and vegetation (Pearson et al., 2013; Gonzalez-Ollauri & Mickovski, 2016; Li et al., 2017), causing a direct significant impact on the dynamics of the ecosystems primarily by the removal of all existing vegetation, upper fertile layer of soil, and nutrients (Geertsema et al., 2009). They are the main erosional force which modifies the terrain and topography of the mountains (Geertsema et al., 2009; Parker et al., 2011). Moreover, due to the removal of vegetation, the rate of surface runoff increases on slopes which results in the washout of soil nutrients. Consequently, it resulted in the decrease of crop production (Lal, 2001).

The quality of land is closely interlinked with the quality of the soil. Typically, assessments of soil quality concentrate on determining soil nutrients, which are

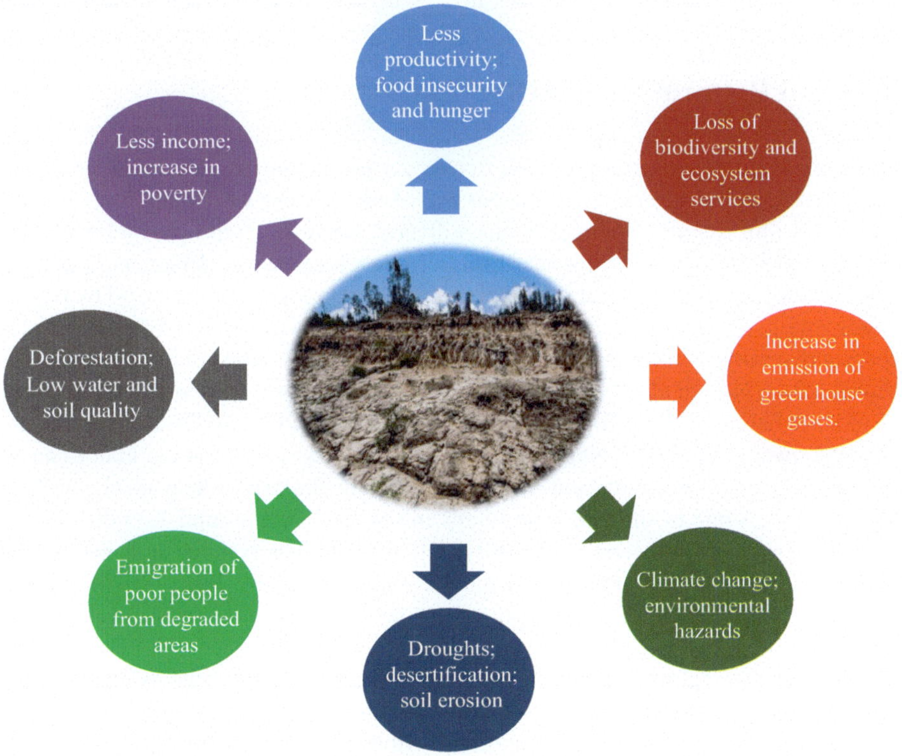

Fig. 2 Various impacts of land degradation

chemical indicators that satisfy agronomic needs and offer guidance for management practices. Land degradation is clearly visible by various signs such as decrease in soil organic carbon, lack of nutrients, the disintegration of soil structure and other physical properties, and impaired biological activities, all of which indicate a deterioration in soil quality (Basak et al., 2016, 2021; Choudhary et al., 2018). Beyond just the production of food, healthy soils provide a range of "ecosystem services," like water filtration, biodiversity preservation, and nutrient cycling (UK National Ecosystem Assessment, 2011). Thus, during the evaluation of soil health, it is important to consider the management objectives associated with different soil types and crop systems, followed by an assessment of pertinent physical, chemical, and biological soil indicators. Therefore, sustainable management of this essential natural resource is of utmost importance.

To enhance the resilience of national food security, it is crucial to devote concentrated effort to rehabilitating soils in mountainous areas as well apart from highly cultivated regions. Among the SDGs, Goal 15 is concentrated on "Life on land: Protect, restore and promote sustainable use of terrestrial ecosystems, sustainably manage forests, combat desertification, and halt and reverse land degradation and halt biodiversity loss" (United Nations, 2015). To achieve SDG 15, land degradation neutrality (LDN) Target 15.3 is focussed.

3 Land Restoration in IHR

The Indian Himalayan Region (IHR) covers a prominent part of India, with 11 states and 2 union territories, and about 6% of the country's population resides in the Himalayan Region. The region has a diverse landscape, with over 41.5% of it covered in forests, representing one-third of India's forest cover (Bhatt et al., 2020). The Himalayas has the special identity in the mountain ranges of the world by being young and geo-dynamic in nature. This region, with a wide variety of genetic and species diversity, is listed among the "Mega Biodiversity Hotspot" areas (Palni & Rawal, 2013; Negi et al., 2022). However, the decline in the forest areas and the loss of wildlife habitats are posing significant threat to this valuable environmental heritage (Semwal et al., 2013; Negi et al., 2018). Land degradation is the major cause for the loss of biodiversity and wildlife habitat in the IHR (Negi et al., 2022). Moreover, it is one of the major problems in the IHR, and around 1,33,06,736 ha land is under degraded land (SAC-ISRO, 2016). This has resulted in a variety of ecological and socioeconomic consequences as reported by Joshi (2018). These include various issues like drying up of springs/streams, soil erosion and downstream flooding, declining natural resources and livelihood options, etc. The natural resources in this region are a source of income and livelihood for the dwelling communities, making ecological restoration of degraded areas an essential activity to promote biodiversity conservation and improve livelihood opportunities.

3.1 Plant Species for Restoration

The potential for adaptation to and mitigation of climate change is enormous through the re-vegetation of a significant portion of degraded landscapes and natural forests, as highlighted by Lewis et al. (2019). According to research by Xu et al. (2019), ecological restoration can prevent landslides and stabilize slopes, as well as provide additional benefits to communities. For example, in Nepal, restoration efforts successfully stabilized degraded slopes and prevented landslides (Devkota et al., 2019), while in the Central Himalayas of India, the restoration of wastelands helped reduce deforestation (Dhyani & Dhyani, 2016).

The plant species selection for restoration processes depends on the adaptation of plant species to the eco-physiological conditions of the region along with the requirement of the local people (Negi et al., 2015; Bhatt et al., 2020). *Pennisetum purpureum*, a type of fodder grass, and *Cymbopogon citratus*, a medicinal grass, have the potential to serve multiple purposes, including stabilizing slopes, preventing soil erosion, reducing runoff, enhancing percolation, and increasing fodder production. *Alnus nepalensis*, *Quercus leucotrichophora*, *Quercus glauca*, *Grewia optiva*, *Pyrus pashia*, *Bombax ceiba*, etc. are preferred in Uttarakhand for their fuel, fodder, medicinal, and social values. Similarly, in the state of Himachal Pradesh, people prefer multipurpose species like *Toona ciliata*, *Vitex negundo*, *Prinsepia utilis*, *Prunus puddum*, etc. (Negi et al., 2022).

The "G.B. Pant National Institute of Himalayan Environment" (GBP-NIHE) developed the "Sloping Watershed Environmental Engineering Technology" (SWEET) model, an exemplary instance of NbS that manages natural resources, incorporating easy and relevant technological developments to mitigate the landslide hazards in the Indian Himalayas (Anonymous, 1994; Maikhuri et al., 2011). The model has been successfully implemented in several locations in the IHR and improved by the "Integrated Eco-development Research Programme" (IERP) of the GBP-NIHE (Bhatt et al., 2020). In North-Eastern Himalayas, "Contour Hedgerow Farming System Technology" (CHFST) was devised. This technology incorporates planting two lines of indigenous nitrogen-fixing hedgerow species (such as *Crotalaria tetragona*, *Desmodium rensonii*, *Leucaena leucocephala*, etc.) parallel to the contour lines (Sundriyal & Jamir, 2005).

3.2 Land Restoration for Achieving SDGs

India's progress towards achieving the SDGs is evident in the improved overall score of its SDG Index, rising from 60 in 2019–2020 to 66 in 2020–2021. This indicates that the country has made significant strides towards achieving its SDG targets. For SDG 15 (Life on land), India has a composite score of 66 (NITI Aayog, 2021).

According to the International Resource Panel (IRP, 2019), the approach to restoring land has the huge capability to offer mutual benefits to all Sustainable Development Goals (SDGs), either directly or indirectly. The SDGs, along with various global policy frameworks, discuss land and soil in direct and indirect ways. Healthy and fertile soils and sustainable land use management are important for achieving several SDGs. Therefore, a paradigm shift in techniques and perceptions is necessary to achieve the 2030 Agenda and the SDGs. SDG 15 cannot be tackled in isolation as it is intricately linked to all other SDGs with cause-and-effect relations (Fig. 3). Development efforts at the local, national, and international levels have a significant impact either directly or indirectly on the results of this SDG. To ensure that land fulfills its critical role in creating a sustainable environment, a comprehensive approach that encompasses management of information and data, research, policy formulation, action planning, governance, communication, capacity building, skill development, economic requirements, conflict resolution, and political and social issues is essential. The adoption of integrated, cooperative, and transdisciplinary approaches is necessary to achieve a 360-degree perspective. The existing policies and institutions require enhancement, and all stakeholders must collaborate to establish a compact that integrates land deterioration avoidance and restoration into developmental initiatives.

3.3 *Restoration Programs Initiated by the Government of India*

A global movement has arisen to counteract the human-caused changing climate, advocating for the restoration of forest areas as a means of improving ecosystem services, supporting local dwelling population, and mitigating the effects of changing climate. India has also implemented reforestation and restoration initiatives to

Fig. 3 Role of SDG 15 (Life on land) in achieving the other sustainable development goals

address soil erosion, water drainage, and desertification. According to Gopalakrishna et al. (2022), approximately 1.58 million hectares of area in India is appropriate for restoration, offering the potential for 61.3 Tg of carbon sink.

The government has a long history of making efforts to restore degraded and deforested land. The National Forest Policy (NFP) of 1952 was the first forest policy of independent India which facilitated forests to support the economy and livelihood of rural people, increasing the productivity of forests and fostering the development of forest-dependent industries. This policy was the first to emphasize that a minimum of 33% of the total national land area should be covered by forests (Bhattacharya, 2015). Later on, to regulate forest clearing, the Forest Conservation Act (1980) and National Forest Policy (NFP) of 1988 were enforced. The state governments were obligated to ask for prior permission from the central government, and tree felling was limited in fragile and eco-sensitive areas. The NFP of 1988 aimed to establish connections between forests and people, strengthening the involvement of locals in restoration efforts (Ahuja, 2014; Bhattacharjee, 2020). The term "restoration" was incorporated in this forest policy which focusses on the restoration of mining areas (section 4.4.2 of NFP, 1988). Another initiative of the authorities is Joint Forest Management (JFM) that was made official in the NFP of 1988 which got momentum in the 1998 guidelines of JFM which was focussed on protecting forests and forest-depending people and preventing soil erosion and deforestation. Other major initiatives that incorporated the forest restoration in India includes Forest Rights Act (2006), Biological Diversity Act (2002), and the National Mission for Green India (NMGI) that was launched in 2008. The Compensatory Afforestation Fund Management and Planning Authority (CAMPA) was established by the NMGI and was considered as a statuary body in 2016.

Various restoration models and activities have been implemented, and various movements and missions have also been initiated in the IHR (Bhatt et al., 2020; Dhyani et al., 2020), involving the dwelling population and community. Some agroforestry and silvi-pastoral models were developed in various villages and districts of Uttarakhand. Some noteworthy examples are as follows: Badrivan Restoration Programme (BRP) at Badrinath in 1993; eco-restoration and biodiversity conservation at Kolidhaik Village, Lohaghat, 2004; and the development of an agroforestry model for degraded community in Banswara Village, Rudraprayag District (Negi & Joshi, 2001). GBP-NIHE's restoration models implemented across various regions of the Indian Himalayan Region (IHR) have significantly contributed to fulfilling the demand for forest resources, including fodder, fuel wood, leaf litter, and non-timber forest products, while simultaneously enhancing local livelihoods and generating global environmental advantages. For example, Contour Hedgerow Farming System Technology (CHFST), in the North-East region, has achieved a remarkable 40–50% increase in crop and vegetable yields while effectively mitigating soil erosion by approximately 50%.

In the state of Uttar Pradesh (now Uttarakhand), during the years 1972–1979, a famous event known as the "Chipko Andolan" occurred. In response to the order from the government to acquire land for business purposes, mostly women from economic-weaker sections hugged the trees when logging started, as reported by

Haigh (1988). Restoration Opportunities Assessment Methodology (ROAM) is an adaptable and efficient methodology for countries to recognize and assess the FLR potential at a national or sub-national level (IUCN and WRI, 2014). G.B. Pant National Institute of Himalayan Environment and Sustainable Development (GBPNIHESD) is the partner of this framework in the Uttarakhand state.

3.4 *Land Restoration Programs (LRPs) at Global Scale*

Land degradation is a matter of immense global significance. The loss of ecosystem services and biodiversity resulting from degradation, as highlighted by Negi et al. (2022), has significant impacts on human livelihoods, as noted by Campbell et al. (2002). Globally, land degradation is estimated to cause losses exceeding 10% of the annual GDP (gross domestic product), affecting more than 3.2 billion people worldwide (Negi et al., 2022).

The Bonn Challenge, launched in 2011, was initiated to tackle the pressing problem of land restoration efforts on a worldwide level (Laestadius et al., 2015; Negi et al., 2022). The aim of the Bonn Challenge was to restore 0.15 billion hectares of deteriorated land by 2020, which was later extended to 2030 after the New York Declaration on Forests at the 2014 UN Climate Summit (InfoFLR, 2019). The new target aims to restore 0.35 billion hectares of degraded land. In 2015, India signed the Bonn Challenge during the United Nations Framework Convention on Climate Change (UNFCCC) held at Paris, pledging to restore 130 lakh hectares of deforested and deteriorated land by 2020 and additionally 80 lakh hectares by 2030 (Bonn Challenge, 2019), highlighting the country's commitment to restoring the environment. In 2018, the first country progress report towards the Bonn Challenge, jointly prepared by the Government of India (GoI) and the IUCN, reveals that India has successfully restored 98 lakh hectares of land (Bhattacharjee, 2020).

The efforts to restore degraded land through land restoration programs (LRPs) have gained momentum in the last decade and are at the base of the three Rio Conventions. These Rio Conventions are legal-bind agreements comprising of the "Convention on Biological Diversity" (CBD), the "United Nations Framework Convention on Climate Change" (UNFCCC), and the "United Nations Convention to Combat Desertification" (UNCCD) (Gichuki et al., 2019). At the worldwide level, decision and policymakers have exhibited stimulating political commitment to gain targets in several international conventions, challenges, and treaties through nature-based restoration processes (Fig. 4). The declaration of the "United Nations Decade on Ecosystem Restoration (2021–2030)" is a good example of global interest in LRPs (Negi et al., 2022). Moreover, the concept of NbS is gaining momentum as the IUCN and other organizations, such as the European Commission, are actively promoting and incorporating it. The IUCN adopted the concept of NbS in its 2009 position paper on the "United Nations Framework Convention on Climate Change (UNFCCC) CoP 15" (Cohen-Shacham et al., 2016).

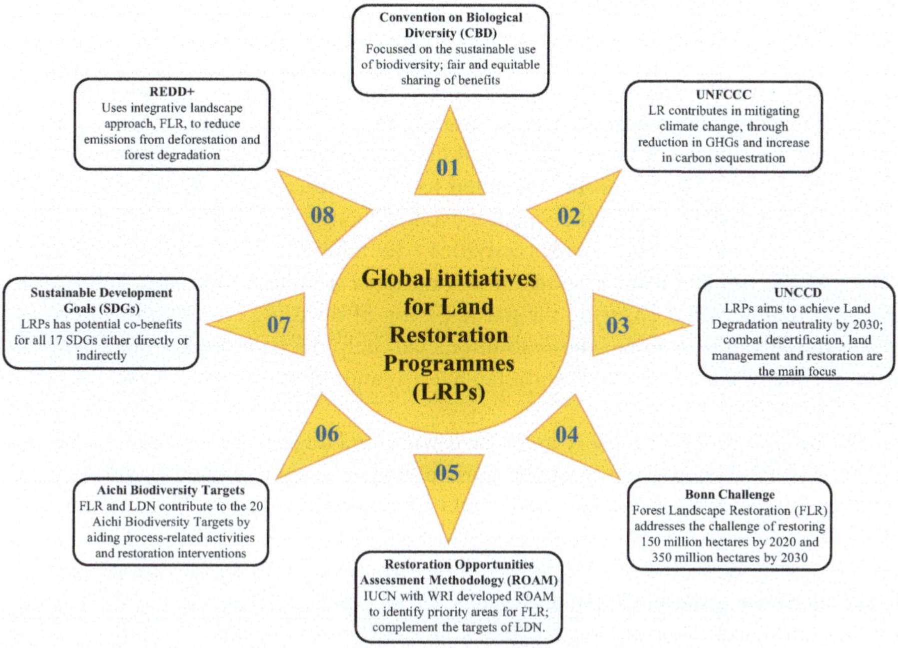

Fig. 4 Global conventions targeted to achieve through land restoration programs

In the context of mountainous areas, several global initiatives focussed on the restoration of mountains, recognizing their ecological, cultural, and socioeconomic importance. Recently in 2020, a brief—"Elevating Mountains in the Post-2020: Global Biodiversity Framework 2.0"—was prepared by GRID-Arendal and the UN Environment Programme, which highlighted that mountains are rich in biodiversity and are exposed to numerous threats and stresses. This brief outlines concrete policy recommendations for the inclusion of mountains in the Post-2020 Global Biodiversity Framework (UNEP, 2020). Similarly, at the UN Biodiversity Conference (2022), in Montreal, the UN recognizes multi-country initiatives to preserve mountain ecosystems in Europe and Africa with special award. This multi-country flagship on ecosystem restoration in mountain regions is coordinated by the UN Environment Programme, the Carpathian Convention, and the Mountain Partnership. Moreover, there are international alliances and organizations that promote the restoration of mountains: The "Mountain Partnership" is a UN voluntary alliance that aims to promote sustainable mountain development worldwide. It brings together governments, civil society organizations, and other stakeholders to address the challenges faced by mountain regions and implement strategies for their sustainable development and restoration. The "Global Environment Facility (GEF)" is an international financial mechanism that provides funding to address global environmental challenges. It supports various projects related to the restoration of mountains, including biodiversity conservation, sustainable land management, climate change adaptation, and the promotion of ecosystem services.

4 Conclusions

The Indian Himalayan Region is prone to frequent and devastating landslides that cause severe land degradation. Landslides damage local forests and soil characteristics, and the release of large volumes of landslide deposits can further exacerbate the problem. Landslides can be attributed to several factors, including rugged terrain, active denudation, extreme weather events, change in climate, excessive deforestation, and various man-made activities. To address this issue, nature-based solutions (NbSs) can offer practical and long-term solutions to protect and restore degraded lands, while ensuring the participation of dwelling communities in restoration initiatives is critical. These solutions can help decision-makers find ways to manage the risks associated with landslides and other various forms of land degradation.

The quality of soil is closely associated with land degradation. Thus, rehabilitating these soils in areas with intensive cultivation is crucial to enhance national food security. It is important to realize that SDG 15 cannot be addressed in isolation as it is intertwined with other SDGs in complex ways. Therefore, countries should focus on realigning their efforts to tackle deforestation through increased cooperation in restoring forest and non-forest landscapes across borders. To ensure the sustainability of forests and other natural resources, it is necessary to implement and achieve SDGs through participatory and collaborative forest governance.

Acknowledgments The authors are thankful to the Dean of the University School of Environment Management, Guru Gobind Singh Indraprastha University, for the support and for providing the facilities.

References

Ahuja, N. (2014). Comeback of community-based forest management: The need to revamp strategies to promote decentralized environmental governance in India and Brazil. *Florida A & M University Law Review, 9*(2), 310–342. https://commons.law.famu.edu/famulawreview/vol9/iss2/4

Alexander, S., Nelson, C. R., Aronson, J., Lamb, D., Cliquet, A., Erwin, K. L., Finlayson, C. M., de Groot, R. S., Harris, J. A., Higgs, E. S., Hobbs, R. J., Robin Lewis, R. R., Martinez, D., & Murcia, C. (2011). Opportunities and challenges for ecological restoration within REDD+. *Restoration Ecology, 19*(6), 683–689. https://doi.org/10.1111/j.1526-100x.2011.00822.x

Anonymous. (1994). *Sustainable development and rehabilitation of degraded village lands in the Himalaya* (GBPIHED Himvikas Publication, No. 8). Bishen Singh Mahendra Pal Singh.

Aronson, J., & Alexander, S. (2013). Ecosystem restoration is now a global priority: Time to roll up our sleeves. *Restoration Ecology, 21*(3), 293–296. https://doi.org/10.1111/rec.12011

Bai, Z. G., Dent, D. L., Olsson, L., & Schaepman, M. E. (2008). *Global assessment of land degradation and improvement: 1. Identification by remote sensing* (Report 2008/01). ISRIC-World Soil Information.

Basak, N., Datta, A., Biswas, S., Mitran, T., & Mandal, B. (2016). Organic amendment influences soil quality in farmers' field under rice-based cropping systems in Indo-Gangetic plains of India. *Journal of the Indian Society of Soil Science, 64*(2), 138–147.

Basak, N., Mandal, B., Datta, A., Kundu, M. C., Rai, A. K., Basak, P., & Mitran, T. (2021). Stock and stability of organic carbon in soils under major agro-ecological zones and cropping systems of sub-tropical India. *Agriculture, Ecosystems & Environment, 312*, 107317. https://doi.org/10.1016/j.agee.2021.107317

Bhadouria, R., Singh, S., Tripathi, S., & Singh, P. (Eds.). (2023). *Understanding soils of mountainous landscapes: Sustainable use of soil ecosystem services and management*. Elsevier.

Bhardwaj, A., Wasson, R. J., Ziegler, A. D., Chow, W. T., & Sundriyal, Y. P. (2019). Characteristics of rain-induced landslides in the Indian Himalaya: A case study of the Mandakini Catchment during the 2013 flood. *Geomorphology, 330*, 100–115.

Bhatt, I. D., Negi, V. S., & Rawal, R. S. (2020). Promoting nature-based solution (NbS) through restoration of degraded landscapes in the Indian Himalayan region. In S. Dhyani, A. Gupta, & M. Karki (Eds.), *Nature-based solutions for resilient ecosystems and societies. Disaster resilience and green growth*. Springer. https://doi.org/10.1007/978-981-15-4712-6_12

Bhattacharjee, A. (2020). Forest landscape restoration as a NbS strategy for achieving Bonn challenge pledge: Lessons from India's restoration efforts. In S. Dhyani, A. Gupta, & M. Karki (Eds.), *Nature-based solutions for resilient ecosystems and societies. Disaster resilience and green growth*. Springer. https://doi.org/10.1007/978-981-15-4712-6_8

Bhattacharya, S. (2015). Wood energy in India: Status and prospects. *Energy, 85*, 310–316. https://doi.org/10.1016/j.energy.2015.03.043

Bhattacharyya, R., Ghosh, B. N., Mishra, P. K., Mandal, B., Rao, C. S., Sarkar, D., Das, K., Anil, K. S., Lalitha, M., Hati, K. M., & Franzluebbers, A. J. (2015). Soil degradation in India: Challenges and potential solutions. *Sustainability, 7*(4), 3528–3570. https://doi.org/10.3390/su7043528

BIS. (2002). *Indian standards code of practice for earthquake resistant design of structures*.

Bonn Challenge. (2019). http://www.bonnchallenge.org/. Accessed 18 March 2023.

Campbell, B., Sayer, J. A., Frost, P., Vermeulen, S., Pérez, M. R., Cunningham, A., & Prabhu, R. (2002). Assessing the performance of natural resource systems. *Conservation Ecology, 5*(2), 22. https://www.jstor.org/stable/26271815

Choudhary, M., Datta, A., Jat, H. S., Yadav, A. K., Gathala, M. K., Sapkota, T. B., Das, A. K., Sharma, P. C., Jat, M. L., Singh, R., & Ladha, J. K. (2018). Changes in soil biology under conservation agriculture based sustainable intensification of cereal systems in Indo-Gangetic Plains. *Geoderma, 313*, 193–204. https://doi.org/10.1016/j.geoderma.2017.10.041

Cohen-Shacham, E., Walters, G., Janzen, C., & Maginnis, S. (2016). *Nature based solutions to address global societal challenges* (p. xiii + 97). IUCN. https://doi.org/10.2305/IUCN.CH.2016.13.en

Crozier, M. J. (2010). Deciphering the effect of climate change on landslide activity: A review. *Geomorphology, 124*(3–4), 260–267. https://doi.org/10.1016/j.geomorph.2010.04.009

Devkota, S., Shakya, N. M., & Sudmeier-Rieux, K. (2019). Framework for assessment of eco-safe rural roads in Panchase geographic region in Central–Western Nepal Hills. *Environments, 6*(6), 59. https://doi.org/10.3390/environments6060059

Dhyani, S., & Dhyani, D. (2016). Strategies for reducing deforestation and disaster risk: Lessons from Garhwal Himalaya, India. In F. G. Renaud, K. Sudmeier-Rieux, M. Estrella, & U. Nehren (Eds.), *Ecosystem-based disaster risk reduction and adaptation in practice. Advances in natural and technological hazards research* (pp. 507–528). Springer International Publishing.

Dhyani, S., Karki, M., & Petwal, A. (2018). Localizing SDGs in India using nature based solutions (NbS). *Current Science, 115*(8), 1442–1443.

Dhyani, S., Gupta, A. K., & Karki, M. (Eds.). (2020). *Nature-based solutions for resilient ecosystems and societies*. Springer.

Escobedo, F. J., Giannico, V., Jim, C. Y., Sanesi, G., & Lafortezza, R. (2019). Urban forests, ecosystem services, green infrastructure and nature-based solutions: Nexus or evolving metaphors? *Urban Forestry & Urban Greening, 37*, 3–12. https://doi.org/10.1016/j.ufug.2018.02.011

Food and Agriculture Organization. (2014). *Land degradation assessment in dryland*. Retrieved from: http://www.fao.org/nr/lada/ and http://www.fao.org/nr/land/degradation/en/

Gariano, S. L., & Guzzetti, F. (2016). Landslides in a changing climate. *Earth-Science Reviews, 162*, 227–252. https://doi.org/10.1016/j.earscirev.2016.08.011
Geertsema, M., Highland, L., & Vaugeouis, L. (2009). Environmental impact of landslides. In K. Sassa & P. Canuti (Eds.), *Landslides – Disaster risk reduction* (pp. 589–607). Springer.
Gichuki, L., Brouwer, R., Davies, J., Vidal, A., Kuzee, M., Magero, C., Walter, S., Lara, P., Oragbade, C., & Gilbey, B. (2019). *Reviving land and restoring landscapes: Policy convergence between forest landscape restoration and land degradation neutrality* (p. viii + 34). IUCN.
Gilbey, B., Davies, J., Metternicht, G., & Magero, C. (2019). Taking land degradation neutrality from concept to practice: Early reflections on LDN target setting and planning. *Environmental Science and Policy, 100*, 230–237. https://doi.org/10.1016/j.envsci.2019.04.007
Gonzalez-Ollauri, A., & Mickovski, S. B. (2016). Using the root spread information of pioneer plants to quantify their mitigation potential against shallow landslides and erosion in temperate humid climates. *Ecological Engineering, 95*, 302–315. https://doi.org/10.1016/j.ecoleng.2016.06.028
Gopalakrishna, T., Lomax, G., Aguirre-Gutiérrez, J., Bauman, D., Roy, P. S., Joshi, P. K., & Malhi, Y. (2022). Existing land uses constrain climate change mitigation potential of forest restoration in India. *Conservation Letters, 15*(2), e12867. https://doi.org/10.1111/conl.12867
Guariguata, M. R. (1990). Landslide disturbance and forest regeneration in the upper Luquillo Mountains of Puerto Rico. *The Journal of Ecology*, 814–832. https://doi.org/10.2307/2260901
Gupta, G. S. (2019). Land degradation and challenges of food security. *Review of European Studies, 11*(1), 63. https://doi.org/10.5539/res.v11n1p63
Haigh, M. J. (1988). Understanding "Chipko": The Himalayan people's movement for forest conservation. *International Journal of Environmental Studies, 31*(2–3), 99–110. https://doi.org/10.1080/00207238808710418
Huggel, C., Clague, J. J., & Korup, O. (2012). Is climate change responsible for changing landslide activity in high mountains? *Earth Surface Processes and Landforms, 37*(1), 77–91. https://doi.org/10.1002/esp.2223
Hung, L. Q., Van, N. T. H., Duc, D. M., Ha, L. T. C., Son, P. V., Khanh, N. H., & Binh, L. T. (2016). Landslide susceptibility mapping by combining the analytical hierarchy process and weighted linear combination methods: A case study in the upper Lo River catchment (Vietnam). *Landslides, 13*, 1285–1301. https://doi.org/10.1007/s10346-015-0657-3
ICIMOD. (1994). Constraints and opportunities. In *Proceedings of the International Symposium on Mountain Environment and Development*. International Centre for Integrated Mountain Development, Kathmandu, Nepal.
InfoFLR. (2019). https://infoflr.org/. Accessed 18 March 2023.
Intergovernmental Platform on Biodiversity and Ecosystem Services. (2013). *Deliverable 3(b) (i): Thematic assessment on land degradation and restoration*. PBES.
IRP. (2019). *Land restoration for achieving the Sustainable Development Goals: An International Resource Panel think piece*. Herrick, J.E., Abrahamse, T., Abhilash, P.C., Ali, S.H., Alvarez-Torres, P., Barau, A.S., Branquinho, C., Chhatre, A., Chotte, J.L., Cowie, A.L., Davis, K.F., Edrisi, S.A., Fennessy, M.S., Fletcher, S., Flores-Díaz, A.C., Franco, I.B., Ganguli, A.C., Speranza, C.I, Kamar, M.J., Kaudia, A.A., Kimiti, D.W., Luz, A.C., Matos, P., Metternicht, G., Neff, J., Nunes, A., Olaniyi, A.O., Pinho, P., Primmer, E., Quandt, A., Sarkar, P., Scherr, S.J., Singh, A., Sudoi, V., von Maltitz, G.P., Wertz, L., Zeleke, G. A think piece of the International Resource Panel. United Nations Environment Programme, Nairobi, Kenya.
IUCN. (2014). *Land degradation neutrality: Implications and opportunities for conservation* (Technical brief, Second edition 27/08/2015, 19p). IUCN.
IUCN. (2015). *Land degradation neutrality: Implications and opportunities for conservation*. International Union for the Conservation of Nature. https://www.iucn.org/sites/dev/files/content/documents/tech_brief_land_degradation_neutrality_revised_2017.pdf
IUCN, WRI. (2014). *A guide to the Restoration Opportunities Assessment Methodology (ROAM): Assessing forest landscape restoration opportunities at the national or sub-national level* (Working Paper, Road-test edition, 125pp). IUCN.

Joshi, B. (2018). Recent trends of rural out-migration and its socio-economic and environmental impacts in Uttarakhand Himalaya. *Journal of Urban and Regional Studies on Contemporary India, 4*(2), 1–14.

Laestadius, L., Buckingham, K., Maginnis, S., & Saint-Laurent, C. (2015). Before Bonn and beyond: The history and future of forest landscape restoration. *Unasylva, 66*(245), 11–18.

Lal, R. (2001). Soil degradation by erosion. *Land Degradation & Development, 12*(6), 519–539. https://doi.org/10.1002/ldr.472

Lewis, S. L., Wheeler, C. E., Mitchard, E. T. A., & Koch, A. (2019). Restoring natural forests is the best way to remove atmospheric carbon. *Nature, 568*, 25–28. https://doi.org/10.1038/d41586-019-01026-8

Li, B., Zeng, T., Ran, J., Yue, B., Zhang, M., Shang, T., & Zhu, D. (2017). Characteristics of the early secondary succession after landslides in a broad-leaved deciduous forest in the south Minshan Mountains. *Forest Ecology and Management, 405*, 238–245. https://doi.org/10.1016/j.foreco.2017.09.020

Maikhuri, R., Rawat, L. S., Negi, V., Purohit, V. K., Rao, K., & Saxena, K. G. (2011). Managing natural resources through simple and appropriate technological interventions for sustainable mountain development. *Current Science, 100*(7), 992–997. https://www.jstor.org/stable/24076516

Mishra, P. K., Rai, A., Abdelrahman, K., Rai, S. C., & Tiwari, A. (2022). Land degradation, overland flow, soil erosion, and nutrient loss in the eastern Himalayas, India. *Land, 11*(2), 179. https://doi.org/10.3390/land11020179

Montanarella, L., Scholes, R., & Brainich, A. (2018). *The IPBES assessment report on land degradation and restoration* (744pp). Secretariat of the Intergovernmental Science-Policy Platform on Biodiversity and Ecosystem Services.

National Bureau of Soil Survey & Land Use Planning. (2004). *Soil map (1:1 million scale)*. NBSS&LUP.

Negi, G. C., & Joshi, V. (2001). Agroforestry trees restore degraded land in the Himalayas. *Agroforestry Today, 13*(1–2), 19–21.

Negi, V. S., Bhatt, I. D., Phondani, P. C., & Kothyari, B. P. (2015). Rehabilitation of degraded community land in Western Himalaya: Linking environmental conservation with livelihood. *Current Science, 109*(3), 520–528. https://www.jstor.org/stable/24906106

Negi, V. S., Joshi, B. C., Pathak, R., Rawal, R. S., & Sekar, K. C. (2018). Assessment of fuel wood diversity and consumption patterns in cold desert part of Indian Himalaya: Implication for conservation and quality of life. *Journal of Cleaner Production, 196*, 23–31. https://doi.org/10.1016/j.jclepro.2018.05.237

Negi, V. S., Pathak, R., Dhyani, V., Durgapal, M., Joshi, R. K., & Bhatt, I. D. (2022). Land restoration in the Himalayan region: Steps towards biosphere integrity. *Land Use Policy, 121*, 106317. https://doi.org/10.1016/j.landusepol.2022.106317

NITI Aayog. (2021). *SDG India index and dashboard 2020–21 partnerships in the decade of action* (pp. 1–347). NITI Aayog, Government of India.

Oldeman, L. R., Hakkeling, R. T. A. & Sombroek, W. G. (1991). World Map of the Status of Human-Induced Soil Degradation: An explanatory Note (rev. ed.), UNEP and ISRIC, Wageningen.

Olsson, L., Barbosa, H., Bhadwal, S., Cowie, A., Delusca, K., Flores-Renteria, D., Hermans, K., Jobbagy, E., Kurz, W., Li, D., Sonwa, D. J., & Stringer, L. (2019). Land degradation. In *Climate change and land: An IPCC special report on climate change, desertification, land degradation, sustainable land management, food security, and greenhouse gas fluxes in terrestrial ecosystems*. IPCC.

Pal, S. C., Das, B., & Malik, S. (2019). Potential landslide vulnerability zonation using integrated analytic hierarchy process and GIS technique of Upper Rangit Catchment Area, West Sikkim, India. *Journal of the Indian Society of Remote Sensing, 47*, 1643–1655. https://doi.org/10.1007/s12524-019-01009-2

Palni, L. M. S., & Rawal, R. S. (2013). *The Himalayan biodiversity: Richness, representativeness, uniqueness and life-support values* (p. 84). G. B. Pant Institute of Himalayan Environment and Development (GBPIHED).

Parker, R. N., Densmore, A. L., Rosser, N. J., De Michele, M., Li, Y., Huang, R., Whadcoat, S., & Petley, D. N. (2011). Mass wasting triggered by the 2008 Wenchuan earthquake is greater than orogenic growth. *Nature Geoscience, 4*(7), 449–452. https://doi.org/10.1038/ngeo1154

Pearson, R., Gibson, A., & Inkpen, R. (2013). Biodiversity and species succession of the Black Ven-Spittle's landslide complex, Dorset. *Geoscience in South-West England, 13*(2), 228–231.

Prein, A. F., Liu, C., Ikeda, K., Trier, S. B., Rasmussen, R. M., Holland, G. J., & Clark, M. P. (2017). Increased rainfall volume from future convective storms in the US. *Nature Climate Change, 7*(12), 880–884. https://doi.org/10.1038/s41558-017-0007-7

Rawat, M. S., Uniyal, D. P., Dobhal, R., Joshi, V., Rawat, B. S., Bartwal, A., Singh, D., & Aswal, A. (2015). Study of landslide hazard zonation in Mandakini Valley, Rudraprayag district, Uttarakhand using remote sensing and GIS. *Current Science*, 158–170. https://www.jstor.org/stable/24905700

Reddy, V. R. (2003). Land degradation in India: Extent, costs and determinants. *Economic and Political Weekly, 38*(44), 4700–4713. https://www.jstor.org/stable/4414225

SAC-ISRO (2016). Desertification and land degradation atlas of India (Based on IRS AWiFS data of 2011-13 and 2003-05). Ahmedabad: Space Applications Centre, ISRO, Ahmedabad, India, 219.

Semwal, R. L., Nautiyal, S., Maikhuri, R. K., Rao, K. S., & Saxena, K. G. (2013). Growth and carbon stocks of multipurpose tree species plantations in degraded lands in central Himalaya, India. *Forest Ecology and Management, 310*, 450–459. https://doi.org/10.1016/j.foreco.2013.08.023

Sreenivas, K., Sujatha, G., Mitran, T., Suresh, K. G., Ravisankar, T., & Rao, P. V. N. (2021). Decadal changes in land degradation status of India. *Current Science, 121*(4), 539–550.

Sundriyal, R. C., & Jamir, S. (2005). Contour hedgerow intercropping for improving crop yields and ecosystem function. In P. S. Ramakrishnan, K. G. Saxena, M. J. Swift, K. S. Rao, & R. K. Maikhuri (Eds.), *Soil biodiversity, ecological processes and landscape management* (pp. 211–218). Oxford & IBH Publication.

UK National Ecosystem Assessment. (2011). *The UK National Ecosystem Assessment: Synthesis of the key findings*. UNEP-WCMC. Available at: uknea.unep-wcmc.org. Accessed 17 March 2023.

UNEP, GRID-Arendal, GMBA and MRI. (2020). *Elevating Mountains in the post-2020 Global Biodiversity Framework 2.0.*

United Nations. (2015). *Transforming our world: The 2030 Agenda for Sustainable Development*. https://sdgs.un.org/publications/transforming-our-world-2030-agenda-sustainable-development-17981. Accessed 15 March 2023.

Walker, L. R., Velazquez, E., & Shiels, A. B. (2009). Applying lessons from ecological succession to the restoration of landslides. *Plant and Soil, 324*(1), 157–168. https://doi.org/10.1007/s11104-008-9864-1

Xu, R., Li, X., Yang, W., Jiang, C., & Rabiei, M. (2019). Use of local plants for ecological restoration and slope stability: A possible application in Yan'an, loess plateau, China. *Geomatics, Natural Hazards and Risk, 10*(1), 2106–2128. https://doi.org/10.1080/19475705.2019.1679891

Community-Managed Forests and Their Effectiveness in SDG Implications in the Western Himalayan Region

Honey Bhatt and Harshit Pant Jugran

Abbreviations

CMF	Community-managed forest
SDGs	Sustainable Development Goals
VP	Van Panchayat

1 Introduction

The ongoing scenario of forest degradation and resource depletion prior to natural assets form sustainable development the only solution left for policymakers (Mensah, 2019). The notion of sustainable development focuses on human well-being, economic development, and ecological sustenance (WCED, 1987; Paul, 2008). Sustainable developmental has always been a political goal and guiding policy ever since the Brundtland Commission report (Bardal et al., 2021). In 2015, the UN General Assembly to a greater extent looked into the Brundtland report and picked up the Sustainable Development Goals (SDGs) and 2030 Agenda. The agenda, comprising 17 SDGs and 169 targets, is a general step of action to abolish poverty, safeguard the planet, and assure that all individuals achieve prosperity and peace by the end of 2030 (De Jong et al., 2018). The adoption in 2015 of the 17 UN SDGs and 2030 Agenda, as a consequence, led to a growing interest among various nations to outline the component that either challenge or ease the execution of the SDGs in holistic way.

H. Bhatt (✉)
Centre for Socio-Economic Development (CSED), G.B. Pant National Institute of Himalayan Environment (GBPNIHE), Almora, Uttarakhand, India

College of Horticulture, VCSG, Uttarakhand University of Horticulture and Forestry, Bharsar, Uttarakhand, India

H. P. Jugran
Centre for Socio-Economic Development (CSED), G.B. Pant National Institute of Himalayan Environment (GBPNIHE), Almora, Uttarakhand, India

S. Tripathi et al. (eds.), *Warming Mountains*,
https://doi.org/10.1007/978-3-031-62197-0_22

As of 2022, India is the 121st country in the globe on universal SDG score, measuring countries' overall progress in the direction of accomplishing all 17 SDGs (Bhattacharya et al., 2022). The necessity of forests in the implementation of the SDGs has thus been fairly acknowledged. One of the significant sets of forest actors encompasses a rural group who resides in or near the forests and participates in forestry activities as a sustenance segment of their livelihood activities. The community-managed forests are historical in the method of their long-term administration of their forest assets in perpetuity with social dependency and provide multifaceted gains in terms of ecosystem services (ES) plus socio-economic benefits. It's a well-known fact that community-based forest management (CBFM) is an exercise of forestry related to general economic, legal, administrative, and social domains, as well as technical and scientific domains of forest management. This exercise includes the administration of forest assets by managing the accumulation plus dissemination of forest products by the consumer groups (Bisht et al., 2021). The Himalayan forests vary greatly from each other respectively in terms of structure and functioning and the magnitude and seasonality of the flow of (ES) (Joshi & Negi, 2011). The major SDGs that can be assessed in accordance with CBFM are SDGs 5, 7, 10, 12, 13, and 15 which are the vital point of present study. The major components which have been addressed under this are the assessment of community participation towards resource management, gender-based distribution of resource management, and decision-making along with the sustainable utilization of resources towards climate change.

Community-based forest management is majorly favored in developing nations, with 31% of the total forests themselves being sustained and managed by the community in various nations (Wangchuk et al., 2018; Maraseni et al., 2019). Numerous prior studies on community-based forests have examined both global (Porter-Bolland et al., 2012) and local (Thapliyal, 2013) resource management approaches. These studies have consistently shown that community forestry not only demonstrates excellence in forest management but also ensures sustainable provisioning of resources to local inhabitants. This highlights the effectiveness and viability of community forestry in meeting the needs of local communities.

Preparing a set of guidelines, guarantees that the forest administration meets the native community's requirements, with the natives, in return, supporting forest preservation. Furthermore, studies have shown that more than 500 million individuals rely on community-managed forests (Niraula & Pokharel, 2016).

Biodiversity management by native groups is an age-old task around the world, especially in developing nations. The concept of phyto-resource use contributes significantly to the conservation of certain species, including threatened species (Berkes, 2007; Negi et al., 2018). Customary proficiency systems and practices are more common in the Indian Himalayan Region (IHR) due to its extreme dependence on natural assets. The native inhabitants of this region have evolved a symbiotic relationship with the environment and have a long history of cultural and aesthetical cultural heritage (Jaryan et al., 2010; Kala & Sharma, 2010; Vidyarthi et al., 2013). Forest dwellers have resided in harmony with nature, protecting and stabilizing the forests and forest resources, by securing the flow of ES through customary ecological knowledge. Ancient exercises of biodiversity conservation and forest management include the portrayal of sacred forests, sacred groves, and dedicated forest patches. It is one of the necessary exercises to conserve forest areas. Dedicating forests to an indigenous deity directly contributes to biodiversity conservation in the region (Gadgil & Vartak, 1976; Negi & Maikhuri, 2013). These are usually associated with temples, which are also culturally important to the communities. These sites function as refuges for a vast group of faunal and floral biodiversity resources and are therefore a vital source of biodiversity management and conservation.

The northern Indian state of Uttarakhand occupies a special and distinctive position in participatory management of customary natural forest resources because of its Van Panchayat (VPs). In the nineteenth century, the British colonial forest department's rejection of the communities' customary rights led to disputes with the local people. VP was established in 1931.

After the Independence in 1947, VP rules were run through and amended to support evolutionary changes in the alliance between native villages and the state forest department (Nagahama et al., 2016). India became the first nation in the world to bring an administrative policy between the government and the native community. This is another example of forest management exercise that reflects the crucial role of the Indigenous Knowledge System (IKS), as these community-based forests (VPs) are an important means of biodiversity conservation and poverty alleviation (Agrawal, 2007; Newton et al., 2015; Germain et al., 2018; Pandey et al., 2017).

Traditional management of forest resources in the IHR is the most widely used and important means of forest resource conservation and utilization. Alongside others, Uttarakhand is frequently recognized for adequate administration of forests via the management and regulation of elected bodies such as VP (Agrawal, 2007; Negi et al., 2012a, b). In Uttarakhand state, the VPs are one of the ancient examples of co-administration of forest assets by the community themselves. Most of the forests in the state are managed by the Forest Department, although 15% of the forests is under the village's management community, commonly known as VP or Forest Council (Negi et al., 2021) (Fig. 1).

Community forest management (CFM) remains an issue that is constantly in the spotlight for forest resource managers, policy planners, and researchers. The majority of the global studies points towards the fact that the involvement of indigenous groups in the forest asset management in addition to CFM for guaranteeing their environmental sustainability and the SDGs (Pathak et al., 2021).

India's National Forest Policy (NFP) of 1988 created an opportunity to build the nation's community forestry. The NFP highlighted the importance to state the quality of life of the estimated 275 million people who depend on forests for grasses, fodder, fuel wood, and other forest products (Bahuguna & Upadhay, 2002; Germain et al., 2018; Pathak et al., 2021). It also reinforced the premise that community-based bodies should provide the thrust to change the usual view of resource governance and prioritize the well-being and ecological stability of indigenous communities by halting forest degradation (Pari, 1998; Germain et al., 2018). As a result, the NFP uplifted the part of local consumers, especially female, in assisting forest assets (Bahuguna & Upadhay, 2002).

In the case of community-based forests, specifically*Van Panchayats* (VPs), the valuation of ES has been conducted for the major tree species of the region, especially pine and oak (Chakraborty et al., 2018). As these community-managed

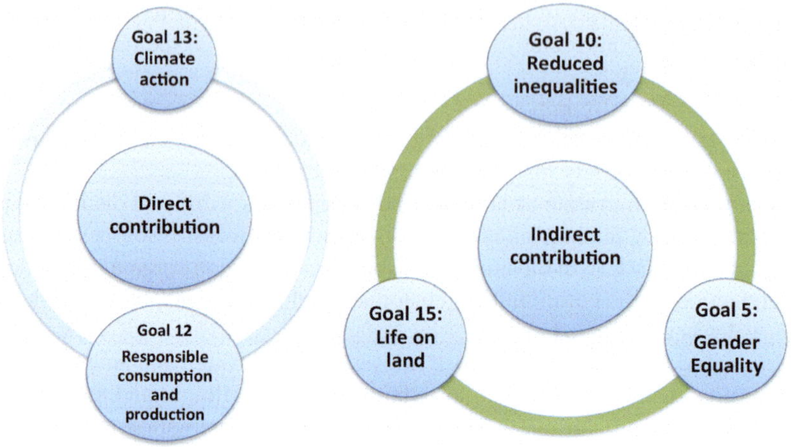

Fig. 1 Contribution of community managed forest in achieving Sustainable Development Goals (SDGs). (Modified from Negi et al. (2021))

forests produce an extensive scale of forest ES and goods, they eminently contribute to the livelihood of the local inhabitants and give rise to employment and income. The key groups and stakeholders involved in forestry occupation are the local people, within the parameters of modern societies and in countries with modest per capita gross domestic product (GDP) (De Jong et al., 2018). Community-managed forests (CMFs), which focus specifically on the qualitative management of natural forests, and efforts to support such CMFs are therefore extremely pertinent to the worldwide pursuit of the SDGs (RSI, 2017).

The research questions explored through the present study are as follows:

1. What components can be identified as promoting the execution of the SDGs in Indian Himalayan Region CMF planning?
2. How are the expediting factors conditioned by the different local and regional institutional contexts with respect to forest resource management?

Considering the problems confronted by the VPs, the key goal set for the study was to recognize the local people's perception of several problems related to:

(i) Threats and challenges across VPs
(ii) Examining the potential of protection and sustainable management of VPs with the people's involvement and influence of vital management resolution on the livelihood of native communities
(iii) Ecological and social impacts of Van Panchayats

It can be addressed that the VP system can, with suitable modifications, be recognized as an institutional mechanism to implement and vest above the usual types of rights under the FRA.

1.1 Status of Local Governance Under the Present Scenario

This study focuses on identifying challenges in forest governance at the local level and suggestive solutions for the implementation of SDGs in the VPs in recognition of their effective contribution to participatory natural resource management in the state of Uttarakhand, India. VP forests are self-managed by the native villagers. The key aim behind the selection of VPs was to gain an intensive understanding of the present condition of these forests under various climatic and anthropogenic pressures.

The Forest Councils or VPs of Kumaun are one of the primitive existing examples of CBFs in India (Guha, 2001) and cover 33% of the overall forest area in the Kumaun region. The Van Panchayats (VPs), comprised of villagers residing near the forest fringes, has only 5500 practicable ones out of 11,522. Although the VPs were established in the hills in 1931 under the Indian Forest Act, 1927, to authorize villagers to take care of and safeguard the forests in their native area, their powers have been frequently curtailed with consecutive amendments to the Act in 1976, 2001, 2005, and 2012 (TOI, 2021).

The present study was conducted in Hawalbagh developmental block of Almora District in Uttarakhand, India. The district comprises 2017 VPs, which is 16.7% of the total VPs in Uttarakhand (though the number has reportedly declined to 11217 VPs by 2024), which ranks second in India after the state of Madhya Pradesh, with a total of 14,428 forest villages (Negi et al. 2012a, b; Thapliyal, 2013; Pathak et al., 2021), and these VPs constitute about 42% of the total forest area in Almora District of Uttarakhand (UFD, 2017).

Almora District is comprised of 13 developmental blocks, in which the highest number of VPs is in Chaukhutiya developmental block and the lowest is in the Sult developmental block. According to the records of UKFD (2017), Talai, Mahrauli talli, and Sated VPs were well established in 1889 and hold the oldest VPs (130 years old) in this landscape.

The present study also provides a notion of the different local management practices opted by the stakeholders of two different administrative divisions in Uttarakhand and their role in the sustainable use and preservation of forest assets in their *Van Panchayats* or community-managed forests and their significance in large-scale landscape restoration in the context of the Himalayan region.

In general, the major role of community forests (Van Panchayats) in the region includes the following points:

(i) It delineates one of the earliest instances of decentralization initiatives across the globe (Agrawal, 2001; Pathak et al., 2021) and provides a fruitful case study for evaluating the outcomes of decentralization initiatives in the management of natural assets.

(ii) It portrays a prominent role in controlling poaching and illicit logging of trees. It illustrates a notable role in the prevention of forest fires in and around the community forests.

(iii) It ensures sustainable use of forest resources to meet the community's needs for fodder, timber, fuel wood, leaf litter, and other forest products.

In Almora District, the range of community-managed forests (CMFs) or VPs is occurring between <1000 and >2000 masl across the developmental blocks. The highest number of VPs located between 1000 and 1500 msl is in Hawalbagh developmental block, while the lowest is in Sult developmental block (Fig. 2). Most VPs fall under the 1000–1500 msl level and have a higher number of villages dependent on them. The present study sites also lie in the reported values of VPs in Almora District (Table 1).

1.2 Significance of Van Panchayats

A crucial aim of Van Panchayat is to revive and endure parts of the civil-soyam forests for the community. It also avoids neighboring hamlets from encroaching on this territory, which was once precisely bounded as VP Forest. Preserving the forests from illicit felling and fire and avoiding harm to trees due to lopping along with grazing are also the sole authority of VPs (Malik et al., 2015). The VPs' powers are legally designed to ensure that forest land in villages is not used for other purposes (e.g., agriculture and construction) and is maintained as a prime forest resource for the village community.

The performance of VPs is normally managed by the community through guarding the forest on a rotational basis or appointing a security guard and reimbursing the salary or wages through community contribution. VP also permits consent for grazing, cutting grass, and collecting fallen wood and imposing fine with government approval. The power to extract NTFPs, resin, and trees is also allotted to VPs

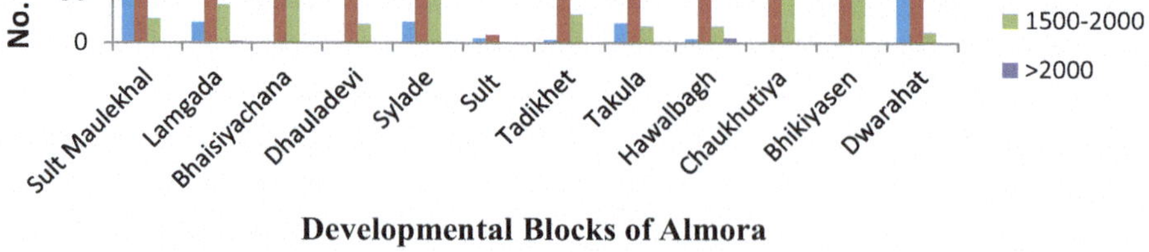

Fig. 2 Elevation distribution of CMFs (VPs) in various developmental blocks in Almora district. (Source: UKFD (2017))

Table 1 Distribution of the studied CMFs (VPs) along the altitudinal gradient and the dominant species in Almora District, Uttarakhand, Western Himalaya

S. no.	Elevation	Van Panchayat	Developmental block (district)	Dominant tree species	Associated species
1.	1702	Champa	Hawalbagh (Almora)	*Quercus leucotrichophora* A. Camus	*Pinus roxburghii* Sarg., *Lantana camara* L., *Alnus nepalensis* D. Don
2.	1510	Kayala	Hawalbagh (Almora)	*Quercus leucotrichophora* A. Camus	*Myrica esculenta, Rubus ellipticus* Sm., *Searsia parviflora Roxb.*
3.	1855	Kafali	Dhauladevi (Almora)	*Cedrus deodara* (Roxb. ex D. Don) G. Don	*Quercus leucotrichophora A. Camus, Myrica esculenta* Buch.-Ham. ex D. Don
4.	1900	Koturi	Dhauladevi (Almora)	*Cedrus deodara* (Roxb. ex D. Don) G. Don	*Rhododendron arboreum* Sm.
5.	1450	Kaneli	Hawalbagh (Almora)	*Pinus roxburghii* Sarg.	*Myrica esculenta* Buch.-Ham. ex D. Don, *Rubus ellipticus* Sm.
6.	1400	Jyoli	Hawalbagh (Almora)	*Pinus roxburghii* Sarg.	*Myrica esculenta* Buch.-Ham. ex D. Don, *Rubus ellipticus* Sm.

on the suggestion of the government (LEAD, 2007). It also has the authority to impose fines or punish for contravening the rules. Apart from that, each VP should create its own rules and regulations as per the requirement and wisdom.

Due to the prominent role of VPs in managing community forests, the state government commences a guide for the creation of new VPs in 2002, resulting in a total number of 12,089 VPs in the total 15,000 villages of the state by 2007. The VPs face new threats and struggles brought about by objective-driven policies sanctioned from above (Mukherjee, 2003). Due to the absence of policy support, financial resources, and weakening management practices, VPs in current years are recorded as an incompetent institution, especially in contrast to its renowned past.

2 Methodology

The current study was conducted in Almora District of Uttarakhand, India, and covered two developmental blocks of the district, namely, Hawalbagh and Dhauladevi (Table 1) (Fig. 3). The data was collected after conducting the primary survey in the selected four VPs. The VPs were selected to meet the following criteria, i.e., dominant vegetation (oak dominant and pine dominant), establishment year (i.e., 2000 and before; after 2000), and size of VPs. A semi-structured questionnaire was used to assess the opinion of participants regarding the forest health status, disturbance factors, and management scenario following Lise (2000).

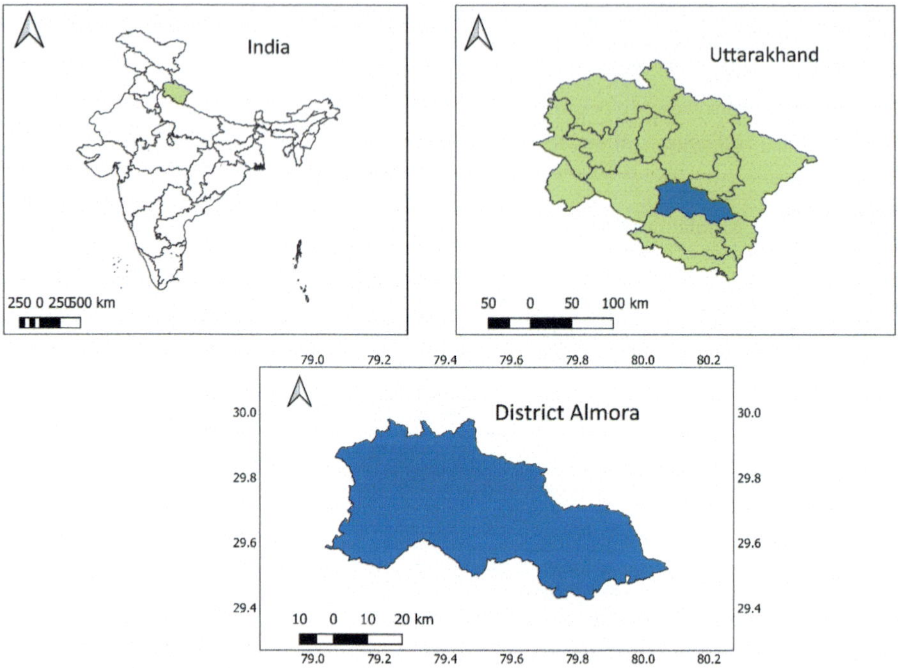

Fig. 3 The map of the study area

3 Pertinent SDGs for the Himalayan Region Vis-à-Vis VPs

The mountains cover about 30% of the world's land surface, 23% of the world's forests, and 14% of the earth's mankind population, supplying a broad range of FES to the communities (Dangwal et al., 2022). Among the world's mountains, Asia comprises the highest, largest, and majorly populated mountain system—the Himalayas (Ehrlich et al., 2021). As a provider of ecosystem goods and a range of services, the Himalaya is considered an important source of livelihood for millions of people in the highlands and billions in the lowlands (Rawal et al., 2021). The forests of the Indian Himalayan Region (IHR) are reported to sequester nearly 3000 million tonnes of carbon, which is about 40% of the total carbon stock of Indian forests (Dangwal et al., 2022). The Himalayan forests are reported to provide nearly 65 million tonnes of carbon annually (Tolangay & Moktan, 2020), with sufficient potential to mitigate global warming and climate change. The Indian Himalayan Region hosts significant natural capital and provides indispensable ES in the region and beyond, assisting livelihoods and additional daily requirements (Malik et al., 2014). The region also holds a prominent role in the provision and fulfillment of several SDGs which have been proposed by the United Nations in 2015. At the same time, the 17 SDGs and the 169 targets aim to protect the planet, end poverty, and ensure that all people enjoy prosperity and peace by 2030. The SDGs have gone through significant academic attention covering several disciplines such as development economics and sustainability science and also within disciplines like public administration, international or global governance, and the science of international relations. It can be even proclaimed that it holds the management and conservation scenario together, with the SDGs embodying the socio-ecological system.

As human societies and forests interact with each other and have been connected with certain SDGs, the relevant SDGs in mountainous regions are particularly SDG 5 (Gender equality) and SDG 10 (Reduce inequalities). As the mountain inhabitants are often associated to indigenous groups and minorities that are socially, politically, and economically excluded, their major portion is comprehensive of women and children who are at greater risk of malnutrition and hunger, and they are often excluded from decisions influencing the management and use of local assets. The management and restoration of VPs can be done through compliance with SDG 5 by empowering women and the weaker section towards VP conservation and decision-making, which can also relate to SDG 10 by stipulating equal chances for them to actively participate in the decision-building process and to be constantly present in the meetings of VPs. The VPs can also work towards strengthening SDG 12 (Ensure responsible consumption and production pattern) and SDG 13 (Climate action) related capacity building through protection and management of their existing forest assets and their sustainable utilization as the ecosystems in mountains are easily disturbed by climate anomalies. SDG 15 (Life on land, which focuses on managing forests sustainably) can also be taken for compliance towards the management of forests through enhancing its cover and variety of tree diversity through community participation. Moreover, SDG 15 (Restore, protect, and sustainably administer forests, encourage sustainable use of terrestrial ecosystems, combat desertification, and restrict and hinder loss of biodiversity and land degradation) can also be taken into consideration for the management of these VPs, as the promotion and conservation of these age-old management institution are of prime priority.

It is well known that the forests in the mountain encompass 23% of the globe's total forest cover. They supply fuel, timber, fodder, medicinal and aromatic plants, and food. Mountain forests safeguard the earth by lessening CO_2 emissions and protecting against natural hazards such as excessive weather events. India ranks first out of the top ten countries in terms of forest area as a percentage of the country's total geographical area and as a percentage of the earth's forest area, representing 2% of the earth's total forest area in 2022 (FSI, 2021); therefore, the VPs specific to the Uttarakhand state of India execute a prominent role in the preservation, promotion, and restoration of forest areas with equal participation from the communities that are highly dependent on them.

4 Results

The results are interpreted in two sections under management scenario and impact of mismanagement on VPs' situation. Under the management scenario, the details of women's active involvement in meetings and decision-making were explored on the basis of questionnaire survey (Annexure 1). The gaps in management are studied on the basis of questionnaire survey and ground-based reports by observation during the collection of primary data. The detailed results are described as follows.

4.1 Management Scenario

It has been reported that in Uttarakhand, the forest council (VPs) rules declare that in the management committee, out of a total of nine members, the reserved seats for female participants shall be four. The research has revealed that among the four forest councils (VPs) examined, only two villages (Kafali Khan and Kayala) had a sufficient four female members in its governing committee, while other remaining VPs had two or sometimes three female members. It was reported that all female interviewees had faced adversity in engaging in the survey. They were concerned regarding how their responses would be depicted and what would be the after-impacts of these responses. Several respondents prefer to keep their feedback "off the record" and favor to talk in the absence of a man in the room. Almost 50% of the female participants revealed that forest council (VPs) meetings are not held at times convenient for them.

They also reported that they did not have enough opportunity to speak at meetings, and when they did, their recommendations were frequently not taken seriously. Therefore, few of the female committee members often negotiate only with the thoughts of their spouses, who are the hidden decision-makers. In addition, the present study found that women lack solidarity. Forty percent of the female respondents complained that other female members do not take their recommendations seriously.

4.2 Gaps in Management

The major reasons that have been ascertained in the present study can be defined as grazing, the influence of forest fire, lopping, illicit felling, and illegal or uncontrolled construction inside the periphery of VPs.

4.2.1 Forest Fire

The incidences of forest fire are less in the VPs, and it was tackled best due to the involvement of local communities. As people were having a sense of responsibility towards the forest and were involved actively in preventing and tackling forest fires (Picture 1). Though some incidents have been reported in late winter, the fire was controlled after the intervention of local villagers.

Many other studies have examined the level of involvement in fighting forest fires, but community intervention in fighting forest fires depends on the direct utilization of forest resources. It can be suggested that awareness and educational activities can be constructive by involving the community and several other committees in a fire management program and holding the community as accountable for fire management.

4.2.2 Grazing and Lopping

The grazing and lopping incidences in the VPs were high, and they are directly impacting the regeneration capacity of the VP forest (Picture 2). As it takes on average almost 4 hours to stack a bundle of firewood (approx. 35 kg), the greatest pressure is on the trees located close to the villages. The lopping incidents were equal in pine-, oak-, and deodar-dominated forests which are lopped for fuel wood (pine), fodder (oak), and deodar (construction material). It was also noted that even small trees (<20 cm) were also harvested for construction poles (pine) and furniture (deodar) which have good market value. Moreover, the grazing hours are not limited as the forests are open throughout the year, which leads to the degradation of forest. Uncontrolled grazing reduces the regeneration capacity of the forests, specifically in oak forests, where the seedlings take a long time to grow and develop into mature saplings.

Picture 1 Forest fire incidents in the studied VPs

Picture 2 Grazing and lopping incidences in VPs

Picture 3 Illicit felling incidences in VPs

4.2.3 Illicit Felling

Households in the study region were completely dependent on nearby community forests for leaf fodder, firewood, and grazing. Firewood acts as a crucial source for cooking and is also the sole source for heating. The alternative for the main cooking fuel is liquefied petroleum gas (LPG), which about 11% of rural households use as their primary cooking energy source in summer (3% in winter).

Around 90% of timber and firewood and 62% of leaf fodder were collected from nearby community forests. Timber harvesting was least important with regard to biomass, accounting for only one-tenth of the wood extracted for firewood utilization in the households across the studied VPs (Picture 3).

4.2.4 Illegal/Uncontrolled Construction

The development progression comes with its perks, and the impacts can be seen in the VPs where the road cutting in an uncontrolled way has impacted the floral and faunal diversity. Moreover, these uncontrolled constructions have also taken a major chunk of the forest. Similarly, some illegal constructions have also been reported to be unregistered in the offices of VP register due to various obvious reasons. These

Picture 4 Illegal/uncontrolled construction incidences in VPs

constructions not only impact the forest health but also impact the scenic and aesthetical ecosystem services of these VPs (Picture 4).

5 Status of the Present Measures for the Assessment of Governance

The health of the forest depends on the state of regeneration and the abundance of indigenous tree species for its subsequent management. The health status of the forest represents the administration status of the community forest and the reliability of the surrounding community for assets. Our study reveals that the management status of the community forest is dwindling because of various reasons such as out-migration, the lack of funds in the VPs' accounts, and disenchantment from the forest-based resources due to a decline in the large cattle stock and simultaneously increases in small cattle stock which is impacting the forest health through grazing. The status of the present measures for the assessment of forest governance is described as follows.

5.1 Practices for Resource Management

CMF/VP plays a prominent role in the conservation and production of forest resources. In the present study, oak (*Quercus leucotrichophora* A. Camus spp.) species was highly considered suitable for fuel wood and fodder, whereas pine (*Pinus roxburghii* Sarg.) is used for leaf litter and resin by the community people and the Forest Department respectively. Several other species such as *Rhododendron arboreum* Sm. (buransh), *Myrica esculenta* Buch.-Ham. ex D. Don (kaphal), *Prunus armeniaca* Marshal (apricot), and *Prunus domestica* L. (plum) are utilized as additional sources of raw material and income by the native communities.

5.2 Governance at the Administrative Level

The CMF/VP management consists of a total of nine members (five males and four females including one Sarpanch) in its management committee. Van Panchayat members conduct meeting monthly along with self-help group (SHG), Van Suraksha Samiti (VSS), etc. and suggest their valuable recommendations for the upliftment of VPs. This represents that community is well-known about forest conservation and management. Van Mahotsav is celebrated every year by the VPs in which they organize plantation programs for the betterment of forest assets. Females are actively engaged for the upliftment of forests and also prepare thematic songs on the management and conservation of Van Panchayats (VPs). Community people realized that if they strictly follow the Van Panchayat rules and regulations, the forest's assets can be preserved and sustained efficiently for the future generation.

5.3 Management Efficiency

The VP in Uttarakhand, repeatedly quoted as one of the primitive examples of community forests, is known to be influential and constructive bodies for forest management and regulation. The analysis has revealed that the VPs having a larger area are having good management status, as the management meetings are regular in these VPs. For others it was fair and poor. The CMF conducts a meeting with all its members once a week or sometimes once a month.

5.4 Out-Migration by the Youth

As far as Uttarakhand is concerned, our state government has introduced National Livelihood Mission (also called Aajivika Mission), Mahatma Gandhi National Rural Employment Guarantee Act (MGNREGA), State Rural Livelihood Mission, Swarnjayanti Gram Swarojgar Yojna (SGSY), Integrated Livelihood Support Project (ILSP), Uttarakhand Livelihoods Improvement Project for the Himalayas (ULIPH), Veer Chandra Singh Garhwali Paryatan Swarozgar Yojana, etc. was administered by the state government to develop self-reliance in rural areas and control or stop out-migration, Although, these kind of initiatives are not widespread and could have a greater impact on upgrading income levels of rural well-beings in the mountain Region.

Despite being concerned with the rural areas, these programs have so far not shown commendable results in self-employment ventures, because self-employment is not part of the lifestyle of the people, so no one is interested in the risk involved in the investment, and if someone wants to initiate interest, the banks show no interest in giving them easy loans and without the financial inclusion the self-employment is hardly adopted. All these schemes have been started for the development of rural communities in the mountainous areas by the government; nevertheless, they are unable to curb the pace of migration and drastically change the living conditions of the rural population (Bora, 2000).

A study conducted by Sati (2021) reported that the semi-permanent out-migration reported after 2011 is 6338 (40.25% of the total villages), from the total number of villages in Uttarakhand. Of these, most villages were from Almora, Pauri, and Tehri, accounting for more than 10%. In five districts—Chamoli, Rudraprayag, Pithoragarh, Champawat, and Nainital—the proportions of villages where semi-permanent migration has taken place are 5–10%. In Haridwar, Bageshwar, Uttarkashi, Dehradun, and Udham Singh Nagar (USN) districts, villages have reported semi-permanent migration at a rate of <5%.

After 2011, around 383,726 migrants (3.8% of the total population) emigrated semi-permanently. Most semi-permanent migrants (>10%) were reported from, Tehri Almora, Chamoli, and Pauri. Three districts mainly—Rudraprayag, Champawat, and Pithoragarh—recorded 5–10% semi-permanent migrants. The least number of migrants (<5%) were observed dominantly in Nainital, Dehradun, Bageshwar, Uttarkashi, Haridwar, and USN districts, from higher to lower in order, though the last two districts, namely, USN and Haridwar, don't have any VPs.

It is important to notice that permanent migration was mainly in the rural hamlets of Uttarakhand. Permanent migration overtook a total of 3946 villages (25.1% of the total villages). Nevertheless, most villages, where more than 10% people migrated, were recorded in Almora, Pauri, and Tehri districts. Chamoli, Pithoragarh, Nainital, Rudraprayag, and Champawat districts, the number of villages from which rural-urban migration has taken place, are in the middle range (5–10), so these districts also have a higher number of VPs, and the impact of out-migration is also higher. As reported by Mehta and Maikhuri (2018), the Census of India, 2011, had reported that in Uttarakhand, two districts, i.e., Almora and Pauri, displayed a negative trend in population growth of 1.73% and 1.51%, respectively, against the national average of 17.

Additional districts like Haridwar, Bageshwar, USN, Dehradun, and Uttarkashi have fewer villages (<5%) where permanent migration was recorded. The number of permanent migrants after 2011 was 118,981 (1.18% of the total population).

Higher than 10% permanent out-migration appeared in the rural hamlets of Pauri (21.5%), Tehri (15.82%), Almora (13.62%), and Chamoli (12%) districts. On the other hand, Pithoragarh (8.3%), Rudraprayag (6.59%), and Champawat (6%) districts followed permanent migration (5–10%) at a medium pace. Nainital,

Bageshwar, Uttarkashi, Dehradun, USN, and Haridwar districts had the lowest pace of permanent migration at <5%.

Awasthi and Mehta (2020) have assessed the status of in-migration after the COVID-19 pandemic in Uttarakhand state and have reported that amplifying the livelihood opportunities and economic base by aiming on niche activities, where the state has relative benefits, like niche products, horticulture, amenity services, eco-tourism, and microhydel plants integrated with ameliorated provision of health and education infrastructure and services, can help stem out-migration from Uttarakhand. These interventions, if kept in the mind with the management of the VPs, can contribute to sustainable management and upliftment of the VPs in the state.

It has impacted the management of these VPs as only elderly or women folks have remained in the villages and are the primary consumers of the forest resources.

5.5 Implications of Non-coordination Between the Managing Institutions

The primary factor is the low participation of the people in a community in maintaining the forest with at times little support from the government or NGOs that have a passive role in forest management. The government may provide for funds when sought if there is some infrastructure-related work to be done in the forest or for other help like seeking enforcement of their legal rights in the court of law or any other function which the government may perform. NGOs may furnish them with volunteers and other people who may help them perform a task in a better and more efficient way. Even if these two external factors are absent, the community is self-sufficient in maintaining the forest. It is only when there is some serious problem that they may seek external aid, but that is very infrequent.

Also, the involvement of government authorities and the local communities varies in different areas depending upon the local rules and regulations prevailing in that place. This form of forest can exist only when the community of an area is interested in the management of the forest. Their interest can be ascertained due to various prevailing factors like their dependence on forests; the higher the dependency of the community on the resources of forests, the more they are likely to be interested in the management and conservation of forests and vice versa. Also, there are many places in the world where people worship forests and take it in aesthetical consideration as their God, and owing to that, a sense of fear combined with an intention to always please their God ensures that they take care of it and manage it leading to aesthetical and spiritual responsibility for forest resource management. There can be various other factors, and also these above-explained factors can vary greatly according to the community residing in forest vicinity, but more or less the condition is comparable.

It can be noted that the prime reasons of the management lapse may be divided into the following points in the context of Almora District which is a portion of the larger scenario of VP management in Uttarakhand:

- Community's less participation, specifically of women, in VP procedures and a move away from forest-based sustenance.
- Institutional (VP)—low adaptability to changes in gender composition and a departure from community-specific needs.
- Policy-reduced community use and access to forest resources and curtailed autonomy of VPs. The results suggest that out-migration is one of the several factors and that its impact on VPs is mediated by the broader political economy around VPs, so that a linear causal relationship is not possible. The findings are therefore useful for policy and future research as they highlight the links between various contextual factors at the regional and community level and the issues that matter for VPs' social resilience, with a focus on out-migration (Prateek, 2017).

6 Conclusion and Possible Recommendation

In the socio-cultural context, Van Panchayats (VPs) are unable to achieve the envisaged participation of female counterparts compared to Gram Panchayat (GPs), suggesting a great possibility of factors apart from patriarchal norms restricting involvement in the case of VPs. The Uttarakhand Panchayati Forest Rules, 2001, have moreover weakened VPs and strengthened the administrative control of the District Administration and the Forest Department. The rights given to VPs in these rules are insignificant as they are enforced to take consent either from the Forest Department or the District Administration for their functioning.

It is mandatory for the VPs to prepare a micro plan for a time period of 5 years for the advancement and management of the forest. The developed micro plan has to be sanctioned by the concerned authorities and approved by the Panchayat Van Vikas Adhikari (DFO). Once the micro plan is approved, the VPs have mandatory practice to develop the Annual Implementation Plan, which needs to be sanctioned from Sahayak Panchayat Vikas Adhikari (Forest Ranger). However, VPs execute a subordinate role. In general, the constitution of the micro plan is not supported by the whole rural community; instead, only few villagers participate in its preparation. Most of the rural communities are not informed about the developmental activities to be initiated under the micro plan. The Forest Department also claims 20% of the micro plan cost from the native community. It is also crucial to note that the intervention of the district forest department is also high, which makes the VP committee dependent on the Forest Department. The trees that have been marked and officially nominated by the Divisional Forest Officer (DFO) for felling on the basis of rotation age can be harvested by VPs.

The VPs must endure the instructions and implement the DFO's orders concerning the preservation and upgradation of VPs' forests and have to take prior consent of the DFO to ensure the local sale of the forest assets. The Uttarakhand State Government has developed the Area Advisory Committee, District Advisory Committee, and State-Level Advisory Committee to support VPs in their progressive activities. The Forest Department nominates the members of these committees which showcase administrative authority over VPs.

The amendment in the 73rd Constitution needs decentralization of power to Gram Panchayats; however, the power of VPs has been reduced by the government through the revised rules, which restricts their involvement in the management and use of assets. There is no procedure to restrict government functionaries responsible for the non-functionary of VPs. The higher rate of degradation in VPs is due to the excessive extraction of forest assets. Fascinating management practices have developed (i.e., devoting forest to native deity). The personal interaction with native communities has disclosed a reduced importance of native communities specifically the younger generations in the management of the VPs.

The major indicators from the outcomes of our study are:

- Ensuring the sustainability of forest management outcomes requires community involvement in management and rule-making.
- Participation in community-based governance and rule-making can reduce inter- and intra-conflicts.
- According to the previous peer-reviewed studies, there are many good examples of well-managed CFs in Bhutan and Nepal, which has significant contribution in livelihood upliftment and forest resource management through the community-based interventions, and these successful examples can be replicated in this region to have a successful management regime for the VPs in collaboration with the state's Forest Department.
- In the Himalayan region, policy frameworks to attract investment have been managed with little effect and need to be revisited to enable more effective and sustainable industrial growth in the region.
- The existing forest management policies still require an in-depth intervention from the lens of community-based forest management, and higher pro-poor and Himalayan (mountain)-oriented policies are required to be drafted while conserving and safeguarding the biological diversity of the region.
- State governments, in conjunction with ongoing programs of the central government, needs to initiate the development of commodity-based small-scale industries that utilize available biological resources and apply traditional methods at the local level to discourage immigration. It is necessary to promote the traditional handicraft industry.
- It can be concluded that a proactive approach of coupling of the environment and economical management of forest-based natural resources in these VPs is an eminent necessity to ensure employment generation for the community dependent on these VPs while ensuring the sustainable development of the mountainous region.

Acknowledgment The authors would like to thank Prof. (Dr.) Sunil Nautiyal, Director, G.B. Pant National Institute of Himalayan Environment (GBPNIHE), for imparting the work opportunities. The authors are also grateful to the village resource persons and van panchayat sarpanch(s) for providing their valuable inputs during the questionnaire and village survey.

Study Design HB conducted the field-based assessment and writing, and HPJ has conceptualized, visualized, edited, and revised the manuscript.

Annexure

Date:

Questionnaire No.:

Community managed forests and their effectiveness in SDGs implications in Western Himalayan Region

State		**District**			
Tehsil		**Taluk**			
Forest Division		**Forest Range**			
Longitude (Degree Decimal)		**Latitude (Degree Decimal)**		**Altitude (m)**	
Major Tree Species					
Anthropogenic Disturbances		High/Medium/Low/Least			
Wildlife Disturbances		High/Medium/Low/Least			

1. **Name of Respondent**: ……………………………………………………………………
2. a) **Gender** M / F b) **Age**……… c) **Education**……………
3. **Mobile no**:
4. Age distribution of family members

Member	**Up to 18**	**19–55**	**Above 55**
Male			
Female			
Migrated for earning			
Migrated for education			

Type of house Pakka/ ☐ Semi-Pakka/ Kaccha ☐

Educational level	**Illiterate**	**Primary**	**Secondary**	**Intermediate**	**Bachelor and above**
Male					
Female					

5. **Economic status**

Primary Profession	Agriculture	Wage labour	Service	Skilled job	Business/ Entrepreneur	Others
Dependence (%)						
	Power tiller	Two-wheeler	Four-wheeler		Refrigerator	

Major cooking energy sources	Fuelwood	Kerosene	LPG/cylinder	Heater	Cow Dung
If fuelwood then collection	Daily	Weekly	Monthly		
If fuelwood who collects	Male	Female	Children (5–10 Yrs)		

6. **Number of livestock owned:**

Cow	**Buffalo**	**Calves**	**Goat**	**Ox**	**Others**

INFORMATION ABOUT FODDER FROM FORESTS (Last One Year Only)

Particulars	Grass/Shrub Collection	Tree Fodder Collection	Grazing
Applicable	Yes/No	Yes/No	Yes/No
Collection Source**	GF/VF/CL/OL	GF/VF/CL/OL	GF/VF/CL/OL
Who collects fodder	Male	Female	Children (10–18 yrs)

**GF-Govt. Forest, VF-Village Forest, CL-Community Land, OL-Own Land

INFORMATION ABOUT THE USE OF TANGIBLE SERVICES FROM FORESTS

A. **Food Use (Last One Year)**

Forest resource		Purpose of collection	Frequency of collection (in number of days in a year)
Fruit 1		Self use/Selling	
Fruit 2			
Wild edibles		Self use/Selling	
Lichen		Self use/Selling	
Honey		Self use/Selling	
Nuts		Self use/Selling	

C. **NTFP (Last One Year)**

Forest resource	Purpose of collection	Frequency of collection (in number of days in a year)	Quantity collected (kg)	Price per unit (Rs/kg)
Gums	Self use/Selling			
Resins	Self use/Selling			
Tooth-cleaning twigs	Self use/Selling			
Ropes/fibres/string	Self use/Selling			
Thatching grass	Self use/Selling			
Brooms (grass/shrub)	Self use/Selling			

D. Regulating (Last 10 years)

Intangible Service	Respondent Opinion	Regular/ Intermittent
Do you obtain water from the nearby forests?	Yes/No/NA	
Do the nearby forests affect the air quality?	Yes/No/Don't know	
Do the nearby forests support soil retention on your agriculture field?	Yes/No/Don't know	
Do the nearby forests increase the fertility of your agriculture field?	Yes/No/Don't know	

Will you pay an amount (Rs.) for the conservation of your forest?	Yes/No	If yes how much 10–100 Rs. ☐ 101–100 Rs. ☐ 100–1000 Rs. ☐ >1000 Rs. ☐
Do you want to plant the broadleaved species in your forest?	Yes/No	If yes; specify the species
Do you have any tree/plant/animal species which is not reported since last 10 years?	Yes/No	Tree species | Plant species | Animal species |

E. **Cultural services**

What do you think that cultural and religious ES are affected by which climatic factors	Climate change/land conversion for residential purposes/pollution/invasive alien species/
Is there any forest caves/temple/place situated inside your van panchayat	Yes/No if yes then how often cultural activities performed there Once in a year ☐ 6 months ☐ 3 months ☐ Every month ☐
Are you enjoying the greenery or forest beauty around your nearby VP Surroundings	Yes ☐ NO ☐ Somewhat agree ☐
Is there any policy for tourism/homestay in your village	YES ☐ NO ☐ E.g.: Himalaya darshan, traditional Pahadi food, homestay, bird watching
Is there any fair or folk festival celebrated in your Village?	YES ☐ NO ☐ If YES then what
Do any of your family member is/was affected from any disease (Water & Air borne disease)	YES ☐ NO ☐ If YES then what-------------------------------------- (Ex. Diahorrea, Food poisoning Typhoid, asthma etc.)
Do you feel peace in your nearby surrounding?	Completely agree ☐ Agree ☐ Somewhat agree ☐ Somewhat disagree ☐ Completely disagree ☐

Signature and date

References

Agrawal, A., (2001). State formation in community spaces? Decentralization of control over forests in the Kumaon Himalaya, India. *The Journal of Asian Studies, 60*(1), 9–40. https://doi.org/10.2307/2659503

Agarwal, B., (2007). Gender inequality, cooperation and environmental sustainability. Inequality, Cooperation And Environmental Sustainability. Princeton University Press, Princeton, New Jersey, pp. 274–313.

Awasthi, I., & Mehta, B. S. (2020). Forced out-migration from hill regions and return migration during the pandemic: Evidence from Uttarakhand. *The Indian Journal of Labour Economics, 63*, 1107–1124. https://doi.org/10.1007/s41027-020-00291-w

Bahuguna, V. K., & Upadhay, A. (2002). Forest fires in India: Policy initiatives for community participation. *International Forestry Review, 4*(2), 122–127. https://doi.org/10.1505/IFOR.4.2.122.17446

Bardal, K. G., Reinar, M. B., Lundberg, A. K., & Bjørkan, M. (2021). Factors facilitating the implementation of the sustainable development goals in regional and local planning—Experiences from Norway. *Sustainability, 13*(8), 4282. https://doi.org/10.3390/su13084282

Berkes, F. (2007). Community-based conservation in a globalized world. *Proceedings of the National Academy of Sciences, 104*(39), 15188–15193. https://doi.org/10.1073/pnas.0702098104

Bhattacharya, R., Kumari, A., & Bose, D. (2022). Impact of Covid-19 on SDG 6 and integrated approaches for clean water access and sanitation. *Sustainability and Climate Change, 15*(5), 298–306. Published online: 21 Oct 2022. https://doi.org/10.1089/scc.2022.0040

Bisht, D., Joshi, V. C., Yadava, A. K., Sundriyal, R. C., & Jugran, P. H. (2021). Community forest management in Uttarakhand: A comparative case study of Western Himalayan Van Panchayats. *ENVIS Bulletin Himalayan Ecology, 28*, 55–59. https://gbpihed.gov.in/PDF/Publication/ENVIS_Bulletin_2020.pdf

Bora, R. S. (2000). *Out-migration: Some policy implications for Uttarakhand*. Indus Publishing Company. CSO. (2013). Annual Survey of Industries, New Delhi.

Chakraborty, A., Joshi, P. K., & Sachdeva, K. (2018). Capturing forest dependency in the central Himalayan region: Variations between Oak (*Quercus* spp.) and Pine (*Pinus* spp.) dominated forest landscapes. *Ambio, 47*(4), 504–522. https://doi.org/10.1007/s13820-017-0947-1

Dangwal, B., Rana, S. K., Negi, V. S., & Bhatt, I. D. (2022). Forest restoration enhances plant diversity and carbon stock in the sub-tropical forests of western Himalaya. Trees, Forests and People, 7, p.100201. https://doi.org/10.1016/j.tfp.2022.100201

De Jong, W., Pokorny, B., Katila, P., Galloway, G., & Pacheco, P. (2018). Community forestry and the sustainable development goals: A two way street. *Forests, 9*(6), 331. https://doi.org/10.3390/f9060331

Ehrlich, D., Melchiorri, M., & Capitani, C. (2021). Population trends and urbanisation in mountain ranges of the world. *Land, 10*, 255. https://doi.org/10.3390/land10030255

FSI. (2021). *India state of forest report*. Forest Survey of India (Ministry of Environment Forest and Climate Change).

Gadgil, M., & Vartak, V. D. (1976). The sacred groves of Western Ghats in India. *Economic Botany, 30*(2), 152–160.

Germain, R., Ghosh, C., & Jayasuriya, M. (2018). Community forestry in the state of Uttarakhand, India: Not meeting the needs of the villagers. *Small-Scale Forestry, 17*(2), 225–242. https://doi.org/10.1007/s11842-017-9384-z

Guha, R. (2001). The prehistory of community forestry in India. *Environmental History, 6*(2), 213–238. http://envhis.oxfordjournals.org/content/6/2/213.extract

Jaryan, V., Uniyal, S. K., Singh, R. D., Lal, B., Kumar, A., & Sharma, V. (2010). Role of traditional conservation practice: Highlighting the importance of Shivbari sacred grove in biodiversity conservation. *Environmentalist, 30*(2), 101–110. https://doi.org/10.1007/s10669-009-9249-x

Joshi, G., & Negi, G. C. S. (2011). Quantification and valuation of forest ecosystem services in the western Himalayan region of India. *International Journal of Biodiversity Science, Ecosystem Services & Management, 7*(1), 2–11. https://doi.org/10.1080/21513732.2011.598134

Kala, M., & Sharma, A. (2010). Traditional Indian beliefs: A key toward sustainable living. *Environmentalist, 30*(1), 85–89. https://doi.org/10.1007/s10669-009-9247-z

LEAD. (2007). *Local environmental governance in India: Case studies*. Lead India.

Lise, W. (2000). Factors influencing people's participation in forest management in India. *Ecological Economics, 34*(3), 379–392. https://doi.org/10.1016/S0921-8009(00)00182-8

Malik, Z. A., Jahangeer, A. B., & Bhatt, A. B. (2014). Forest resource use pattern in Kedarnath Wildlife Sanctuary and its fringe areas (a case study from Western Himalaya, India). *Energy Policy, 67*, 138–145. https://doi.org/10.1016/j.enpol.2013.12.016

Malik, A., Iqbal, K., Mahato, S., & Pala, N. (2015). Van Panchyat Institutions at crossroad: A prospective perception of forest dependent stakeholders from Indian Himalaya. 9, 14–20. https://doi.org/10.5829/idosi.gjer.2015.9.02.94164

Maraseni, T. N., Bhattarai, N., Karky, B. S., Cadman, T., Timalsina, N., Bhandari, T. S., Apan, A., Ma, H. O., Rawat, R. S., Verma, N., & San, S. M. (2019). An assessment of governance quality for community based forest management systems in Asia: Prioritization of governance indicators at various scales. *Land Use Policy, 81*, 750–761. https://doi.org/10.1016/j.landusepol.2018.11.04

Mehta, K., & Maikhuri, R. K. (2018). Assessing Reasons of Outmigration from Garhwal Himalayan Region A case study of three hill district. https://ijcrt.org/papers/IJCRTRIETS032.pdf

Mensah, J. (2019). Sustainable development: Meaning, history, principles, pillars, and implications for human action: Literature review. *Cogent Social Sciences, 5*, 1653531. https://doi.org/10.1080/23311886.2019.1653531

Mukherjee, P. (2003). *Community forest management in India: The Van Panchayats of Uttaranchal*. FAO.

Nagahama, K., Satya, L., Saito, K., & Desk, S. (2016). The Van Panchayat movement and struggle for achieving sustainable management of the forest: A case study of Uttarakhand in North India. *SDRP Journal of Earth Sciences & Environmental Studies, 1*(3), 1–11. https://doi.org/10.15436/JESES.1.3.1

Negi, V. S., & Maikhuri, R. K. (2013). Socio-ecological and religious perspective of agrobiodiversity conservation: Issues, concern and priority for sustainable agriculture, Central Himalaya. *Journal of Agricultural and Environmental Ethics, 26*(2), 491–512. https://doi.org/10.1007/s10806-012-9386-y

Negi, B., Chauhan, D., & Todaria, N. (2012a). Administrative and policy bottlenecks in effective management of Van Panchayats in Uttarakhand, India. *Law, Environment and Development Journal, 8*(1), 141–148. http://www.lead-journal.org

Negi, V. S., Maikhuri, R. K., & Rawat, L. S. (2012b). Paradigm and ecological implication of changing agricultural land-use: A case study from Govind Wildlife Sanctuary, Central Himalaya, India. *Journal of Mountain Science, 9*(4), 547–557. https://doi.org/10.1007/s11629-012-2216-x

Negi, V. S., Pathak, R., Sekar, K. C., Rawal, R. S., Bhatt, I. D., Nandi, S. K., & Dhyani, P. P. (2018). Traditional knowledge and biodiversity conservation: A case study from Byans Valley in Kailash Sacred Landscape, India. *Journal of Environmental Planning and Management, 61*(10), 1722–1743. https://doi.org/10.1080/09640568.2017.1371006

Negi, V. S., Pathak, R., Thakur, S., Joshi, R. K., Bhatt, I. D., & Rawal, R. S. (2021). Scoping the need of mainstreaming indigenous knowledge for sustainable use of bioresources in the Indian Himalayan region. *Environmental Management, 72*, 135–146. https://doi.org/10.1007/s00267-021-01510-w

Newton, P., Schaap, B., Fournier, M., Cornwall, M., Rosenbach, D. W., DeBoer, J., Whittemore, J., Stock, R., Yoders, M., Brodnig, G., & Agrawal, A. (2015). Community forest management and REDD+. *Forest Policy and Economics, 56*, 27–37. https://doi.org/10.1016/j.forpol.2015.03.008

Niraula, R. R., & Pokharel, B. K. (2016). Community forest management as climate change adaptation measure in Nepal's Himalaya. In N. Salzmann, C. Huggel, S. Nussbaumer, & G. Ziervogel (Eds.), *Climate change adaptation strategies—An upstream-downstream perspective* (pp. 101–120). Springer. https://doi.org/10.1007/978-3-319-40773-9_6

Pandey, S. S., Maraseni, T. N., Reardon-Smith, K., & Cockfield, G. (2017). Analysing foregone costs of communities and carbon benefits in small scale community based forestry practice in Nepal. *Land Use Policy, 69*, 160–166. https://doi.org/10.1016/j.landusepol.2017.09.007

Pari, B. (1998). The persistence of population in Indian forest policy. *Journal of Peasant Studies, 5*, 96–123.

Pathak, R., Thakur, S., Negi, V. S., Rawal, R. S., Bahukhandi, A., Durgapal, K., Barola, A., & Bhatt, I. D. (2021). Ecological condition and management status of Community Forests in Indian western Himalaya. *Land Use Policy, 109*, 105636. https://doi.org/10.1016/j.landusepol.2021.105636

Paul, B. D. (2008). A history of the concept of sustainable development: Literature review. *Annals of the University of Oradea: Economic Science, 17*, 576–580. a1930a9b09f1c486fb66260c001b2606344f1dd0.

Porter-Bolland, L., Ellis, E. A., Guariguata, M. R., Ruiz-Mallén, I., Negrete-Yankelevich, S., & Reyes-García, V. (2012). Community managed forests and forest protected areas, an assessment of their conservation effectiveness across the tropics. *Forest Ecology and Management, 268*, 6–17. https://doi.org/10.1016/j.foreco.2011.05.034

Prateek, G. (2017). *Out-migration in Kumaon: Are Van Panchayats (forest councils) socially resilient?* Arizona State University.

Rawal, R. S., Negi, V. S., & Bhatt, I. D. (2021). Changing outlook on harnessing biodiversity values–A special focus on Indian Himalaya. *Journal of Graphic Era University, 9*, 55–82. https://doi.org/10.13052/jgeu0975-1416.914

RSI. (2017). *Securing community land rights: Priorities and opportunities to advance climate and sustainable development goals*. Rights and Resources Initiative.

Sati, V. P. (2021). Out-migration in Uttarakhand Himalaya: Its types, reasons, and consequences. *Migration Letters, 18*(3), 281–295.

Thapliyal, J. (2013). *Van panchayats script a success story*. Jharkhand Panchayat Women Resource Centre. http://jpwrc4.blogspot.com/2013/01/

TOI. (2021). Over 50% Van Panchayats dysfunctional in Uttarakhand. *Times of India*, New Delhi. Published on: 9 Apr 2021. Accessed: 15 Mar 2023.

Tolangay, D., & Moktan, S. (2020). Trend of studies on carbon sequestration dynamics in the Himalaya hotspot region: A review. *Journal of Applied and Natural Science, 12*(4), 647–660. https://doi.org/10.31018/jans.v12i4.2426

UFD. (2017). *Uttarakhand forest statistics 2016–2017*. Uttarakhand Forest Department, Dehradun. uk.gov.in/pages/display/174-van-panchayat

UKFD. (2017). *Van Panchayats in Uttarakhand*. Government of Uttarakhand. https://www.forest.uk.gov.in/van-panchayat

Vidyarthi, S., Samant, S. S., & Sharma, P. (2013). Traditional and indigenous uses of medicinal plants by local residents in Himachal Pradesh, North Western Himalaya, India. *International Journal of Biodiversity Science, Ecosystem Services & Management, 9*(3), 185–200. https://doi.org/10.1080/21513732.2013.823113

Wangchuk, S., Phuntsho, S., & Wangdi, T. (2018). Management issues in community forests management, a case from Bumthang, Bhutan. *Forests Trees and Livelihoods, 27*(1), 54–60. https://doi.org/10.1080/14728028.2017.1353444

WCED. (1987). World Commission on Environment and Development. *Our Common Future, 17*, 1–91.

Index

S. Tripathi et al. (eds.), *Warming Mountains*,
https://doi.org/10.1007/978-3-031-62197-0

U

V

W

The manufacturer's authorised representative in the EU is Springer Nature Customer Service Centre GmbH, Europaplatz 3, 69115 Heidelberg, Germany. If you have any concerns regarding our products, please contact ProductSafety@springernature.com

Printed and bound by CPI Group (UK) Ltd, Croydon, CR0 4YY
01/10/2026
02225965-0014